THE CELL IN CONTACT

THE NEUROSCIENCES INSTITUTE
of the Neurosciences Research Program

Gerald M. Edelman, *Director*
W. Einar Gall, *Research Director*
W. Maxwell Cowan, *Chairman,*
Scientific Advisory Committee

The Neurosciences Institute was founded in 1981 by the Neurosciences Research Program to promote the study of scientific problems within the broad range of disciplines related to the neurosciences. It provides visiting scientists with facilities for planning and review of experimental and theoretical research with emphasis on understanding the biological basis for higher brain function.

The Institute has initiated an active publishing program. This volume, which results from a conference held at the Institute, addresses a central issue in developmental biology that is of particular relevance to the formation of the nervous system. It focuses on cell adhesion molecules, substrate adhesion molecules, and cell junctional molecules, on the regulation of their expression, and on their roles in key events during embryological development.

Support for the Neurosciences Research Foundation, Inc., which makes the Institute's programs possible, has come in part from generous gifts by The Vincent Astor Foundation, Lily Auchincloss, Francois de Menil, Sybil & William T. Golden Foundation, Lita Annenberg Hazen, Lita Annenberg Hazen Charitable Trust, The IFF Foundation, Inc., Johnson & Johnson, Harvey L. Karp, John D. & Catherine T. MacArthur Foundation, Josiah Macy, Jr. Foundation, Rockefeller Brothers Fund, Alfred P. Sloan Foundation, van Ameringen Foundation, The G. Unger Vetlesen Foundation, and The Vollmer Foundation.

The Neurosciences Institute Publications Series

Neurophysiological Approaches to Higher Brain Functions
Edward V. Evarts, Yoshikazu Shinoda, and Steven P. Wise

Protein Phosphorylation in the Nervous System
Eric J. Nestler and Paul Greengard

Dynamic Aspects of Neocortical Function
Gerald M. Edelman, W. Einar Gall, and W. Maxwell Cowan, Editors

Molecular Bases of Neural Development
Gerald M. Edelman, W. Einar Gall, and W. Maxwell Cowan, Editors

The Cell in Contact: Adhesions and Junctions as Morphogenetic Determinants
Gerald M. Edelman and Jean-Paul Thiery, Editors

THE CELL IN CONTACT
Adhesions and Junctions as Morphogenetic Determinants

Edited by

GERALD M. EDELMAN
The Rockefeller University

JEAN-PAUL THIERY
Institut d'Embryologie, CNRS

A Neurosciences Institute Publication
JOHN WILEY & SONS
New York • Chichester • Brisbane • Toronto • Singapore

Library of Congress Cataloging in Publication Data:

Main entry under title:

The cell in contact.

(The Neurosciences Institute publications series)
Includes index.
1. Cell adhesion.   2. Cell junctions.   3. Morpho-
genesis.   I. Edelman, Gerald M.   II. Thiery, Jean-Paul.
III. Series
QH623.C45   1985   599′.0188   85-91284
ISBN 0-471-83872-1
Printed in the United States of America
10 9 8 7 6 5 4 3 2 1

Contents

SECTION 3   CELL ADHESION MOLECULES

Section 1

Morphology, Molecules, and Membrane Structure

General Introduction

GERALD M. EDELMAN
JEAN-PAUL THIERY

Developmental biology currently has no adequate theory in the sense that there are adequate theories of evolution and genetics. There are a number of reasons for the lack. The first is that we do not have a sufficiently complete description of the molecular bases of the spatial dependencies within the embryo of the primary processes of development—cell division, cell movement, cell death, cell adhesion, and cell differentiation. The second reason has to do with time—in epigenetic sequences of development, the action of a gene is followed more or less precisely by that of a completely unrelated gene in time periods that are long compared to these regulating intracellular events. To analyze the highly parallel nature of the events coordinating the primary processes and also determine the dependence of such control events upon higher order cellular patterns established in epigenetic sequences poses a strong challenge to the ingenuity of the molecular embryologist.

For these reasons and others, regulative development resulting in animal form appears to be staggeringly complex. In the midst of this complexity, however, there are important regularities and invariants, as can be seen by the construction of fate maps (Vogt, 1929; Keller, 1975; Ballard, 1982; Vakaet, 1984). With this information in hand, a number of large questions can be formulated to guide the search for morphogenetically significant molecules. These include (1) *The developmental genetic question.* How can a one-dimensional genetic code specify a three-dimensional animal? (2) *The evolutionary question.* How can the mechanisms proposed to answer the developmental genetic question be reconciled with relatively large changes in form that can occur in relatively short periods of evolutionary time? (3) *The question of functional complexity.* How does cell differentiation during histogenesis take place within the context set by three-dimensional pattern formation or morphogenesis? (4) *The scaling question.* How does growth regulation occur within this same context?

The answers to these questions must come from analyses at the molecular level, and these analyses must be formulated in terms of much smaller, more focused questions. While this is obvious for early specification and determination and for later cell differentiation, it is perhaps less obvious for the establishment of the body plan. During embryonic induction, cells of different histories are brought together by morphogenetic movements resulting in mutual and milieu-dependent gene expression (Saxen and Toivonen, 1962; Spemann, 1938; Jacobson, 1966). The movements not only involve cells in loose mesenchyme, but also tissue folding and remolding (Holtfreter, 1943; Sengel, 1976). In searching for molecules and mechanisms regulative in embryonic induction (Tiedemann, 1983), we must span several levels of organization: genetic, molecular, cellular, and histological.

We are led to the conclusion that the problem of the molecular regulation of form is mechanochemical as well as genetic in nature. This gives us a strong hint with respect to the guiding hypotheses that might be employed in searching for molecules directly mediating the determination of animal form: (1) Such molecules are likely to be proteins, that is, molecules directly alterable by genetic mutation; (2) they should appear and act at the cell surface; (3) they should be shown by well-defined assays to mediate primary processes of development; initially, attention should be focused upon those processes that bear directly upon the initiation of cell motion and the establishment of cell position; (4) they should show defined sequences of appearances and regulation that can ultimately be connected with alterations in the macroscopic form of tissues and organs.

Of the candidate molecules linking the gene program to mechanisms of mechanochemical regulation, the three protein families mediating cell–cell adhesion, cell–substrate adhesion, and the formation of cell junctions seem to us to warrant the most attention at this time. While knowledge of these molecules is only beginning to accumulate, it appears likely that a number of insights will emerge from considering their possible interactions and the temporal relationships of their expression sequences during development. If this surmise is correct, one should try to explore not only the properties of these molecules but also construct additional hypotheses on how they might interact with each other in several modes: regulatory at the gene level, modulatory at the cell level, and mechanical at the tissue level.

The fact that we have set aside detailed consideration of molecules concerned with cell movement is not a denial of their central importance. It is a reflection rather of the conjecture that such molecules seem to be unlikely candidates for *direct* mediation of the flow of information from gene to pattern. A large number of genes contribute to the formation of cytoskeletal structures and the apparatus of cell motion (see the special issue of the *Journal of Cell Biology*, Volume 99, Number 1, Part 2, July 1984, on the cytoplasmic matrix). Moreover, even single genes specifying morphologically significant molecules cannot contain prior information on the local space–time coordinates or position that a moving cell will eventually occupy.

This turns our attention back to the cell in contact and to the question of what mechanisms might regulate the expression pattern of gene products such as cell adhesion molecules (CAMs), substrate adhesion molecules (SAMs), and cell junctional molecules (CJMs). Embryonic induction via morphogenetic movements and foldings of tissue sheets and neural histogenesis provide apt examples. Two main types of mechanism have been proposed to account for the invariances and cell orderings seen during these events. The first is the existence of large numbers of prespecified cell surface markers mediating cell recognition. In certain cases, for example, the nervous system, such gene products have been proposed to mediate specificity down to the level of single cells (Sperry, 1963). While this is feasible in systems consisting of small numbers of cells, in more complex systems it poses great problems of regulation of expression, particularly when this regulation is considered in connection with the variance of cell motion. Furthermore, this cell recognition hypothesis is difficult to reconcile with the plasticity of regulative development as shown by the classic experiments of Spemann (1938), Holtfreter (see Townes and Holtfreter, 1955), and Weiss (1939, 1961). It is also difficult to see how it could provide a satisfactory answer to the evolutionary question: Mutations in a few key markers among a very large set could be disastrous, and it is hard to envision how multiple complementing mutations of such cell recognition markers that might rescue certain variants could be stabilized in a population.

The alternative view is that a small set of surface molecules of different specificity might lead to pattern by regulating the driving forces of cell motion that lead through induction to differential gene expression. This view bears directly upon the expression of the three families of molecules considered in this book. The requirements for a defined epigenetic sequence of expression and for a topological invariance in cell patterns during morphogenesis translate directly: We must look for causal counterparts both in the expression and in the function of these families of surface molecules.

The known roles of these molecules are somewhat divergent. CAMs form a family of different proteins with definite specificities that link cells in epithelia and condense mesenchyme within and across germ layers (Edelman, 1983). SAMs include extracellular matrix components together with their specific cell surface receptors. They provide bases for mesenchymal cell movement, provide geometric isolation through formation of basal laminae, and set up matrices for hard tissue (Hay, 1982). CJMs provide direct communication pathways among cytoplasms of linked cells (Revel and Karnovsky, 1967), provide permeability barriers, and also provide linkages that contribute to the mechanical strength of tissues (Farquhar and Palade, 1962; Gilula, 1978; Hull and Staehelin, 1979). These differences in properties suggest that the three sets of molecules will distribute differently in time and space during development. Just as strongly, however, it suggests that at key times, these molecules are likely to link the primary processes during development and thus show coordinated patterns of regulation and mutual action.

Consideration of this possibility brings us to a pivotal issue and to some questions for future research. The issue is that CAMs, SAMs, and CJMs are cell surface molecules. They are thus responsive to the properties of the cell membrane and they communicate in two directions—to the outside and the inside of the cell. This raises the possibility that local modulation of these molecules, with consequent global modulation of the cell surface involving the cytoskeleton, may be the transducing process that mediates the connection between genetic and epigenetic mechanical events in morphogenesis (Edelman, 1976, 1983). Evidence is accumulating to indicate that such modulation events occur. The task remains to link them and to show how they might apply in particular systems.

As seen in this collection of chapters beginning with George Palade's introductory assessment of the role of the cell membrane, salient features related to this issue have been uncovered in many biological systems ranging from *Chlamydomonas* to higher brains. The meeting from which these chapters were gleaned led to a number of pertinent questions that remain to be answered in detail; each is related in some way to the four major questions concerning genetics, evolution, complexity, and scaling with which we began this introduction. We list them here in the hope that they may prompt thought and further research by those reading this collection:

1. What is the structure, function, and genetic control of CAMs, SAMs, and CJMs?
2. Are the molecules in the three families structurally or evolutionarily related?
3. What are their sequences of expression in time and space during morphogenesis, and how are these sequences related to key events of specification, determination, and induction?
4. What signals are responsible for expression, turnover, and modulation of each of these families of molecules, and is there evidence for mutual regulatory interaction or intermodulation?
5. Are the regulatory genes mediating these expressions under separate control from those concerned with cell differentiation during histogenesis?
6. How do these molecular families relate to growth and size regulation?
7. Do the roles and functions of these molecules differ in embryonic and adult life, that is, in early morphogenesis as compared to regeneration?

As the following chapters show, each of these questions can be put in a variety of contexts. The answers that will be forthcoming in the next decade should transform our view of developmental biology and bring us one step closer to the construction of an adequate theory of morphogenesis.

REFERENCES

Ballard, W. W. (1982) Morphogenetic movements and fate maps of the cypriniform teleost, *Catostomus commersoni*. *J. Exp. Zool.* **219**:301–321.

Edelman, G. M. (1976) Surface modulation in cell recognition and cell growth. *Science* **192**:218–226.

Edelman, G. M. (1983) Cell adhesion molecules. *Science* **219**:450–457.

Farquhar, M. G., and G. E. Palade (1962) Junctional complexes in various epithelia. *J. Cell Biol.* **17**:375–412.

Gilula, N. B. (1978) Structure of intercellular junctions. In *Intercellular Junctions and Synapses: Receptors and Recognition*, Series B., Vol. 2, J. Feldman, N. B. Gilula, and J. D. Pitts, eds., pp. 1–19, Chapman and Hall, London.

Hay, E. D., ed. (1982) *Cell Biology of Extracellular Matrix*, Plenum, New York.

Holtfreter, J. (1943) A study of the mechanics of gastrulation. *J. Exp. Zool.* **94**:261–318.

Hull, B. E., and L. A. Staehelin (1979) The terminal web: A reevaluation of its structure and function. *J. Cell Biol.* **81**:67–82.

Jacobson, A. G. (1966) Inductive processes in embryonic development. *Science* **152**:25–34.

Keller, R. E. (1975) Vital dye mapping of the gastrula and neurula of *Xenopus laevis*. I. Prospective areas and morphogenetic movements of the superficial layer. *Dev. Biol.* **42**:222–241.

Revel, J.-P., and M. J. Karnovsky (1967) Hexagonal array of subunits in intercellular junctions of the mouse heart and liver. *J. Cell Biol.* **33**:C7–C12.

Saxen, L., and S. Toivonen (1962) *Primary Embryonic Induction*, Academic, London.

Sengel, P. (1976) *Morphogenesis of Skin*, Cambridge Univ. Press, Cambridge, England.

Spemann, H. (1938) *Embryonic Development and Induction*, Yale Univ. Press, New Haven. Reprinted in 1962 by Hafner, New York.

Sperry, R. W. (1963) Chemoaffinity in the orderly growth of nerve fiber patterns and connections. *Proc. Natl. Acad. Sci. USA* **50**:703–710.

Tiedemann, H. (1983) Neural embryonic induction. *NATO Adv. Study Inst. Ser. A (Life Sci.)* **77**:89–105.

Townes, P. L., and J. Holtfreter (1955) Directed movements and selective adhesion of embryonic amphibian cells. *J. Exp. Zool.* **128**:53–120.

Vakaet, L. (1984) Early development of birds. In *Chimaeras in Developmental Biology*, N. M. Le Douarin and A. McLaren, eds., pp. 71–88, Academic, London.

Vogt, N. (1929) Gestaltunganalyse am Amphibienkiem mit ortlicker Vitalfarbung. II. Gastrulation und Mesodermbildung bei Urodelen und Anuren. *W. Roux' Arch. Entwicklungsmech. Org.* **120**:384–706.

Weiss, P. (1939) *Principles of Development*, Henry Holt, New York.

Weiss, P. (1961) Guiding principles in cell locomotion and cell aggregation. *Exp. Cell. Res.* **8**:260–281.

Chapter 1

Differentiated Microdomains in Cellular Membranes: Current Status

GEORGE E. PALADE

BASIC PREMISES DEFINED

Chemical Diversity of Cellular Membranes

It is now generally understood that eukaryotic cells use large amounts of chemically diversified membranes as critical components in the construction of their own bodies. At present, the number of well-defined, chemically differentiated membranes is approximately 10 in an animal eukaryotic cell and 13 in its plant counterpart. However, these figures should be considered minimal, temporary estimates; we have not yet achieved a complete inventory of all subcellular membranes because many membrane systems considered homogeneous until a few years ago have proved to be heterogeneous on a relatively large dimensional scale (the Golgi complex is an outstanding example), and because differentiated domains ranging in size from greater than 10 μm^2 to less than 100 nm^2 have been found, and continue to be found, in the continuity of many (but not all) cellular membranes.

Chemical differentiation affects both the lipid and protein components of cellular membranes, but with a well-established and important distinction: Differences in lipid composition are essentially quantitative; the same individual polar lipids, or families of polar lipids, are found in all cellular membranes in different relative concentrations. In contrast, differences in protein composition are both quantitative and qualitative in nature, and the qualitative differences are well defined and clearly recognized as being functionally important. Chemical specificity is reflected in the structural characteristics of each cellular membrane system and represents, of course, the basis for the functional specialization of each type of membrane and each type of compartment outlined by its cognate membrane within the cell.

The Fluid Lipid Bilayer

Notwithstanding their diversity, cellular membranes have in common a structural, chemical, and functional denominator, the polar lipid bilayer, which acts as a builder of closed vesicular structures (irrespective of geometric complexities), as a solvent for the hydrophobic domains of integral membrane proteins, and as a diffusion barrier for hydrophilic solutes. These basic functions are not apparently affected by the quantitative differences, already mentioned, in the polar lipid composition of the bilayers.

In all cellular systems so far investigated, the lipid bilayer is fluid at the temperature of the immediate ambient, irrespective of whether it is that of a polar ocean, a hot spring, or a homeothermic multicellular organism. Moreover, extensive studies carried out primarily in prokaryotic cells have shown that cell growth depends on cell membrane fluidity. Past a certain ratio of solid phase to fluid phase in the bilayer of their membranes, cells cease growing and eventually die (Cronan, 1975, 1978). In addition, such studies have documented, to an extent, the molecular mechanisms by which cells succeed in maintaining bilayer fluidity in a variety of environments, primarily by modifying the structure of the fatty acyl groups of their phospholipids and the nature (ester vs. ether) of the glycerol–fatty acid bonds (Cronan, 1975).

In principle, lipid and integral protein molecules should be able to diffuse laterally in the plane of a fluid bilayer and thereby randomize the chemical composition of any membrane. In fact, experimental data derived primarily from photobleaching recovery studies have established that both lipid and protein molecules diffuse rapidly in the plane of eukaryotic cell membranes, though in the case of proteins at considerably lower rates than in solution (Schlessinger et al., 1977; Webb et al., 1981; Jacobson et al., 1982). Evidence on the ubiquity of lipid bilayers in cellular membranes and on their fluidity had been accumulated over the years before Singer and Nicolson (1972) used it to formulate their fluid mosaic model of membrane structure and to systematize the interactions of membrane proteins with the bilayer. The model came, however, at the right time to have a considerable impact on general ideas in membrane biology.

Functional Implications of Bilayer Fluidity

At present, we are beginning to understand why cells use fluid bilayers in the construction of their membranes. Bilayer fluidity makes possible the insertion of new molecular components, lipids as well as proteins, into a preexisting membrane matrix without creating discontinuities in its hydrophobic diffusion barrier. All studies published to date indicate that, in cellular systems, membranes grow by expansion of preexisting membranes, not by assembly *de novo* (Palade, 1977, 1983). Expansion of solid bilayers may not be possible without the loss of permeability barriers.

Membrane fluidity is a prerequisite for membrane–membrane fusion at the cell surface as well as inside the cell, an essential process in cell division and zygote formation (fecundation). It is also necessary in many activities carried out at high frequency between cell divisions during the life span of any cell, including any type of vesicular transport involved in endocytosis, exocytosis, transcytosis, membrane recycling, and the intracellular movement of secretory, lysosomal, and many (but not all) membrane proteins. The underlying requirement in all these cases is the maintenance of the diffusion barriers created by the bilayers between different cell compartments and the cytosol, and between the cytosol and the external environment. The fluidity of the bilayers makes possible the fusion–fission of the interacting membranes, which leads to the formation of a continuous bilayer for the fused membranes, and thereby ensures the maintenance of a continuous hydrophobic barrier between the external environment and the cytosol (in exocytosis, endocytosis, and transcytosis) or between compartmental contents and the cytosol (in intracellular vesicular transport). Viewed in the light of the critical importance of all these processes for the cell's metabolic and reproductive activities, it becomes clear that membrane fluidity should be considered an obligatory condition—a *condicio sine qua non*—for cellular life. Cells must have fluid membranes, otherwise they could not exist.

Internal Contradictions Exist but Are Resolved

Membrane fluidity brings with it, however, its own problems. In addition to the randomization possible and proven within the plane of a single membrane accessible to direct investigation, such as the plasmalemma (Webb et al., 1981), fluidity could lead to the randomization of the chemistry of interacting membranes when the interaction involves the continuity established between their bilayers, a condition regularly encountered in all processes that involve membrane fusion–fission, as previously mentioned. Yet, experimental data recorded thus far indicate that chemical randomization does not occur (Palade, 1982).

It is clear then that cells have succeeded in reconciling constructively two sets of conditions that in principle are mutually exclusive. They have fluid membranes expected to randomize their molecular components; yet they succeed in maintaining the chemical specificity of their interacting membranes, and in generating, within the continuity of many of their membranes, differentiated domains that are clearly and sharply different in structure, chemistry, and function from the surrounding membrane matrix.

Our understanding of the solutions cells use to solve the problems posed by such "internal contradictions" was advanced by two independent lines of research from different fields—virology and cell biology—that developed in nearly complete isolation from one another over the past two decades.

THE ENVELOPED VIRUS MODEL

Interactions Involved in the Formation of Specific Viral Envelopes

The first model of differentiated microdomains was provided by studies of the process of replication of enveloped viruses in animal cells. A newly replicated virion leaves the infected cell by budding, usually at the cell surface. In the process, it acquires an envelope at the expense of differentiated microdomains that it generates in the plasmalemma of the infected cell. The viral genome encodes for one or a few specific glycoproteins that are processed by the cell and transported to the site of budding by the equipment the host cell uses for its own glycoproteins. In addition, the viral genome encodes for one or more nucleocapsid or structural proteins that behave like peripheral membrane proteins. They reach the target membrane, where they recognize, and interact with, the endodomains of the viral glycoproteins. The interactions lead to the creation of a microdomain in which the viral proteins are concentrated and from which the membrane proteins of the host cell are, as a rule, excluded (Lenard and Compans, 1974; Choppin and Compans, 1975). The lipids of the bilayer of the viral envelope are, however, either a random sample or, in some cases, a partially selected sample of the lipids of the host membrane bilayer.

The principles involved, therefore, are the creation of a chemically specific microdomain through specific interactions between viral integral transmembrane glycoproteins and an infrastructure constructed from nucleocapsid or structural viral proteins. This infrastructure ensures the stability of the microdomain by preventing the lateral diffusion of viral proteins out of the domain and of host proteins into the domain. The lipids are apparently of secondary importance for the biology of the virus. Hence, they are probably allowed to move by lateral diffusion in and out of the microdomain. The assembly creates not only a chemically specific microdomain within the continuity of the plasmalemma; it controls, in addition, the budding process and eventually the form of the virions by the number and types of protein molecules involved and by the geometry of their interactions.

Coated Pits and Coated Vesicles

The eukaryotic cell equivalent of the viral envelope-type of differentiated microdomain is the coated pit–coated vesicle system. It was first discovered by Roth and Porter (1964) as a morphologically recognizable differentiation of the plasmalemma of mosquito oocytes. Then coated vesicles were isolated from mammalian brains as a distinct cell fraction by Pearse (1975), who identified the main component of the vesicular coat as a 180-kD protein she named clathrin. It has since become clear that coated pits and coated vesicles are involved in adsorptive endocytosis (Goldstein et al., 1979), and in conjunction with this activity become sites of transient or permanent locations of receptors for peptide hormones and for a number of other physiologically significant ligands such as low-density lipoprotein particles and immunoglobulins (IgG, polymeric IgA). The coat is the equivalent of a stabilizing infrastructure and has a complex but regular geometry reminiscent of the geodesic domes constructed by Buckminster Fuller (1975).

More recently, additional information has been collected concerning the proteins that form the characteristic, cagelike infrastructure of coated pits and vesicles. In addition to the 180-kD protein (heavy-chain clathrin), two light chains (30–35 kD) and two groups of approximately 100 kD and 50 kD have been added to the list of the cage's components (Keen et al., 1981; Pfeffer et al., 1983). Much less progress has been made in the study of the resident proteins and lipids of the coated vesicles and in the analysis of the interactions of the core proteins (100 kD and 50 kD groups) of the coat with the integral proteins of the pit or vesicle membrane.

Organizational similarities between budding viral envelopes and coated vesicles are clear, but a number of distinct differences are also evident. To begin with, the coated pit is constructed for internalization, not for budding. Its cage is a much more versatile structure than a nucleocapsid, as evidenced by variations in its size, form, and distribution. And finally, the number of receptors already localized in the membrane of coated pits and vesicles is large and is fast growing larger, to the point that at present it is difficult to reconcile the diversity of receptors for physiologically significant or opportunistic ligands with the existence of an apparently common infrastructure for all coated pits. Either the specificity of the interactions is unusually broad or adaptor molecules are involved in the building of these half-diversified and half-uniform constructions.

Noteworthy are two other aspects of the coated pit–coated vesicle system. First, pits and vesicles of similar types appear in a variety of locations on intracellular membrane systems, especially in association with elements of the Golgi complex, as originally discovered by Friend and Farquhar (1967), but also with condensing vacuoles, lysosomes, and endosomes. More rudimentary cages are found on the transitional elements of the rough endoplasmic reticulum and on peripheral Golgi vesicles. Other different receptors are expected to be found in these additional intracellular locations, and an example is already available in the case of the mannose-6-phosphate receptor for lysosomal enzymes (Brown and Farquhar, 1984). Second, to the extent known, the spectrum of receptors, that is, integral membrane proteins associated with coated pits and vesicles, is not only diverse, it is also variable. Many receptors are randomly distributed in the plane of the cognate membrane and become preferentially segregated into coated pits only upon binding their physiological ligands. Since in many cases the binding leads to the phosphorylation of the receptor and other cellular proteins, it is possible that such, or similar, modifications are the signal that triggers aggregation and subsequent internalization.

It should be clear that the coated pit–coated vesicle system is much more versatile than the budding enveloped-virus system with which it shares only some basic principles of organization. It is built for the internalization of specific macromolecules, it is responsive to specific extracellular or intracellular ligands, and it has a built-in signal transducing system which remains to be analyzed and understood.

Coated Vesicle Relations to Vesicular Transport

At present, it is assumed that the coated pit–coated vesicle system is essentially dynamic, in the sense that the geodesic cages of the vesicles are assembled and disassembled at each transport operation. It should be clear, however, that we are dealing with a form of vesicular transport in which carrier vesicles ply between two termini, for example, between the plasmalemma and endosomes (Helenius et al., 1983), or between Golgi cisternae and lysosomes (Brown and Farquhar, 1984). In fact, there is no convincing evidence that the coats of these carrier vesicles are fully dismantled when the vesicles fuse with one or the other terminus. If the coat persists, at least in part, it may stabilize the chemistry of the carrier membrane for the duration of the continuity of its bilayer with that of any of the two termini between which it operates. This type of stabilization is needed not only to prevent randomization by lateral diffusion in the plane of the fused membranes, but to allow nonrandom removal of the carrier membrane from the membranes of its two termini (Bergeron et al., 1973; Palade, 1976, 1982). Random removal is incompatible with the retention of chemical membrane specificity.

Other Types of Stabilizing Infrastructures

Clathrin cages are undoubtedly the infrastructures eukaryotic cells most widely use to stabilize the differentiated microdomains of their membranes. But they are not the only option cells have to achieve this purpose. During exocytosis, for instance, a fibrillar network appears on the cytoplasmic aspect of the membrane of discharging secretion granules; nondischarging granules do not have a fibrillar coat (Palade, 1975). Cytochemically, this coat consists of actin filaments, but the presence of other proteins such as actin-binding proteins and myosin in the same meshwork is to be expected. Since the membranes of secretion granules appear to be nonrandomly recovered from the plasmalemma (De Camilli et al., 1976; Dowd et al., 1983), the fibrillar meshwork probably functions as a stabilizing infrastructure which prevents chemical randomization by restricting lateral diffusion and intermixing of the proteins of the two membranes and by making possible the nonrandom removal of secretion granule membranes from the apical plasmalemma.

Differentiated Microdomains in the Plasmalemma of the Vascular Endothelium

Work done over the last few years has led to the unexpected discovery that the luminal plasmalemma of the capillary endothelium is a mosaic of differentiated microdomains, especially in the case of fenestrated visceral capillaries. In addition to rare coated pits and vesicles, these endothelial cells have numerous plasmalemmal vesicles, some with stomatal diaphragms. They also have many fenestrae covered by fenestral diaphragms and a smaller number of transendothelial channels fitted, like the plasmalemmal vesicles, with stomatal diaphragms. All these elements appear as differentiated microdomains situated in the continuity of the plasmalemma proper. Their differentiation is due primarily to an uneven distribution of anionic sites (detectable by perfusion with such cationic probes as cationized ferritin). The density of anionic sites is high on the plasmalemma proper, extremely high on fenestral diaphragms, and low or nil on plasmalemmal vesicles, transendothelial channels, and associated stomatal diaphragms (Simionescu et al., 1981a). The nature as well as the density of the anionic sites varies from one type of microdomain to another: On fenestral diaphragms the sites were found (by enzyme digestion) to be contributed primarily by heparan sulfate proteoglycans (Simionescu et al., 1981b). In endothelium, stabilizing infrastructures appear to be associated primarily with the plasmalemma proper rather than with plasmalemmal vesicles, although the latter appear to have their own specific structure consisting of a set of meridional ridges (Peters et al., 1984) on the cytoplasmic aspect of their membranes. Cells have, therefore, two main options regarding stabilizing infrastructures: (1) they can attach their infrastructures to mobile elements, as in coated vesicles, or (2) they can assemble them under the membrane of one or both termini. The latter appears to be the case for vascular endothelium, in which the termini for transcytotic carriers are represented by the two aspects, luminal and abluminal, of the endothelial plasmalemma proper. [Preliminary results suggest that the stabilizing infrastructures of the plasmalemma proper consist of actin (Herman and D'Amore, 1983), nonmuscle myosin (Joyce et al., 1985), spectrin, and fodrin (Heltianu et al., 1984).] The second option may represent the cells' response to the physical forces—pressure, shearing, distortion—to which endothelial cells are subjected during normal blood circulation. In this respect, their physical habitat is similar to that of circulating red blood cells. Like blood cells, they apparently need a strong, extensive infrastructure capable of imparting tensile strength to their continuously stressed plasmalemma. In mature erythrocytes, the functional role of the spectrin–actin infrastructure is essentially mechanical; in endothelial cells, it is also responsible for the maintenance of differentiated microdomains.

The precise association of chemically defined, structurally differentiated microdomains during blood plasma–interstitial fluid exchanges led to the postulate (Simionescu et al., 1981b) that macromolecules of different charges (pI) follow different pathways across the endothelium, with such anionic proteins as albumin and most other plasma proteins being preferentially routed through plasmalemmal vesicles and transendothelial channels because their passage through fenestral diaphragms was expected to be restricted by charge repulsion. Recent work indicates that this is indeed the case: Albumin, introduced as albumin–colloidal gold complexes, binds preferentially to the membranes of plasmalemmal vesicles that move the tracer across the endothelium (Simionescu et al., 1985). Binding implies binding sites. Therefore it is possible that chemical specificity, in addition to charge interaction, plays a role in endothelial transcytosis, and it is also possible that adsorptive endocytosis or transcytosis based on differentiated microdomains may be a process operating on a broader scale than currently assumed.

Stabilizing Interactions in the Plane of the Membrane

All the examples mentioned thus far cover differentiations created and stabilized by interactions of integral membrane proteins with an infrastructure built by peripheral membrane proteins on the cytoplasmic side of the membrane. But other types of interactions with a different topography seem to be used in creating differentiated domains in the continuity of other membranes. Examples of interest are ribophorins (Kreibich et al., 1978, 1983) and signal-recognition particle receptors or docking proteins (Meyer et al., 1982), both restricted in their distribution to the membrane of the rough endoplasmic reticulum, and 5′-nucleotidase, detected cytochemically only on the distended rims of Golgi cisternae (Farquhar et al., 1974). Since in such cases morphological studies do not provide evidence for the existence of associated infrastructures, it is possible that the stabilizing interactions that restrict the distribution of these proteins to the rough part of the endoplasmic reticulum (they are not present in the smooth part of the system) occur in the membrane proper, perhaps in the bilayer or immediately adjacent to it. This type of interaction is probably responsible for the concentration of acetylcholine receptors on the postsynaptic membrane of electric organ junctions, where the receptors often appear to be organized in a planar crystalline lattice (Heuser and Salpeter, 1979).

THE JUNCTIONAL MODEL

The second line of research that led to the recognition and subsequent characterization of differentiated plasmalemmal domains began with electron microscope studies of various epithelia. The first results led to the description of both attachment devices between epithelial cells and of sealing devices of different efficiency for epithelial intercellular spaces.

The Junctional Complex

The simplest and clearest examples of such devices are seen in the junctional complexes of simple columnar or cuboidal epithelia (Farquhar and Palade, 1963). Each complex consists of: (1) an intercellular-space sealing device located at the transition from the apical to the lateral aspect of the plasmalemma and disposed as a belt around each cell (hence the name occluding zonule); (2) an adhering device immediately behind the occluding zonule, again organized in most cases as a continuous belt (hence the name adhering zonule); and (3) a variable set of more elaborate but discontinuous, buttonlike adhering devices (called adhering maculae or desmosomes). The presence of some of these elements (desmosomes, terminal bars) had been recorded in the light microscope literature, and occluding zonules had been described as "tight junctions" in the epithelium of the toad bladder (Peachey and Rasmussen, 1961), but a systematic description of these devices and a first generalization of the related concepts appeared only in 1963 (Farquhar and Palade, 1963). It was soon realized that these occluding and adhering devices represented a new class of differentiated domains within the plasmalemma involving interactions among the membranes of two adjacent cells that resulted in symmetric, in-phase differentiations in the plasmalemma of each partner. The stability of this new type of microdomain was ensured by interactions between membrane proteins and infrastructures within each cell, and by interactions between the membrane proteins of each cell with similar or complementary molecules of its neighbor cells. The latter interactions can be direct, but in most cases they appear to be mediated by extracellular proteins that can be viewed as secretory cell products.

In time, the initial morphological studies were followed by chemical characterization of the junctional components either by immunocytochemistry or cell fractionation work or by other, more indirect, means. The stabilizing infrastructure of the occluding zonules was found to consist, in part at least, of actin (Meza et al., 1980). That of the adhering zonules proved to be considerably more complex: it comprises α-actinin and vinculin (130 kD) in addition to actin (Geiger et al., 1981). That of the desmosomes proved to be special: it consists of dense plates made up of two large (215 kD and 250 kD) proteins called desmoplakins (Mueller and Franke, 1983). The glycoproteins of the desmosomal membrane itself and of the extracellular plates (Grobsky and Steinberg, 1981) have also been identified, and a partial characterization of the glycoprotein involved in the construction of the characteristic strands of the occluding zonules has apparently been achieved (Stevenson and Goodenough, 1984).

Occluding Zonules as Boundaries Between Apical and Basolateral Plasmalemmal Domains

Soon after the discovery of the occluding zonule, it was realized that this zonule could function as a restrictive boundary for the molecular components of the apical and basolateral domains of the plasmalemma. Polarized epithelial cells have an asymmetric distribution of integral membrane proteins (enzymes, transporters, and receptors) with distinct components for each domain. In intestinal and nephronic epithelium, for instance, terminal digestive hydrolases and transporters are restricted to the apical domain (cf. Louvard et al., 1973), whereas sodium and potassium ion pumps are concentrated in the basolateral domain (Kyte, 1976a,b). Occluding zonules can be reversibly dismantled by removing Ca^{2+} from the incubation medium of isolated, cultured epithelia (Martinez-Palomo et al., 1980). Concomitant with their disorganization, partial randomization of the apical and basolateral membrane components has been recorded (Meldolesi et al., 1978; Zlomek et al., 1980). Intact occluding zonules appear, therefore, to be able to limit lateral diffusion of membrane proteins to one or the other domain, but apparently they do not restrict lipid molecule diffusion from one domain to the other (Dragsten et al., 1981).

Structural Functions of Junctional Elements

Adhering zonules and maculae were visualized from the beginning as anchoring sites of elaborate intracellular fibrillar systems (Farquhar and Palade, 1963), and more recent studies have analyzed in considerable detail the chemistry and architecture of these associated fibrillar elements. The system associated with the adhering zonule is known as the terminal web. It consists of actin in at least two different forms, actin-associated proteins, spectrin, fodrin, and myosin (Hirokawa et al., 1983; Mooseker et al., 1984) and probably functions as a shaker-motor for the enzymes and transporters of the apical domain, although the type of motion and the way in which it is generated and controlled are still obscure. The fibrils associated with desmosomes are cytokeratins of different kinds characteristic for each cell type (Franke et al., 1981). They form bundles that crisscross the cytoplasm between desmosomes on opposite sides of the cell's body and are anchored at each end in desmosomal plaques. Their number and size (as well as the density and size of the desmosomes) appear to correlate directly with the degree of pressure and shear to which the epithelia are exposed at different locations. Because of such observations, the differentiated junctional microdomains acquire a special functional significance: They are anchoring sites for intracellular muscles that achieve stability and mechanical strength through cell–cell interactions, and they are anchoring sites for noncontractile, intermediary filament systems that function as bracing devices between rigid, solid desmosomal plates on opposite sides of the cell's body. The resulting complex is comparable to a tensegrity system (Fuller, 1975; Ingber, 1984) built of solid knots (the desmosomal plates) and flexible elements (the bundles of cytokeratin fibrils) that is maintained under tension by physical forces operating in the environment. The system can withstand pressure and shear without collapsing, successfully resisting applied forces by distributing and dissipating them over large distances throughout the entire construction.

Junctional elements are found not only between epithelial cells, but also between these cells and their natural substrate, the basement membranes, which are complex extracellular structures formed by the assembly of secretory proteins (collagens, laminin, fibronectin, proteoglycans) produced by the cells themselves (Hay, 1981). The attachments to the basement membrane are called hemidesmosomes, because when well developed they have indeed the structural appearance of half a desmosome. Via these hemidesmosomes the tensegrity system extends beyond the epithelium to the basement membranes, which in turn are linked to other components of the extracellular matrix and through them to other basement membranes and to other cells by interactions specific to cells and components of the extracellular matrix (Kleinman et al., 1981; Hynes and Yamada, 1982). Tissues, organs, and eventually whole organisms thus become mechanically continuous systems made up of cells and polymerized cell products (secretory glycoproteins and proteoglycans).

Under experimental conditions, cells cultured *in vitro* acquire partially differentiated plasmalemmal domains in the area of attachment to an artificial substrate (Chen and Singer, 1982). Compared to hemidesmosomes, these attachment sites appear rather rudimentary, yet they are already functional in the sense that they lead to a preferential distribution of membrane proteins by interactions with the substrate and, in time, with the extracellular matrix proteins produced by the cells themselves. The type of substrate and the type of attachment appear to influence the behavior of cells (migration, differentiation), especially during embryonic development. Several chapters in this volume discuss the specificity of such interactions in ontogeny.

GAP JUNCTIONS

In addition to the junctional differentiations already described, which function primarily as mechanical devices for distributing and dissipating physical forces that impinge on cells in the structural context of a multicellular organism, animal eukaryotic cells have developed an entirely different type of intercellular junction based, again, on interactions between macromolecular assemblies organized in-phase in the apposed membranes of two adjacent cells. These "gap junctions" (Revel and Karnovsky, 1967) consist of two (one for each cell) coupled planar lattices of hexamers or connexons of an approximately 26-kD protein called connexin (Caspar et al., 1977a; Hertzberg and Gilula, 1979; Nicholson et al., 1981). Each connexon has a central channel approximately 1.5 nm in diameter which, when aligned with the channel of an apposing connexon, creates a patent, water-filled passage across the two membranes. Ions and molecules up to 600 Daltons can move freely from cell to cell along these channels (Caspar et al., 1977b; Unwin and Zampigi, 1980; Loewenstein, 1981), which in their aggregate constitute areas of low-resistance electrical coupling (Loewenstein, 1981) between connected cells. The opening of the channels is controlled by intracellular Ca^{2+} concentrations (Rose and Loewenstein, 1975) and pH. In the case of the gap junctions, the differentiated microdomains, which are the planar lattices themselves, are stabilized by strong interactions from connexon to connexon between adjacent cells and by weaker inter-actions among the connexons of each plate, as suggested by the fact that the lattices are often disordered. So far, no stabilizing cytoplasmic infra-structure has been detected for gap junctions. These junctions act by necessity as cell–cell attachment devices, but their main function is cell–cell communication at the intracellular level. Physiologically, they establish a common intracellular environment for large cell populations and make possible the propagation of membrane depolarization waves from cell to cell. For this reason they are found extensively developed among cells of the myocardium and those of visceral smooth muscle.

IN LIEU OF CONCLUSION

Eukaryotic cells are able to create differentiated domains or microdomains in most of their membrane systems by a variety of means for a variety of purposes. These domains appear to be the result of stabilizing interactions among the components of the microdomains themselves, or between these components and a relatively small spectrum of characteristic infrastructures, among which clathrin cages predominate. The main function of these stabilizing interactions is to prevent diffusion of functionally important components out of the corresponding microdomains (however, they ap-parently allow diffusion into and out of the microdomains under specific conditions). In this respect, they can be viewed as correctives for the drawbacks of membrane fluidity.

Many of the microdomains that have been identified and studied are connected with vesicular transport. They may represent carrier vesicles in transient residence at one of their termini, stabilized to ensure maintenance of membrane specificity by preventing randomization either by lateral diffusion or by nonrandom removal during recycling.

Some differentiated microdomains are essentially structural devices con-nected to intercellular junctions and rely extensively on cell–cell interactions for their stability. These domains can also be regarded as correctives for the drawbacks of membrane fluidity, but their primary function is different: They convert aggregates of individual cells into mechanically and functionally integrated systems. In the process they produce effective epithelial barriers and create the boundaries that separate apical from basolateral plasmalemmal domains in polarized epithelial cells. This separation becomes the basis for vectorial transport across epithelia that evidently is a development of major importance for whole organisms. For the cells involved in the process, however, polarization introduces additional elements into the complexity of mechanisms involved in the control of vesicular carrier traffic and of chemical membrane specificity. It means an additional terminus and probably additional sets of target-specific vesicular carriers.

REFERENCES

Bergeron, J. J. M., P. Siekevitz, and G. E. Palade (1973) Golgi fractions prepared from rat liver homogenates. II. Biochemical characterization. *J. Cell Biol.* **59**:73–88.

Brown, W. J., and M. G. Farquhar (1984) The mannose-6-phosphate receptor of lysosomal enzymes is concentrated in *cis* Golgi cisternae. *Cell* **36**:295–307.

Caspar, D. L. D., D. A. Goodenough, L. Makowski, and W. C. Phillips (1977a) Gap junction structures. I. Correlated electron microscopy and X-ray diffraction. *J. Cell Biol.* **74**:605–628.

Caspar, D. L. D., D. A. Goodenough, L. Makowski, and W. C. Phillips (1977b) Gap junction structures. II. Analysis of X-ray diffraction data. *J. Cell Biol.* **74**:629–645.

Chen, W. T., and S. J. Singer (1982) Immunoelectron microscopic studies of the sites of cell–substratum and cell–cell contacts in cultured fibroblasts. *J. Cell Biol.* **95**:205–222.

Choppin, P. W., and R. W. Compans (1975) Reproduction of paramyxoviruses. *Comp. Virol.* **4**:95–178.

Cronan, J. E., Jr. (1975) Thermal regulation of membrane lipid composition in *Escherichia coli*. *J. Biol. Chem.* **250**:7074–7077.

Cronan, J. E., Jr. (1978) Molecular biology of bacterial membrane lipids. *Annu. Rev. Biochem.* **47**:163–189.

De Camilli, P., D. Peluchetti, and J. Meldolesi (1976) Dynamic changes of the luminal plasmalemma in stimulated parotid acinar cells. A freeze-fracture study. *J. Cell Biol.* **70**:59–74.

Dowd, D. J., C. Edward, D. Englert, J. E. Mazurkiewicz, and H. Z. Ye (1983) Immunofluorescent evidence for exocytosis and internalization of secretory granule membrane in isolated chromaffin cells. *Neuroscience* **10**:1025–1033.

Dragsten, P. R., R. Blumental, and J.S. Handler (1981) Membrane asymmetry in epithelia: Is the tight junction a barrier to diffusion in the plasma membrane? *Nature* **294**:718–722.

Farquhar, M. G., and G. E. Palade (1963) Junctional complexes in various epithelia. *J. Cell Biol.* **17**:375–412.

Farquhar, M. G., J. J. M. Bergeron, and G. E. Palade (1974) Biochemistry of Golgi fractions isolated from rat liver. *J. Cell Biol.* **60**:8–25.

Franke, W. W., D. L. Schiller, R. Moll, E. Schmid, I. Englebrecht, H. Denk, R. Krepler, and B. Platzer (1981) Diversity of cytokeratins. Differentiation-specific expression of cytokeratin polypeptides in epithelial cells and tissues. *J. Mol. Biol.* **153**:933–959.

Friend, D. S., and M. G. Farquhar (1967) Functions of coated vesicles during protein absorption in the rat vas deferens. *J. Cell Biol.* **35**:357–376.

Fuller, R. Buckminster (1975) *Synergetics,* Macmillan, New York.

Geiger, B., A. H. Dutton, K. T. Tokuyasu, and S. J. Singer (1981) Immunoelectron microscope studies of membrane-microfilament interactions. The distribution of α-actinin, tropomyosin, and vinculin in intestinal epithelial brush borders and in chicken gizzard smooth muscle cells. *J. Cell Biol.* **91**:614–628.

Goldstein, J. L., R. G. W. Anderson, and M. S. Brown (1979) Coated pits, coated vesicles, and receptor-mediated endocytosis. *Nature* **279**:679–685.

Grobsky G., and M. S. Steinberg (1981) Isolation of the intercellular glycoproteins of desmosomes. *J. Cell Biol.* **90**:243–248.

Hay, E. D. (1981) Extracellular matrix. *J. Cell. Biol.* **91**:205s–223s.

Helenius, A., I. Mellman, A. Hubbard, and D. Wall (1983) Endosomes. *Trends Biochem. Sci.* **8**:245–250.

Heltianu, C., E. Constantinescu, I. Bogdan, and M. Simionescu (1984) Immunological identification of a spectrin-like protein in endothelial cells and platelets. *J. Cell Biol.* **99**:302a.

Herman I., and P. A. D'Amore (1983) Discrimination of vascular endothelium, pericytes and smooth muscle with affinity-fractionated antiactin IgGs. *J. Cell Biol.* **97**:278a.

Hertzberg, E. L., and N. B. Gilula (1979) Isolation and characterization of gap junctions from rat liver. *J. Biol. Chem.* **254**:2138–2147.

Heuser, J. E., and S. R. Salpeter (1979) Organization of acetylcholine receptors in quick-frozen, deep-etched, and rotary-replicated *Torpedo* postsynaptic membranes. *J. Cell Biol.* **82**:150–173.

Hirokawa, N., R. E. Cheney, and M. Willard (1983) Localization of a protein of the fodrin–spectrin-TW 260/240 family in the mouse intestinal brush border. *Cell* **32**:953–965.

Hynes, R. O., and K. M. Yamada (1982) Fibronectins: Multifunctional modular glycoproteins. *J. Cell Biol.* **95**:369–377.

Ingber, D. E. (1984) Basement membrane polarizes epithelial cells and its loss can result in neoplastic disorganization. Unpublished doctoral dissertation, Yale University School of Medicine, New Haven.

Jacobson, K., E. Elson, D. Koppel, and W. Webb (1982) Fluorescence photobleaching in cell biology. *Nature* **295**:283–284.

Joyce, N., M. Haire, and G. E. Palade (1985) Contractile proteins in pericytes. II. Immunocytochemical evidence for the presence of two isomyosins in graded concentrations. *J. Cell Biol.* (in press).

Keen, J. H., M. C. Willingham, and J. H. Pastan (1981) Clathrin and coated vesicle proteins. *J. Biol. Chem.* **256**:2538–2544.

Kleinman, H. K., R. J. Klebe, and G. R. Martin (1981) Role of collagenous matrices in the adhesion and growth of cells. *J. Cell Biol.* **88**:473–485.

Kreibich, G., B. L. Ulrich, and D. D. Sabatini (1978) Proteins of rough microsomal membranes related to ribosome binding. I. Identification of ribophorins I and II: Membrane characteristics of rough microsomes. *J. Cell Biol.* **77**:464–487.

Kreibich, G., E. E. Marcantonio, and D. D. Sabatini (1983) Ribophorins I and II: Membrane proteins characteristic of the rough endoplasmic reticulum. *Methods Enzymol.* **96**:520–530.

Kyte, J. (1976a) Immunoferritin determination of (Na$^+$, K$^+$) ATPase over the plasma membrane of renal convoluted tubules. I. Distal segment. *J. Cell Biol.* **68**:287–303.

Kyte, J. (1976b) Immunoferritin of (Na$^+$, K$^+$) ATPase over the plasma membrane of renal convoluted tubules. II. Proximal segment. *J. Cell Biol.* **68**:304–313.

Lenard, J., and R. W. Compans (1974) The membrane structure of lipid-containing viruses. *Biochim. Biophys. Acta* **344**:51–94.

Loewenstein, W. R. (1981) Junctional intercellular communication: The cell-to-cell membrane channel. *Physiol. Rev.* **61**:829–913.

Louvard, D., S. Maroux, J. Barrati, P. Desnuelle, and S. Mutaftschiev (1973) On the preparation and some properties of closed membrane vesicles from hog duodenal and jejunal brush borders. *Biochim. Biophys. Acta* **291**:747–763.

Martinez-Palomo, A., I. Meza, G. Beaty, and M. Cereijido (1980) Experimental modulation of occluding junctions in a transporting epithelium. *J. Cell Biol.* **87**:736–745.

Meldolesi, J., G. Castiglioni, R. Parma, N. Nassivera, and P. De Camilli (1978) Ca^{2+}-dependent disassembly and reassembly of occluding junctions in guinea pig pancreatic acinar cells. *J. Cell Biol.* **79**:156–172.

Meyer, D. I., E. Krause, and B. Dobberstein (1982) Secretory protein translocation across membranes—the role of the "docking protein." *Nature* **297**:647–650.

Meza, I., G. Ibarra, M. Sabanero, A. Martinez-Palomo, and M. Cereijido (1980) Occluding junctions and cytoskeletal components in a cultured transporting epithelium. *J. Cell. Biol.* **87**:746–754.

Mooseker, M. S., E. M. Bonder, K. A. Conzelman, D. J. Fishkind, C. L. Howe, and T. C. S. Keller, III (1984) Brush border cytoskeleton and integration of cellular functions. *J. Cell. Biol.* **99**:104s–112s.

Mueller, H., and W. W. Franke (1983) Biological and immunological characterization of desmoplakin I and II, the major peptides of the desmosomal plaque. *J. Mol. Biol.* **163**:647–671.

Nicholson, B. J., M. W. Hunkapiller, L. B. Grim, L. E. Hood, and J.-P. Revel (1981) Rat liver gap junction protein: Properties and partial sequence. *Proc. Natl. Acad. Sci. USA* **78**:7594–7598.

Palade, G. E. (1975) Intracellular aspects of the process of protein secretion. *Science* **189**:347–358.

Palade, G. E. (1976) Interactions among cellular membranes. In *Pontificae Academiae Scientiarum Scripta Varia, Semaine d'Étude sur le Thème Membranes Biologiques et Artificielles et la Désalinisation de l'Eau*, R. Passino, ed., pp. 85–109, Pontifical Academy of Sciences, Rome.

Palade, G. E. (1977) Membrane biogenesis. In *Molecular Specialization and Symmetry in Membrane Fraction*, A. K. Solomon and M. Karnowski, eds., pp. 3–30, Harvard Univ. Press, Cambridge.

Palade, G. E. (1982) Problems in membrane traffic. *Ciba Found. Symp.* **92**:1–14.

Palade, G. E. (1983) Membrane biogenesis: An overview. *Methods Enzymol.* **96**:xxix–lv.

Peachey, L. D., and H. Rasmussen (1961) Structure of the toad's urinary bladder as related to its physiology. *J. Biophys. Biochem. Cytol.* **10**:529–533.

Pearse, B. M. F. (1975) Coated vesicles from pig brain: Purification and biochemical characterization. *J. Mol. Biol.* **97**: 93–98.

Peters, K. R., W. W. Carley, and G. E. Palade (1984) Plasmalemmal vesicles have a characteristic structure different from that of coated vesicles. *J. Cell Biol.* **99**:365a.

Pfeffer, S. R., D. G. Drubin, and R. B. Kelly (1983) Identification of three coated vesicle components as α- and β- tubulin linked to a phosphorylated 50,000 mol. wt. polypeptide. *J. Cell. Biol.* **97**:40–47.

Revel, J.-P., and M. J. Karnovsky (1967) Hexagonal array of subunits in intercellular junctions of the mouse heart and liver. *J. Cell Biol.* **33**:C7–C12.

Rose, B., and W. R. Loewenstein (1975) Permeability of cell junction depends on local cytoplasmic calcium activity. *Nature* **254**:250–252.

Roth, T., and K. R. Porter (1964) Yolk protein uptake in the oocyte of the mosquito *Aedes aegypti L. J. Cell Biol.* **20**:313–332.

Schlessinger, J., D. E. Axelrod, D. E. Koppel, W. W. Webb, and E. L. Elson (1977) Lateral transport of a lipid probe and labelled proteins on a cell membrane. *Science* **195**:307–309.

Simionescu, M., N. Simionescu, J. E. Silbert, and G. E. Palade (1981a) Differentiated microdomains on the luminal surface of capillary endothelium. II. Partial characterization of their anionic sites. *J. Cell. Biol.* **90**:614–621.

Simionescu, N., M. Simionescu, and G. E. Palade (1981b) Differentiated microdomains on the luminal surface of capillary endothelium. I. Preferential distribution of anionic sites. *J. Cell Biol.* **90**:605–613.

Simionescu, M., N. Simionescu, and L. Ghitescu (1985) Transendothelial transport of albumin. *J. Cell Biol.* (in press).

Singer, S. J., and G. Nicolson (1972) The fluid mosaic model of the structure of cell membranes. *Science* **175**:720–731.

Stevenson, B. R., and D. A. Goodenough (1984) Zonulae occludentes in junctional complex-enriched fractions from mouse liver: Preliminary morphological and biochemical characterization. *J. Cell Biol.* **98**:1209–1221.

Unwin, P. N. T., and G. Zampigi (1980) Structure of the junctions between communicating cells. *Nature* **283**:545–549.

Webb, W. W., L. S. Barak, D. W. Tank, and E. S. Wu (1981) Molecular mobility on the cell surface. *Biochem. Soc. Symp.* **46**:191–205.

Zlomek, C. A., S. Schulman, and M. Edidin (1980) Redistribution of membrane proteins in isolated mouse intestinal epithelial cells. *J. Cell Biol.* **86**:849–857.

Section 2

Some Determinants of Animal Form

Chapter 2

Three Types of Cell Interaction Regulate the Generation of Cell Diversity in the Mouse Blastocyst

MARTIN H. JOHNSON

ABSTRACT

The development of the mouse embryo from the fertilized egg to the 32-cell blastocyst takes three days and involves the generation of two distinct cell lineages. It is proposed that three types of cellular events are adequate to explain the generation of cell diversity, each involving cell interaction. These are (1) polarization of nonpolar cells in response to asymmetric cell contact; (2) regulation by cell contact of the position and orientation of the cleavage plane in polarized cells; and (3) maximization of contact between adjacent cells to minimize the degree of asymmetry of contact for apolar cells. The involvement of the extracellular matrix, the cell surface, and the cytoskeleton in mediating the cell interactions is considered in detail.

The preimplantation embryo of the mouse is peculiarly suited for studies on the mechanisms underlying the generation of cell diversity, for its constituent cells are resilient to a variety of experimental insults. One consequence of this resilience is that a wealth of experimental data has been published regarding the effects of both spatial relocation and temporal asynchrony on the ultimate developmental fate of cells. These studies have yielded three broad conclusions: (1) cells are developmentally labile, even after acquiring differentiated features; (2) the division order of cells can influence the relative position in which their progeny find themselves; (3) relative cell position influences subsequent cell fate.

However, an obsession with establishing the lineage relationships of cells that are clearly reluctant to make final decisions has obscured the real virtues of the mouse embryo as a developmental system in which precise questions about the decision-making process itself can be studied. For

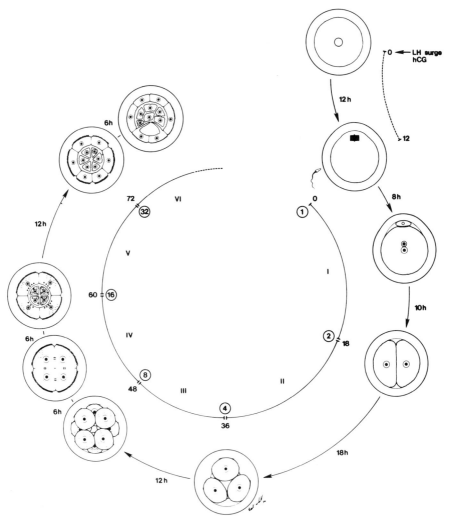

Figure 1. *Development of the mouse blastocyst.* Development is initiated with the activation of the preovulatory follicle by endogenous luteinizing hormone (LH) or exogenous human chorionic gonadotrophin (hCG). The germinal vesicle, containing chromosomes in first meiotic prophase, breaks down and meiosis proceeds to second meiotic metaphase over a 12-hour period. During this period the egg is ovulated. Insemination at zero hours leads to development through the six developmental cell cycles indicated by Roman numerals, cleavage occurring at approximately 18, 36, 48, 60, and 72 hours postinsemination. Polarization of blastomeres, intercellular flattening, and junction formation occur in the IVth cell cycle, and the blastocyst forms in the VIth cell cycle, the blastocoele, inner cell mass, and trophectoderm being visible.

example, the capacity to manipulate individual cells, or small groups of cells, makes it possible to inquire about the nature of the positional signals, and the responses to them, that deflect cells along particular pathways. Moreover, the pace of early mammalian development is comparatively slow, such that, should developmental decisions be linked to cell cycle-dependent events, there is adequate time for observation or for experimental intrusion. Since developmental decisions are evidently made incrementally and/or flexibly, the nature and timing of the signals, and the responses characterizing the decision-making process, are of more interest, and indeed are more accessible, than the final decision that results. Most other types of early embryos develop rapidly, with short cell cycles, and are not easily manipulated, so that, although in contrast to the mammal a useful system of developmental genetics may be established from the study of these embryos, it is not always easy to interpret the mechanisms underlying this genetics at the time when the genes are actually being expressed. Moreover, it is becoming clear that many of the details of the developmental decision-making process may not be regulated at the level of the genes at all, but rather at a posttranscriptional level (for a review, see Johnson, et al., 1984).

In this chapter, the mechanisms, rather than the description, of cell diversification are considered, as illustrated by the formation of the trophectoderm (TE) and inner cell mass (ICM) of the 3.5-day-old blastocyst (Figure 1). Three categories of mechanism appear sufficient to specify a blastocyst, and each involves cell interaction. First, however, the routes to cell diversification are reviewed briefly (a more detailed consideration of lineage experiments is given in Johnson and Pratt, 1983, and Johnson, 1985).

ROUTES TO CELL DIVERSIFICATION

Two distinct cell subpopulations characterized by a variety of criteria are first detected at the 16-cell stage of development. A population of larger, polar, outside cells surrounds a population of smaller, apolar, inside cells. Corresponding cell subpopulations exist at the 32-cell stage when blastocyst formation begins, the outer cells becoming TE and the inner ICM. Two routes by which the cells of each type may arise have been described.

The Division of a Polarized Eight-Cell Blastomere

Late eight-cell blastomeres are polarized and elements of this polarity are retained through division to give two 16-cell blastomeres. Depending on the orientation of the division plane, the progeny may either both be polar or one may be polar and the other apolar. If all cells were to divide along their axis of polarity, all progeny would be polar. Were all cells to divide

along an axis perpendicular to their axis of polarity, eight polar and eight
apolar cells would result. In the latter case (or indeed in intermediate
cases) cell diversity would be generated by a process of differential inheritance
(Johnson and Ziomek, 1981b).

Diversity *de novo* at the 16- and 32-Cell Stages

A polar 16-cell blastomere also retains its polarity at division to yield two
32-cell blastomeres. Again, therefore, the orientation of its cleavage plane
will influence whether it generates two polar 32-cell blastomeres (2 × TE),
or one polar and one apolar cell (1 × TE + 1 × ICM). Thus, apolar cells
can arise from polar 16-cell blastomeres.

An apolar 16-cell blastomere will retain its apolar properties for only as
long as it is totally surrounded by other cells. If exposed to an asymmetry
of cell contacts, it will polarize and behave thereafter like a polar cell.
Thus, polar 16-cell blastomeres can arise directly from apolar 16-cell blastomeres (Johnson and Ziomek, 1983).

Balance of Contribution by Each Route

While there is general agreement that these two routes to cell diversification
exist, there is disagreement over the relative contribution made by each
route in a "typical" embryo. This disagreement is, however, of little consequence since the same cellular mechanisms underlie each route. An
understanding of three types of cellular event is required to explain either
one: the process and maintenance of cell polarization (whether in 8- , 16-,
or 32-cell blastomeres; Figure 2); the way in which division planes are
oriented with respect to a cell's axis of polarity (whether at the 8- to 16-,
or the 16- to 32-cell transitions); and the process by which cells flatten on
or surround one another, thereby influencing the extent of their contact
patterns (Figure 2). We consider each of these cellular events and the
factors that influence them in the context of the generation of cell diversity.

POLARIZATION AND CELL INTERACTION

The Evidence

The blastomeres of the early eight-cell embryo are not obviously polarized.
They are spherical and covered evenly with microvilli and surface receptors,
with a fairly homogeneous internal dispersal of Golgi apparatus, endosomes,
clathrin-coated vesicles, and cytoplasmic actin. By the late eight-cell stage,
the blastomeres have become reorganized radically such that their surfaces
have clear apical and basolateral domains (Ducibella and Anderson, 1975;
Handyside, 1980; Reeve and Ziomek, 1981). Intracellular actin is concentrated

(1) Non-polar cells polarize in response to asymmetric cell contact

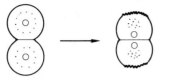

(2) Non-microvillous surface membranes maximize contact and thereby reduce asymmetry
of contact

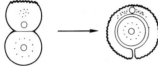

(3) Cell interaction influences orientation of division plane in polarized cells

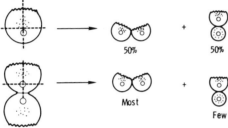

Figure 2. *Schematic summary of principal cellular mechanisms involved in generating cell diversity in the blastocyst.* 1: Non-polar cells at the 8-, 16-, and 32-cell stage polarize in response to asymmetric cell contact, their nuclei migrate basally, endosomes aggregate apically, and a surface pole develops opposite to the point of contact. 2: Polarization is inhibited if the nonpolar cell (in this case a 1/16 cell) is totally surrounded either by many cells or, as shown here, by a single polar cell enveloping it. In the absence of total cell contact, polarity is generated as shown in 1. 3: The proportion of polar cells dividing conservatively or differentiatively varies and depends on contact patterns, as summarized here for polar 1/16 cells.

apically (Johnson and Maro, 1984), as are coated vesicles and endosomes (Reeve, 1981a; Fleming and Pickering, 1985; Maro et al., 1985), whereas the nuclei of the cells are located basally (Reeve and Kelly, 1983; Johnson and Maro, 1984). This extensive reorganization occurs over a period of 8–9 hours and coincides with a change in cell surface properties that results in the maximization of cell contact and the generation of tight junctional and gap junctional contact between cells. These events—polarization, cell flattening, and junctional communication—are together called compaction.

The changes at compaction are remarkable, and their control and potential interdependence have been the subject of considerable study. The three events do not appear to be interdependent, as it is possible to inhibit each selectively. For example, polyclonal antisera to embryonal carcinoma cells will inhibit junction formation and reduce the extent of cell flattening but do not inhibit polarization. Reduction of Ca^{2+} levels inhibits both flattening and junction formation, but not polarization (Pratt et al., 1982). Conversely, there are some circumstances, for example, aggregation of large numbers of eight-cell blastomeres or of small groups of heterogeneous cell types, in which flattening occurs but polarization does not (Johnson and Ziomek, 1981a). Moreover, each event takes place at a different time during the 11–12 hours of the eight-cell's life, gap junction formation and intracellular reorganization of organelles occurring over the period 3–5 hours, cell flattening between five and seven hours, and surface polarity between six and nine hours. Thus, each of these events appears to be an independent expression of a common underlying developmental program rather than part of a causally related sequence.

Cellular reorganization at the time of polarization may arise by any of three routes, no one of which is mutually exclusive: (1) Overt polarity in the eight-cell blastomere may be the manifestation of a covert polarity inherited from earlier stages; (2) polarity may be generated *de novo* at the eight-cell stage with the remnant of the cleavage furrow, the so-called midbody, as an orienting focus; or (3) polarity *de novo* may arise by cell interaction via contact-dependent processes. These is good evidence for the last mechanism, and little for the others.

Newly formed eight-cell blastomeres may be isolated and cultured *in vitro* free from other contacts. Under such conditions they polarize poorly if at all (Ziomek and Johnson, 1980). However, if isolated cells are reaggregated and cultured in pairs, trios, or quartets, polarity develops, and the axis of polarity in each cell is determined by the position of its cell contacts (Ziomek and Johnson, 1980; Johnson and Ziomek, 1981a). If after aggregation and culture for 3–5 hours pairs of cells are disaggregated and reaggregated in a new orientation with respect to each other, the axes of polarity in these cells develop with respect to the *new* points of contact. However, if this procedure is performed after more than five hours of coculture, then the *old* point of contact is "remembered" and the polarity that develops is "off-axis" with respect to the new contact point (Johnson and Ziomek, 1981a). In all of these experiments, both the original orientation of the blastomere within the embryo and the position of the midbody (indicating the position of the previous cleavage furrow) are irrelevant to the orientation of the polarity that develops. From these results we conclude that whether or not an intrinsic polarity or the position of the previous cleavage plane influence development of polarity, such influences can be overridden by changing cell contacts. In summary, an asymmetry of contacts induces polarization, the axis of polarity being determined by the point(s) of contact. This process of induction takes a maximum of about five hours, and thereafter the axis of polarity is stable in the face of changing contact relationships.

The capacity to polarize in response to an asymmetry of cell contacts is not restricted to early eight-cell blastomeres. Cells from the apolar (inside) cell population at the 16-cell stage will, if isolated and cultured together in pairs, polarize over a period of 7–9 hours (Ziomek and Johnson 1981) and achieve a stable, polarized phenotype (Johnson and Ziomek, 1983). Isolated inner apolar cells at the 32- and probably the 64-cell stage will similarly polarize when exposed to an asymmetry of cell contacts, but will take around 24 hours to complete this process (Fleming et al., 1984). This inherent capacity to polarize emphasizes its importance for normal regulative development.

The Mechanism

We do not know the nature of the surface signal or of the immediate polarizing cellular response to it. The inducing signal shows some specificity, since newly fertilized eggs fail to induce polarity, this property developing at the two-cell stage. It is the capacity to respond to an inducing signal that appears to be limiting, developing first at the eight-cell stage. The complete process of polarization is also highly resilient to experimental manipulation: It is resistant to the effects of agents that disrupt gap junction formation (Goodall and Johnson, 1982, 1984; McLachlin et al., 1983), prevent cell flattening (Johnson et al., 1979; Pratt el al., 1982), prevent transcription from at least the four-cell stage (Johnson and Pratt, 1983; M. H. Johnson and H. P. M. Pratt, unpublished observations), prevent DNA replication at the four- and eight-cell stages (Smith and Johnson, 1985), and inhibit glycosylation from the two-cell stage (Surani et al., 1981, 1983; Pratt et al., 1982, Sutherland and Calarco-Gillam, 1983). Four agents have been shown to have effects on polarization and these are described in more detail.

Monoclonal antibodies to a cell surface antigen derived from embryonal carcinoma cells, named variously uvomorulin (Hyafil et al., 1980, 1981) and cadherin (Ogou et al., 1982; Shirayoshi et al., 1983; Yoshido-Noro et al., 1984), bind to early mouse embryos and both prevent and reverse intercellular flattening at compaction. The antigen(s) recognized may be the same in each case, has a maximum molecular weight of 124 kD, is a fucosylated glycoprotein, and is present in a similar form on many other cell types where it is also involved in adhesion (Takeichi et al., 1981, and this volume). Shirayoshi et al. (1983) have reported that the presence of this antibody prevents not only intercellular flattening but also cell polarization. We have reinvestigated this interesting observation (M. H. Johnson, B. Maro, and M. Takeichi, unpublished observations), but are unable to find conditions in which polarization is affected as assessed by the polarized distribution of concanavalin A receptors, intracellular actin, and clathrin-coated vesicles (Table 1, line 2). At present we are investigating the basis for this discrepancy in results.

Cytochalasin D (CCD) modifies interactions between microfilaments and blocks actin turnover within microfilaments, leading to their dissociation (discussed in detail in Maro et al., 1984). CCD has been used on whole embryos to study the role of microfilaments in polarization, but interpretation of the results is difficult because the drug was usually present throughout

Table 1.   Incidence of Polarity in 2/8 Natural Pairs Exposed to a Variety of
            Experimental Protocols[a]

Culture condition	Number of cells analyzed	With surface polarity (%)	With cytoplasmic polarity (%)
Control	110	70	74
ECCD-1[b]	107	66	73
Cytochalasin D (0.5 μg/ml)	42	71	44[c]
Nocodazole (10 μM)	113	55	13

[a] Four-cell embryos were disaggregated to 4 × 1/4 blastomeres and placed in culture. Each hour all 1/4 blastomeres dividing to 2/8 were harvested and cultured as natural 2/8 pairs for nine hours before scoring for polarity as described in Johnson and Maro (1984). About half of the 1/4 cells recorded from ECCD-1 cultures were placed in ECCD-1 prior to division to 2/8.
[b] Monoclonal antibody reactive with cadherin (Shirayoshi et al., 1983).
[c] Clathrin only; for actin the percentage is zero.

periods of cytokinesis, during which turnover and relocation of actin occurs (Ducibella and Anderson, 1975; Surani et al., 1980; Pratt et al., 1981, 1982; Sutherland and Calarco-Gillam, 1983), and because some cell features used to identify polarity such as microvilli are modified by the drug. Therefore, we restricted CCD exposure to the eight-cell stage and applied the drug throughout the process of polarization, or after its completion but before the division to 2/16 cells, in order to examine the role of microfilament turnover in the generation and stabilization of polarity (Johnson and Maro, 1984). The incidence of surface polarity was not reduced regardless of the period of exposure to the drug (Table 1, line 3). However, the polar distribution of cytoplasmic actin that is normally observed was lost completely and that of clathrin-coated vesicles reduced. Moreover, the surface poles were less well organized, containing long and irregular microvilli (as observed under the scanning electron microscope), and when the drug was present throughout, poles were frequently not opposite to the point of intercellular contact as is the case in control embryos. From these results, we conclude that microfilaments probably do play a role in the detailed organization of the surface pole and possibly in orienting the pole with respect to the point of intercellular contact, but the precise nature of that role remains unclear.

Drugs such as nocodazole, colcemid, vinblastine, or colchicine, which shift the equilibrium from microtubular to monomeric tubulin, have also been used to investigate polarity, but again most studies involved a period of drug presence including one or more mitotic divisions prior to analysis

(Ducibella and Anderson, 1975; Surani et al., 1980; Pratt et al., 1982; Sutherland and Calarco-Gillam, 1983). When colcemid or nocodazole was applied either to newly formed, whole eight-cell embryos (Ducibella, 1982; Maro and Pickering, 1984) or to newly formed pairs of eight cells (Johnson and Maro, 1985), and embryos or cells were examined before entry into the next mitotic cycle had occurred, the drugs were shown to disrupt the intracellular microtubule network (Ducibella, 1982; Maro and Pickering, 1984; Johnson and Maro, 1985). The drugs also exerted profound effects on the organization of cytoplasmic features of polarization (Table 1, line 4). For example, the incidence of polar actin and clathrin patterns was reduced almost to zero, whether the drug was present throughout or added only after polarization had developed (Johnson and Maro, 1985). However, once again surface polarity developed, albeit at a reduced incidence, and once established was relatively stable in the presence of the drug (Ducibella, 1982; Maro and Pickering, 1984; Johnson and Maro, 1985). Surface polarity is most easily scored in pairs of cells (Johnson and Maro, 1985) or in thin sections (Ducibella, 1982) rather than by scanning electron microscopy, since the poles of microvilli/concanavalin A binding tend to be less clearly demarcated at their periphery and spread over a broader area of the cell surface in drug-treated embryos than in controls. From these results we conclude that disruption of microtubules has profound effects on both the development and stability of intracellular polarity, but that its effects on both the development and stability of surface polarity are less marked.

Culture of newly formed eight-cell embryos in a fourth drug, taxol, which induces a noncontrolled polymerization of microtubules into randomly distributed bundles in eight-cell blastomeres (Maro and Pickering, 1984), also fails to prevent either development of surface polarity or loss of established surface polarity during interphase (Johnson and Maro, 1985), although again cytoplasmic polarity is lost.

The one clear conclusion that might be drawn from the data on drug effects is that the development and stability of cell surface polarity is relatively resistant to the action of drugs, when compared to intracellular cytoplasmic polarity. This conclusion raises questions about the nature of the causal sequence of events leading to development of polarity. Is it possible that the earliest elements of polarity are generated first in the form of a surface heterogeneity, which serves as a focus for the organization of both overt surface polarity and for cytoskeletal polarity? Cytoplasmic polarity might then develop in secondary association with the cytoskeleton (Johnson and Maro, 1985). Such a sequence would also be consistent with the observation that elements of cytocortical, but not of cytoplasmic, polarity are conserved during division of polarized eight-cell blastomeres (Johnson and Ziomek, 1981b; Reeve, 1981b). The notion that heterogeneity within the cytocortex might provide a focus for secondary cytoplasmic organization is one familiar to nonmammalian embryologists (see Johnson and Pratt,

1983), although the precise basis for such foci has yet to be established (see, however, Blikstad et al., 1983; Lazarides et al., 1984; Moon et al., 1984). If such a conclusion is indeed correct, then the cell contact-mediated signal that leads to polarization must either elicit its response within the plane of the membrane, for example by bulk flow or lateral diffusion of membrane constituents in relation to the point of contact (Pratt, 1978; Wolf, 1983), or via signals that function independently of the detailed structural organization of the cytoplasm, such as ion currents derived from focal changes in ionic permeability (Nuccitelli and Wiley, 1983).

CLEAVAGE PLANES AND CELL INTERACTIONS

The position and orientation of the cleavage plane within a cell will determine the relative size of the two progeny cells, and if the cell is not organized symmetrically, may also influence the qualitative distribution of resources between the progeny. An excellent general review of the influence of the cleavage plane on cell diversification has been published recently by Freeman (1983).

The mammalian embryo develops from an oocyte generated by two serial nonequivalent divisions, during which the small first and second polar bodies are discarded (Figure 1). The mechanism underlying the second of these divisions has been investigated recently (Maro et al., 1984). It should be stressed that the cleavage process that follows fertilization is not in itself essential for successful progress through the *temporal* program of molecular and cytological maturation to the blastocyst, since it occurs to give "one- or two-cell blastocysts" in the presence of CCD, which inhibits cytokinesis (Surani et al., 1980; Kimber and Surani, 1981; Pratt et al., 1981; Petzoldt et al., 1983).

Cleavage to two, four, and eight cells does not appear to result in any systematic size differences among blastomeres, although by the four- to eight-cell transition there can be an asynchrony of 1–3 hours among blastomeres within an embryo (Kelly et al., 1978). The division from the polarized eight-cell stage to the 16-cell stage is marked by the generation of subpopulations of apolar and polar cells. The latter are larger than the former, although there is considerable overlap between the two groups and size *per se* is an inadequate diagnostic criterion for the state of polarity (Johnson and Ziomek, 1981b). However, when an individual polar 1/8 blastomere divides so as to generate one polar and one apolar cell, it is the apolar cell that is smaller in 95% of the couplets in which a size difference is evident (Johnson and Ziomek, 1981b). The systematic difference in cell size generated at this division appears to be related to the eccentric position of the nucleus, and hence of the mitotic spindle, in the more basal region of the polarized eight-cell blastomere (Reeve and Kelly, 1983). This basal position is determined by the cell interaction that induces polarity

(Reeve and Kelly, 1983), and thus it is a cell interaction that determines the generation of cells of different sizes.

The cleavage plane of an *isolated*, polarized eight-cell blastomere appears to be random with respect to the surface pole of microvilli. Division may cut the pole and generate two cells, each possessing part of the surface pole that extends across the midbody. These conservative divisions (i.e., that "conserve" the polar phenotype in both progeny) constitute, on average, one to two out of every eight. The remaining divisions are differentiative (and lead to progeny that may be "differentiated" from each other by their polar and apolar phenotypes). Thus, eight isolated 1/8 blastomeres tend to generate nine or ten polar and six or seven apolar 1/16 cells (Johnson and Ziomek, 1981b).

Is the cleavage plane similarly random when 1/8 cells are in contact with other cells? Initially, we investigated this question by disaggregating 16-cell embryos and assigning their constituent cells to either polar or apolar categories; we found that in most cases (93%) embryos had six, seven, or eight apolar cells (Johnson and Ziomek, 1981b). This result suggested that contact had little effect on the orientation of the cleavage plane. Others have used a range of techniques to investigate cell ratios at the 16-cell stage and find considerable variation among embryos, occasionally reporting two or fewer inside apolar cells and more usually four or more inside apolar cells (Barlow et al., 1972; Graham and Lehtonen, 1979; Handyside, 1981; Balakier and Pedersen 1982; Surani and Barton, 1984). It is clear that this variation among individual embryos makes large sample sizes important. Recently, we reexamined this question by analyzing blastomeres in intact embryos during the transition from the 8- to the 16-cell stage for the orientation of their spindles in relation to the axis of polarity (M. H. Johnson, S. J. Pickering, and B. Maro, unpublished observations). Our results suggest that about four cells divide differentiatively and that our earlier estimates of 6–8 differentiative divisions may have been high. If correct, this result implies that cell contact within the intact embryo can influence the orientation of the spindle away from alignment along the axis of polarity. We are now investigating how this is achieved.

The division of polarized *16-cell* blastomeres has been shown clearly and unequivocally to be influenced by cell interaction (Johnson and Ziomek, 1983). A polar cell alone can divide differentiatively or conservatively; a polar cell aggregated with a second polar cell divides conservatively at higher frequency; a polar cell aggregated with an apolar cell will only divide conservatively. The increase in the frequency of conservative divisions in this sequence of combinations appears to be related directly to the extent of intercellular flattening. This situation is analogous to that described by Meshcheryakov (cited in Freeman, 1983) for *Lymnaea* blastomeres, in which the degree of intercellular flattening places constraints on the possible orientations of the cleavage spindle. Thus, cell interaction influences the orientation of cleavage from 16 to 32 cells by virture of the adhesive properties of the interacting cells.

CELL ADHESION AND CELL INTERACTION

Blastomeres tend to flatten on each other to a limited extent throughout cleavage (Magnuson et al., 1977), but the phenomenon is particularly marked at the mid-eight-cell stage during compaction. At this time, cell outlines become so indistinct that the embryo appears as a single cell at the light microscope level (Ducibella and Anderson, 1975; Lehtonen, 1980; Ziomek and Johnson, 1980). Associated with the intercellular flattening at compaction is the first appearance of gap junctions and focal tight junctions (Ducibella et al., 1975; Magnuson et al., 1977; Lo and Gilula, 1979; Goodall and Johnson, 1982, 1984; McLachlin et al., 1983). However, although extensive intercellular flattening and the formation of specialized junctions develop together, it is not clear that they are interdependent. Manipulation of Ca^{2+} levels, use of antisera to embryonal carcinoma cells, or use of agents that interfere with the cytoskeleton can dissociate the two sets of events (H. Goodall, in preparation).

Several observations combine to suggest that the striking morphological changes observed at compaction are merely the first expression of a progressive change in surface properties manifested by changing patterns of cell adhesion, cell spreading, and junctional organization. First, it is more difficult to decompact embryos with a variety of agents at the 16-cell stage than at the eight-cell stage, and embryos cultured continuously in such agents from precompact stages nonetheless tend to show some flattening behavior from the 16-cell stage onward (Shirayoshi et al., 1983; Surani et al., 1983). Second, when pairs of blastomeres are examined for their cell-flattening properties, distinctive patterns of adhesiveness may be identified that relate to the progressively more advanced developmental stages of the blastomeres and to their position within the embryo (Ziomek and Johnson, 1981; Kimber et al., 1982). Third, the organization of junctional complexes matures during the 16- and 32-cell stages such that desmosomes appear, and the focal tight junctions become zonular on outer cells and mark a boundary between the outer pole of relatively nonadhesive microvilli and the inner, more adhesive surface (Ducibella et al., 1975; Magnuson et al., 1977). The presence of these zonular junctions, in conjunction with the vectorial transport of fluid in the polarized outer trophectoderm cells, permits the formation of a blastocoele containing blastocoelic fluid (McLaren and Smith, 1977; Wiley and Eglitis, 1981).

The surface adhesive properties of the constituent cells of embryos from the 16-cell stage onward, and especially the differences between inner apolar and outer polar cells, are of particular importance for the regulation of the inner cell–ICM lineage, since exposure of the inner apolar cells to asymmetric cell contact generates polarity and thus a crossing of lineage. An analysis of the nature and regulation of surface adhesive properties is therefore essential for an understanding of the generation of cell diversity (Kimber et al., 1982). In order to analyze the processes of intercellular flattening, it is useful to discriminate between changes in the extracellular matrix, in the surface properties of cells, and in their underlying cortical structure.

The Intercellular Matrix

The principal and most important extracellular matrix of the early embryo is its zona pellucida, derived from the granulosa cells of the follicle and composed of three major glycoprotein species (Bleil and Wassarman, 1980,b; East and Dean, 1984). The zona pellucida performs important regulatory roles in embryogenesis, because in many species it reduces the incidence of polyspermy (Bedford, 1983), prevents both the aggregation of adjacent cleaving embryos to form chimeras and the dispersal of cells from a single embryo to form twins (Mintz, 1965; Tsunoda and McLaren, 1983), and may also affect the packing of blastomeres and thereby influence distribution of cells within the lineages of the embryo (Graham and Lehtonen, 1979).

The only extracellular matrix glycoprotein to be detected at preblastocyst stages is laminin. Immunocytochemical analyses have provided evidence for the presence of laminin within and between apposing surfaces of blastomeres at the 8- and 16-cell stages, but not earlier (Leivo et al., 1980; Wu et al., 1983). Concentrations of laminin at sites of developing intercellular junctions were particularly marked. Immunocytochemical detection does not distinguish between the presence of intact laminin molecules and the presence of its isolated constituent A, B_1, and B_2 chains (Cooper et al., 1981). Metabolic labeling of laminin subunits, and their specific immuno-precipitation, has revealed that coordinate synthesis of all three subunits cannot be detected until the 16-cell stage (Cooper and MacQueen, 1983). Prior to that stage, oocytes and fertilized eggs synthesize only B_1 chains, two-cell embryos synthesize little detectable laminin, and four- and eight-cell embryos synthesize only B_1 and B_2 chains. All three chains are required for synthesis and secretion of the complete and functional laminin molecule (Cooper et al., 1981), so it is possible that the extracellular immunoreactive material detected at the eight-cell stage is nonfunctional. Evidence consistent with this possibility is provided by the distinctive pattern of reactivity at these early stages, observed after use of certain monoclonal antibodies to laminin (Wan et al., 1984). What does seem clear is that whether or not laminin has a role to play in mediating intercellular adhesion at the eight-cell stage, it may well be involved in the more marked adhesion found between cells from the 16-cell stage onward.

The Cell Surface

Adhesion and recognition tend to involve glycosylated molecules, particularly integral membrane glycoproteins and glycolipids. The mouse preimplantation embryo shows characteristic changes in its surface profile of sugars and glycosylated molecules, as evidenced by surface labeling and fractionation, internal labeling and immunoprecipitation, immunoblotting, and probing with antisera and lectins under a variety of conditions (Solter and Knowles, 1978; Willison and Stern, 1978; Surani, 1979; Gooi et al., 1981; Hyafil et al., 1981; Surani et al., 1981; Takeichi et al., 1981; for a review, see Webb, 1983). Prolonged tunicamycin treatment prevents some of the changes from occurring (Magnuson and Epstein, 1981; Surani et al., 1981) and also prevents or retards the process of intercellular flattening, although the effect is rarely total and the presence of undesirable effects on protein synthesis is not always controlled effectively (Maylie-Pfenninger and Bennett, 1979; Surani, 1979; Atienza-Samols et al., 1980; Surani et al., 1981; Webb and Duksin, 1981; Pratt et al., 1982; Sutherland and Calarco-Gillam, 1983). The confusingly named "compactin" (ML236B) also inhibits glycosylation via its effects on HMG CoA reductase, and the inhibition of flattening that it produces is substantially reversed by the addition of mevalonic acid (Surani et al., 1983). Unlike tunicamycin, compactin did not inhibit intercellular flattening when embryos were cultured in the reagent from the two-cell stage. However, with continuing incubation in the drug, decompaction occurred at the 16-cell stage, cell outlines becoming visible. It is important to note, as indicated earlier, that cell polarization was not found to be inhibited in any of the studies in which it was assessed.

These results are consistent with, but not proof of, a role for glycosylated molecules in the process of intercellular flattening. More direct evidence relating to this question comes from an analysis of the effects of sugars or sugar-binding molecules on cell spreading. Thus, attempts to mimic intercellular flattening by aggregating a single 1/8 blastomere to a lectin-coated agarose bead (Kimber and Surani, 1982) revealed that only beads coated with peanut lectin, wheat germ agglutinin, and concanavalin A were suitable substrates for flattening and engulfment. Moreover, these processes were inhibited by sugars with high affinity for the lectin. However, when free sugars, alone or in various combinations, and a range of sugar-containing molecules were examined for their effects on cell flattening in the intact embryo, little or no physiological effect was evident (Kimber and Surani, 1982), with the exception of certain fucosylated molecules such as lacto-N-fucopentoase III (alone or in combination with lacto-N-fucopentoase II) (Bird and Kimber, 1984). In the latter cases, the sugars did not *prevent* cell flattening, even when present from the two- to four-cell precompact stages. However, presence of the fucopentoases subsequently did reverse intercellular flattening. These results provide evidence that a subset of surface fucosylated molecules might be involved in the later stages of intercellular flattening.

It is not clear from these results with tunicamycin and oligosaccharides whether glycosylated determinants are merely permissive for intercellular flattening, or whether changes in the glycosylation status determine the onset and subsequent development of flattening. It is possible that the appropriate surface profile of sugars is displayed early in development and that some other change triggers flattening. Evidence consistent with this possibility is provided by an analysis of the antigenic determinant(s) uvomorulin/cadherin.

Uvomorulin/cadherin (or a similar molecule) is detectable on many types of epithelial cells and is involved in intercellular adhesion (Ogou et al., 1983; Yoshida-Noro et al., 1984). Its adhesive properties (and its capacity to bind antibody) are inhibited by low Ca^{2+}, which also renders the molecule susceptible to proteolytic digestion by trypsin (Hyafil et al., 1980; Takeichi et al., 1981). Monoclonal antibodies (designated DEI and ECCD-1) to uvomorulin/cadherin bind the antigen only in the presence of Ca^{2+} (Yoshida-Noro et al., 1984). Moreover, when individual blastomeres that have been exposed to trypsin plus low Ca^{2+} and thereby deprived of their uvomorulin/cadherin are aggregated together or with normal blastomeres, flattening is inhibited (Bilozur and Powers, 1983), suggesting that mutual expression of the molecule on interacting cells is required. Taken together, these observations suggest that the uvomorulin/cadherin molecular complex of itself can provide an adequate explanation for the cell surface-mediated components of cell adhesion in the embryo. The monoclonal antibody to cadherin is less effective at blocking cell flattening and interaction at the later 16-cell to early 32-cell stage, and it fails to block the formation of zonular tight junctions that takes place at this time (Shirayoshi et al., 1983). Thus, uvomorulin/cadherin would seem to be involved primarily in the early stage of cell flattening. However, uvomorulin/cadherin is also detected on the surface of blastomeres from the one-cell stage onward and does not show evidence of a marked change in expression at the time of compaction (Hyafil et al., 1981; Ogou et al., 1982). Moreover, the limited flattening of one blastomere upon another observed at the two- and four-cell stages (Magnuson et al., 1977) is also inhibited by ECCD-1 (M.H. Johnson, unpublished observations). Thus, we are again forced to examine the possibility that the changes in cell flattening observed at the time of compaction may not occur as a result of the appearance of uvomorulin/cadherin, but rather as a result of changes either in its organizaton within the membrane or in the organization of the underlying cytocortical structure.

The Cytocortex

The apposed cell surfaces of adjacent blastomeres prior to, during, and after compaction are characterized by a subcortical zone deficient in both myosin and actin (Lehtonen and Badley, 1980; Sobel, 1983; Johnson and Maro, 1984). If the extent of the zone of contact is decreased, or if its location is altered by rotating cells to new contact points, then the regions deficient in myosin and actin are correspondingly affected. If CCD is added to blastomeres flattened against each other, the blastomeres round up

(Ducibella and Anderson, 1975; Surani et al., 1980; Kimber and Surani, 1981; Pratt et al., 1981, 1982; Sutherland and Calarco-Gillam, 1983) and the cytoplasmic actin is dispersed (Johnson and Maro, 1984). Hence, the disposition of actin and myosin is related intimately to the capacity of cells to flatten upon each other. In contrast, spectrin is disposed throughout the cortex of blastomeres regardless of contact points (Sobel and Alliegro, 1983).

The redistribution of microtubules associated with cell flattening (Ducibella and Anderson, 1975; Ducibella et al., 1977) is *not* matched by a corresponding sensitivity of the process to drugs that depolymerize microtubules. Thus, as long as drugs such as nocodazole or colcemid are not applied to embryos (or pairs of cells) during a mitotic period (when blockage in mitosis with a rounded morphology occurs), they do not inhibit or reverse flattening (Ducibella, 1982; Sutherland and Calarco-Gillam, 1983; Maro and Pickering, 1984; Johnson and Maro, 1985). Indeed, nocodazole reduces the variation in the amount of time it takes embryos at the mid-eight-cell stage to complete flattening (Maro and Pickering, 1984), almost as though, once a signal to flatten is received, the presence of microtubules slows down the flattening process. This view is supported by the effects of taxol, which by generating multiple disorganized microtubule bundles inhibits and reverses flattening (Maro and Pickering, 1984).

It seems clear that an important change at the mid-eight-cell stage triggers the development of the capacity for flattening as mediated by extension of cytoplasmic processes under the regulatory control of the microtubule system and involving actin–myosin interaction. The basis for this change is not clear, but it is not obviously dependent upon proximate genetic activity, as it is resistant to inhibitors of transcription and DNA replication applied as early as the newly formed four-cell stage (Smith and Johnson, 1985; M.H. Johnson, unpublished data). Therefore, it is presumably triggered by some component of a cytoplasmic program for development; in this context, it is of interest that this period is characterized by changes in the phosphorylation status of certain unidentified polypeptides (Lopo and Calarco, 1982). Moreover, synthesis of intermediate filament proteins is first detected around the time of compaction (Oshima et al., 1983), and the appearance of intermediate filaments themselves may be detected at this time (Lehtonen et al., 1983). Evidence from immunoblotting studies suggests the existence of a considerable pool of presynthesized, soluble, cytokeratinlike intermediate filament proteins (Lehtonen et al., 1983). Since in mature cells newly formed intermediate filament proteins are usually modified posttranslationally to an insoluble, filament form immediately after synthesis (Moon and Lazarides, 1983), the finding of a pool of soluble monomers in early blastomeres might provide another pathway for the rapid and posttranscriptional regulation of cell form. This phase of development also coincides with major changes in the patterns of lipid metabolism, which has led to the suggestion that altered fluidity within intra-membranous domains might play a regulatory role in flattening (discussed in detail in Pratt, 1978, 1982).

CONCLUSION

The results obtained from manipulation of the cells of the early mouse embryo have made it possible to define three basic patterns of cell interactions that mold the developing blastocyst. Each interaction has distinct consequences for the intracellular organization of the interacting cells, and for the disposition of this organization at division. The setting up of cell diversity depends upon the processes of polarization and cell division. The subsequent regulation of the ratio of inside-to-outside cells depends on the process of cell flattening. Both polarization and cell flattening to maximize cell contact appear to involve specific cell surface molecules that, in the latter case at least, have been identified as uvomorulin/cadherin. However, the changing pattern of organization observed at the eight-cell stage does not seem to be explicable simply in terms of the appearance *de novo* of specific recognition molecules. Rather, these are present from at least the two-cell stage onward. The trigger for the various events that constitute compaction requires a reorganization of preexisting cell components, much of which occurs independent of proximate transcriptional activity.

ACKNOWLEDGMENTS

I wish to thank Bernard Maro, Virginia Bolton, Harry Goodall, Sue Kimber, Hester Pratt, and Brigid Hogan for helpful comments and discussion, and Anne M. Wright for preparation of the manuscript. The work described here was supported by grants from the Medical Research Council and the Cancer Research Campaign.

REFERENCES

Atienza-Samols, S: B., P. R. Pine, and M. I. Sherman (1980) Effects of tunicamycin upon glycoprotein synthesis and development of early mouse embryos. *Dev. Biol.* **79**:19–32.

Balakier, H., and R. A. Pedersen (1982) Allocation of cells to inner cell mass and trophectoderm lineages in preimplantation mouse embryos. *Dev. Biol.* **90**:352–362.

Barlow, P., D. A. J. Owen, and C. F. Graham (1972) DNA synthesis in the preimplantation mouse embryo. *J. Embryol. Exp. Morphol.* **27**:431–445.

Bedford, J. M. (1983) Fertilization. In *Reproduction in Mammals*, Vol. 1, C. R. Austin and R. V. C. Short, eds., pp. 128–164, Cambridge Univ. Press, Cambridge.

Bilozur, M., and R. D. Powers (1983) Effect of calcium-free trypsinization on interaction of blastomeres from compact mouse embryos. *J. Cell Biol.* **97**:37a.

Bird, J. M., and S. J. Kimber (1984) Oligosaccharides containing fucose linked $\alpha(1-3)$ and $\alpha(1-4)$ to N-acetylglucosamine cause decompaction of mouse morulae. *Dev. Biol.* **104**:449–460.

Bleil, J. D., and P. M. Wassarman (1980a) Structure and function of the zona pellucida: Identification and characterization of the proteins of the mouse oocyte's zona pellucida. *Dev. Biol.* **76**:185–202.

Bleil, J. D., and P. M. Wassarman (1980b) Synthesis of zona pellucida proteins by denuded and follicle-enclosed mouse oocytes during culture *in vitro. Proc. Natl. Acad. Sci. USA.* **77**:1029–1033.

Blikstad, I., W. J. Nelson, R. T. Moon, and E. Lazarides (1983) Synthesis and assembly of spectrin during avian erythropoiesis: Stochiometric assembly but unequal synthesis of α and β spectrin. *Cell* **32**:1081–1091.

Cooper, A. R., and H. A. MacQueen (1983) Subunits of laminin are differentially synthesized in mouse eggs and early embryos. *Dev. Biol.* **96**:467–471.

Cooper, A. R., M. Kurkinen, A. Taylor, and B. L. M. Hogan (1981) Studies on the biosynthesis of laminin by murine parietal endoderm cells. *Eur. J. Biochem.* **119**:189–197.

Ducibella, T. (1982) Depolymerization of microtubules prior to compaction. *Exp. Cell Res.* **138**:31–38.

Ducibella, T., and E. Anderson (1975) Cell shape and membrane changes in the eight-cell mouse embryo. Prerequisites for morphogenesis of the blastocyst. *Dev. Biol.* **47**:45–58.

Ducibella, T., D. F. Albertini, E. Anderson, and J. D. Biggers (1975) The preimplantation mammalian embryo: Characterization of intercellular junctions and their appearance during development. *Dev. Biol.* **45**:231–250.

Ducibella, T., T. Ukena, M. Karnovsky, and E. Anderson (1977) Changes in cell surface and cortical cytoplasmic organization during early embryogenesis in the preimplantation mouse embryo. *J. Cell Biol.* **74**:153–167.

East, I. J., and J. Dean (1984) Monoclonal antibodies as probes of the distribution of ZP-2, the major sulphated glycoprotein of the murine zona pellucida. *J. Cell Biol.* **98**:795–800.

Fleming, T. P., and S. J. Pickering (1985) Maturation and polarization of the endocytotic system in outside blastomeres during mouse preimplantation development. *J. Embryol. Exp. Morphol.* (in press).

Fleming, T. P., P. D. Warren, J. C. Chisholm, and M. H. Johnson (1984) Trophectodermal processes regulate the expression of totipotency within the inner cell mass of the mouse expanding blastocyst. *J. Embryol. Exp. Morphol.* **84**:63–90.

Freeman, G. (1983) The role of egg organization in the generation of cleavage patterns. In *Time, Space and Pattern in Embryonic Development,* W. R. Jeffery and R. A. Raff, eds., pp. 171–196, Alan R. Liss, New York.

Goodall, H., and M. H. Johnson (1982) The use of carboxyfluorescein diacetate to study formation of permeable channels between mouse blastomeres. *Nature* **295**:524–526.

Goodall, H., and M. H. Johnson (1984) The nature of intercellular coupling within the preimplantation mouse embryo. *J. Embryol. Exp. Morphol.* **79**:53–76.

Gooi, H. C., T. Feizi, A. Kapadia, B. B. Knowles, D. Solter, and M. J. Evans (1981) Stage specific embryonic antigen involves 1–3 fucosylated type 2 blood group chains. *Nature* **292**:156–158.

Graham, C. F., and E. Lehtonen (1979) Formation and consequences of cell patterns in preimplantation mouse development. *J. Embryol. Exp. Morphol.* **49**:277–294.

Handyside, A. H. (1980) Distribution of antibody- and lectin-binding sites on dissociated blastomeres from mouse morulae: Evidence for polarization at compaction. *J. Embryol. Exp. Morphol.* **60**:99–116.

Handyside, A. H. (1981) Immunofluorescent techniques for detection of the numbers of inside and outside cells from mouse morulae. *J. Reprod. Immunol.* **2**:339–350.

Hyafil, F., D. Morello, C. Babinet, and F. Jacob (1980) A cell surface glycoprotein involved in the compaction of embryonal carcinoma cells and cleavage-stage embryos. *Cell* **21**:927–934.

Hyafil, F., C. Babinet, and F. Jacob (1981) Cell-cell interactions in early embryogenesis: A molecular approach to the role of calcium. *Cell* **26**:447–454.

Johnson, M. H. (1985) Manipulation of early mammalian development: What does it tell us about cell lineages? In *Developmental Biology: A Comprehensive Synthesis*, Vol. 4., R. B. L. Gwatkin, ed., Plenum, New York (in press).

Johnson, M. H., and B. Maro (1984) The distribution of cytoplasmic actin in mouse 8-cell blastomeres. *J. Embryol. Exp. Morphol.* **82**:97–117.

Johnson, M. H., and B. Maro (1985) A dissection of the mechanisms generating and stabilizing polarity in mouse 8- and 16- cell blastomeres: The role of the cytoskeleton. *J. Embryol. Exp. Morphol.* (in press).

Johnson, M. H., and H. P. M. Pratt (1983) Cytoplasmic localizations and cell interactions in the formation of the mouse blastocyst. In *Time, Space and Pattern in Embryonic Development*, W. R. Jeffery and R. A. Raff, eds., pp. 287–312, Alan R. Liss, New York.

Johnson, M. H., and C. A. Ziomek (1981a) Induction of polarity in mouse 8-cell blastomeres: Specificity, geometry, and stability. *J. Cell Biol.* **91**:303–308.

Johnson, M. H., and C. A. Ziomek (1981b) The foundation of two distinct cell lineages within the mouse morula. *Cell* **24**:71–80.

Johnson, M. H., and C. A. Ziomek (1983) Cell interactions influence the fate of mouse blastomeres undergoing the transition from the 16- to the 32-cell stage. *Dev. Biol.* **95**:211–218.

Johnson, M. H., A. H. Handyside, J. Chakraborty, K. Willison, and P. Stern (1979) The effect of prolonged decompaction on the development of the preimplantation mouse embryo. *J. Embryol. Exp. Morphol.* **54**:241–261.

Johnson, M. H., J. McConnell, and J. Van Blerkon (1984) Programmed development in the mouse embryo. *J. Embryol. Exp. Morphol.* (Suppl.) **83**:197–231.

Kelly, S. J., J. G. Mulnard, and C. F. Graham (1978) Cell division and cell allocation in early mouse development. *J. Embryol. Exp. Morphol.* **48**:37–51.

Kimber, S. J., and M. A. H. Surani (1981) Morphogenetic analysis of changing cell associations following release of 2-cell and 4-cell mouse embryos from cleavage arrest. *J. Embryol. Exp. Morphol.* **61**:331–345.

Kimber, S. J., and M. A. H. Surani (1982) Spreading of blastomeres from 8-cell mouse embryos on lectin-coated beads. *J. Cell Sci.* **56**:191–206.

Kimber, S. J., M. A. H. Surani, and S. C. Barton (1982) Interactions of blastomeres suggest changes in cell surface adhesiveness during the formation of inner cell mass and trophectoderm in the preimplantation mouse embryo. *J. Embryol. Exp. Morphol.* **70**:133–152.

Lazarides, E., W. J. Nelson, and T. Kasamatsu (1984) Segregation of two spectrin forms in the chicken optic system: A mechanism for establishing restricted membrane-cytoskeletal domains in neurons. *Cell* **36**:269–278.

Lehtonen, E. (1980) Changes in cell dimensions and intercellular contacts during cleavage-stage cell cycles in mouse embryonic cells. *J. Embryol. Exp. Morphol.* **58**:213–249.

Lehtonen, E., and R. A. Badley (1980) Localization of cytoskeletal proteins in preimplantation mouse embryos. *J. Embryol. Exp. Morphol.* **55**:211–225.

Lehtonen, E., U. P. Lehto, T. Vartio, R. A. Badley, and I. Virtanen (1983) Expression of cytokeratin polypeptides in mouse oocytes and preimplantation embryos. *Dev. Biol.* **100**:158–165.

Leivo, L., A. Vaheri, R. Timpl, and J. Wartiovaara (1980) Appearance and distribution of collagens and laminin in the early mouse embryo. *Dev. Biol.* **76**:100–114.

Lo, C. W., and N. B. Gilula (1979) Gap junctional communication in the preimplantation mouse embryo. *Cell* **18**:399–409.

Lopo, A. C., and P. G. Calarco (1982) Stage-specific changes in protein phosphorylation during preimplantation development in the mouse. *Gamete Res.* **5**:283–290.

McLachlin, J. R., S. Caveney, and G. M. Kidder (1983) Control of gap junction formation in early mouse embryos. *Dev. Biol.* **98**:155–164.

McLaren, A. and R. Smith (1977) A functional test of tight junctions in the mouse blastocyst. *Nature* **267**:351–353.

Magnuson, T., and C. J. Epstein (1981) Characterization of concanvalin A precipitated proteins from early mouse embryos: A 2-dimensional gel electrophoresis study. *Dev. Biol.* **81**:193–199.

Magnuson, T., A. Demsey, and C. W. Stackpole (1977) Characterization of intercellular junctions in the preimplantation mouse embryo by freeze-fracture and thin-section electron microscopy. *Dev. Biol.* **61**:252–261.

Maro, B., and S. J. Pickering (1984) Microtubules influence compaction in preimplantation mouse embryos. *J. Embryol. Exp. Morphol.* **84**:217–232.

Maro, B., M. H. Johnson, S. J. Pickering, and G. Flach (1984) Changes in actin distribution during fertilization of the mouse egg. *J. Embryol. Exp. Morphol.* **81**:211–237.

Maro, B., M. H. Johnson, S. J. Pickering, and D. Louvard (1985) Changes in the distribution of membranous organelles during mouse early development. *J. Embryol. Exp. Morphol.* (in press).

Maylie-Pfenninger, M. F., and D. Bennett (1979) The effect of tunicamycin on the development of mouse preimplantation embryos. *J. Cell Biol.* **83**:216a.

Mintz, B. (1965) Experimental genetic mosaicism in the mouse. *CIBA Found. Symp.* **1**:194–209.

Moon, R. T., and E. Lazarides (1983) Canavanine inhibits vimentin assembly but not its synthesis in chicken embryo erythroid cells. *J. Cell Biol.* **97**:1309–1314.

Moon, R. T., I. Blikstad, and E. Lazarides (1984) Regulation and assembly of the spectrin-based membrane skeleton in chicken embryo erythroid cells. In *Cell Membranes: Methods and Reviews*, Vol. 2, E. L. Elson, W. A. Frazier and L. Glaser, eds., pp. 197–218, Plenum, New York.

Nuccitelli, R., and L. M. Wiley (1983) Polarity of isolated blastomeres from mouse morulae: Detection of transcellular ion currents. *J. Cell Biol.* **97**:32a.

Ogou, S-I., T. S. Okada, and M. Takeichi (1982) Cleavage-stage mouse embryos share a common cell adhesion system with teratocarcinoma cells. *Dev. Biol.* **92**:521–528.

Ogou, S-I., C. Yoshida-Noro, and M. Takeichi (1983) Calcium-dependent cell–cell adhesion molecules common to hepatocytes and teratocarcinoma stem cells. *J. Cell Biol.* **97**:944–948.

Oshima, R. G., W. E. Howe, F. G. Klier, E. D. Adamson, and L. H. Shevinsky (1983) Intermediate filament protein synthesis in preimplantation murine embryos. *Dev. Biol.* **99**:447–455.

Petzoldt, U., K. Burki, G. R. Illmensee, and K. Illmensee (1983) Protein synthesis in mouse embryos with experimentally produced asynchrony between chromosome replication and cell division. *Roux's Arch. Dev. Biol.* **192**:138–144.

Pratt, H. P. M. (1978) Lipids and transitions in embryos. In *Development in Mammals*, Vol. 3, M. H. Johnson, ed., pp. 83–130, Elsevier, Amsterdam.

Pratt, H. P. M. (1982) Preimplantation mouse embryos synthesize membrane sterols. *Dev. Biol.* **89**:101–111.

Pratt, H. P. M., J. Chakraborty, and M. A. H. Surani (1981) Molecular and morphological differentiation of the mouse blastocyst after manipulations of compaction with cytochalasin D. *Cell* **26**:279–292.

Pratt, H. P. M., C. A. Ziomek, W. J. D. Reeve, and M. H. Johnson (1982) Compaction of the mouse embryo: An analysis of its components. *J. Embryol. Exp. Morphol.* **70**:113–132.

Reeve, W. J. D. (1981a) Cytoplasmic polarity develops at compaction in rat and mouse embryos. *J. Embryol. Exp. Morphol.* **62**:351–367.

Reeve, W. J. D. (1981b) The distribution of injected horseradish peroxidase in the 16-cell mouse embryo. *J. Embryol. Exp. Morphol.* **66**: 191–207.

Reeve, W. J. D., and F. Kelly (1983) Nuclear position in cells of the mouse early embryo. *J. Embryol. Exp. Morphol.* **75**:117–139.

Reeve, W. J. D., and C. A. Ziomek (1981) Distribution of microvilli on dissociated blastomeres from mouse embryos: Evidence for surface polarization at compaction. *J. Embryol. Exp. Morphol.* **62**:339–350.

Shirayoshi, Y., T. S. Okada, and M. Takeichi (1983) The calcium-dependent cell–cell adhesion system regulates inner cell mass formation and cell surface polarization in early mouse development. *Cell* **35**:631–638.

Smith, R. K. W., and M. H. Johnson (1985) DNA replication and compaction in the cleaving embryo of the mouse. *J. Embryol. Exp. Morphol.* (in press).

Sobel, S. (1983) Cell–cell modulation of myosin organization in the early mouse embryo. *Dev. Biol.* **100**:207–213.

Sobel, S., and M. A. Alliegro (1983) Localization of a spectrin-like protein in the preimplantation mouse embryo. *J. Cell Biol.* **97**:36a.

Solter D., and B. B. Knowles (1978) Monoclonal antibody defining a stage-specific mouse embryonic antigen (SSEA1). *Proc. Natl. Acad. Sci. USA* **75**:5565–5569.

Surani, M. A. H. (1979) Glycoprotein synthesis and inhibition of glycosylation by tunicamycin in preimplantation mouse embryos: Compaction and trophoblast adhesion. *Cell* **18**:217–227.

Surani, M. A. H., and S. C. Barton (1984) Spatial distribution of blastomeres is dependent on cell division order and interactions in mouse morulae. *Dev. Biol.* **102**:335–343.

Surani, M. A. H., S. C. Barton, and A. Burling (1980) Differentiation of 2-cell and 8-cell mouse embryos arrested by cytoskeletal inhibitors. *Exp. Cell Res.* **125**:275–286.

Surani, M. A. H., S. J. Kimber, and A. H. Handyside (1981) Synthesis and role of cell surface glycoproteins in preimplantation mouse development. *Exp. Cell Res.* **133**:331–339.

Surani, M. A. H., S. J. Kimber, and J. C. Osborn (1983) Mevalonate reverses the developmental arrest of preimplantation mouse embryos by compactin, an inhibitor of HMG Co A reductase. *J. Embryol. Exp. Morphol.* **75**:205–223.

Sutherland, A. E., and P. G. Calarco-Gillam (1983) Analysis of compaction in the preimplantation mouse embryo. *Dev. Biol.* **100**:328–338.

Takeichi, M., T. Atsumi, C. Yoshida, K. Uno, and T. S. Okada (1981) Selective adhesion of embryonal carcinoma cells and differentiated cells by Ca^{2+}-dependent sites. *Dev. Biol.* **87**:340–350.

Tsunoda, Y., and A. McLaren (1983) Effect of various procedures on the viability of mouse embryos containing half the normal number of blastomeres. *J. Reprod. Fertil.* **69**:315–322.

Wan, Y.-J., T-C. Wu, A. E. Chung, and I. Damjanov (1984) Monoclonal antibodies to laminin reveal the heterogeneity of basement membranes in developing and adult mouse tissues. *J. Cell Biol.* **98**:971–979.

Webb, C. A. G. (1983) Glycoproteins on gametes and early embryos. In *Development in Mammals*, Vol. 5, M. H. Johnson, ed., pp. 155–185, Elsevier, Amsterdam.

Webb, C. A. G., and D. Duksin (1981) Involvement of glycoprotein in the development of early mouse embryos: Effect of tunicamycin and α, α' dipyridyl *in vitro*. *Differentiation* **20**:81–86.

Wiley, L. M., and M. A. Eglitis (1981) Cell surface and cytoskeletal elements: Cavitation in the mouse preimplantation embryo. *Dev. Biol.* **86**:493–501.

Willison, K. R., and P. L. Stern (1978) Expression of a Forssman antigenic specificity in the preimplantation embryo. *Cell* **14**:785–793.

Wolf, D. E. (1983) The plasma membrane in early embryogenesis. In *Development in Mammals*, Vol. 5, M. H. Johnson, ed., pp. 187–208, Elsevier, Amsterdam.

Wu, T-C., Y-J. Wan, A. E. Chung, and I. Damjanov (1983) Immunohistochemical localization of entactin and laminin in mouse embryos and fetuses. *Dev. Biol.* **100**:496–505.

Yoshida-Noro, C., N. Suzuki, and M. Takeichi (1984) Molecular nature of the calcium-dependent cell–cell adhesion system in mouse teratocarcinoma and embryonic cells studied with a monoclonal antibody. *Dev. Biol.* **101**:19–27.

Ziomek, C. A., and M. H. Johnson (1980) Cell surface interaction induces polarization of mouse 8-cell blastomeres at compaction. *Cell* **21**:935–942.

Ziomek, C. A., and M. H. Johnson (1981) Properties of polar and apolar cells from the 16-cell mouse morula. *Roux's Arch. Dev. Biol.* **190**:287–296.

Chapter 3

Adhesion and Movement of Cells
May Be Coupled to Produce Neurulation

ANTONE G. JACOBSON

ABSTRACT

The neural plate of the newt embryo consists of two domains of cells. One domain, called the notoplate, overlies the notochord and elongates and narrows along the midline during neurulation. Cells in the notoplate often change neighbors and thus the region behaves like a fluid. Cells in the rest of the neural plate do not change neighbors; this domain behaves like an elastic solid. Cells of the early neural plate are also programmed to shrink their apical surfaces. Experiments and computer simulation show that shrinkage of the apical surfaces and elongation of the notoplate along the midline are both necessary, and are together sufficient, for early neurulation. The anisotropic reshaping of the neural plate is a result of the elongation of the notoplate.

The period of neural tube formation coincides exactly with that of rapid elongation of the nervous system in newt and chick embryos, and experiments that stop tube formation also stop nervous system elongation. I propose a model to explain elongation of the neural plate by the notoplate. Cells in the two domains of the neural plate may differ in adhesiveness in such a way that the cells of the notoplate attempt to mix with the cells of the rest of the neural plate. Since the rest of the plate is a solid, actual mixing is not possible, and so the boundary between the notoplate and the rest of the neural plate extends, lengthening the midline and buckling the plate into a tube. The contraction of actin filaments moves the cells about, but adhesive differences at the boundary of the two domains order cell movements into a lengthening of the midline. Adhesion and contraction are thus coupled, and both must function to produce neurulation.

Neurulation, and possibly some other anisotropic reshapings in embryos, may be explained by a model in which adhesive differences between groups of cells are coupled to cell movements produced by the contractions of actin filaments. In this model, cell contractility provides the force to move cells about and change the shapes of cells, but adhesive differences at the boundaries of groups of cells guide the reshuffling of the cells in specific directions or patterns such as the shaping and elongation of the neural plate and tube.

Neurulation has been described most thoroughly in amphibian embryos. I focus here mostly on the West Coast newt, *Taricha torosa*. Like most amphibians, *Taricha torosa* has no growth in any embryonic region until at least the larval stage when the heart begins to beat. At that time, some parts such as the nervous system begin to grow at the expense of other cells, mostly the large yolk-laden cells of the gut. Cell divisions in the embryo result in ever-smaller daughter cells. Since there is no growth (no increase in mass) in such embryos, growth can have no role in the early morphogenesis of the organs of these animals. Volume measurements of the entire nervous system of *Taricha torosa*, made from the earliest inception of the neural plate to the larval stage when the heart begins to beat, confirm that no growth occurs in the nervous system during this period (Jacobson, 1978).

In amphibians, the commencement of morphogenesis of the nervous system is the first noticeable consequence of embryonic induction. During gastrulation in the newt embryo, all the cells of the animal hemisphere, from which the neural plate will form, move toward the lips of the blastopore in the movement called epiboly (Figure 1A). Surface cells at the equator and in the vegetal hemisphere of the egg involute over the dorsal lip of the blastopore and extend forward beneath the prospective neural plate. These underlying cells induce the neural plate, and there is an immediate, dramatic change in the behavior of the responding cells. The movements of epiboly toward the blastopore cease abruptly in the cells of the forming plate, and these cells reverse direction and displace toward the animal pole and the midline (Figure 1B). I have observed these movements in scores of time-lapse cinematographic records.

Cellular displacements have been followed in detail during the subsequent transformation of the neural plate into a keyhole shape (Figure 2). In the newt, individual cells have been identified by variegations in the amounts of egg pigment they contain, and their movements have been followed frame by frame in time-lapse movies (Burnside and Jacobson, 1968).

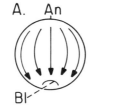
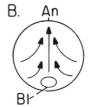

Figure 1. *View of the prospective dorsal surface of the newt embryo. A:* Arrows indicate the direction of movement of surface cells during gastrulation. *B:* Arrows indicate the direction of movement of cells of the forming neural plate at the end of gastrulation. An, animal pole; Bl, blastopore.

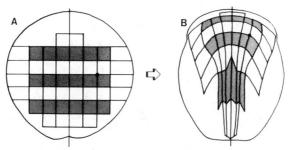

Figure 2. *A: Coordinate grid drawn on the neural plate of a late gastrula-stage newt embryo. B:* At midneurula stage, the coordinate grid is shown distorted by the movements or displacement of the cells of the neural plate. Cells at the intersections of the grid were followed frame by frame in time-lapse movies. (Modified from Burnside and Jacobson, 1968.)

During its transformation from a circular disk shape to a flattened keyhole shape, the neural plate reduces its surface area by 39% and the length of its midline increases by 29%. Burnside (1971, 1973) related the surface shrinkage of the neural plate to the apical constriction of the cells of the plate, which is accomplished by contraction of a purse string of actin filaments just below the apical surface of each cell (first noted, in anuran embryos, by Baker and Schroeder, 1967). The plate cells, which lie perpendicular to the surface of the plate, also elongate, making the plate thicker. Cell elongation during this period was found to be proportional to the amount of apical shrinkage; thus the cells remain cylindrical. Burnside (1973) has described the microtubules, oriented on the long axes of these cells, that are essential for normal elongation.

To measure such changes in cell shape, it is essential that the system be mapped. Since groups of cells are displaced during late gastrulation/ neurulation, it is necessary to measure a group of cells at the beginning stage and then locate later, from the map, the same group of cells and measure them again. A number of studies that claim to describe changes in cell shape have been made, but they were made from "typical" sections at different stages. During the time period studied, cells moved through the sections, and as a result different cells were described at the beginning and at the end of the period (e.g., Baker and Schroeder, 1967; Schroeder, 1970).

From the varied changes in area in the distorted coordinate grid of the Burnside and Jacobson map (Figure 2), it was clear that the surface areas of cells in different regions of the plate must be reduced by different amounts between late gastrula and midneurula stages. Jacobson and Gordon (1976) measured groups of cells over the entire neural plate surface at the late gastrula stage, and the same groups, displaced, at the midneurula stage. This gave us an empirical map of prospective surface shrinkage of the late-gastrula neural plate (Figure 3).

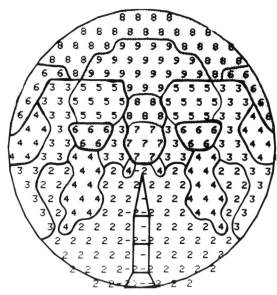

Figure 3. *Representation of the neural plate of a late gastrula-stage newt embryo.* The numbers on the disk are an indication of the amount of apical surface shrinkage the cells of a region will undergo by midneurula stages. Cells labeled 9 shrink the most, and those labeled 2 the least; the shaded areas are regions of greatest shrinkage. The rocket-shaped area represents the notoplate; the blastopore would be located at its base. (Modified from Jacobson and Gordon, 1976.)

The cells of the neural plate that overlie the notochord behave quite differently than do the rest of the plate cells. These cells engage in little or no contraction of their apical surfaces; instead, the area encompassing these cells narrows and elongates along the midline (Figure 4). The behavior of this area was analyzed from time-lapse movies. In most specimens, the neural plate cells over the notochord have less pigment than do other cells of the plate, and their close attachment to the notochord causes them to reflect light differently. This makes it possible to discern and follow them

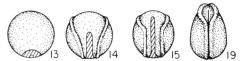

Figure 4. *Four stages of neurulation of a newt embryo (dorsal view, blastopore site at the bottom)* show changes in the shape of the notoplate (diagonal stripes). The stage numbers are arbitrary. Neurulation begins in the late gastrula-stage embryo (13). In the early neurula-stage embryo (14), neural folds have formed and the notoplate is narrowing and elongating along the midline. By midneurula stage, notoplate elongation distorts the neural plate into a keyhole shape (15). Toward the end of neurulation (19), the plate has rolled into a tube and the notoplate is strung out in a line along the midline.

in movies of neurulation. We have named the region of the neural plate that overlies the notochord the "notoplate."

COMPUTER SIMULATION OF EARLY NEURULATION

Using our empirical map of apical shrinkage and our observations of displacement of the notoplate region, Gordon and I did a computer simulation of neurulation from the time the plate has just begun to form to the midneurula stage (Jacobson and Gordon, 1976; Gordon and Jacobson, 1978). We stopped the simulation at a stage at which the plate is still flat and shaped like a keyhole, and is about to roll into a tube.

I mention here but a few of the many things learned from this simulation and from the efforts to make it work. We demonstrated that the pattern of apical contraction of the plate cells and the elongation of the midline of the plate resulting from displacement of the notoplate cells were both necessary, and together were sufficient, to bring about the change in shape from a disk to a keyhole. Apical surface shrinkage alone (observed either by computer simulation or by experimental isolation of the early plate)

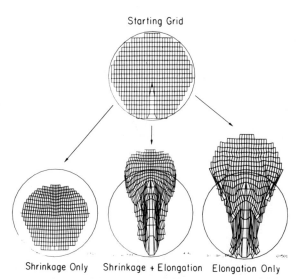

Figure 5. *Computer simulations of changes in the shape of the neural plate.* The starting grid (*top*) represents a late gastrula-stage newt embryo. A simulation of shrinkage of the apical surfaces of the plate cells and of elongation along the midline of the notoplate produces the keyhole shape seen in midneurula-stage embryos (*bottom center*). Shrinkage of apical surfaces of plate cells without notochord elongation reduces plate size (*bottom left*). Midline elongation with no shrinkage of cell surfaces (*bottom right*) makes a keyhole-shaped plate that is too large. Both shrinkage and elongation are necessary, and together are sufficient, to produce the change in shape of the neural plate of the early neurula-stage embryo. (Modified from Jacobson and Gordon, 1976.)

produced a smaller disk, and the transformation appeared to be a conformal remapping. The displacement of the notoplate region elongated the plate and was the source of anisotropy (distortion in some directions more than in others). A computer "experiment" in which the apical shrinkage did not occur, but in which notoplate displacement proceeded normally, demonstrated that elongation of the plate and most of the shaping into a keyhole form results from notoplate displacement along the midline (Figure 5).

NOTOPLATE ELONGATION DURING NEURAL TUBE FORMATION

Models of neural tube formation often invoke as their driving force the change of shape of neural plate cells from cylindrical to wedge-shaped (Karfunkel, 1974). Continued constriction of the apical surfaces of newt-embryo plate cells has been observed to occur extensively between the keyhole-shaped stage of the midneurula embryo and the time of tube closure. The cells become wedge-shaped, that is, their basal ends become broader than their apical ends as the plate rolls up into a tube (Burnside, 1971). If the plate cells become wedge-shaped of their own accord, it is possible that this contributes to tube formation; however, it is also possible that the cells become wedge-shaped passively as the tube rolls up, driven by some other mechanism.

Elongation of the neural plate is another mechanism that may roll the plate into a tube. We measured elongation of the newt nervous system

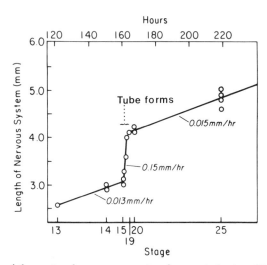

Figure 6. *Length of the newt embryo nervous system from gastrula stage 13 to tailbud stage 25. The nervous system elongates 10 times faster during those stages when the plate is rolling into a tube (stages 15–19). (Modified from Jacobson, 1981.)*

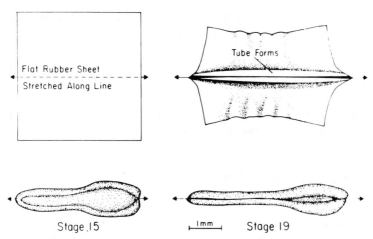

Figure 7. *A rubber sheet stretched along a line forms a tube (top).* The neural plate of the newt embryo (drawn to scale, *bottom*) may also be buckled into a tube as it is stretched along the midline by notoplate elongation. (Modified from Jacobson, 1980.)

during tube formation and found a dramatic correlation between the rate of elongation and the period of tube formation (Jacobson and Gordon, 1976; Jacobson, 1978): During those stages when the neural tube forms, nervous system elongation occurs 10 times more rapidly than it does before or afterward (Figure 6). This remarkable correlation between rapid elongation of the nervous system and neural tube closure led me to suggest that the neural plate is buckled into a tube when the plate is stretched along the midline, much as a sheet of rubber or fabric will form a tube when stretched along a line (Figure 7; Gordon and Jacobson, 1978; Jacobson, 1978, 1980, 1981).

To ascertain that the notoplate is responsible for the elongation of the nervous system during tube formation, the neural plate of the midneurula-stage embryo was split from head to tail along a line at one edge of the notoplate (Figure 8). Elongation of the two parts of the split plate was then measured after the period during which neural tubes of control embryos closed (Jacobson, 1981). The part of the neural plate that included

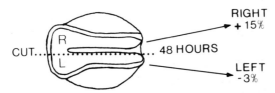

Figure 8. *Neural plate of a newt embryo severed along a craniocaudal line to the left of the notoplate.* The change in length of each severed half was measured 48 hours later, when the plates of intact control embryos had closed into tubes. Only the section containing the notoplate elongated. Percentages of length change are averaged from 12 cases.

the notoplate elongated, but the part lacking the notoplate did not. Furthermore, the part containing the notoplate closed into a tube; the other did not.

These experiments strongly support the midline elongation model for neural tube formation and clearly show the necessary role of the midline tissues in this process. The question remained as to whether elongation resulted from the activities of the notoplate or whether the underlying notochord, present in these experiments, drove elongation.

DOES NOTOPLATE ELONGATION REQUIRE NOTOCHORD ELONGATION?

Experiments show that the notoplate, from the midneurula stage to the time of tube closure, will elongate the midline of the plate in the absence of the notochord. Kitchin (1949) removed the notochord from beneath the open neural plate of otherwise intact midneurula-stage axolotl embryos. The nervous system continued elongation during neurulation and the tube closed normally.

I explanted the open neural plate of midneurula-stage newt embryos with and without the underlying notochord and measured their elongation (Jacobson, 1981). Plates elongated and neural tubes formed both in explanted plates without notochord and in those with notochord (Figure 9). Although notochord elongation may normally assist and augment notoplate elongation, the notochord is not required after the midneurula stage. The ability to elongate the neural plate is thus intrinsic to the notoplate region of the neural plate during tube formation.

AN ADHESION MODEL TO ACCOUNT FOR NOTOPLATE ELONGATION OF THE NEURAL PLATE

The period of neural tube formation during which the embryonic newt nervous system rapidly elongates occurs long before growth takes place.

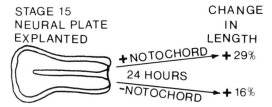

Figure 9. *Neural plates explanted at midneurula stage 15 with and without underlying notochord.* The amount of plate elongation over the next 24 hours was recorded. In both cases enough elongation occurs to close the plates into tubes, but more elongation occurs if the notochord is present.

Nervous system elongation must, therefore, result from rearrangement of the existing mass of nervous system cells. The neural plate elongates by increasing length at the expense of cross-sectional area. I have measured cross sections of the plate through the neurulation period and have counted the numbers of cells present (Jacobson, 1978). There is an exact correlation between the decreasing cross-sectional area (and the decreasing number of cells in the cross sections) and the increasing length of the forming neural tube. Thus cells that are initially positioned in a line running across and perpendicular to the midline may end up strung out along the midline and contributing to plate elongation. Such a rearrangement has been directly observed in time-lapse movies of neurulation (Jacobson and Gordon, 1976).

The cells that rearrange are principally those of the notoplate (Figure 4). The rest of the plate is also distorted, but like an elastic sheet. The cells of the notoplate must change neighbors extensively, and toward the end of neurulation all or nearly all of the notoplate cells are strung like a string of beads along the midline, each notoplate cell virtually surrounded by cells of the rest of the plate. It is of incidental interest that the notoplate region maps exactly with the floor plate of the brain and spinal cord. The ultimate fate of almost the entire floor plate is to become a raphe, which is consistent with the model I propose.

Since notoplate cells make constant changes of neighbors, they behave like a liquid. The rest of the neural plate cells, in contrast, make few, if any, changes of neighbors. In our computer simulation work (Jacobson and Gordon, 1976), we pointed out that cells in the plate change neighbors only along the notoplate boundary, within the notoplate, and between the neural plate and epidermis (along the neural folds). Therefore, except along the midline and its outer edges, the neural plate behaves like an elastic solid. The contraction of the cells' apical surfaces and the displacement of the notoplate distort this elastic sheet into new shapes.

Possibly the behavior of the notoplate can be accounted for by adhesive differences between notoplate cells and the cells of the rest of the plate. The relative adhesiveness of notoplate cells and other neural plate cells must be such that mixing should result (Figure 10). Otherwise the two cell domains would tend to sort out (two-dimensionally within the plate) and the length of the boundary between them would minimize (Steinberg, 1970). Since the boundary is increasing, the conditions must be right for mixing. Actual mixing may be prevented because the rest of the plate is in the solid state so mixing cannot occur and the boundary between cell types lengthens.

Cells of the plate play out a complicated pattern of apical surface shrinkage (Figure 3). The basal ends of the cells are also capable of movement and are endowed with filopodia and lamellipodia (Jacobson, 1981); the lateral surfaces of plate cells have broad lamellipodia as well. The cells' actin filaments are responsible for producing movements in the plate. These movements, which are quite apparent in time-lapse movies, vary from

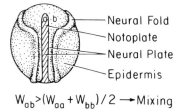

Neural Fold
Notoplate
Neural Plate
Epidermis

$$W_{ab} > (W_{aa} + W_{bb})/2 \longrightarrow \text{Mixing}$$

Figure 10. *Dorsal view of a midneurula-stage newt embryo.* The adhesive relationships between notoplate cells (a) and cells of the rest of the neural plate (b) that could lead to attempted mixing and midline elongation are indicated. W, work of adhesion between cell types.

being random centers of contraction to being rather organized waves of contraction that sweep down the plate. The effects of the movements are to jiggle the whole system such that surfaces of notoplate cells are brought into new contacts with other cells of the plate. When such contacts occur, the cells stick to one another and the system ratchets, elongating the midline. The movements produced by actin contraction provide the force for neurulation, but these forces are organized to extend the midline of the plate by the nature of the adhesive differences between notoplate cells and the cells of the rest of the plate. Thus cell adhesion and cell movement are coupled to shape the neural tube.

One can ask why the notoplate extends along the midline rather than in some other direction or directions. Initially, the notoplate area is a broad crescent over the dorsal lip of the blastopore (Figure 4). The bias to start extending along the midline may come from the shape of the shrinkage pattern of the cells in the rest of the neural plate. A long tongue of cells that greatly contract their apical surfaces extends down along the midline (Figure 3). We have shown that this shrinkage pattern is programmed and is carried out even when groups of these cells are transplanted elsewhere (Jacobson and Gordon, 1976). Contraction of these cells could cause the notoplate to begin extending and narrowing along the midline; once this process has started, considerable directed force would have to be applied to change it in another direction.

I mentioned above that the main anisotropy of the neural plate is caused by the extension of the notoplate up the midline. When the apical surfaces

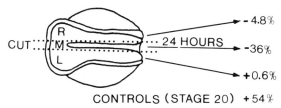

CONTROLS (STAGE 20) +54%

Figure 11. *Neural plates of midneurula-stage newt embryos severed along both sides of the notoplate.* The length of each section was measured 24 hours later (when the plates of intact control embryos had closed into tubes). No elongation occurs when notoplate cannot interact with the rest of the plate.

of plate cells contract in early neurulae, they pull equally in every direction because the apical cell surfaces at those stages are isodiametric (Jacobson, 1981). In later neurulae, apical surfaces of many cells of the plate, especially those toward the neural folds, become stretched in a direction perpendicular to the midline axis. These cells may be stretched until they are nine times longer than they are wide (Jacobson and Gordon, 1976). When these stretched cells contract, they will do so in the direction of stretch, thus augmenting the closing of the tube. The most likely reason for the stretching of these cells is the distortion of the plate caused by the elongation of the notoplate along the midline (resisted by the elastic resistance of the epidermis). Thus there is a transfer of anisotropy from the activities of the notoplate to the other cells of the neural plate. This transfer helps form the tube.

We have done a number of experiments to test this adhesion model of neurulation. One such test of the proposed model was to sever the neural plates of midneurula-stage newt embryos along both sides of the notoplate, thus preventing contact between the notoplate and the rest of the neural plate. Because the two domains of cells could no longer interact, it would be expected that none of the severed fragments would elongate; that indeed was the result of the experiment (Figure 11). We have also examined the embryos of chicks and mice to see if the model will hold for other species.

THE ROLE OF ELONGATION IN NEURULATION OF THE CHICK EMBRYO

The entire neural plate of the newt embryo closes into a tube at about the same time. In contrast, the chick embryo tube closes first in the brain region. Closure then progresses caudally while more embryo is still being added by the regressing primitive streak. During the stage at which the brain has just closed into a tube, there is a plate region in the spinal cord that is in the process of closing and a more caudal plate region that is open flat. I marked the boundaries of the closed tube, the closing plate, and the open flat plate with lines of Nile blue sulfate and then measured the elongation of each region through the four hours required for the closing region to become a tube (Figure 12). (During the period of the experiment, the "open" region converts to "closing.") The results show a dramatic correlation between nervous system elongation and the closing of the plate into a tube (Jacobson, 1981). The closing region elongated at seven times the rate of the closed region and at four times the rate of the open region.

To be certain that these results were not peculiar to the region of the neural plate tested, I measured the length of the brain, from its tip to the anterior edge of the first somite, during and after the time of closure of

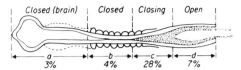

Figure 12. *Chick embryos at the stages shown were marked with Nile blue sulfate (dashed lines) to measure regional elongation.* Four hours later the "closing" region had formed a tube, and the distances between the lines of dye were measured again to determine the relative elongation (%) of each region. The most elongation occurs where the plate is closing into a tube. (Derived from Jacobson, 1978, 1980.)

the brain plate into a tube (Figure 13). The period of most rapid elongation coincides exactly with the time of closure (Jacobson, 1984). In the chick embryo, it appears that rapid elongation of the plate and formation of the neural tube begin in the brain region and then pass like a wave together down the length of the embryo.

If the elongation that accompanies tube formation is indeed the cause of tube formation as I propose, then treatments that stop neural tube formation should also stop elongation of the nervous system. I carried out experiments to see if this is true.

It has been known for years that the application of very low doses of ultraviolet irradiation to the open neural plate of the chick embryo will

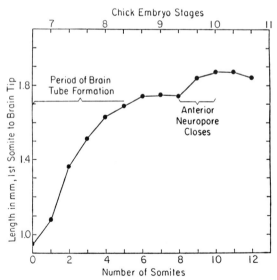

Figure 13. *Groups of chick embryos measured from the tip of the brain to the anterior edge of the first somite at different stages.* Each point is the average of 7–38 embryos. The increase in brain length is most rapid during the time when the plate forms a brain tube. Closure of the anterior neuropore is somewhat delayed and a small spurt of elongation coincides with it. (From Jacobson, 1984.)

stop neural tube formation (Davis, 1944). I repeated these experiments to see whether plate elongation is also halted (Jacobson, 1984). A fluence of 60 J/m^2 of ultraviolet light (predominantly 254 nm wavelength) was delivered to the vitelline membrane covering the open brain plates. The membrane transmits only about 7% of the ultraviolet light, so the dosage absorbed by the brain plate was very low. After 13 hours of incubation in the dark, when the brains of the control embryos had closed into tubes, more than half of the experimental embryos still had open or partly open brain plates. The lengths of the brains were measured: Brains of control embryos were 1.74 mm long; brains of experimental embryos with closed tubes were 1.73 mm long; those with partly open plates were 1.42 mm long; and those with open plates were 1.30 mm long. I concluded that ultraviolet irradiation simultaneously stops brain elongation and closure of the brain plate into a tube.

It occurred to me that photoreactivation, that is, the application of visible light after ultraviolet treatment, might reverse the effects of the irradiation. It did. Embryos were subjected to ultraviolet irradiation, followed by two hours of irradiation by visible light, and an additional 11 hours of incubation in the dark. Control embryos received no ultraviolet irradiation but were otherwise treated similarly. In both the experimental and the control groups, the nervous systems elongated and the brain plates closed into tubes.

Others (e.g., Karfunkel, 1972) have shown that cytochalasin B stops neurulation in the chick embryo. I repeated these experiments and also asked whether elongation of the nervous system is stopped simultaneously (Jacobson, 1984, unpublished results). A small quantity (0.2 ml) of 0.5 mg/ ml cytochalasin B was injected under the vitelline membrane of chick embryos with open brain plates (stage 8). After additional incubation, all control embryos (injected with the carrier, but without the cytochalasin B) had closed brain plates. The average length of the control brains was 1.68 mm. In more than three-fourths of the embryos injected with cytochalasin B the brain plates did not close into tubes, and the brain plates measured 1.28 mm. The experimental embryos in which the brains did close into tubes had brains that averaged 1.63 mm in length. These experiments, which again demonstrate the correlation between nervous system elongation and tube formation, are of special interest because the drug used is believed to affect primarily the actin-contraction system. That elongation of the nervous system was also stopped is consistent with my model, in which cell movements resulting from actin contraction are coupled to the adhesion system. Both systems must function if the nervous system is to elongate and the tube form.

My model also predicts that a drug such as the lectin concanavalin A, whose effects should be primarily on the adhesion system, should stop neurulation and nervous system elongation. Lee (1976) has already reported that this lectin stops neural tube formation in the chick embryo. I am

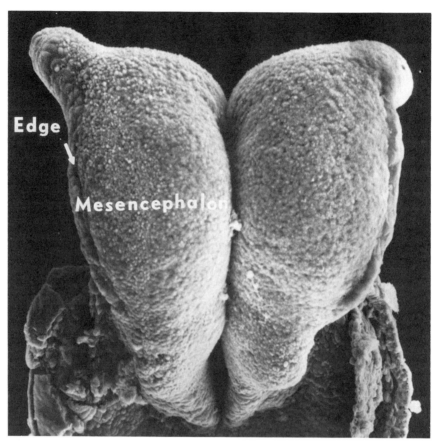

Figure 14. *Scanning electron micrograph of the open neural plate of a mouse embryo brain at seven somites of development.* Each lateral half of the brain plate bulges upward convexly, opposite to the way it must fold to form a tube. Tube formation is beginning at the lateral edges of the mesencephalon; the folds there begin to reverse the convex bulge. This process continues toward the midline in later stages. (Modified from Jacobson and Tam, 1982.)

repeating these experiments to see whether nervous system elongation is simultaneously halted.

NERVOUS SYSTEM ELONGATION AND NEURULATION IN THE MOUSE EMBRYO

Using scanning electron microscopy and morphometry, we (Jacobson and Tam, 1982) have examined neurulation in the brain of the mouse embryo to see whether a mammal exhibits the same correlations between nervous system elongation and neurulation that have been found in newt and chick embryos.

Neural plates of mouse embryos are quite different from those of newt and chick embryos. The early brain plate of the mouse embryo is very broad, and each lateral half bulges upward convexly, just opposite to the way it must fold to become a tube. The process of plate closure begins at the lateral edges (Figure 14).

We measured the lengths of the midline and of the lateral edges of the plate in the mesencephalon region during the period of tube formation. The midline elongated very little, but the lateral edges elongated considerably. Stretching the lateral edges could roll the plate into a tube, and the rolling does begin at, and proceed from, the lateral edges in these brains. In the mouse embryo, elongation of the lateral edges may play a similar role to that played by the notoplate in the newt embryo. This finding has led us to ask whether lateral elongation also has a role in the closure of the tube in newt and chick embryos. We are examining this question.

OTHER DEVELOPING SYSTEMS THAT STRETCH ALONG A LINE

During the first day of development of the chick blastoderm, the primitive streak is established and becomes stretched out to its maximum length along the midline of the embryo. The primitive streak begins as a broad crescent at the posterior end of the blastoderm and then narrows and elongates along the midline. During the period of extension of the primitive streak, the blastoderm is transformed from a disk shape to a keyhole shape (Figure 15). This transformation looks remarkably like the change from disk to keyhole shape observed in the early neural plate of the newt embryo. It seems likely that the cells of the primitive streak and the cells of the rest of the blastoderm may have adhesive relationships much like those between the notoplate and the rest of the neural plate in the newt embryo.

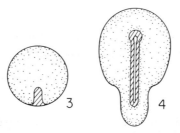

Figure 15. *When the primitive streak (diagonal lines) is forming at the posterior end of the early chick blastoderm (stage 3), the clear area of the blastoderm is disk-shaped.* By the time the streak has extended up the midline to its greatest length (stage 4), the blastoderm has been distorted into a keyhole shape. These changes of shape are similar to those of the early neural plate of the newt embryo.

The cells of the primitive streak behave like a fluid, even to the extent that cells in the streak sink below the surface of the epithelial epiblast and migrate away beneath the epiblast. The cells of the rest of the blastoderm seem to act more like an elastic solid, and the whole blastoderm becomes distorted into a keyhole shape by the lengthening of the primitive streak up the midline.

This is another important morphogenetic system that deserves analysis. Adhesive differences between domains of cells may have an important role in gastrulation as well as neurulation. Genes could have a role in organizing such morphogenetic systems by providing for the timely synthesis and modulation of adhesion molecules in the different cell domains (Edelman, 1983).

ACKNOWLEDGMENTS

My research is supported by National Institutes of Health Grant NS-16072. I thank Gertrud Threm for help with some of the experiments.

REFERENCES

Baker, P. C., and T. E. Schroeder (1967) Cytoplasmic filaments and morphogenetic movement in the amphibian neural plate. *Dev. Biol.* **15**:432–450.

Burnside, B. (1971) Microtubules and microfilaments in newt neurulation. *Dev. Biol.* **26**:416–441.

Burnside, B. (1973) Microtubules and microfilaments in amphibian neurulation. *Am. Zool.* **13**:989–1006.

Burnside, B., and A. G. Jacobson (1968) Analysis of morphogenetic movements in the neural plate of the newt *Taricha torosa. Dev. Biol.* **18**:537–552.

Davis, J. O. (1944) Photochemical spectral analysis of neural tube formation. *Biol. Bull.* **87**:73–95.

Edelman, G. M. (1983) Cell adhesion molecules. *Science* **219**:450–457.

Gordon, R., and A. G. Jacobson (1978) The shaping of tissues in embryos. *Sci. Am.* **238**:106–113.

Jacobson, A. G. (1978) Some forces that shape the nervous system. *ZOON* **6**:13–21.

Jacobson, A. G. (1980) Computer modeling of morphogenesis. *Am. Zool.* **20**:669–677.

Jacobson, A. G. (1981) Morphogenesis of the neural plate and tube. In *Morphogenesis and Pattern Formation*, T. G. Connelly, L. L. Brinkley, and B. M. Carlson, eds., pp. 233–263, Raven, New York.

Jacobson, A. G. (1984) Further evidence that formation of the neural tube requires elongation of the nervous system. *J. Exp. Zool.* **230**:23–28.

Jacobson, A. G., and R. Gordon (1976) Changes in the shape of the developing vertebrate nervous system analyzed experimentally, mathematically, and by computer simulation. *J. Exp. Zool.* **197**:191–246.

Jacobson, A. G., and P. P. L. Tam (1982) Cephalic neurulation in the mouse embryo analyzed by SEM and morphometry. *Anat. Rec.* **203**:375–396.

Karfunkel, P. (1972) The activity of microtubules and microfilaments in neurulation in the chick. *J. Exp. Zool.* **181**:289–302.

Karfunkel, P. (1974) The mechanisms of neural tube formation. *Int. Rev. Cytol.* **38**:245–272.

Kitchin, J. C. (1949) The effects of notochordectomy in *Ambystoma mexicanum*. *J. Exp. Zool.* **112**:393–415.

Lee, H.-Y. (1976) Inhibition of neurulation and interkinetic nuclear migration by concanavalin A in explanted early chick embryos. *Dev. Biol.* **48**:392–399.

Schroeder, T. E. (1970) Neurulation in *Xenopus laevis*. An analysis and model based upon light and electron microscopy. *J. Embryol. Exp. Morphol.* **23**:427–462.

Steinberg, M. S. (1970) Does differential adhesion govern self-assembly processes in histogenesis? Equilibrium configurations and the emergence of a hierarchy among populations of embryonic cells. *J. Exp. Zool.* **173**:395–434.

Chapter 4

Contact Regulation
of Neuronal Migration

PASKO RAKIC

ABSTRACT

Electron microscope, immunocytochemical, and autoradiographic analyses of cell motion in the developing central nervous system indicate that contact interaction between migrating neurons and the surfaces of neighboring cells may play a crucial role in the selection of migratory pathways and in the orientation, displacement, and final positioning of neurons. With regard to pathway selection, migrating neurons fall into three major categories: (1) gliophilic cells that follow glial fibers and ignore neuronal surfaces; (2) neurophilic cells that follow neuronal, particularly axonal, surfaces and ignore glial shafts; and (3) biphilic cells that display temporal or regional affinities towards either glial or neuronal surfaces. Selective displacement of migrating neurons along preferred surfaces can be explained by the presence of two types of affinities mediated by at least three classes of adhesion molecules that are inserted at the growing tips of leading processes. The proposed model of contact guidance based on differential adhesion is supported by several lines of evidence, including the ultrastructural characteristics of migrating cells, the pattern of cell surface growth, the speed of cell displacement, and the consequences of genetic and acquired alteration of cell locomotion.

Although contact interactions play an important role in morphogenetic movements in most organs of multicellular organisms, these interactions assume special significance during the migration of neurons through the developing central nervous system. With few exceptions, neurons in the mammalian brain, especially those of humans, originate at a considerable distance from the place where they reside in the adult (Rakic, 1971a, 1978, 1985b; Sidman and Rakic, 1973). The intervening process of migration differs in several important respects from the movement of cells in most other organs. First, migration is initiated only after completion of the final cell division. Second, individual postmitotic cells move actively along preferred surfaces rather than being passively shifted as a group of cells. Third, the final location of each individual neuron, which depends on both the form of its migratory pathway and the speed of its migration, determines its synaptic relations and therefore specifies its function. Since the phenotype of an individual neuron has proved in several regions of the brain to be specified at the time of the last cell division, failure of the migrating cell to reach the correct destination leads to either neuronal death or inappropriate synaptic relationships, in either case producing functional deficits (Rakic, 1984).

It is safe to state that the vast majority of neurons in the mammalian central nervous system migrate from the site of their last mitotic division to their final position. In the rhesus monkey, the length of the migratory pathway between those two sites may range from less than 100 to more than 3000 μm. The migratory cell usually has three well-defined components: (1) the leading process, which may range in length from 50 to 200 μm; (2) a bipolar soma consisting of an ovoid nucleus and surrounding cytoplasmic ring 10–20 μm long; and (3) an attentuated trailing process of variable length. Although this description is based on neurons of embryonic monkey brains, the dimensions of migrating cells are similar in other mammals. In most instances, the leading process is more voluminous and contains a larger amount and higher diversity of cytoplasmic organelles—in particular, free ribosomes, rough and smooth endoplasmic reticulum, Golgi apparatus, and an extensive cysternal and vesicular system. In contrast, the trailing process is relatively poor in organelles as it rapidly tapers into an elongated cylinder free of ribosomes but well endowed with microtubules.

Since most migratory neurons, particularly those generated at the later developmental stages, must penetrate nervous tissue that is densely populated with previously generated cells and their processes, the two most obvious questions are: (1) How are cells propelled through this terrain? And (2), since each neuron has the opportunity to select many pathways on the way to its destination, how are the appropriate choices made? Examination of the migratory behavior of postmitotic cells strongly suggests that both the locomotion and selectivity of cell movement may depend on surface-mediated interactions. The nature of this interaction and the molecular mechanisms involved are not understood, although our present knowledge allows some reasonable propositions and several molecules have been suggested to play a role.

In this chapter I provide selected examples illustrating three basic categories of neuronal migration. All are derived from the studies carried out in my laboratory on neurogenetic processes in the fetal rhesus monkey which, because of its size and slow development, allows favorable temporal and spatial resolution of migratory events. In the final section, I discuss possible cellular and molecular mechanisms that may be involved in the process of neuronal migration.

THERE ARE THREE BASIC MODES OF NEURONAL MIGRATION

In the densely packed cerebral tissue, the leading process, cell soma, and trailing process of the migrating cell are invariably apposed to adjacent cells. Detailed electron microscope analysis of migration zones in a variety of brain structures shows that the leading process of a given class of postmitotic cells selectively follows either glial or neuronal surfaces. In terms of their relation to other cells in the developing brain, migrating neurons and their leading processes can be classified into three basic categories: (1) *gliophilic* cells that follow glial membrane surfaces and ignore other neurons; (2) *neurophilic* cells that follow neuronal surfaces, particularly axons, and ignore glial cells; and (3) *biphilic* cells that possess two types of processes—one that follows neurons and the other that follows glial fibers exclusively. The gliophilic mode of migration, at least in primates, is the most prominent and is described first.

Migrating Cells That Follow Glial Guides

Initially I observed this type of migration in the neocortical region of the fetal monkey cerebral wall (Rakic, 1971a, 1972). Later studies in our and other laboratories showed that this form of migration is the most prevalent type in the mammalian brain (for a review, see Rakic, 1985b). Translocation of postmitotic cells may not be easily noticeable in small structures or in species with smaller brains, where migratory pathways may be short in relation to the total length of the migratory cell itself. Such migration along glial guides is nevertheless present even in regions where postmitotic cells move relatively short distances, as in the pyramidal sector of the hippocampal formation (Nowakowski and Rakic, 1979), or in the dentate gyrus (Eckenhoff and Rakic, 1984) and diencephalon (Rakic, 1977).

Perhaps because of the extraordinary length of their trajectory, the translocation of neocortical neurons in primates is both the most explicit and most useful example of neuronal migration (Figure 1). Studies conducted in the developing rhesus monkey during the past fifteen years (Rakic, 1971a, 1972, 1978; Rakic et al., 1974; Rakic, 1985b) indicate that the radial pathways of migrating cells are most probably established by a guidance mechanism that depends on cellular interaction between membranes of migrating neurons and adjacent radial glial cells. In the developing telencephalon, these transient populations of nonneuronal cells have remarkably elongated fibers that stretch radially across the full thickness of the developing telencephalic wall (Figure 1). They show the early cytological and biochemical properties of astroglial cells (Rakic, 1972; Schmechel and Rakic, 1979a; Levitt and Rakic, 1980). More recently, immunohistochemical studies at the ultrastructural level (Levitt et al., 1981) revealed that a separate glial cell line is established relatively early, before the first neurons

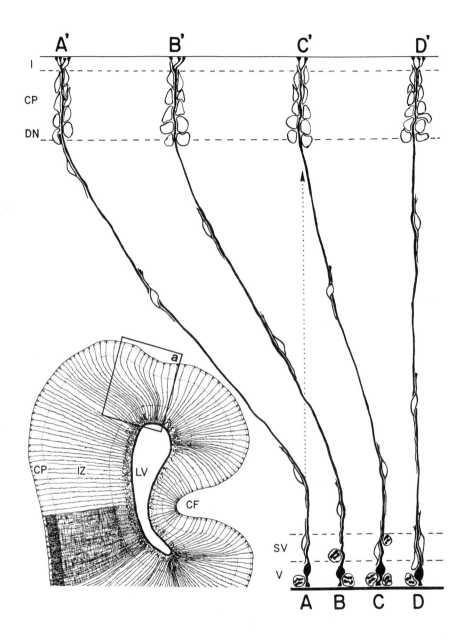

destined for a given structure have left their proliferative zone. Many radial glial cells, which appear to form a scaffolding on the developing cerebral wall, stop dividing for about two months during the midgestational period (Schmechel and Rakic, 1979b). Eventually, however, they reenter the cell division cycle, loose their attachments to the cerebral surface, and become astrocytes. During the transformation into typical astrocytes, radial glial cells may change their biochemical properties, as evidenced by the loss of immunoreactivity to specific antibodies (McKay and Hockfield, 1983; Pixley and deVellis, 1984). The newly formed astrocytes also assume different functions (Rakic, 1985b). Thus, a line of glial cells in the mammalian telencephalon have a unique developmental history as they pass through two distinct phases in the course of their evolution.

Corticogenesis in the rhesus monkey lasts for about two months and occupies approximately the middle 60 days of the 165-day gestational period (Rakic, 1974). During these two months, progenitors of cortical neurons are generated first in the ventricular and later in the subventricular zones. Young neurons migrate to the cortical plate only after their last cell division (Rakic, 1975, 1978). A point relevant to the discussion of proposed mechanisms of migration is that postmitotic neurons which migrate to the monkey neocortex appear to be relatively simple bipolar cells with leading processes of lengths that are only a fraction of the total migration pathway. On the basis of electron microscope findings (Figure 2), I proposed that migrating neurons may find their way to the distant cortex by using radially oriented glial fibers as guides (Rakic, 1971a, 1972). It appeared reasonable that such structurally defined guidelines could provide mechanisms that postmitotic neurons use to penetrate the large cerebrum of larger mammals that consists of a mixture of cellular elements including neuronal and glial cells and their various cytoplasmic processes, as well as blood vessels of various sizes and orientations. Indeed, throughout the

---

Figure 1.   *Schematic drawing of the relationship between radial glial cells and cohorts of associated migrating neurons in developing monkey telencephalon.* A low-power illustration of the Golgi image of a transverse section of monkey telencephalon at midgestation is in the left lower corner. A selectively impregnated, transient population of radial glial cells stretch from the surface of the lateral ventricle (LV) across the intermediate zone (IZ) to the cortical plate (CP). The area located within rectangle *a* is enlarged and simplified to emphasize the point that all neurons produced in the same sites of the ventricular (V) and subventricular (SV) zones (proliferative units A–D) migrate in succession along the same fibers to the developing cortical plate (CP), where they establish ontogenetic radial columns (A'–D'). Within each column the newly arrived neurons bypass more deeply situated neurons (DN) generated earlier and come to occupy the most superficial position at the borderline between the developing cortical plate and the first cortical layer (I). Radial glial grids preserve the topographic relationships between successive generations of neurons produced in the proliferative units at their final positions in the cortical plate and prevent mismatching, which could occur, for example, if the cells from unit A were to take a direct, straight route to column C' (this abnormal pathway is indicated by the *dotted line*).

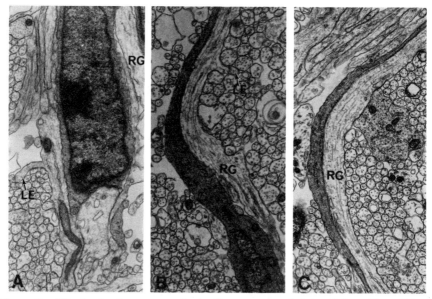

Figure 2. *Ultrastructural appearance of migrating neurons and their cellular milieu within the intermediate zone of an 80-day-old monkey fetus.* The areas selected for illustration show the soma (*A*), the leading process (*B*), and its attenuated tip (*C*). As revealed by ^{3}H-thymidine labeling of their DNA (Rakic, 1974), this class of cells is migrating to the superficial strata (layers II and II) of the visual cortex.

course of its migration, the plasma membrane of a young neuron and its elongated radial guide remain in contact in sectors where the glial fiber twists or curves (Figure 2) even though there are a number of other elements that are geometrically aligned with the initial direction of movement.

These observations collectively suggest the existence of some kind of recognition and differential affinity either between the surface components of the two cell types or between cells and an intercellular coating and/or binding matrix substance. As is elaborated in the last section of this chapter, such contact interaction may provide both a basis for selective affinities between these two cell types and the mechanism for directed cell movement.

Migrating Cells that Follow Neuronal Guides

Although neuronal movement along glial guides is the prevalent mode of migration in the mammalian central nervous system, some cell classes preferentially follow the surface of other neurons. An instructive example in this category is the migration of neurons from the rhombic lip to the gray nucleus of the pons in the brain stem. Neuronal migration in this region of the brain stem was described by Essick (1912), but until recently nothing was known about its possible mechanisms. In our laboratory, a series of monkey embryos were injected with [³H]thymidine at selected times during the second month of gestation and sacrificed shortly after treatment. Analysis of autoradiograms resolved that after the last cell division, pontine neurons migrate from the proliferative center (rhombic lip) tangentially over the surface of the brain stem (Rakic, 1985c).

Examination of 1-μm-thick serial sections cut across the brain stems of embryos sacrificed at the end of the second fetal month reveals cohorts of migrating cells forming a superficial bundle situated at the ventral surface of the developing pons (Figure 3). This bundle, termed the corpus pontobulbare (Essick, 1912), is a transient structure that in the monkey fetus exists for a period of less than one month. After leaving the rhombic lip, postmitotic cells assume a bipolar shape and become oriented lateromedially as they stream toward the ventral portion of the pons (Figure 3). At the peak of neuronal migration, the corpus pontobulbare is 3–5 cells thick, but toward the end of neurogenesis, around the 70th embryonic day, one can see only solitary individual cells situated below the pial surface (Rakic, 1985c).

Electron microscope examination of the corpus pontobulbare reveals the ultrastructural features of typical postmitotic cells, with nuclei and cytoplasm that differ little morphologically from that of migrating neurons destined for the neocortex, hippocampus, or cerebellar granule layer in the same species (Rakic, 1971a, 1972; Nowakowski and Rakic, 1979; Eckenhoff and Rakic, 1984). However, the orientation of these migratory cells and their relationship to the elements in the surrounding tissue is different. First, the orientation of the bipolar cells is orthogonal to the radial axis of the brain stem wall (Figure 3). Second, migrating cells are aligned with fascicles of axons that run at the surface of the developing pons (Figure 4). Third, although portions of individual migrating cells may have direct surface contact with radial glial cells or their end feet, the main axis of cell movement is orthogonal to the orientation of individual glial fibers or radial glial fascicles (Figures 3, 4). Therefore, migrating cells in the corpus pontobulbare are in contact with a series of glial processes, as if they move from the one to the other. Obviously, in this case, one would have to assume that a different mechanism or at least a different set of adhesive affinities may be responsible for cell translocation than the one suggested for neocortical cells.

The remarkably consistent, well-timed, and precisely oriented migratory pathway of the pontine neurons requires some explanation in terms of their guidance. Since pontine cells have a single source or origin, a unique pathway, and selective destinations, indiscriminate, random movement is not a satisfactory answer. In analogy to the migratory behavior of cortical neurons in the telencephalon described above, there should be some cues distributed over the brain stem surface that guide the pontine cells to the appropriate position across the ventral surface of the pons. Electron mi-

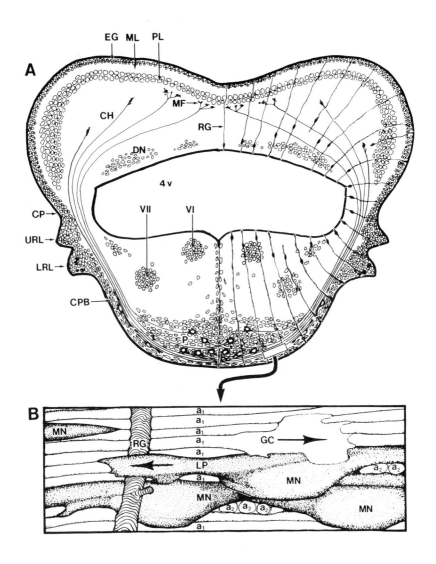

74

croscope analysis indicates that pontine neurons may follow preexisting guidelines consisting of the axons of previously generated neurons that run at the surface of the developing brain stem (Rakic, 1985c). This hypothesis is based on the observation that migrating neurons in the corpus ponto-bulbare are invariably aligned along the axonal fascicles.

As illustrated in Figure 3A, most of the axons at the base of the developing pons originate from the cells situated on the opposite side of the rhom-bencephalon—they grow over the lateral surface of the pons and enter the cerebellar hemisphere via middle cerebellar peduncles and form mossy fiber terminals. Thus migrating neurons move along axonal shafts but in a direction opposite to the growth of these axons (Figure 3B). Differential affinity between the surfaces of migrating neurons and superficially dis-tributed axons is suggested by their close and parallel apposition as well as by the fact that the neurons' leading tips or processes do not follow radially oriented glial shafts (RG in Figures 3B, 4A) or axons of corticobulbar and corticospinal tracts that run in a rostrocaudal direction (e.g., a_2 in Figure 3B). Thus, unlike the leading tips of telencephalic cortical neurons, pontine neurons display preferential alignment with the surfaces of axons belonging to the same class of cells that have originated on the opposite side and moved to their final positions earlier. The combined findings from our autoradiographic and electron microscope analyses demonstrate that migrating cells of the pontine nuclei are aligned with axons and growth cones that lie in the same plane (Figure 4) but run in the opposite direction as predicted from the scheme illustrated in Figure 3A.

Figure 3. *A: Semischematic drawing of a transverse section of the cerebellar hemispheres (CH) and pons (P) of a 55-day-old monkey embryo.* By this age, cranial nerve nuclei (e.g. VI, VII) as well as Purkinje cells (PL) and most of deep cerebellar neurons (DN) have completed their genesis near the fourth ventricle (4v) and migrated along radial glial fibers (RG) to their positions. However, cells generated later in the lower rhombic lip (LRL) migrate via the corpus pontobulbare (CPB) along the brain stem surface, perpendicular to the orientation of radial glial fibers, and accumulate at the bottom of the pons (P). These cells form the pontine gray nuclei, which eventually project to the opposite cerebellar hemisphere beneath the Purkinje cell layer (PL) as mossy fiber terminals (MF). Note that germinative cells orginating in the upper rhombic lip (URL) simultaneously form the external granular layer (EG) that coats the external surface of the cerebellum. Progeny of cells from the external granular layer migrate inward after their last cell division, across the molecular layer (ML), and eventually contact "waiting" mossy fiber terminals. B: Enlargement of the area encompassed by the rectangle in A displays at higher magnification the cytological organization of the corpus pontobulbare. Note that cohorts of migratory neurons (MN) and their leading processes (LP) move along axons presumably originating from previously generated pontine neurons (a_1) that project to the contralateral hemisphere; they bypass radial glial fibers (RG). Simultaneously, the growth cones (GC) of the most recently generated cells situated in the contralateral pontine neurons move in the opposite direction, toward the cerebellar peduncle (CP in A). It may be significant that migrating neurons do not follow axons of the corticospinal and corticobulbar system that run in a rostrocaudal direction (a_2) and in this plane of section appear as round profiles. (From Rakic, 1985c.)

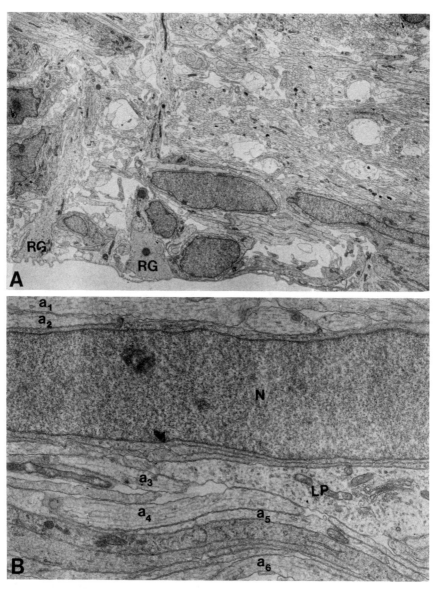

Figure 4. *A: Low-power electron micrograph showing several neurons of the corpus pontobulbare migrating from the left to right.* Note that the direction of migration is orthogonal to the orientation of radial glial fibers and their end feet (RG). *B: Higher-power electron micrograph showing close apposition of migrating neuron (N) and its leading processes (LP) to axonal fascicles (a_{1-5}).*

Cells with Regional and Temporal Affinities to Neuronal or Glial Surfaces

This category of migrating cells is the least frequent as it includes those postmitotic cells that first develop cytoplasmic extensions that grow selectively along the neuronal surface and either simultaneously or later develop second processes that extend only along the glial surface. Because such cells display a region- and/or time-dependent affinity to glial and neuronal surfaces, they are classified as biphilic. The most remarkable example of this type of migratory behavior is the morphogenetic transformation and translocation of cerebellar granule cells across the developing molecular layer (Figure 5). These cells require well-coordinated interactions with both the neurons and the glial cells that compose their immediate cellular milieu. In the adult cerebellar cortex, granule cell somas lie in the granular layer situated under the Purkinje cell somas, but the granule cell axons pass radially upward and branch at 90° to form fibers that lie parallel to one another in the molecular layer, above the layer of Purkinje cell somas (Figure 5D). The basic developmental problems relevant to the subject of neuronal migration presented here are: How do granule cells move from the surface of the cerebellum, where they originate in the transient external granular layer, to the position below the Purkinje cell layer, where they synapse upon mossy fiber terminals? How are the very thin (about 0.2 μm in diameter) and very long (3–7 mm in length) horizontal segments of granule cell axons inserted into proper position in the molecular layer? How do they achieve appropriate orientation and succeed in recognizing postsynaptic targets on the dendritic spines of the relevant sectors of Purkinje cells?

These developmental problems can be explained by the juxtaposition of the simultaneously growing granule cell axons with other parallel fibers and the alignment of the leading tips with glial guides at the outer border of the incipient molecular layer while, and even before, the granule cell soma becomes translocated inward (Rakic, 1971b). The proposed concept is based on the combined Golgi, electron microscope, and [^{3}H]thymidine autoradiographic observations summarized diagrammatically in Figure 5. Precursors of granule cell neurons originate in the upper rhombic lip (near, but dorsal to, the origin of pontine cells). However, presumptive granule cells continue to divide and spread upward, eventually covering the entire surface of the developing cerebellum (Figure 3A). The early postmitotic granule cell undergoes a remarkable series of changes in shape that transform it from a round form at the time of the last cell division through bipolar (Figure 5A) and tripolar forms (Figure 5B), to the mature cell neuron that is situated in the granular layer (GL in Figure 5). An initial event in this sequence is the genesis of two horizontal cytoplasmic processes that originate on opposite sides of the cell. These processes lie in the plane oriented longitudinal to the folium and grow exclusively along the surfaces of other

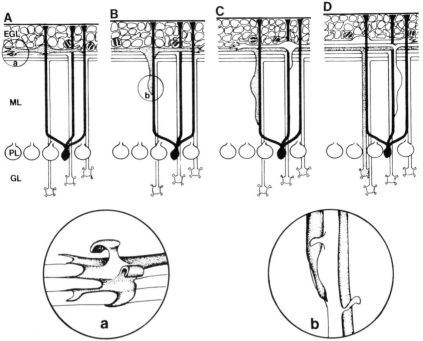

Figure 5. *Semidiagrammatic illustration of the morphogenetic transformation of a cerebellar granule cell (stippled) during its translocation across the molecular layer (ML) and Purkinje cell layer (PL) to the granular layer (GL).* The morphological difference between the leading tips of the two horizontal processes and the descending process are illustrated in enlargements *a* and *b,* respectively. Note that the horizontal processes invariably align with axons of previously generated granule cells, while the vertical process descends along radial glial shafts. (From Rakic, 1985a.)

axons (*a* in Figure 5) and therefore express a strong neurophilia. Again, as in the two types of neuronal migration described previously, the crucial observation came from electron microscope analysis of the behavior of these horizontal axons *in vivo* (Rakic, 1971b). However, even if removed from their normal position, these axons nevertheless grow preferentially in a straight-line fashion along other axons (Mangold et al., 1984). *In situ,* these processes invariably follow the surface of the so-called parallel fibers; that is, the horizontal portions of T-shaped axons of previously generated granule cells (Figure 6). The position of such an axon, relative to the position of the Purkinje cell somas, depends on when the granule cell differentiates (Rakic, 1971b). As a consequence, parallel fibers in the deepest strata of the mature molecular layer belong to granule cells generated earliest while those situated in the most external strata belong to the last ones that are formed (cf. A–D in Figure 5).

The third descending process of the immature granule cell enters the molecular layer at the next stage (Figure 5). Again, electron microscope

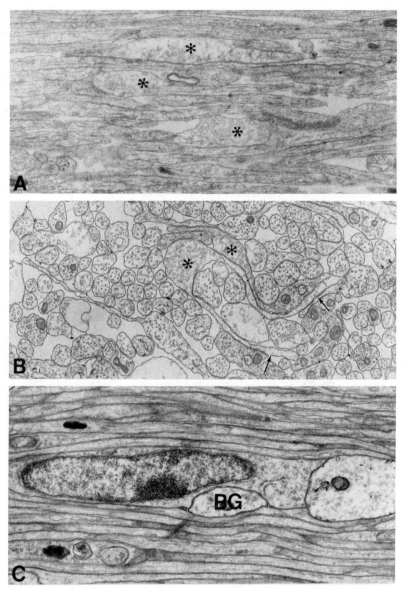

Figure 6. *Electron micrographs of the superficial strata of the molecular layer*, situated at the border external to the granular layer and cut in the three cardinal planes of section. In the plane oriented longitudinally to the folium (*A*) there are several axonal enlargements (*asterisks*) belonging to the growth cones of recently generated parallel fibers. Although in this plane the growth cones appear as bulbous enlargements, the transverse plane of section (*B*) reveals their lamellipodial form (*asterisks and arrows*). In the plane parallel to the pial surface (*C*), the soma of a granule cell in the bipolar stage of its transformation is closely associated with a transversely cut Bergmann glial shaft (BG).

analysis provided the first evidence that the descending process invariably follows a Bergmann glial fiber (Rakic, 1971b). Thus, the third process is strongly gliophilic. It should be pointed out that the Bergmann glial cells of the cerebellum are equivalent to the radial glial cells in the other regions of the vertebrate brain, as determined by their morphological, ultrastructural, and immunocytochemical characteristics (Rakic, 1971b, Levitt and Rakic, 1980; deBlas, 1984). Developmental studies carried out on several mammalian species indicate that Bergmann glial fibers are present early in embryogenesis and that they are the first element in the cerebellum that is geometrically organized, even before the wave of granule cell migration starts (Rakic, 1971b; del Cerro and Swarz, 1976; Zecevic and Rakic, 1976; Levitt and Rakic, 1980).

As a result of a remarkable morphological transformation, the two bipolar horizontal processes and the vertical process that trailed out behind the migrating granule cell are transformed into a T-shaped axon, the horizontal portion of which is termed a parallel fiber (Figure 5). The striking difference between the growth behavior of the vertical process and that of the horizontal processes, all of which are formed by a single granule cell, implies a regional difference in the structural composition of both the cell membrane and the cytoplasm. Indeed, electron microscope examination reveals that the growing tips of the two horizontal processes terminate as typical growth cones in the form of broad, sheetlike lamellipodia (*a* in Figure 5; Figure 6B). These growth cones, which in three-dimensional reconstruction resemble axonal growth cones described in other structures of the primate brain *in vivo* (Williams and Rakic, 1985), contain a relatively light, filamentous matrix and very few other cellular organelles besides an occasional cluster of membrane-bound vesicles (Figure 6). In contrast, smoothly tapered, descending leading tips contain a dense, filamentous cytoplasmic matrix, numerous parallel microtubules, ribosomes, rough endoplasmic reticulum, Golgi apparatus cisternae, and vesicles of variable size (Figure 7; Rakic, 1971b, 1985a). Examination of the plasma membrane surface of developing granule cells in freeze-fracture preparations shows a different quantity and different pattern of distribution of intermembraneous particles on the two classes of processes and on their growing tips (Garcia-Segura and Rakic, 1985).

A general implication of these findings is that each of the two classes of granule cell processes must also have differential affinities for the surrounding cellular elements. Thus, the horizontal processes grow exclusively along the surfaces of other neurons (other parallel fibers) and bypass nearby glial shafts (Bergmann glial fibers), whereas the leading processes descend only along glial surfaces and bypass the myriad of parallel fibers with which they come into direct contact. As discussed in the next section, this finding suggests the presence of at least three sets of adhesion affinities.

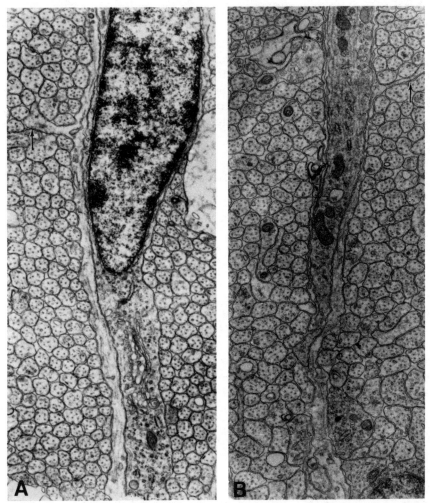

Figure 7. *A: Vertically oriented soma of a migrating granule cell in the middle of the molecular layer.* Note the abundance of cytoplasmic organelles in the leading process, particularly in the elaborate reticulum and ribosomes. On the left side of the cell is an electronluscent Bergmann glial fiber with longitudinally oriented microtubules. The Bergmann process has a lateral expansion (*arrow*) but the surface shared in common with the migrating granule cell is relatively smooth. In the granule cell, the cytoplasm of the leading process close to the nucleus is rich in organelles, including mitochondria, free ribosomes, Golgi apparatus, and multivesicular bodies. x 16,500. *B:* Tip of the leading process of another migrating granule cell enveloped on both sides with Bergmann fibers. On the right is a lateral lamellated expansion (*arrow*). Note the longitudinally oriented microtubules, and numerous free ribosomes, mitochondria, and vesicles, but an absence of growth cones. × 16,500. (From Rakic, 1971.)

SEVERAL MECHANISMS PLAY A ROLE IN NEURONAL MIGRATION

At present, our understanding of the phenomenology of neuronal migration is ahead of our understanding of the underlying cellular and molecular mechanisms. At least two aspects of neuronal migration have to be considered: one relates to the capacity of cells to recognize and follow appropriate substrates, and the other concerns the ability of an individual cell to propel itself along the preferred pathway.

Recognition of Migratory Pathway

The crucial observation relevant to the recognition of a preferred pathway is that the leading tip of a migrating neuron extends selectively along the surface of one class of cells and ignores the contacts with membranes of other cells in the immediate environment (Rakic, 1971a,b). This observation suggests the existence of selective affinity of postmitotic cells for specific surfaces in their environment. If we apply a differential adhesion hypothesis to the three categories of neuronal migration discussed in the previous section of this chapter, two conclusions can be drawn. First, there must be at least two types of bonds: one attaching neurons to neurons and the other binding neurons to glial cells. Second, behavior of the three classes of migratory cells indicates that the two types of bonds must be present selectively in the different cell classes or in some instances even distributed regionally in different components of the same neuron.

The molecules responsible for binding may be an intricate part of the membrane bilayer, or the adhesion between cells may be mediated by the extracellular matrix. Available data from experiments on the central nervous system can not exclude either of these two possibilities, although work in several laboratories indicates that surface molecules may play a primary role. At present, two types of cell adhesion molecules, the neural cell adhesion molecule (N-CAM) and the neuron–glia cell adhesion molecule (Ng-CAM), may provide instructive examples of how to approach this problem experimentally (Edelman, this volume). Initial studies carried out on the developing neural retina showed that the antibodies to N-CAM prevent neuron–neuron binding (Edelman, 1983), while the antibodies to Ng-CAM, which is present only on the neuronal membranes, prevent adhesion between neurons and glia (Grumet et al., 1984). The absence of Ng-CAM on glial cells indicates the possibility that a different but complementary molecule may be present on their surface (Edelman, this volume). It was also shown that an antibody prepared against a membrane protein (L1) of the developing cerebellum can prevent granule cell migration across the molecular layer in cerebellar slices (Lindner et al., 1983; Schachner et al., this volume). These molecules, and the antibodies prepared against their active sites, when used in a tissue culture system of dissociated cells

(Hatten and Liem, 1981; Willinger et al., 1981; Hatten et al., 1984; Trenkner et al., 1984) are highly promising models for further experimental analysis.

If we accept that the interaction between substrate cells and molecules at the leading process of a migrating cell plays a crucial role in selecting the migratory pathway, one would like to find out how the molecule or the membrane component containing such molecules is inserted at the appropriate site within the leading process. The ultrastructural composition of the leading process suggests that the synthesis of membrane proteins, their intracellular transport, and their delivery and insertion at the surface of the leading tips in migrating cells may be similar to the mechanism postulated for other epithelial tissues such as secretory cells (Jamieson and Palade, 1967; Palade, 1975). Although pulse-labeling of proteins, specific antibodies to membrane components, and applications of pharmacological ligands to follow membrane flow have not been utilized in the migratory neurons, electron microscope analysis of the order, shape, and distribution of organelles in the leading process indicate that a similar sequence of events may occur here (Rakic, 1985a). The analogy can be pursued to suggest that membrane patches destined to be inserted in the growing tip of the leading process may be synthesized in polysomes bound to the endoplasmic reticulum near the nucleus of the migrating cell. The abundance of Golgi apparatus, smooth endoplasmic reticulum, and cisternae of various sizes within the leading tip (Figure 7) suggests that membrane containing adhesion molecules or their receptors may be incorporated at selected sites in the leading tip (Figure 8). Electron microscope observations indicate that the process of membrane insertion may be basically similar to the mechanism suggested for recycling of synaptic vesicles (Heuser and Rees, 1973) and for elongation of growth cones in tissue culture (Pfenninger, 1982). In the case of migrating neurons, the cumulative outcome of membrane incorporation may be to allow for the selection between possible migratory pathways. However, the transport of the membrane components may not always be unidirectional. In the case of the biphilic cells, such as the cerebellar granule cells, the proposed model requires that the preassembled molecules be selectively distributed; one set of molecules responsible for neurophilic adhesion should move to the horizontal processes and the set responsible for gliophilic bonds should move to the descending process. One can only speculate that each set of molecules is synthesized in distinct domains of a single neuron or that the cytoskeleton and/or differential distribution of two species of mRNA may regulate the appropriate direction of membrane transport. A compartmentalization of different species of polyribosomes that could lead to differential distribution of proteins in migrating granule cells has not been demonstrated, but it seems clear that the direction of flow must be controlled by the genome of individual cells. For example, postmitotic granule cells that are removed from their cellular environment develop neurites with properties similar to those formed *in vivo* (Mangold et al., 1984). Furthermore, recent tissue culture analysis

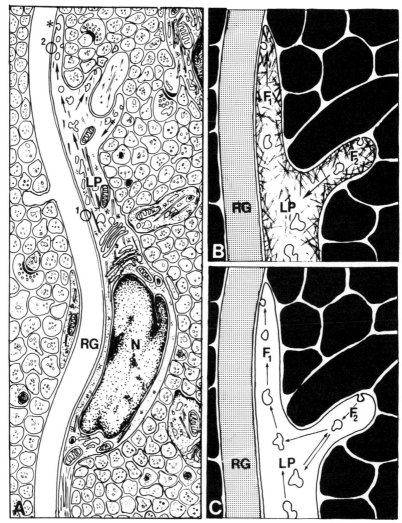

Figure 8. *A: Semischematic drawing of a cross section through a migrating neuron (N) and its leading process (LP) as it moves along a radial glial shaft (RG) through terrain densely packed with various other cells and their processes.* The membrane segments formed near the nucleus are transported toward the tip *(arrows)* within the leading process *via* rough and smooth endoplasmic reticulum and vesicles in order to be inserted *(asterisk)* at the surface of the filopodia. The leading process continues to grow along the surface of radial glial cells, while the filopodia apposed to other surfaces withdraw *(crossed arrow)*. *B:* Sketch of a possible mechanism for the advancement of filopodia (F_1), alignment of the glial shaft (RG), and regression of filopodia (F_2), that have contact with less adhesive surfaces of other cells. The uniform pooling of contractile fibers within the cytoplasm would result in withdrawal of processes with lesser adhesion and preferential elongation along the surface with higher adhesion. *C:* Sketch of how a higher rate of endocytosis in the filopodia associated with less adhesive surfaces (F_2) could result in withdrawal, while a net increase in exocytosis in the filopodia associated with the more adhesive surface (F_1) produces selective outgrowth of the leading process (LP) along a radial glial fiber (RG). Further explanation can be found in the text.

indicates that granule cells have only a relatively short time period during which they are capable of migration, and the length of this period is under strict genetic control (Trenkner et al., 1984).

Another question relevant to the selectivity of migratory pathways concerns the number of different classes of adhesion molecules or their receptors that is needed to allow for the diversity of migratory behaviors and precision exhibited in reaching appropriate destinations in the central nervous system. Although no definitive answer to this question is available, it appears that relatively few molecules could account for all the known migratory events. Even if the binding between neurons and glial cell surfaces is not cell- or region-specific and proves to be mediated by a single molecule (e.g., all migrating neurons have the same affinity for all radial glial cells including the Bergmann glial cells), the spatiotemporal order of cell migration can be achieved provided that each postmitotic neuron becomes attached exclusively to the neighboring glial shaft (Rakic, 1981). This is indeed the case, although we have occasionally observed that a migrating neuron may translocate from one radial glial fiber to another nearby (Rakic et al., 1974; Schmechel and Rakic, 1979a). Lack of regional selectivity between granule cells and glial fibers originating in different regions is supported by the observation that all cerebellar granule cells migrate equally well along any Bergmann and radial glial fiber in tissue culture (Hatten et al., 1984).

Mechanism of Neuronal Displacement

Even if we understood the mechanism involved in the recognition of preferred migratory pathways, we would have to explain how a postmitotic neuron penetrates and subsequently translocates its entire soma across densely packed tissue filled with other cells and their processes running in various directions. To propel itself, the migrating neuron would first need to have the capacity to dissolve the bonds between other cellular elements that are lying in its way, and second, it would need to have the appropriate structural machinery and a sufficient amount of energy to move its soma physically through this terrain. The complexity and magnitude of these two problems come into focus during the late stages of neuronal migration in the primate neocortex, when the migratory pathways become long and the extracellular spaces become reduced to intermembranous clefts, typically 200–300 Å wide.

The mechanism for penetration of the neural tissue by migrating cells is not understood, but the suggestion that leading processes release some proteolytic enzymes at their growing tips seems reasonable. Indeed, it has been shown that plasminogen and plasminogen activator are abundant in developing neural tissues at the period of active cell migration (Kristocek and Seeds, 1981; Soreq and Miskin, 1981; Moonen et al., 1982). These or some other proteolytic enzymes could be present on the outer surface of the growing tips or released into extracellular space along the migratory pathway. Electron microscope analysis of the developing telencephalon showed that the intermediate zone through which all neocortical neurons have to pass is particularly loosely arranged during the peak of neuronal migration and the only close apposition observed is between glial shafts and migrating neurons. The detachment and dispersion of other cellular elements in the vicinity is clearly not a fixation artifact, since cells in other nearby structures that are not involved in cell migration at this period are densely packed and have much smaller amounts of extracellular space (Rakic, 1972).

The understanding of how a migrating cell physically propels itself across a long trajectory is equally tentative, although several important advancements were made in recent years. Among the possibilities that could account for both displacement and selectivity is that the membranes of a migrating cell and a guiding fiber become fixed at all points along their interface (Rakic, 1981). Although individual cells have contact with other cells, the most firmly bound filopodium would be the most likely to succeed in steering the growth of the leading process, while the filopodia less attached to the surrounding elements would retract (Figure 8B). Numerous tissue culture observations indicate that neuronal processes grow preferentially along surfaces to which they adhere more stongly (Letourneau, 1975; Hatten et al., 1984). It seems contraintuitive, but the strong bond between neuronal and glial surfaces *in vivo* would promote rather than interfere with neuronal migration. According to the model, the migrating cell could move by adding new membrane components to its growing tip, and as a consequence the leading process would progressively extend along the preferred substrate while the nucleus and surrounding cytoplasm subsequently become transferred to a new position within the newly formed segment of the leading process (Figure 8A). The bonds formed between the leading process and the glial shaft (e.g., circle 1 in Figure 8A) do not have to be broken since new bonds are formed continuously at the more distal segments (circle 2 in Figure 8A), allowing growth of the leading tip and subsequent translocation of the entire cell.

The rate of movement of migrating neurons in the primate central nervous system, which varies between 0.5 and 10 μ per hour, is compatible with the generation and insertion of new membrane along the leading process. Several other observations give credence to this mechanism of neurite extension, including the ultrastructural composition of the leading tip (Rakic, 1972), the finding of a continuous increase in the surface areas of migrating cells as they approach the target structure (Rakic et al., 1974), and the predominant growth of neurites at their leading tips (Bray, 1973). The additional membrane needed for rapid growth can be provided by the retraction of filopodia that have been extended transiently along less adhesive surfaces.

The retraction of filopodia that grow along less adhesive surfaces may be achieved by two basic mechanisms or by their combination. One mechanism assumes uniform contraction of actin filaments, which are as abundant in the leading tips of migrating neurons as they are in growing neurites (Johnston and Wessels, 1980). It can be expected that indiscriminate traction of the plasma membrane toward the center of the cell would eventually result in involution of those filopodia that have weaker bonds with surrounding cells (e.g., F_2 in Figure 8B) while filopodia with stronger bonds (F_1 in Figure 8B) would be retained. A cumulative effect of this process would result in extension of the filopodia aligned along the most adhesive surface. Another mechanism that could also contribute to the retraction of supernumerary filopodia is a higher rate of endocytosis of membrane patches that lack strong bonds with adjacent cells (Figure 8C). It has been proposed that the endocytic cycle may play a role in the locomotion of fibroblasts (Bretscher, 1984), and a similar mechanism may apply to migrating nerve cells.

A model of neuronal migration that depends on the ratio between insertion and intake of membrane at the leading tip is attractive since it takes into account the rate of membrane biosynthesis that is ultimately controlled by both the genome of individual cells and by their interaction with adjacent cells. The membrane components containing specific binding or receptor molecules may be generated only during the restricted period of time when the cell is displaying migratory behavior. The studies of the migration of cerebellar granule cells *in vitro* indicate that the length of the period during which an individual cell is capable of migration is intrinsically predetermined and, in this case, lasts about four days (Trenkner et al., 1984). After reaching their final destinations, individual cells lose their ability to move, suggesting that neurons may change surface properties by stopping biosynthesis of appropriate membrane components.

Although both the selective and self-propelling capacities of the migrating cell can be explained by contact recognition and selective adhesion, contractile proteins must provide for the displacement of the nucleus and surrounding cytoplasm within the elongated leading process. The presence of such molecules has been detected in immature brain cells (Goldberg, 1982) but their role in migrating neurons has not been clarified. According to the proposed model, the contractile properties of cytoplasm may be essential for both the retraction of filopodia as well as for the completion of nuclear displacement, but they are not expected to play a crucial role in the selection of appropriate pathways or the determination of an individual neuron's destination (Rakic, 1985b).

A neuron ultimately acquires its permanent position in relation to other cells within the target structure; however, the mechanism of this final step of migration may be independent of and unrelated to the preceding cellular events (Rakic, 1975; Caviness and Rakic, 1978). Several lines of evidence indicate that the correct position attained by a migrating young neuron in the cerebral cortex is determined innately before the cell reaches its final position (Caviness, 1982; Jensen and Killackey, 1984). Under normal conditions, each neuronal soma within a given ontogenetic column in the neocortex takes a position distal (external) to its predecessor in all species so far examined (Rakic, 1974). As a result, cortical layers are formed from inside to outside as more superficial cells arrive at progressively later times. In the mutant *reeler* mouse, postmitotic cells appear to migrate normally towards the cortical plate but seem to be unable to bypass previously generated cells (e.g., see Caviness and Rakic, 1978; Goffinet, 1984). It has been suggested that faulty interaction between radial glial cells and migrating neurons at the termination of their journey may be the cause of cell malposition (Pinto-Lord et al., 1982). On the other hand, it is possible that the leading tip of a migrating cell fails to produce a sufficient amount of proteolytic agents, essential for breaking the bond between previously arrived neurons and radial glial shafts. There are several other examples, observed mostly in the mammalian species, which show that alteration at a single genetic locus may alter neuron–glia interaction, resulting in defective migration and malformed brain structure (Rakic and Sidman, 1973; Willinger et al., 1981; Hatten et al., 1984; Nowakowski, 1984; Rakic, 1984, 1985b).

The accumulated evidence on the behavior of migrating neurons in normal and malformed brains of various mammalian species indicates that both pathfinding and displacement of young neurons depends, to a large degree, on contact regulation and cooperation between heterogeneous cell classes. As is evident from the other chapters in this volume, contact regulation of cell movement is a ubiquitous feature of developing organisms, but it assumes a special significance in determining detailed cell positioning in the central nervous system. Disruption or even minimal delay of this regulation leads to abnormal neuronal and synaptic organization that ultimately results in varying degrees of abnormal function in the adult (Rakic, 1978, 1984, 1985b).

REFERENCES

Bray, D. (1973) Model for membrane movements in the neural growth cone. *Nature* **244**:93–96.

Bretscher, M.S. (1984) Endocytosis: Relation to mapping and cell locomotion. *Science* **224**:681–686.

Caviness, V. S., Jr. (1982) Development of neocortical afferent systems: Studies in the *reeler* mouse. *Neurosci. Res. Program Bull.* **20**:560–569.

Caviness, V. S., Jr., and P. Rakic (1978) Mechanisms of cortical development: A view from mutations in mice. *Annu. Rev. Neurosci.* **1**:297–326.

deBlas, A. L. (1984) Monoclonal antibodies to specific astroglial and neuronal antigens reveal the cytoarchitecture of the Bergmann glial fibers in the cerebellum. *J. Neurosci.* **4**:265–273.

del Cerro, M. P., and J. R. Swarz (1976) Prenatal development of Bergmann glial fibers in rodent cerebellum. *J. Neurocytol.* **5**:669–676.

Eckenhoff, M. F., and P. Rakic (1984) Radial organization of the hippocampal dentate gyrus: A Golgi, ultrastructural, and immunohistochemical analysis in the developing rhesus monkey. *J. Comp. Neurol.* **223**:1–21.

Edelman, G. M. (1983) Cell adhesion molecules. *Science* **219**:450–457.

Essick, C. R. (1912) The development of the nuclei points and the nucleus arcuatus in man. *Am. J. Anat.* **13**:25–54.

Garcia-Segura, M., and P. Rakic (1985) Differential distribution of intermembranous particles in the plasmalemma of migrating cerebellar granule cells. *Dev. Brain Res.* (in press).

Goffinet, A. M. (1984) Events governing organizaton of postmigratory neurons: Studies on brain development in normal and *reeler* mice. *Brain Res. Rev.* **7**:261–296.

Goldberg, D. J. (1982) Actin and myosin in nerve cells. In *Axoplasmic Transport in Physiology and Pathology*, D. G. Weiss and A. Gorio, eds., pp. 73–80, Springer, Berlin.

Grumet, M., S. Hoffman, and G. M. Edelman (1984) Two antigenetically related neuronal CAMs of different specificities mediate neuron-neuron and neuron-glia adhesion. *Proc. Natl. Acad. Sci. USA* **81**:267–271.

Hatten, M. E., and R. K. H. Liem (1981) Astroglia provide a template for the positioning of cerebellar neurons *in vitro*. *J. Cell Biol.* **90**:622–630.

Hatten, M. E., R. K. H. Liem, and C. Mason (1984) Two forms of cerebellar glial cells interact differently with neurons *in vitro*. *J. Cell Biol.* **98**:193–204.

Heuser, J. E., and T. S. Rees (1973) Evidence for recycling of synaptic vesicle membrane during transmitter release in the frog neuromuscular junction. *J. Cell Biol.* **57**:315–344.

Jamieson, J. D., and G. E. Palade (1967) Intracellular transport of secretory protein in the pancreatic exocrine cell. I. Role of the peripheral elements of the Golgi complex. *J. Cell Biol.* **34**:577–596.

Jensen, K. F., and H. P. Killackey (1984) Subcortical projections from ectopic neocortical neurons. *Proc. Natl. Acad. Sci. USA* **81**:964–968.

Johnston, R. N., and N. K. Wessels (1980) Regulation of the elongating nerve fiber. *Curr. Top. Dev. Biol.* **16**:165–206.

Kristocek, A., and N. W. Seeds (1981) PI Plasminogen activator release at the neuronal growth cone. *Science* **213**:1532–1534.

Letourneau, P. (1975) Cell-to-substrate adhesion and guidance of axonal elongation. *Dev. Biol.* **44**:92–111.

Levitt, P. R., and P. Rakic (1980) Immunoperoxidase localization of glial fibrillary acid protein in radial glial cells and astrocytes of the developing rhesus monkey brain. *J. Comp. Neurol.* **193**:815–840.

Levitt, P. R., M. L. Cooper, and P. Rakic (1981) Coexistence of neuronal and glial precursor cells in the cerebral ventricular zone of the fetal monkey: An ultrastructural immuno-peroxidase analysis. *J. Neurosci.* **1**:27–39.

Lindner, J., F. G. Rathjen, and M. Schachner (1983) Ll mono- and polyclonal antibodies modify cell migration in early postnatal mouse cerebellum. *Nature* **305**:427–430.

McKay, R., and S. Hockfield (1983) A monoclonal antibody which recognizes radial glial cells in the developing central nervous system. *Soc. Neurosci. Abstr.* **9**:899.

Mangold, V., J. Sievers, and M. Berry (1984) 6-OHDA-induced ectopic of external granule cells in the subarachnoid space covering the cerebellum. II. Differentiation of granule cells: A scanning and transmission electron microscopic study. *J. Comp. Neurol.* **227**:267–284.

Moonen, G., M. P. Grau-Wagemans, and I. Seler (1982) Plasminogen activator–plasmin system and neuronal migration. *Nature* **298**:753–755.

Nowakowski, R. S. (1984) The mode of inheritance of a defect in lamination in the hippocampus of BALB/c mice. *J. Neurogenet.* **1**:249–258.

Nowakowski, R. S., and P. Rakic (1979) The mode of migration of neurons to the hippocampus: A Golgi and electron microscopic analysis in fetal rhesus monkey. *J. Neurocytol.* **8**:697–718.

Palade, G. E. (1975) Intracellular aspects of the process of protein secretion. *Science* **189**:347–358.

Pfenninger, K. H. (1982) Axonal transport in the sprouting neuron: Transfer of newly synthesized membrane components to the cell surface. In *Axoplasmic Transport in Physiology and Pathology*, D. G. Weiss and A. Gorio, eds., pp. 52–61, Springer, Berlin.

Pinto-Lord, C. M., E. Evrard, and V. S. Caviness, Jr. (1982) Obstructed neuronal migration along radial glial fibers in the neocortex of the *reeler* mouse: A Golgi–EM analysis. *Dev. Brain Res.* **4**:379–383.

Pixley, S. R., and J. DeVellis (1984) Transition between immature radial glial and mature astrocytes studied with a monoclonal antibody to dimentia. *Dev. Brain Res.* **15**:201–209.

Rakic, P. (1971a) Guidance of neurons migrating to the fetal monkey neocortex. *Brain Res.* **33**:471–476.

Rakic, P. (1971b) Neuron–glia relationship during granule cell migration in developing cerebellar cortex. A Golgi and electron microscopic study in macaque rhesus. *J. Comp. Neurol.* **141**:283–312.

Rakic, P. (1972) Mode of cell migration to the superficial layers of fetal monkey neocortex. *J. Comp. Neurol.* **145**:61–84.

Rakic, P. (1974) Neurons in rhesus monkey visual cortex: Systematic relation between time of origin and eventual disposition. *Science* **183**:425–427.

Rakic, P. (1975) Timing of major ontogenetic events in the visual cortex of the rhesus monkey. In *Brain Mechanisms in Mental Retardation*, N. A. Buckwald and M. Brazier, eds., pp. 3–40, Academic, New York.

Rakic, P. (1977) Genesis of the dorsal lateral geniculate nucleus in the rhesus monkey: Site and time of origin, kinetics of proliferation, routes of migration and pattern of distribution of neurons. *J. Comp. Neurol.* **176**:23–52.

Rakic, P. (1978) Neuronal migration and contact guidance in primate telencephalon. *Postgrad. Med. J.* **54**:25–40.

Rakic, P. (1981) Neuronal–glial interaction during brain development. *Trends Neurosci.* **4**:184–187.

Rakic, P. (1984) Defective cell-to-cell interactions as causes of brain malformations. In *Malformations of Development: Biological and Psychological Consequences*, E. S. Gollin, ed., pp. 239–285, Academic, New York.

Rakic, P. (1985a) Mechanisms of neuronal migration in developing cerebellar cortex. In *Molecular Basis of Neural Development*, G. M. Edelman, W. E. Gall, and W. M. Cowan, eds., pp. 139–159, Wiley, New York.

Rakic, P. (1985b) Principles of neuronal migration. In *Handbook of Physiology: Developmental Neurobiology*, W. M. Cowan, ed., American Physiological Society, Bethesda, Md. (in press).

Rakic, P. (1985c) Mode of cell migration in the corpus pontobulbare of the embryonic monkey brain stem (in preparation).

Rakic, P., and R. L. Sidman (1973) Sequence of developmental abnormalities leading to granule cell deficit in cerebellar cortex of *weaver* mutant mice. *J. Comp. Neurol.* **152**:103–132.

Rakic, P., L. J. Stenaas, E. P. Sayre, and R. L. Sidman (1974) Computer-aided three-dimensional reconstruction and quantitative analysis of cells from serial electron microscopic montages of foetal monkey brain. *Nature* **250**:31–34.

Schmechel, D. E., and P. Rakic (1979a) A Golgi study of radial glial cells in developing monkey telencephalon: Morphogenesis and transformation into astrocytes. *Anat. Embryol.* **156**:115–152.

Schmechel, D. E., and P. Rakic (1979b) Arrested proliferation of radial glial cells during midgestation in rhesus monkey. *Nature* **277**:303–305.

Sidman, R. L., and P. Rakic (1973) Neuronal migration, with special reference to developing human brain: A review. *Brain Res.* **62**:1–35.

Soreq, H., and R. Miskin (1981) Plasminogen activator in rodent brain. *Brain Res.* **216**:361–374.

Trenkner, E., D. Smith, and N. Segil (1984) Is cerebellar granule cell migration regulated by an internal clock? *J. Neurosci.* **4**:2850–2855.

Williams, R. W., and P. Rakic (1985) Dispersion of growing axons within the optic nerve of embryonic monkey. *Proc. Natl. Acad. Sci. USA* **82**:3906–3910.

Willinger, M., D. M. Margolis, and R. L. Sidman (1981) Neuronal differentiation in cultures of *weaver (wv)* mutant mouse cerebellum. *J. Supramol. Struct.* **17**:79–86.

Zecevic, N., and P. Rakic (1976) Differentiation of Purkinje cells and their relationship to other components of developing cerebral cortex in man. *J. Comp. Neurol.* **167**:27–47.

Chapter 5

An Assay of Cell Interactions
That Play A Role
In Neuronal Patterning

SCOTT E. FRASER
VIRGINIA BAYER

ABSTRACT

During neuronal development, groups of neurons interconnect with one another in a precisely ordered topographic fashion. It has long been proposed that chemical cues play a major role in the guidance of neurons and the formation of patterned synaptic relationships. An attractive proposal has been that adhesive interactions between neurons help to guide the formation of topographic neuronal connections. However, few assays have been available to test for the involvement of such interactions in the normal formation and maintenance of topographic projections. Here we describe a set of experiments that use a new assay for adhesive cell interactions in the development and regeneration of the retinotectal projection of the clawed frog, Xenopus. An agarose implant containing antibodies to cell surface molecules is placed in the optic tectum, and the effects of the antibody on the distribution of the optic nerve fibers are examined. Antibodies to the neural cell adhesion molecule (N-CAM) distort the pattern of the retinotectal projection and decrease its precision. These experiments demonstrate the utility of this in vivo perturbation technique for investigations of the cell interactions important in neuronal ordering and suggest a role of N-CAM in the patterning of the retinotectal projection.

The patterning of cell migration and cell differentiation during development is one of the most striking features of embryogenesis. These patterns are largely responsible for the orderly development of an embryo from an apparently unpatterned egg. In some species, prepatterns in the egg and early embryo may exist that help to organize the later development of the animal; however, much of development is thought to be the product of cell interactions that take place during embryogenesis. Proposals for the mechanism(s) of these interactions range from adhesive mechanisms

(Steinberg, 1970; Edelman, 1983), to the exchange of diffusible messengers (Wolpert, 1971; Meinhardt and Gierer, 1974), to tractional forces exerted on and between cells (Odell et al., 1981; Stopak and Harris, 1982). Some evidence consistent with each of these mechanisms exists, but true tests of the interactions responsible for the ordering of development remain elusive. The challenge posed to modern developmental and cell biology is to develop better assays of cell interactions and to use these assays to define the interactions that work to pattern the embryo.

Nowhere is the problem of patterning of more importance than in the developing nervous system, since there correct function relies not only on proper patterning within each neural center but also on the interconnections between neural centers. These interconnections demonstrate a high degree of organization, usually based on the topography of the neural centers involved. For example, in the visual system of vertebrates a topographic map of the retina is maintained by neural connections to almost all the centers involved in visual processing. As in the development of nonneural structures, the cells in the nervous system undergo stereotyped cell migration and position-dependent differentiation. However, one facet of this migration, neurite outgrowth, takes a different form. The cell body usually remains stationary while the growth cone of the neurite crawls away, leaving an axon or dendrite trailing behind. Many neurons extend long axons that travel via stereotyped pathways toward their target tissue(s) and once at the target, terminate in an ordered pattern on a subset of the cells located in a subregion of their target tissue. These terminations are often elaborate and demonstrate a characteristic anatomy of branches and synapses. Thus, neural development can be viewed as consisting of at least two nested patterning events: (1) the ordering of each neural center; and (2) the ordering of the interconnections between neural centers. As in the case of nonneural development, several different proposals of how the cells of the nervous system interact to form these patterns have been made; unfortunately, few techniques are available to assay for the exact nature or role of these interactions.

The lower vertebrate visual system has served as a major focus of research into the patterning of neuronal connections. The neuronal projection from the eye to the optic tectum (the retinotectal projection) is the major visual pathway, and is topographically organized such that a "map" of the visual field of the eye is conveyed intact to the contralateral optic tectum (for a review, see Fraser, 1985). Cells that originate at neighboring regions of the retina send their axons down the optic nerve to terminate at neighboring loci in the tectum. The projection is oriented so that cells from the anterior (nasal) part of the retina project to the posterior (caudal) portion of the tectum; cells from the dorsal part of the retina project to the ventral (lateral) part of the tectum. Following trauma to the optic nerve, such as a nerve crush, the optic nerve fibers regenerate to re-form a proper retinotectal projection within weeks. This stereotyped projection pattern, the regenerative ability of the optic nerve axons, and the straightforward assays of the projection together are responsible for the popularity of the retinotectal projection for neuronal patterning studies.

The many experiments performed on the lower vertebrate visual system have created a large data base concerning the response of the system to experimental manipulation. Some of these data indicate that the optic nerve fibers possess the ability to recognize their "correct" termination site in the tectum; still others demonstrate the plasticity of the projection pattern, which allows the system to adjust for changes in the numbers of optic nerve fibers or tectal cells. Several models have been developed that attempt to account for the complexity of the available data. Many of these models propose that basic cell interactions, such as cell–cell adhesion, play a central role in nerve patterning.

Our goal was to establish an *in situ* system for studying the cell interactions important in forming the retinotectal map. Below, we describe some additional background on the retinotectal system, outline some of the models proposed to explain the patterning of the projection, and then describe our recently developed technique for investigating the interactions involved in neuronal patterning of the retinotectal system (Fraser et al., 1984).

BACKGROUND

Assays of the Pattern and Precision of the Retinotectal Projection

Typical assays for the ordering of the retinotectal projection are based on extracellular electrophysiological techniques (for a discussion, see Hunt and Jacobson, 1974). A metal microelectrode is lowered into the neuropil of the optic tectum, where it records the activity of neurons in the near vicinity of the electrode tip. Visual stimuli are then presented to the animal, and the regions of the visual field that elicit activity at the electrode tip are determined. The responsive area of the visual field for a single electrode position is termed the receptive field for that position. The electrode is then moved and the procedure is repeated at successive sites until the entire dorsal surface of the optic tectum has been sampled. The data are typically presented as a set of electrode positions on the tectum and as the center of the responsive region for each electrode position (Figure 1A,B). This representation demonstrates the overall polarity and order of the retinotectal projection.

The size of individual receptive fields (Figure 1C) can be used to determine the precision of the projection (Schmidt and Edwards, 1983). The electrodes used for assaying the retinotectal map preferentially record from the terminal arbors of the optic nerve fibers near the electrode tip because of the electrode resistance and tip geometry. Thus, the receptive field directly indicates the region of the retina that sends optic nerve fibers to terminate

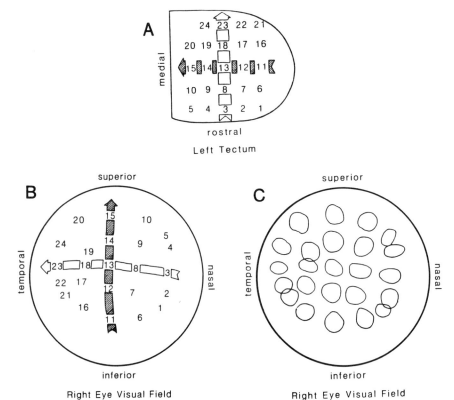

Figure 1.   *The retinotectal projection of* Xenopus. *A:* An electrode was lowered into the tectum at each of the numbered positions and the depth of the electrode was adjusted to maximize the signal recorded. *B* and *C:* The eye of the frog was centered on a projection perimeter and the regions of visual field that caused neural activity at the electrode tip for each electrode position were determined. These were recorded as the center of the responsive area (*B*) and as the entire receptive field for that electrode position (*C*). The arrows on the tectal outline (*A*) and the visual field map (*B*) highlight the two-dimensional ordering of the retinotectal projection.

at the electrode position. If the projection pattern were extremely precise, the receptive field would be very small, limited only by the finite size of the cells in the neural retina and the optic nerve terminals. However, the receptive field sizes are always larger than this lower limit because of short-range disorder in the projection. This disorder or imprecision slightly blurs the projection but maintains its overall topography. The size of the receptive fields therefore offers a direct and fast assay of the projection's precision.

Anatomical assays of the overall pattern and precision of the projection are also possible. For example, to determine the orientation of the projection, a subregion of the eyebud can be filled with tracer molecules such as horseradish peroxidase (HRP). The region of the tectum to which the filled optic nerve fibers project then gives a direct measure of the projection's orientation (cf. Harris, 1980). The precision of the projection can be anatomically assayed by allowing HRP to be retrogradely transported from a small injection site in the tectum to the retina. Only those ganglion cells that innervate the injected region will be labeled with HRP. As the precision of the projection decreases, the cells containing HRP become both less coherent and dispersed over a larger area of the retina (Cook and Rankin, 1984). In another assay of map precision, Meyer (1983) has examined the denervated region in the tectum produced by a small lesion in the retina. If the map is perfectly ordered, the retinal lesion produces an exactly corresponding denervated region in the tectum. Autoradiography after intraocular injection of tritiated proline is used to follow innervation of the tectum by healthy (unlesioned) parts of the retina. As the disorder in the projection is increased, fibers from regions of the retina neighboring the lesion will overlap the projection site of the lesioned retina; therefore, as the precision of the projection decreases, the denervated region will blur and shrink until it is eventually unnoticeable.

The anatomical techniques have generated data largely in agreement with the physiological assay described above, which confirms the utility and accuracy of both the anatomical and the physiological assays. The advantage of an anatomical assay is that it provides a direct measure of the distribution of optic nerve terminals. The advantage of a physiological assay is that it can be performed repeatedly on the same animal and thereby permits the time course of refinement of the projection pattern to be followed directly.

Development and Regeneration of the Retinotectal Projection

Anatomical tracing of developing optic nerve fibers filled with HRP or radiolabel has been used to examine the development of the retinotectal projection. These experiments demonstrate that the optic nerve fibers grow to their correct dorsoventral position from the earliest time that fibers reach the tectum (Holt and Harris, 1983). This early map appears to be independent of: (1) the timing of the arrival of nerve fibers (Holt, 1984); (2) the presence of a normal optic stalk (Harris, 1982); and (3) the presence of normal neuronal activity (Harris, 1980). It remains possible that combinations of these potential interactions cooperate to help organize the projection pattern. Because of the limitations of both physiological and anatomical assays of the projection in small animals such as *Xenopus*, these experiments offer only information on the overall polarity and order of the retinotectal projection, leaving questions about the precision of the projection unanswered.

Following damage to the optic nerve, the retinotectal map can regenerate with proper orientation and near-normal precision within weeks (Meyer,

1983; Schmidt and Edwards, 1983). This regeneration has been used extensively to study nerve patterning since the adult animals are much larger and hardier than embryos, making several classes of experiment much easier to perform. The underlying, though sometimes hidden or forgotten, assumption is that similar interactions mediate both development and regeneration; therefore, it is hoped that these experiments inform on the normal developmental processes responsible for retinotectal map development. Experiments on the regenerating retinotectal projection of goldfish indicate that a normally oriented map can form in the absence of visual experience (cf. Keating and Feldman, 1975) and even in the absence of all neural activity (Meyer, 1983; Schmidt and Edwards, 1983). The size of the adult animals allows a measurement of the precision of the regenerated retinotectal projection by both anatomical and physiological means. The results of these experiments indicate that the map first formed by the regenerating optic nerve fibers is much less precise than that of either the normal adult or the animal several weeks further on in the regeneration process. Interestingly, this process of map refinement is blocked by tetrodotoxin treatments, which abolish all nerve activity (Meyer, 1983; Schmidt and Edwards, 1983). Similarly, rearing the animal in strobe illumination to disrupt normal patterned vision also blocks refinement (Schmidt and Eisele, 1985). Thus, it appears that patterned neuronal activity plays little or no role in the orientation of the retinotectal projection but plays a major role in the refinement of the projection.

Experiments on the regenerating retinotectal projection demonstrate a rich set of responses to surgical rearrangement of the retina or tectum (for a review, see Fraser and Hunt, 1980). If two pieces of tectum are excised and interchanged with one another, the optic nerve fibers can track down the moved pieces and re-form synapses on them (cf. Yoon, 1975). Such experiments have been taken as strong evidence of some form of positional cue on the surface of the tectum. Similarly, if two pieces of eyebud are interchanged, they can map autonomously to the tectum in a manner appropriate for their position of origin (cf. Conway et al., 1980). These experiments suggest that the cells of the developing retina differ from one another in a position-dependent fashion, and that these differences play a role in the patterning of the projection.

In addition to the specificity described above, the retinotectal system is also capable of a good deal of plasticity. After removal of part of the retina or the tectum, the pattern of the projection is modified to at least partially compensate for the deletions (Sharma, 1972). For example, the optic nerve fibers from one half of an eye will spread to innervate most of the tectum if the complementary half of the eye has been deleted (cf. Schmidt et al., 1978). Thus, the retinotectal projection behaves somewhat paradoxically, demonstrating a great deal of both neuronal specificity and neuronal plasticity.

Cell Interactions and Models

Experiments on the regeneration of the retinotectal projection indicate that some form of positional cue plays a role in the patterning of regenerating nerve connections. Furthermore, recent experiments on the initial projection formed in normal embryonic development are also consistent with the idea of positional markers on the retina and tectum playing a role in nerve patterning. Experiments on the patterning of regenerating optic nerve fibers led Sperry (1950, 1963) to propose that chemical markers guide the optic nerve fibers to their correct target in the tectum. Sperry's model appears to be consistent with the mass of data concerning the specificity of the retinotectal projection, but appears to be inconsistent with the plasticity demonstrated by it. In the most extreme version of his model, Sperry came close to proposing unique chemical labels for each cell or small group of cells. When viewed from our perspective (one strongly influenced by and perhaps built upon the chemoaffinity hypothesis), the multitude of labels the hypothesis might require seems implausible. However, it should be remembered that the historical context in which the chemoaffinity model was proposed was dominated by the ideas that mechanical guidance and selective resonance were responsible for patterned behavior or neural connectivity. This context, and the sketchy knowledge of molecular biology at that time, may well have shaped the chemoaffinity hypothesis into such extreme terms. Since the true test of a model is the quality of the work it inspires, it is to its credit that the chemoaffinity model so completely dominated both the experimental design and the thinking of most laboratories.

While many would now reject the details of Sperry's chemoaffinity hypothesis, its ideas have spawned a new class of models in which adhesive interactions between cells play a central role (cf. Edelman, 1984; Fraser, 1985). Two such models (Fraser, 1980, 1985; Whitelaw and Cowan, 1981) propose that optic nerve fibers are patterned by positional adhesive cues in the retina and the tectum, coupled with stronger, position-independent competitive and adhesive interactions. Figure 2 shows the results of a computer simulation of one of these models (Fraser, 1985) and demonstrates the good fit between the model's behavior and the normal retinotectal projection (cf. Figure 1). Instead of a multitude of positional markers required by some models, a smooth gradient or set of gradients is proposed to provide the cells of the retina and the tectum with weak positional preferences. These preferences are sufficient to generate the specificity seen in the retinotectal projection, and yet weak enough to permit the plasticity that is observed. Thus, these models escape two of the pitfalls of the chemoaffinity hypothesis: the multitude of labels required and the inability to predict the plasticity experiments. Interestingly, a key feature of both models is a major, position-independent adhesive interaction be-

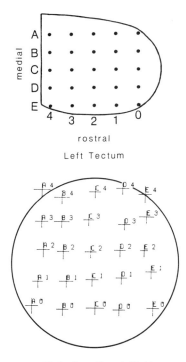

Figure 2. *Computer simulation of a model for neural patterning (Fraser, 1980, 1985).* The model is based on adhesive interactions between the optic nerve fibers and the tectum, adhesive interactions among the optic nerve fibers, and competition between the synapses made by the optic nerve fibers. The dominant interaction is a position-independent adhesion of all optic nerve fibers to the tectum, and the second largest is a competition between synapses. The smallest interaction is a position-dependent adhesion, which gives optic nerve fibers weakly preferred sites on the tectal surface. The computer simulation determined the predicted order of the optic nerve fibers on the surface of the tectum and presented this order by "mapping" the projection as done in the experiment shown in Figure 1. The marked positions in the lower figure represent the receptive fields (*crossed lines* represent the size of the receptive field) for each of the marked "electrode" positions in the upper figure. Comparison with Figure 1 shows good correspondence between the data and the predictions of the model.

tween the optic nerve fibers and the tectum. The models work best if this global adhesion is at least severalfold larger than the graded adhesions; when the magnitude of the global adhesion is diminished the models lose the ability to fit some of the data on the system.

Assays of Cell Interactions

In parallel with these embryological and theoretical considerations of the formation of nerve patterns, several laboratories have been investigating the possible molecular interactions that might pattern the nervous system. One approach is based on the reaggregation of single dissociated embryonic cells when cultured in a shaker bath. The rate of reaggregation is taken as a measure of the strength of the adhesion between cells, although it is uncertain how well this kinetic assay of cell adhesion relates to the behavior of cells in the embryo. Some of these studies have demonstrated regional differences in the kinetics of cell binding (cf. Marchase et al., 1975), but the molecular mechanisms responsible for the patterns remain largely unknown. A related, but perhaps more sound, approach has been to use the reaggregation of dissociated embryonic cells to assay for the presence of a functional cell adhesion system on the cells (cf. Thomas and Steinberg, 1982). Enzymatic and other treatments are then used to help characterize the classes of adhesion mechanisms present.

Other laboratories have approached this question by searching for position-dependent antigens in the retina and the tectum by using monoclonal antibody technology. Here too, a few successful experiments have yielded intriguing results. One antigen, TOP (for toponymic), is distributed in a gradient across the developing chick retina (Trisler et al., 1981); however, the role of either TOP or its gradient during development remains unknown.

A different experimental approach has been developed by Bonhoeffer and his colleagues (cf. Bonhoeffer and Huf, 1982). Their experiments have investigated the behavior of chick optic-nerve growth cones in culture when they are given a choice of growing onto one of two monolayers of tectal cells. By constructing the monolayers from different regions of the tectum, it is possible to determine if the optic nerve fibers have a preference for some regions of the tectum over others. Their experiments show that at least some of the cells can distinguish between tectal cells from different anteroposterior levels in the tectum. The experiments have not yet demonstrated any positional discrimination by optic nerve fibers and have failed to observe any dorsoventral preference. These failures seem troublesome, but may reflect more on the side effects of the cell isolation techniques than on the presence or absence of position-specific behavior.

Neuronal Cell Adhesion Molecule

A functional assay for cell adhesion, coupled with an immunological approach, has succeeded in defining the neural cell adhesion molecule (N-CAM), isolated by Edelman and his coworkers (for a review, see Edelman, 1983). In aggregation studies, N-CAM mediates the dominant adhesion between nerve cells, both embryonic and adult, through an unusual chemistry of homophilic adhesion. That is, each N-CAM molecule binds to another identical N-CAM molecule, instead of to a distinct receptor molecule. Antibodies to N-CAM are capable of blocking the majority of the adhesions between neurons and have been used to demonstrate the widespread distribution of N-CAM in embryonic and adult neural tissues (Edelman,

1983). N-CAM appears to be a heterogeneous molecule in developing systems, producing a wide band on SDS-PAGE because of varying degrees of polysialylation. This heterogeneity is decreased during development through the reduction of the carbohydrate content of the molecule, primarily because of the loss of sialic acid. The large amount of polysialic acid on the embryonic form of N-CAM appears to have profound effects on the effective binding kinetics of the molecule. Thus, the presence of the sialic acid and the heterogeneity of N-CAM may have important developmental consequences by modulating the strength of the adhesive interactions between cells.

The parallels between the properties of N-CAM and some of the interactions proposed in nerve-patterning models bring the possible role of N-CAM in nerve patterning to the forefront. N-CAM represents the major adhesive interaction between neurons in an *in vitro* aggregation assay and is widely distributed in the cells of the visual system. By comparison, two nerve-patterning models require strong general adhesive interactions between the cells of the retina and the tectum (Fraser, 1980; Whitelaw and Cowan, 1981). In one of these (Fraser, 1980), the model works best if the adhesion between cells is mediated by a homophilic adhesive molecule. To this general adhesion, both models add a position-dependent adhesion or a gradient of adhesion. N-CAM therefore offers the advantages of being present in both the retina and the tectum and of demonstrating homophilic adhesion. Furthermore, the heterogeneity of the molecule could conceivably allow positional differences in adhesive strength, thus fulfilling the patterned adhesive interactions proposed by the models.

An *In Vivo* Assay

The parallels between the known chemistry and distribution of N-CAM and the adhesive interactions predicted from nerve-patterning models made it timely to investigate the role of N-CAM in the patterning of the retinotectal projection. This required the development of a new *in vivo* assay for cell interactions in the retinotectal projection. Antibodies to *Xenopus* N-CAM or other cell antigens were incorporated into agar "spikes" and implanted in the optic tectum of young *Xenopus* froglets. In the days following the implantation of the antibodies, the pattern of the retinotectal projection and the precision of the projection were assayed by using electrophysiological techniques.

Details of the Assay

Preparation of Antibody Implants.   All antibodies were introduced into the frog in the form of an agar implant. These implants were made of a 6% solution of low melting-point agarose (Seaprep, FMC Biocolloids). The agarose solution was drawn into a length of polyethylene tubing with a syringe and then chilled for several hours. The solidified agar was forced from the tubing by pressure, cut into lengths, and placed in a multiwell plate where it soaked in a solution of antibody for 24–48 hours. The humidity in the plate was maintained so that the antibody solution was concentrated by about a factor of two during the soaking of the agarose. The agar cylinder was removed from the well, allowed to dry slightly, and then cut into sharpened "spikes" about 500 μm in length. This partial drying of the agar further concentrated the implant and produced a spike with enough inherent strength to be handled with jewelers forceps.

Antibodies Utilized. The antibodies used in these experiments fall into three categories: (1) monoclonal antibodies and Fab' fragments of polyclonal antibodies directed against *Xenopus* N-CAM; (2) Fab' fragments of polyclonal antibodies directed against the cell surface of the *Xenopus* brain but depleted of binding to N-CAM; (3) monoclonal antibodies or Fab' fragments of polyclonal antibodies directed against L-CAM (liver cell adhesion molecule) or preimmune serum. Antibodies in category 1 represent the test antibodies; those in categories 2 and 3 represent control antibodies.

Insertion of the Implants. The implants were inserted into the tectum of *Xenopus* froglets after the skin and bone overlying the tectum was deflected. A puncture wound was made in the tectum with a small, sharpened, metal probe at the site of the implant, and the agarose/antibody spike was then forced, point first, into the puncture wound, leaving approximately 100 μm of the spike projecting from the surface of the tectum. Within seconds, the agar/antibody spike rehydrated and swelled, which held the implant firmly in place. After the implantation was complete, the skull flap and skin were replaced. The frogs were revived and maintained in a solution of Gentamycin sulfate (50 μg/ml; Sigma) to minimize the chances of infection.

Assay of the Retinotectal Projection. Extracellular electrophysiological techniques were used to assay the pattern and precision of the retinotectal projection. The animal was anesthetized with MS-222 and paralyzed with curare; the skin and skull were deflected to expose the dorsal surface of the tecta. A platinum-tipped, platinum–iridium electrode was lowered into the superficial neuropil at a regular grid of positions. The depth of the electrode was adjusted to maximize the responses to the stimuli, which were provided by an Aimark projection perimeter. The signals from the electrode were amplified ($\times$ 1000) and filtered (100 Hz–10 kHz bandpass, 60 Hz rejection filters) before being displayed on an oscilloscope and played over a loudspeaker. For each electrode position, the region of the visual field that elicited any activity at the electrode tip (the receptive field) was determined. Any signal above the background noise of the equipment was taken as a response.

RESULTS

The basic paradigm of these studies was to implant an agar/antibody spike into the optic tectum of the frog and then to follow the effects of the antibody on the ordering of the retinotectal projection. The agar spikes were loaded with monoclonal antibodies or Fab' fragments of rabbit antibodies directed against N-CAM, L-CAM, or other undefined cell surface antigens from *Xenopus* brain. Throughout the experiment, the tissue surrounding the implant remained healthy in appearance and electrophysiological responses could always be recorded. Thus, the implant technique appears to offer a means to introduce the desired antibody with a minimum of trauma.

Both normal frogs and frogs in the midst of regenerating their retinotectal projections were used in the studies. The regenerating animals were used three weeks after an optic nerve crush, when a crudely ordered projection to the tectum had formed. This crude map would normally refine over the next few weeks to near-normal order. The regenerating animals were employed because they were clearly undergoing large synaptic rearrangements at the time that the antibody was introduced into their tecta.

In the days following the implantation of the agar/antibody spikes, the order of the projection was assayed by use of extracellular electrophysiology. In addition, the size of the receptive fields for each electrode position was determined as a measure of the short-range disorder in the projection (the larger the receptive fields, the less precise the ordering). Since the technique of physiologically mapping the retinotectal projection is nondestructive, the same animal can be assayed several times after the administration of the antibody to determine the time course of the antibody effects.

When assayed one week after the implantation, antibodies directed against *Xenopus* N-CAM distorted the projection pattern, whereas control antibodies to the *Xenopus* cell membrane, the chicken L-CAM molecule, or preimmune serum had no significant effect. Figure 3 shows representative maps of frogs regenerating their retinotectal projections one week after insertion of an implant of preimmune serum or rabbit anti-N-CAM Fab' (implant made three weeks after nerve crush). The maps from the animals treated with preimmune serum were indistinguishable from those of normal regenerating animals or animals implanted with agar alone (Figure 3A). In contrast, note the alterations of the projection pattern typical after an implant of antibodies directed against N-CAM (Figure 3B). The antibodies against N-CAM significantly increased the receptive field size ($p = .01-.001$) and produced a distortion in the pattern of the projection. This distortion is larger along the anteroposterior axis of the tectum, as predicted by one model of retinotectal patterning (Fraser, 1980). Normal (nonregenerating) animals showed similar, though slightly smaller, increases in receptive field size when implanted with antibodies directed against N-CAM.

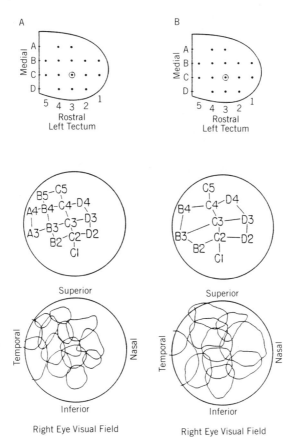

Figure 3. *Effects of antibodies on the retinotectal projection.* An implant containing antibody was inserted at the circled position in the tectum. Either preimmune Fab' fragments (*A*) or anti-N-CAM Fab' fragments (*B*) were used. The animals were implanted with the antibody three weeks after an optic nerve crush, at which time a well-ordered but somewhat imprecise retinotectal map had formed; the order of the retinotectal projection was determined one week later. Preimmune Fab' fragments caused no distortion in the projection pattern (*A*). Anti-N-CAM Fab' fragments caused a distortion in the pattern of the projection (*B, middle*) and a large increase in the size of the receptive fields (*B, bottom*). (Data from Fraser et al., 1984.)

The time course of antibody-induced changes in the precision of the projection was studied in normal animals. Figure 4 presents the receptive field sizes for animals implanted with control antibodies and with antibodies directed against N-CAM. Note that the receptive field size of animals implanted with control antibodies slowly decreases in size, reflecting the normal, continuing refinement of the projection pattern in these young frogs. In contrast, the animals implanted with antibodies to N-CAM demonstrate an increase in receptive field size that becomes maximal about 10 days after the implant is inserted. The receptive field size of these antibody-

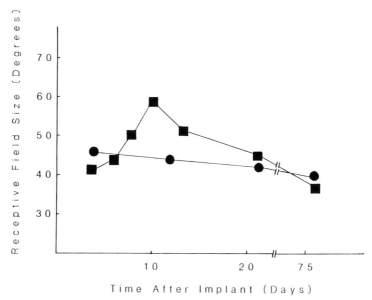

Figure 4. *Time course of the effect of antibody implants.* Implants were inserted into the tectum of newly metamorphosed *Xenopus* froglets and the receptive field sizes were measured at various times after the implant. Preimmune Fab' fragments (*circles*) did not cause an increase in the receptive field sizes, whereas anti-N-CAM Fab' fragments (*squares*) caused a significant increase. The sizes of the symbols are larger than the SEM for the data points.

treated animals decreases over the next several days to return to a value similar to that seen in normal or control animals.

CONCLUSION

The results presented above indicate that the order of the retinotectal projection can be affected by introducing antibodies to N-CAM into the tectal neuropil. Anti-N-CAM causes both a local distortion in the patterning of the projection and a decrease in the precision of the projection, as shown by the enlarged receptive field sizes. The magnitude of the distortion is somewhat larger along the anteroposterior dimension of the tectum. In contrast, the receptive field sizes did not increase more along one dimension. The most profound effects on the projection were produced by a rabbit anti-*Xenopus* N-CAM and a monoclonal antibody directed against *Xenopus* N-CAM. Smaller effects were obtained from another monoclonal antibody against *Xenopus* N-CAM that appears less effective at blocking adhesion in an assay of cell aggregation. Preimmune serum, polyclonal sera against chicken L-CAM, and other monoclonal antibodies against chicken L-CAM had little or no effect.

Theoretical analyses of the retinotectal system have led to a number of models, some of which are based on adhesive interactions between cells. Two of these models show an ability to fit the experimental data generated by a wide range of experiments on the lower vertebrate visual system (Fraser, 1980; Whitelaw and Cowan, 1981). Both models propose a dominant adhesion between the cells of the retina and the tectum that plays a central role in the patterning of connections. In one model, a minor quantitative gradient of adhesion is added to this global adhesion between cells (Whitelaw and Cowan, 1981). In the other, pairs of minor adhesive gradients were added to give each cell a "best-fit" site on the tectum (Fraser, 1980). The distortions found here appear consistent with those predicted by either model if this dominant adhesion were to be diminished. Furthermore, the heterogeneity of the embryonic form of N-CAM offers the possibility that the positional variations in adhesion proposed by both models are provided by N-CAM as well.

Interestingly, very similar effects were obtained when the antibodies were applied to the tecta of normal frogs or of frogs in the process of regenerating their retinotectal projections. This indicates that the retinotectal projection need not be in the midst of massive synaptic reorganization and refinement for the antibodies to have their effects. Thus, the antibodies do not merely block the reestablishment of an ordered projection but can cause the degradation of a well-ordered projection as well. Recent evidence indicates that the frog visual system is continually undergoing minor synaptic rearrangements in order to compensate for the continued growth of the retina and the tectum (cf. Fraser, 1983). The sizable effect of the antibody on the projection of normal frogs may have been caused by interference with this normal dynamic rearrangement of the retinotectal projection. This highlights the issue that many systems previously thought to be somewhat static, such as the adult neuromuscular junction, might continually extend and retract sprouts or synapses (cf. Van Essen, 1982). Therefore, treatments similar to those employed here may have large effects in systems that appear outwardly static, since these systems may undergo a great deal of underlying dynamic synaptic reorganization on a very small scale.

The results discussed above demonstrate the usefulness of an *in vivo* assay for cell interactions involved in the patterning of neural projections. The experiments show that disrupting the cell–cell adhesion between neurons mediated by N-CAM leads to a distortion in the pattern of the retinotectal projection, as well as a decrease in the precision of the projection. The technique offers a bridge between the cell interactions proposed in models of cell patterning and the cell interactions that are experimentally verifiable. The results support the hypothesis that cell surface adhesions are important in establishing neural patterns and suggest that N-CAM plays a central role in the formation of neuronal projections. The heterogeneity and the dynamic changes demonstrated by N-CAM offer the further possibility that N-CAM or a related molecule plays a central role in the fine structure of the the retinotectal projection and of other neural networks (see Edelman, 1984). The approach outlined above offers the possibility of testing directly for the role of defined molecular interactions in the patterning of neurons and their interconnections, and in doing so it should help to refine our understanding of neural patterning.

ACKNOWLEDGMENTS

The assay described was developed in collaboration with Dr. G. M. Edelman and his colleagues, Drs. B. A. Murray and C.-M. Chuong, and I gratefully acknowledge their collaborative efforts. Some of the data presented came from a report on the assay published with these coworkers (Fraser et al., 1984).

REFERENCES

Bonhoeffer, F., and J. Huf (1982) *In vitro* experiments on axon guidance demonstrating an anterior–posterior gradient on the tectum. *EMBO J.* 1:427–431.

Conway, K., K. Feiock, and R. K. Hunt (1980) Polyclones and patterns in developing *Xenopus* larvae. *Curr. Top. Dev. Biol.* 15:216–317.

Cook, J. E., and E. C. C. Rankin (1984) Use of a lectin-peroxidase conjugate (WGA-HRP) to assess the retinotopic precision of goldfish optic terminals. *Neurosci. Lett.* 48:61–66.

Edelman, G. M. (1983) Cell adhesion molecules. *Science* 219:450–457.

Edelman, G. M. (1984) Cell-surface modulation and marker multiplicity in neural patterning. *Trends Neurosci.* 7:48–84.

Fraser, S. E. (1980) A differential adhesion approach to the patterning of nerve connections. *Dev. Biol.* 79:453–464.

Fraser, S. E. (1983) Fiber optic mapping of the *Xenopus* visual system: Shift in the retinotectal projection during development. *Dev. Biol.* 95:505–511.

Fraser, S. E. (1985) Cell interactions involved in neuronal patterning: An experimental and theoretical approach. In *The Molecular Bases of Neural Development*, G. M. Edelman, W. E. Gall, and W. M. Cowan, eds., pp. 481–507, Wiley, New York.

Fraser, S. E., and R. K. Hunt (1980) Retinotectal specificity: Models and experiments in search of a mapping function. *Annu. Rev. Neurosci.* 3:319–352.

Fraser, S. E., B. A. Murray, C.-M. Chuong, and G. M. Edelman (1984) Alteration of the retinotectal map in *Xenopus* by antibodies to neural cell adhesion molecules. *Proc. Natl. Acad. Sci. USA* 81:4222–4226.

Harris, W. A. (1980) The effects of eliminating impulse activity on the development of the retinotectal projection in salamanders. *J. Comp. Neurol.* 194:303–317.

Harris, W. A. (1982) The transplantation of eyes to genetically eyeless salamanders: Visual projections and somatosensory interaction. *J. Neurosci.* 2:339–353.

Holt, C. (1984) Does timing of axon outgrowth influence initial retinotectal topography in *Xenopus? J. Neurosci.* 4:1130–1152.

Holt, C., and W. A. Harris (1983) Order in the initial retinotectal map in *Xenopus:* A new technique for labelling growing nerve fibers. *Nature* **301**:150–152.

Hunt, R. K., and M. Jacobson (1974) Neuronal specificity revisited. *Curr. Top. Dev. Biol.* **8**:203–258.

Keating, M. J., and J. D. Feldman (1975) Visual deprivation and intertectal neuronal connections in *Xenopus laevis. Proc. R. Soc. Lond.(Biol.)* **192**:467–474.

Marchase, R. B., A. J. Barbera, and S. Roth (1975) A molecular approach to retinotectal specificity. *Ciba Found. Symp.* **29**:315–327.

Meinhardt, H., and A. Gierer (1974) Applications of a theory of biological pattern formation based on lateral inhibition. *J. Cell Sci.* **15**:321–346.

Meyer, R. L. (1983) Tetrodotoxin inhibits the formation of refined retinotopography in goldfish. *Dev. Brain Res.* **6**:293–298.

Odell, G. M., G. Oster, P. Alberch, and B. Burnside (1981) The mechanical basis of morphogenesis. I. Epithelial folding and invagination. *Dev. Biol.* **85**:446–462.

Schmidt, J. T., and D. L. Edwards (1983) Activity sharpens the map during the regeneration of the retinotectal projection in goldfish. *Brain Res.* **269**:29–39.

Schmidt, J. T., and L. E. Eisele (1985) Stroboscopic illumination and dark block the sharpening of the regenerated retinotectal map in goldfish. *Neuroscience* **14**:535–546.

Schmidt, J. T., C. M. Cicerone, and S. S. Easter, Jr. (1978) Expansion of the half-retinal projection to the tectum in goldfish: An electrophysiologic and anatomical study. *J. Comp. Neurol.* **177**:257–277.

Sharma, S. C. (1972) Reformation of the retinotectal projection after various tectal ablations in *Xenopus. Exp. Neurol.* **34**:171–182.

Sperry, R. W. (1950) Neuronal Specificity. In *Genetic Neurology,* P. Weiss, ed., pp. 231–248, Univ. of Chicago Press, Chicago.

Sperry, R. W. (1963) Chemoaffinity in the orderly growth of nerve fiber patterns and connections. *Proc. Natl. Acad. Sci. USA* **50**:703–710.

Steinberg, M. S. (1970) Does differential adhesion govern self-assembly processes in histogenesis? Equilibrium configurations and the emergence of hierarchy among populations of embryonic cells. *J. Exp. Zool.* **73**:395–434.

Stopak, D., and A. K. Harris (1982) Connective tissue morphogenesis by fibroblast traction. I. Tissue culture observations. *Dev. Biol.* **90**:383–398.

Thomas, W. A., and M. S. Steinberg (1982) Two distinct adhesion mechanisms in embryonic neural retina cells. II. An immunologic analysis. *Dev. Biol.* **81**:106–114.

Trisler, G. D., M. D. Schneider, and M. Nirenberg (1981) A topographic gradient of molecules in retina can be used to identify neuron position. *Proc. Natl. Acad. Sci. USA* **78**:2145–2149.

Van Essen, D. C. (1982) Neuromuscular synapse elimination. In *Neuronal Development,* N. Spitzer, ed., pp. 333–376, Plenum, New York.

Whitelaw, V. A., and J. D. Cowan (1981) Specificity and plasticity of retinotectal connections: A computational model. *J. Neurosci.* **1**:1369–1387.

Wolpert, L. (1971) Positional information and pattern formation. *Curr. Top. Dev. Biol.* **6**:183–224.

Yoon, M. G. (1975) Topographic polarity of the optic tectum studied by reimplantation of the tectal tissue in adult goldfish. *Cold Spring Harbor Symp. Quant. Biol.* **40**:503–519.

Chapter 6

Chlamydomonas Cells in Contact

URSULA W. GOODENOUGH
W. STEVEN ADAIR
PATRICIA COLLIN-OSDOBY
JOHN E. HEUSER

ABSTRACT

Gametic recognition between Chlamydomonas *cells is mediated by long, fibrous, hydroxyproline-rich glycoproteins called agglutinins, which associate with the flagellar membrane surface in an extrinsic fashion. We review recent research on the biochemistry of the agglutinins and present the structure of these proteins as revealed by quick-freeze, deep-etch electron microscopy. We also describe several shorter, but structurally similar, proteins that coextract with the agglutinins but lack adhesive activity. Monoclonal antibody decoration studies illustrate the feasibility of mapping domains along such proteins. Our studies support the general hypothesis that* Chlamydomonas *produces a family of fibrous glycoproteins, some that form the extracellular matrix and others that mediate sexual cell–cell recognition.*

Eukaryotic protozoa can be assumed to have preceded the metazoa. They therefore must have invented such ubiquitous eukaryotic traits as mitosis, meiosis, centrioles, cilia, and the use of actin and tubulin as intracellular mediators of cell shape. Of the present-day eukaryotic protists, *Chlamydomonas* is usually chosen by textbook writers as the protozoan most closely resembling a progenitor of multicellular plants. Since *Chlamydomonas* cells grow in the dark and since nonphotosynthetic varieties of *Chlamydomonas* exist in nature, a good case can also be made for a primal *Chlamydomonas* as a progenitor of the multicellular animals. It does not follow, of course, that all features of the modern protozoa were invented by their ancestral cells: The modern species have had the same billion years of evolutionary opportunity as have the metazoa. Nevertheless, we have adopted as the premise for our research the notion that the strategies employed by modern *Chlamydomonas* for cell–cell recognition and fertilization may, at the very least, offer insights into the primitive strategies employed by its ancestors, and that at least some of these strategies may have been retained by the metazoa as they came into being. .

BACKGROUND

The *Chlamydomonas* mating system is the subject of three recent reviews (Goodenough and Thorner, 1983; Snell, 1985; van den Ende, 1985). Therefore, we can summarize the system quickly as it pertains to recent data from our laboratory.

Chlamydomonas clones are immortal and resort to sex and meiosis only when deprived of nitrogen. The apparent "purpose" of the sexual option is to form diploid zygotes; these are uniquely capable of secreting a highly cross-linked cell wall, resistant to dessication, which is analogous to the spore coats of yeast and of gram-positive bacteria and cyanobacteria.

Vegetative cells differentiate directly into gametes when starved, a process requiring approximately eight hours and accompanied by a round of mitosis. Each mating type (*plus* and *minus*) can differentiate independently and requires no hormonal stimulus from the other. When *plus* and *minus* gametes are mixed, they undergo an instantaneous agglutination via their flagellar surfaces. Agglutination sends a signal to the cell bodies which results in two identifiable responses: (1) release of autolysin, a factor that catalyzes the breakdown of the glycoproteinaceous matrix surrounding the cell bodies; and (2) activation of mating structures, specialized regions of the plasmalemma that mediate the fusion of the gametes in a mating type-specific fashion. Although the signaling and fusion phases of the mating reaction are of considerable interest to us (Mesland et al., 1980; Goodenough et al., 1982; Detmers et al., 1983), we focus here on the initial recognition/adhesion events.

The pioneering work of Lutz Wiese (Wiese, 1969; Wiese and Wiese, 1978) established that *Chlamydomonas* gametes carry glycoproteins on their flagellar surfaces that mediate recognition/adhesion. Each species of *Chlamydomonas* has two mating types and therefore at least two types of complementary agglutinins. With some 200 species of *Chlamydomonas* described, plus an extensive group of related colonial forms (*Eudorina, Pandorina, Volvox*, etc.) that undertake fertilization in a fundamentally similar fashion, this means that the agglutinins, like all sperm/egg recognition molecules, must be both highly specific *and* capable of modification during the course of speciation. Our research focuses on *Chlamydomonas reinhardii*, the species most amenable to genetic analysis (Levine and Ebersold, 1960), but it is important to note at the outset that parallel studies in the laboratory of van den Ende (Musgrave et al., 1981; Homan, 1982) have demonstrated that similar proteins mediate adhesion in *Chlamydomonas eugamatos*, a species sufficiently different from *C. reinhardii* in certain aspects of its life cycle to predict that such proteins will prove to be utilized at least throughout the Volvocales.

Previous investigations have established four features of sexual adhesion in *Chlamydomonas* that are relevant to our present findings:

1. The agglutinins are not present on the surface of vegetative cells, and activity is lost after the gametes fuse to form quadriflagellated cells (QFCs). Whether the QFCs carry inactivated proteins or shed their agglutinins into the medium is as yet unknown.

2. Gametes lose their ability to agglutinate when treated with a variety of proteolytic enzymes (Wiese and Hayward, 1972). If the enzymes are then washed away, new agglutinins repopulate the flagellar surfaces in a cycloheximide-sensitive, tunicamycin-sensitive process (Snell, 1976, 1981; Solter and Gibor, 1978; Snell and Moore, 1980).

3. Flagella lose their agglutinability during the course of adhesion and rely on a continuous repopulation of protein during prolonged agglutination events. This was first demonstrated by Snell and Roseman (1979), who showed that if isolated *plus* flagella were presented to *minus* gametes, the gametes adhered avidly to the flagella for only about 20 minutes and then disadhered. Addition of fresh flagella elicited a new round of adhesion/disadhesion, whereas addition of fresh gametes yielded no additional adhesion; thus the flagella, not the cells, were dead. Snell and Moore (1980) went on to document this same phenomenon *in vivo*: Using the *imp-1 plus* mutant, which can agglutinate but not fuse and therefore adheres to *minus* gametes for hours, they showed that if an *imp-1/minus* mating mixture were given cycloheximide, the cells disadhered within 20 minutes. It is possible that this phenomenon will prove to be related to the "natural" loss of agglutinability that accompanies QFC formation, but since QFCs appear to lose their adhesiveness at the moment of cell fusion, whereas flagella require 20 minutes to be "used up" during agglutination, some additional event(s) would appear to accompany the loss of QFC agglutinability.

4. Adhesion requires a living cell (Goodenough, 1977; Snell and Roseman, 1979). Thus, as noted above, living gametes will adhere to isolated flagella; they will also adhere to glutaraldehyde-fixed gametes (Goodenough and Weiss, 1975), to membranes derived from flagella (Bergman et al., 1975), and to purified agglutinin proteins adsorbed to glass (Adair et al., 1982) or covalently coupled to agarose beads (Collin-Osdoby et al., 1984). However, none of these "nonliving" preparations gives any indication of interacting with or neutralizing one another: Thus *plus* and *minus* flagella show no tendency to stick together when examined by light microscopy (Snell and Roseman, 1979); beads derivatized with *plus* agglutinin are not inactivated by *minus* agglutinin (Collin-Osdoby et al., 1984), and so on. The physiological/biochemical basis for the requirement of a living cell is also unknown.

AGGLUTININ GENETICS

Mutations at two gene loci, *sag-1* and *sag-2* (sexual *ag*glutination), abolish the ability of *plus* gametes to become sexually agglutinable, and mutations at a single locus, *sad-1* (sexual *ad*hesion), abolish *minus* agglutinability (Goodenough et al., 1978; Hwang et al., 1981). Repeated mutant screens have turned up five alleles at the *sag-1* locus, one at *sag-2*, and two at *sad-1*; therefore, although some additional loci might well emerge with additional effort, such a pattern suggests that the trait may be specified by a relatively small number of genes. The five *sag-1* alleles fail to complement one another in diploids (Adair et al., 1983a), indicating that this locus does not represent a gene cluster.

The mating type (*mt*) locus, on linkage-group VI of *C. reinhardii*, exerts ultimate control over gametic type, and behaves as a single Mendelian locus during meiosis. The mt^+ locus controls expression of *sag-1* and *sag-2*, which are located on different chromosomes from *mt* and from one another. The *sad-1* locus, on the other hand, is tightly linked to mt^- so that we cannot ask whether its expression is mt^- dependent. Nothing is known about the *mt* loci except that when diploid organisms are constructed in the laboratory, they mate as mt^- gametes (Ebersold, 1967); therefore, mt^- is dominant to mt^+ in this situation.

An important feature of all these mutants is that they are apparently defective *only* in recognition/adhesion (Goodenough and Jurivich, 1978). Thus if they are mixed with normal gametes of opposite mating type and a polyclonal antiserum, raised against isolated flagella, is then added to the suspension, the antiserum causes the gametes to agglutinate, and there follows normal autolysin secretion, mating structure activation, cell fusion, QFC maturation, and zygote germination, permitting the acquisition of the genetic data cited above. Presumably, therefore, the system mediating sexual signaling is intact in these mutants and simply lacks its natural "stimulator," a functional agglutinin system. Interestingly, the signaling system appears to be relatively nonspecific: If gametes of a *sag-1* strain are agglutinated to one another via an antiserum, in the absence of any normal *minus* gametes, they still undergo mating structure activation (although not fusion); moreover, the antiserum used can be raised against either gametic or vegetative flagella of either mating type. The flagellar proteins stimulated by these polyclonal antisera await identification.

POLYPEPTIDE A

Agglutinin can be stripped from the surface of *plus* gametes by exposure to 5 mM EDTA for approximately 20 minutes (Adair et al., 1982), and the agglutinin is then purified from contaminating proteins by conventional chromatography, using as an assay the adhesion of gametes of opposite type to aliquots of column fractions dried down on glass slides. The results, which are published elsewhere (Adair et al., 1982, 1983a) and summarized in Figure 1A, indicate that *plus* agglutinin activity cofractionates with a very high molecular weight glycopolypeptide designated polypeptide A, which moves just into the stacker of a 3% SDS-polyacrylamide gel. When the *sag-1* and *sag-2* mutants are analyzed using comparable protocols, they lack this polypeptide (Figure 1B), as do noncomplementing diploid strains between *sag-1* alleles (Adair et al., 1983a).

When *minus* gametes are exposed to EDTA, agglutinin is also stripped from their flagellar surfaces. The process requires a longer time (approximately 45 minutes), but it is not yet known whether this is because the proteins are more abundant, more tightly associated with the membrane, or more rapidly replaced by the cell. Sizing chromatography indicates that the *minus* proteins behave equivalently to the *plus*, and the *minus* polypeptide A migrates in a similar position to that of the *plus* by SDS-polyacrylamide gel electrophoresis. By ion exchange chromatography, on the other hand, the *minus* activity is excluded from an anion exchange (Mono-Q) resin (20 mM Hepes, 5 mM EDTA, pH 7.2), whereas the *plus* material is retained. Therefore, it would appear that the two proteins differ in net charge, and this may facilitate their interaction.

Purified agglutinin proteins can be covalently coupled to agarose beads, and the beads then elicit the adhesion of opposite-type, but not like-type, living gametes (Collin-Osdoby et al., 1984). The beads prove to be a convenient substrate for testing the effects of various reagents on adhesive activity. As summarized in Table 1, agglutinin activity is sensitive to proteolysis by thermolysin and trypsin, but not by chymotrypsin; is sensitive to treatment with periodate or borohydride, but unaffected by glycosidases; and is sensitive to reduction by β-mercaptoethanol or dithiothreitol.

Amino acid analysis of polypeptide A shows the protein to contain approximately 12% hydroxyproline and 10% serine (Cooper et al., 1983). This profile places the agglutinin in a family of hydroxyproline-rich glycopolypeptides that are abundant in the cell walls of algae and higher plants (Lamport, 1980; Fincher et al., 1983). Some of the higher plant proteins, moreover, display lectin activity (Desai et al., 1981; Leach et al., 1982). These proteins, and the agglutinin from *Chlamydomonas eugamatos*, have been shown to carry short-chain oligosaccharides with terminal arabinosides and subterminal galactosides O-linked to the hydroxyproline and serine residues (Miller et al., 1974; Catt et al., 1976; Homan, 1982). The apparent insensitivity of the agglutinin to glycosidases, therefore, may simply be due to the insensitivity of these oligosaccharides to commercially available enzymes. The periodate/borohydride inactivation results do not, of course, document a direct role for sugars in the adhesion reaction, since

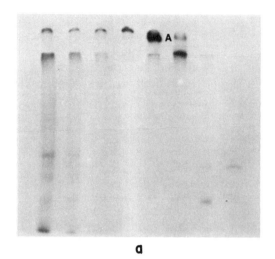

a

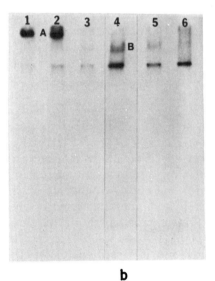

b

c

Figure 1. *Purification of* Chlamydomonas mt$^+$ *agglutinin. a*: Fractogel-75 chromatography of the nonincluded fractions from a Sepharose-6B column presented with a radiolabeled EDTA extract of living cells. Adhesive activity is present only in the two fractions containing polypeptide A. *b*: Polypeptides from wild-type and nonagglutinating mutants. Mutant radiolabeled EDTA extracts were fractionated along with unlabeled wild-type material as an internal marker for adhesive activity. *Lane 1:* Wild-type *plus. Lane 2: imp-1,* a mutant with normal adhesion but defective cell fusion (Goodenough et al., 1982). *Lane 3: imp-2.* Lane 4: *imp-5.* Lane 5: *imp-7.* Lane 6. *imp-8* (the last four strains are nonagglutinating). A and B, polypeptides A and B. *c*: Polypeptides A and B from *mt$^+$*. The purified fraction in *Lane 1,* from a radiolabeled EDTA extract, contains no detectable polypeptide A, whereas some B is present in the fraction in *Lane 2,* from a radiolabeled flagellar preparation extracted with the detergent octylglucoside.

116

Table 1. Inactivation of Bead-Conjugated Agglutinin

Agent or Treatment	Inactivation
Thermolysin	
10 μg/ml	Partial
100 μg/ml	Complete
Trypsin	
100 μg/ml	Partial
1 mg/ml	Complete
Chymotrypsin, 10 mg/ml	No effect
Heating	
45°C	No effect
55°C	No effect
65°C	Complete
Periodate (10 mM)	Complete
Alkaline sodium borohydride (1.0 M)	Complete
Endoglucosaminidase H (0.005 U)	No effect
α-Galactosidase (1.0 U)	No effect
β-Galactosidase (0.85 U)	No effect
α-L-Fucosidase (0.02 U)	No effect
β-N-Acetylglucosaminidase B (0.142 U)	No effect
α-Mannosidase (0.20 U)	No effect
Alkaline phosphatase (0.23 U)	No effect
Neuraminidase (0.05 U)	No effect
Mixed exoglycosidases[a]	No effect
Dithiothreitol (100 mM)	Complete
β-Mercaptoethanol	
10 mM	Partial
120 mM	Complete
Spontaneous reoxidation	None
β-Mercaptoethanol ([*]100 mM) + iodoacetamide (100 mM)	Complete
Iodoacetamide (100 mM)	No effect
β-Mercaptoethanol (100 mM) + N-ethylmaleimide (100 mM)	Complete
N-Ethylmaleimide (100 mM)	No effect
minus Agglutinin	No effect

Source: Collin-Osdoby et al., 1984.

[a] β-Galactosidase (0.4 U); α-L-fucosidase (0.025 U); β-N-Acetylglucosaminidase B (0.5 U).

if sugars are required for maintenance of the protein's overall topology, then their destruction may cause the protein to denature.

POLYPEPTIDE B

EDTA extracts contain a second polypeptide that migrates ahead of polypeptide A in the 3% stacker of SDS-polyacrylamide gels (Figure 1C). Its abundance varies considerably from one extract to the next. In some preparations, *two* bands can be resolved in this region, but for present purposes it suffices to say that at least one additional band, hereafter designated polypeptide B, is encountered.

Two observations about polypeptide B are important to our study. First, it is possible to fractionate polypeptide A away from polypeptide B, and adhesive activity invariably correlates with the presence of A: Fractions containing B alone are inactive. Second, proteins that migrate in the polypeptide B region are present in SDS-polyacrylamide gels of the *sag-1* and *sag-2* mutant strains (Figure 1B) which, as noted above, lack polypeptide A.

Additional insight into the nature of polypeptide B is given in the following sections.

CANES, SHORT CANES, AND LOOPS

The purified agglutinin protein has been visualized by the quick-freeze, deep-etch technique after adsorption to pulverized mica flakes (Heuser, 1983). Figure 2 shows representative molecules and Table 2 summarizes their lengths.

Preparations of pure *mt*$^+$ polypeptide A are found to contain very long (approximately 225 nm) fibrous proteins that we designate canes (Figure 2A, B). Each cane exhibits four distinct domains: a terminal head, a straight rod section, a flexible section, and a terminal hook. In some molecules a

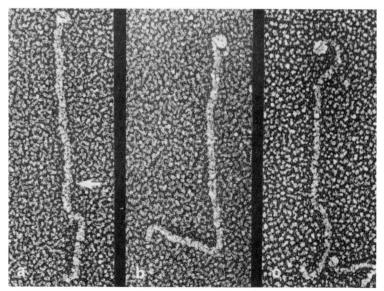

Figure 2. *Agglutinin proteins. a* and *b*: *mt*$^+$ *agglutinins. c*: *mt*$^-$ *agglutinin. Arrow* indicates discontinuity along shaft. × 370,000.

Table 2. Lengths of Agglutinin Proteins

	plus Proteins	minus Proteins
Canes	228 ± 7 nm (n = 68)	226 ± 9 nm (n = 49)
Short canes	128 ± 10 nm (n = 78)	118 ± 11 nm (n = 85)
Loops	129 ± 5 nm (n = 41)	134 ± 16 nm (n = 67)
imp-5 Loops	145 ± 14 nm (n = 66)	
imp-12 Loops		134 ± 12 nm (n = 32)
imp-12 Short canes		127 ± 7 nm (n = 27)

discontinuity appears along the shaft; examples are indicated by the arrows in Figures 2A and 3. These prove, however, to be localized at various positions on the molecule and are therefore not reliable morphological landmarks.

Preparations of pure mt^- polypeptide A contain proteins that are similar in length (approximately 225 nm) and in overall construction to the mt^+ agglutinin (Figure 2C). The major morphological difference is seen at the head/shaft junction which, in the minus species, can adopt the shape of a shepherd's crook (Figure 2C).

The proteins found in fractions enriched in the polypeptide B family prove to be comparable for both plus and minus preparations. Two predominant species are encountered. Some assume the configuration of short canes, approximately 120 nm in length (Table 2), with a head at one end (Figure 3, upper panels). Others assume the configuration of loops, ap-

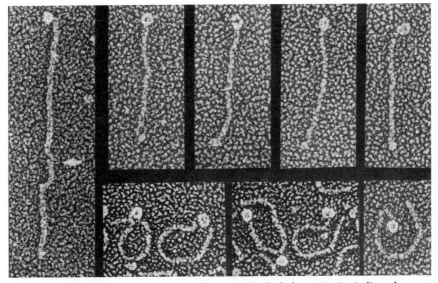

Figure 3. Canes, short canes, and loops. mt^+ cane with shaft discontinuity indicated at arrow is shown at left. Short canes (all mt^- are shown in the top four frames. Loops are shown in the lower small frames (all mt^+ except the last, which is mt^-). × 310,000.

proximately 130 nm in length, some of which carry a globular domain at one end (Figure 3, lower panels). The morphological similarity of these species to the full canes is clear, but we do not yet know how, if at all, the three species are related to one another. All, however, are gamete-specific, being absent from vegetative cells, so a role in the mating reaction is suggested.

THERMOLYSIN DIGESTION OF THE HEAD

A clue about the location of important domains in the mt^+ cane protein has come with the examination of canes that have been digested with thermolysin to destroy agglutinative activity (cf. Table 1). Controls are shown in the upper panel of Figure 4, treated proteins in the lower panel. The digested proteins retain their full length, but many carry visibly abnormal heads: some are totally decapitated (Figure 4, arrow); others carry small

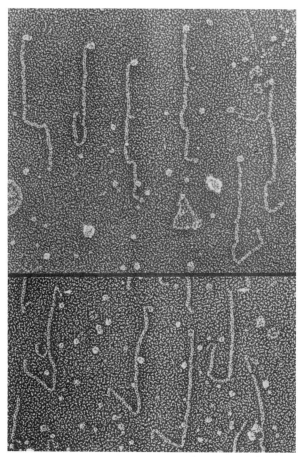

Figure 4.   *Thermolysin digestion. Upper*: Control mt^+ agglutinins. *Lower*: Thermolysin-digested mt^+ agglutinins. *Arrow*, decapitated protein; *arrowhead*, deranged head. × 180,000.

or deranged heads (Figure 4, arrowhead). There are three possible interpretations of this result: (1) the head is an important adhesive domain; (2)
disruption of the head causes some overall change in topology that destroys
activity; (3) thermolysin produces some additional nonvisible changes in
the protein that destroy activity, and the head alteration is irrelevant. Of
these, we regard the first interpretation as the most likely.

THE SURFACE OF GAMETIC FLAGELLA

The obvious next step is to ask how these proteins are displayed on the
flagellar surface. Again we have used the quick-freeze, deep-etch technique.
When deep-etched replicas of gametic flagella are compared with vegetative
flagella, they appear indistinguishable. Both are covered by a dense coat
of short, blunt protuberances (Monk et al., 1983), which, we believe,
correspond to the major integral protein of the *Chlamydomonas* flagellar
membrane, a 350-kD glycopolypeptide (Witman et al., 1972; Bergman et
al., 1975). To better visualize these surfaces, we have mixed intact cells or
isolated flagella with flakes of pulverized mica so that the flagellar surfaces
can be displayed in the same fashion as the isolated agglutinin proteins.
Figures 5 and 6 show representative images. It is clear, first, that fibrous
proteins indeed project from the membrane surface, and these prove to
be absent from comparable preparations of vegetative flagella. Second,
there are occasions when these fibers can be identified unambiguously as
canes (Figure 5, arrows). Third, most of the proteins remain close to the
flagellar surface and adopt a curved configuration.

It is difficult to analyze these curves in preparations of whole flagella,
since portions of them are often obscured by the membrane. More informative are small vesicles derived from the flagellar membrane that stick to
the mica and display their associated fibers (Figure 7). We also encounter
cases wherein a mat of such curved fibrils has adsorbed to a mica surface
(Figure 8). These images indicate that the gametic surface carries not only
the long canes, anchored to the membrane via their hook ends and displaying
their heads, but also a system of curved proteins of unknown function.
The curved arrays carry no globular heads, so that they are unlikely to be
constructed from the loops described earlier (Figure 3). Loops and short
canes, we should note, cosediment with isolated flagella and appear near
the flagella on the mica surface (not shown), but have not as yet been
observed attached to the flagellar membrane. Possibly a loose attachment
exists that is destabilized by the presence of mica.

MONOCLONAL ANTIBODY PROBES OF AGGLUTININ
STRUCTURE

Our present collection of monoclonal antibodies raised against purified
plus agglutinin (Adair et al., 1983b) yields additional information about

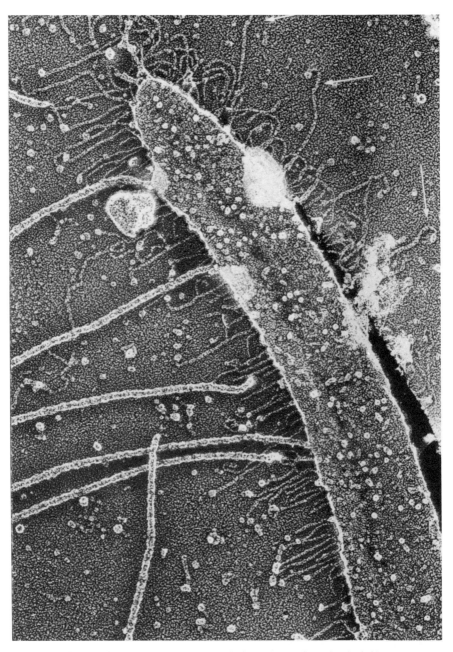

Figure 5. *Flagellum from living* mt⁻ *gamete adsorbed to a mica surface.* The thick fibrous elements to the right are mastigonemes, which appear to play no direct role in adhesion (Bergman et al., 1975). *Arrows* indicate slender fibrous proteins with the shepherds's-crook morphology of canes. The fibers in between circle back on the membrane to form closed loops. × 140,000.

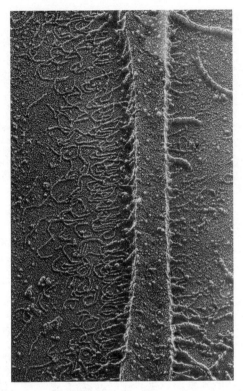

Figure 6. *Flagellum from living* mt⁺ *gamete adsorbed to a mica surface.* Fibrils form closed loops along the flagellar surface. × 79,000.

the domains of the protein. One antibody, denoted A16A, is an IgG that recognizes both agglutinin and other flagellar proteins (as determined by both immunoautoradiographic analysis and immunofluorescence studies). Mixtures of this antibody with agglutinin give patterns of the sort shown in Figure 9: IgGs localize to the head and the hook sections of the protein, but fail to bind to the shaft. An identical pattern is found for antibody A8C, an IgM (Figure 10). By contrast, IgGs from clone A12E show a very different pattern: They interact with the shaft of the protein, so obscuring it that one cannot tell whether the ends are also recognized (Figure 11). Such images indicate that the determinant recognized by antibody A12E repeats along the length of the shaft.

Antibody A12E proves to be uniquely specific for the A and B polypeptides: It recognizes no other flagellar proteins by immunoautoradiography, and stains gametes, but not vegetative cells or QFCs, by immunofluorescence. Therefore, determinants that repeat along the shaft are unique to canes; determinants that reside in the head and hook are shared with other proteins. Antibody A12E, we should note, agglutinates both *plus*

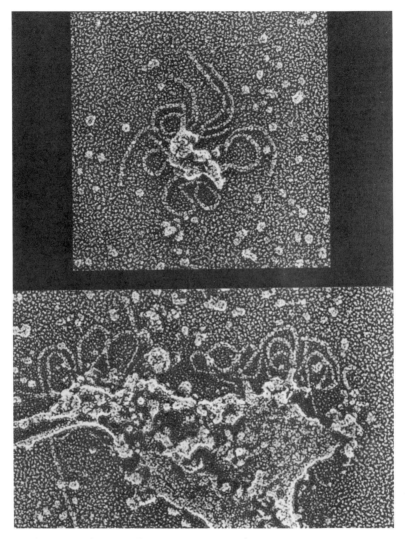

Figure 7. *Membrane fragments from* mt⁻ *(top) and* mt⁺ *(bottom) flagella.* Fibrils form closed loops. The scattered particulate material in the background and on the *mt⁺* membrane is a component of the flagellar surface that readily dissociates as the membrane adsorbs to mica. × 250,000.

and *minus* gametes; therefore, its antigen is agglutinin-specific but not mating type-specific.

Figures 10 and 11 document that the monoclonal antibodies bind to short canes and loops in the same fashion that they bind to intact canes: A8C binds to the heads, and A12E binds along the shaft. These patterns support the proposal that the smaller proteins are related to the full canes.

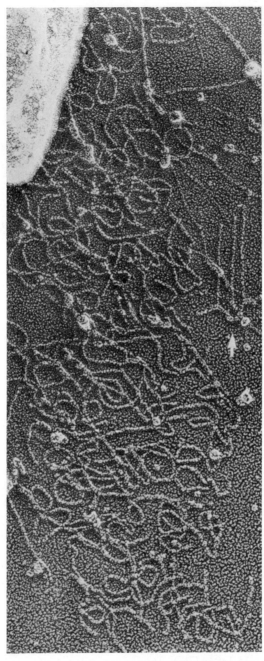

Figure 8. *Closed loops that have adsorbed to the mica surface* from the flagellum just seen in upper left corner. *Arrow* indicates a group of cell wall proteins. × 150,000.

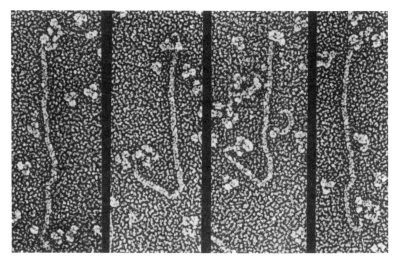

Figure 9. *Agglutinin canes* (mt⁺) decorated with monoclonal antibody A16A, an IgG (note free IgG proteins in background). × 250,000.

AGGLUTININ-RELATED PROTEINS PRODUCED BY NONAGGLUTINATING MUTANTS

We have to date examined the proteins produced by two of our nonagglutinating mutant strains (see above). The *imp-5 mt⁺* mutant and the *imp-12 mt⁻* mutant each produce short canes and loops (Figures 12, 13; Table 2) but no normal canes. These results complement our observations with wild-type proteins, for they again indicate that it is only the full-length cane that has adhesive activity.

The left panel of Figure 13 shows the fibrils displayed on the surface of an *imp-12* flagellum. In addition to the mat of curved fibrils in the center of the field, there are long, straight, approximately 225-nm fibrils that lack terminal heads. If these are agglutinin shafts devoid of heads, which is yet to be proved, then we are again led to conclude that the head is essential for adhesive activity.

The fact that the mutants examined to date produce proteins that are morphologically related to the agglutinin presumably explains why all of our nonagglutinating mutants can be isoagglutinated by all of our monoclonal antibodies.

ADHESION AND DISADHESION

To visualize the adhesion reaction itself we must examine living preparations, since, as noted earlier, the purified proteins appear incapable of forming stable interactions. When mating cells are quick-frozen and deep-etched, regions of adhesion are recognized by the presence of numerous small vesicles blebbing off the flagellar tips. Such a field is shown in Figure 14. The vesicles are connected to one another and to the two flagellar tips via stout fibers, which, when they adsorb to the mica flakes also included in the mating mixture, splay out into agglutinin fibrils (righthand side of Figure 14). We conclude, therefore, that agglutination entails the interaction of agglutinin proteins such that they associate laterally to form thicker fibers. The membranes bearing these fibers then vesiculate off the flagellar tips, presumably explaining, at least in part, the need for continuous production of new agglutinin proteins during a prolonged adhesive interaction (Snell and Moore, 1980; Cooper et al., 1983).

The morphological basis for the disadhesion that follows cell fusion is currently under study. We have, however, obtained an intriguing clue. When agglutinating cells are exposed to 2% glutaraldehyde and then

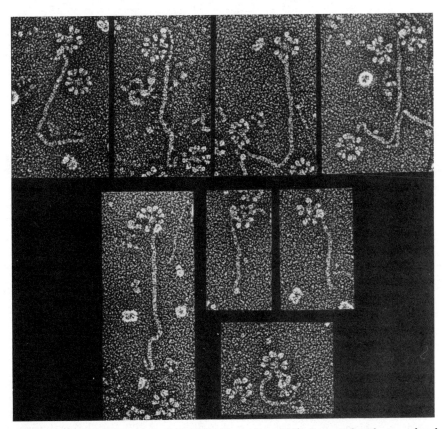

Figure 10. *Agglutinin canes*, two short canes, and a loop (mt^+) decorated with monoclonal antibody A8C, an IgM (note free IgM proteins in background). × 250,000.

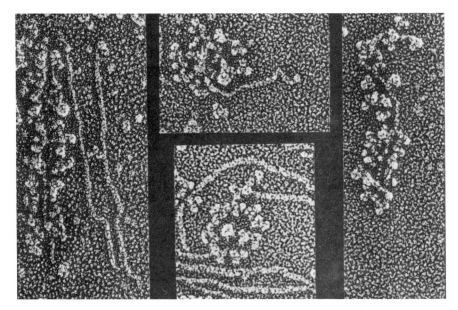

Figure 11. *Agglutinin canes and loops* (mt⁺) *decorated with monoclonal antibody A12E, an IgG.*
For orientation, the *left panel* shows two canes, cross-bridged by a single IgG, lying next
to two canes that are coated with IgGs along their length. Decoration with this antibody
is usually an all-or-none process—a protein is either free of decoration or completely
coated—suggesting that the epitope may be buried within the protein and exposed once
antibody binding initiates an unraveling process. × 290,000.

centrifuged and washed, the clumps disperse into single cells, indicating
that exposure to glutaraldehyde has caused the cells to disadhere (Forest
et al., 1978). The flagella of these cells, by deep-etching, display a strikingly
uniform fringe of 60-nm fibers (Figure 15), which we interpret to be proximal
sectors of agglutinin molecules and/or curved fibrils. Such images suggest
that glutaraldehyde-induced disadhesion is effected by cutting the agglutinin
at a proximal locus. Since every protein on the flagellum is so foreshortened,
the cutting event appears to be extremely efficient and specific. Whether
the same mechanism is used for "natural" disadhesion is now being
investigated.

PERSPECTIVES

The extracellular matrix that surrounds the *Chlamydomonas* soma is composed
of hydroxyproline-rich glycoproteins (Miller et al., 1974; Catt et al., 1976,
1978) that prove to be long, fibrous species when viewed by the deep-etch
technique (Monk et al., 1983; see also Figures 8, 12). We have therefore

proposed that the two protein classes are evolutionary relatives (Cooper et al., 1983). Returning, then, to our opening evolutionary theme, *Chlamydomonas* presents us with a family of fibrous proteins, some of which assemble into an extracellular matrix and others of which mediate a reversible cell–cell recognition event. It is not difficult to imagine how such proteins *could* have evolved to generate extracellular matrices that join cells together into metazoan assemblages, or recognition proteins that "self-destruct" after recognition has taken place. Whether evolution in *fact* followed this course is another question.

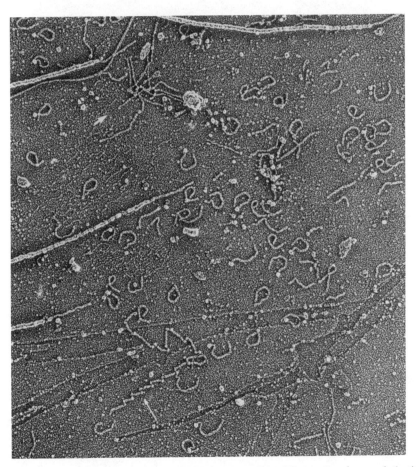

Figure 12. *Short canes and loops released from the flagellar surface of the* imp-5 mt[+] *nonagglutinating mutant. Long arrow* denotes a specialized form of mastigoneme, whose polypeptide subunits interact with one another in a head-to-tail, piggyback fashion. *Short arrow* denotes a cluster of cell wall proteins. Large fibers are "regular" mastigonemes. × 87,000.

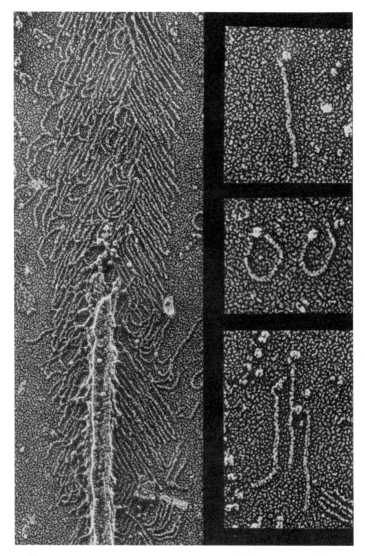

Figure 13. *Short canes and loops associated with or released from the flagellar surface of the* imp-12 mt⁻ *nonagglutinating mutant. Left panel*: The "imprint" left after a flagellum has adsorbed to mica and most of its membrane is then fractured away. × 140,000. *Right panels*: Molecules shed from the *imp-12* flagellum. × 250,000.

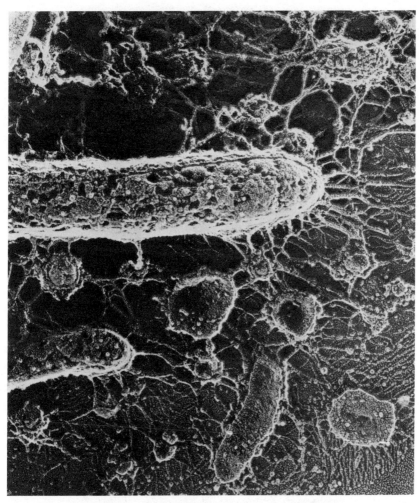

Figure 14. *Adhesion between* mt$^+$ *and* mt$^-$ *flagella.* Vesicles derived from some flagella are seen to connect to the two intact flagellar tips via fibrous material. At the right, the fibers have adsorbed to mica and splay out into agglutinin fibers. × 100,000.

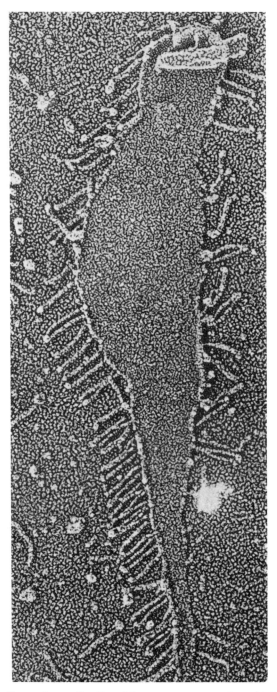

Figure 15. *Flagellar membranes* disadhered by exposure to 2% glutaraldehyde. × 130,000.

REFERENCES

Adair, W. S., B. C. Monk, R. Cohen, and U. W. Goodenough (1982) Sexual agglutinins from the *Chlamydomonas* flagellar membrane. Partial purification and characterization. *J. Biol. Chem.* **257**:4593–4602.

Adair, W. S., C. J. Hwang, and U. W. Goodenough (1983a) Identification and visualization of the sexual agglutinins from mating-type plus flagellar membranes of *Chlamydomonas*. *Cell* 33:183–193.

Adair, W. S., J. Long, W. B. Mehard, J. E. Heuser, and U. W. Goodenough (1983b) Monoclonal antibodies directed againt the sexual agglutinins of *Chlamydomonas reinhardii*. *J. Cell Biol.* **97**:93a.

Bergman, K., U. W. Goodenough, D. A. Goodenough, J. Jawitz, and H. Martin (1975) Gametic differentiation in *Chlamydomonas reinhardii*. II. Flagellar membranes and the agglutination reaction. *J. Cell Biol.* **67**:606–622.

Catt, J. W., G. J. Hills, and K. Roberts (1976) A structural glycoprotein containing hydroxyproline, isolated from the cell wall of *Chlamydomonas reinhardii*. *Planta* **131**:165–171.

Catt, J. W., G. J. Hills, and K. Roberts (1978) Cell wall glycoproteins from *Chlamydomonas reinhardii*, and their self-assembly. *Planta* **138**:91–98.

Collin-Osdoby, P., W. S. Adair, and U. W. Goodenough (1984) *Chlamydomonas* agglutinin conjugated to agarose beads as an *in vitro* probe of adhesion. *Exp. Cell Res.* **150**:282–291.

Cooper, J. B., W. S. Adair, R. P. Mecham, J. E. Heuser, and U. W. Goodenough (1983) *Chlamydomonas* agglutinin is a hydroxyproline-rich glycoprotein. *Proc. Natl. Acad. Sci. USA* **80**:5898–5901.

Desai, N. N., A. K. Allen, and A. Neuberger (1981) Some properties of the lectin from *Datura stramonium* (thorn-apple) and the nature of its glycoprotein linkages. *Biochem. J.* **197**:345–353.

Detmers, P. A., U. W. Goodenough, and J. Condeelis (1983) Elongation of the fertilization tubule in *Chlamydomonas*: New observations on the core microfilaments and the effects of transient intracellular signals on their structural integrity. *J. Cell Biol.* **97**:522–532.

Ebersold, W. T. (1967) *Chlamydomonas reinhardii*: Heterozygous diploid strains. *Science* 157:447–449.

Fincher, G. B., B. A. Stone, and A. E. Clarke (1983) Arabinogalactan-proteins: Structure, biosynthesis, and function. *Annu. Rev. Plant Physiol.* **34**:47–70.

Forest, C. L., D. A. Goodenough, and U. W. Goodenough (1978) Flagellar membrane agglutination and sexual signaling in the conditional *gam-1* mutant of *Chlamydomonas*. *J. Cell Biol.* **79**:74–84.

Goodenough, U. W. (1977) Mating interactions in *Chlamydomonas*. In *Microbial Interactions*, J. L. Reissig, ed., pp. 323–350, Chapman and Hall, London.

Goodenough, U. W., and D. Jurivich (1978) Tipping and mating-structure activation induced in *Chlamydomonas* gametes by flagellar membrane antisera. *J. Cell Biol.* **79**:680–693.

Goodenough, U. W., and J. Thorner (1983) Sexual differentiation and mating strategies in the yeast *Saccharomyces* and the green alga *Chlamydomonas*. In *Cell Interactions and Development: Molecular Mechanisms*, K. M. Yamada, ed., pp. 29–75, Wiley, New York.

Goodenough, U. W., and R. L. Weiss (1975) Gametic differentiation in *Chlamydomonas reinhardii*. III. Cell wall lysis and microfilament-associated mating structure activation in wild-type and mutant strains. *J. Cell Biol.* **67**:623–637.

Goodenough, U. W., C. J. Hwang, and A. J. Warren (1978) Sex-limited expression of gene loci controlling flagellar membrane agglutination in the *Chlamydomonas* mating reaction. *Genetics* **89**:235–243.

Goodenough, U. W., P. A. Detmers, and C. J. Hwang (1982) Activation for cell fusion in *Chlamydomonas*: Analysis of wild-type gametes and non-fusing mutants. *J. Cell Biol.* **42**:378–386.

Heuser, J. E. (1983) A method for freeze-drying molecules adsorbed to mica flakes. *J. Mol. Biol.* **169**:155–196.

Homan, W. (1982) An analysis of the flagellar surface of *Chlamydomonas eugamatos*. Unpublished doctoral dissertation, University of Amsterdam, Holland.

Hwang, C. J., B. C. Monk, and U. W. Goodenough (1981) Linkage of mutations affecting *minus* flagellar membrane agglutinability to the *mt⁻* mating-type locus of *Chlamydomonas*. *Genetics* **99**:41–47.

Lamport, D. T. A. (1980) Structure and function of plant glycoproteins. In *The Biochemistry of Plants*, Vol. 3, J. Preiss, ed., pp. 501–541, Academic, New York.

Leach, J. E., M. A. Cantnell, and L. Sequira (1982) Hydroxyproline-rich bacterial agglutinin from potato. *Plant Physiol.* **70**:1353–1358.

Levine, R. P., and W. T. Ebersold (1960) The genetics and cytology of *Chlamydomonas*. *Annu. Rev. Microbiol.* **14**:197–216.

Mesland, D. A. M., J. L. Hoffman, E. Caligor, and U. W. Goodenough (1980) Flagellar tip activation stimulated by membrane adhesions in *Chlamydomonas* gametes. *J. Cell Biol.* **86**:656–665.

Miller, D. H., J. S. Mellman, D. T. A. Lamport, and M. Miller (1974) The chemical composition of the cell wall of *Chlamydomonas gymnogama* and the concept of cell wall protein. *J. Cell Biol.* **63**:420–429.

Monk, B. C., W. S. Adair, R. A. Cohen, and U. W. Goodenough (1983) Topography of *Chlamydomonas*: Fine structure and polypeptide components of the gametic flagellar membrane surface and the cell wall. *Planta* **158**:517–533.

Musgrave, A., E. van Eijk, R. te Welscher, R. Broekman, P. Lens, W. Homan, and H. van den Ende (1981) Sexual agglutination factor from *Chlamydomonas eugamatos*. *Planta* **153**:362–369.

Snell, W. J. (1976) Mating in *Chlamydomonas*: A system for the study of specific cell adhesion. II. A radioactive binding assay for quantitation of adhesion. *J. Cell Biol.* **68**:70–79.

Snell, W. J. (1981) Flagellar adhesion and deadhesion in *Chlamydomonas* gametes: Effects of tunicamycin and observation on flagellar tip morphology. *J. Supramol. Struct. Cell Biochem.* **16**:371–376.

Snell, W. J. (1985) Cell–cell interactions in *Chlamydomonas*. *Annu. Rev. Plant Physiol.* **36**:287–315.

Snell, W. J. and W. S. Moore (1980) Aggregation-dependent turnover of flagellar adhesion molecules in *Chlamydomonas* gametes. *J. Cell Biol.* **84**:203–210.

Snell, W. J., and S. Roseman (1979) Kinetics of adhesion and deadhesion of *Chlamydomonas* gametes. *J. Biol. Chem.* **254**:10820–10829.

Solter, K. M., and A. Gibor (1978) Removal and recovery of mating receptors on flagella of *Chlamydomonas reinhardii*. *Exp. Cell Res.* **115**:175–181.

van den Ende, H. (1985) Sexual agglutination in *Chlamydomonas*. *Adv. Microb. Physiol.* **27** (in press).

Wiese, L. (1969) Algae. In *Fertilization: Comparative Morphology, Biochemistry and Immunology*, Vol. 2, C. B. Metz and A. Monroy, eds., pp. 135–188, Academic, New York.

Wiese, L., and P. C. Hayward (1972) On sexual agglutination and mating-type substances in isogamous dioecious *Chlamydomonads*. III. The sensitivity of sex cell contact to various enzymes. *Am. J. Botany* **59**:530–536.

Wiese, L., and W. Wiese (1978) Sex cell contact in *Chlamydomonas*: A model for cell recognition. *Symp. Soc. Exp. Biol.* **32**:83–104.

Witman, G. B., K. Carlson, J. Berliner, and J. L. Rosenbaum (1972) *Chlamydomonas* flagella. I. Isolation and electrophoretic analysis of microtubules, matrix, membranes, and mastigonemes. *J. Cell Biol.* **54**:507–539.

Section 3

Cell Adhesion Molecules

Chapter 7

Specific Cell Adhesion in Histogenesis and Morphogenesis

GERALD M. EDELMAN

ABSTRACT

The relationship of the primary processes of development (cell division, movement, adhesion, differentiation, and death) to morphogenesis and histogenesis is a central problem in cell biology. Recent progress in the discovery, isolation, and determination of the structural and functional properties of cell adhesion molecules (CAMs) has shed light on key aspects of this problem. Primary CAMs are found at very early stages of embryogenesis and derive from more than one germ layer. Secondary CAMs appear during histogenesis and are derived from only one germ layer. So far, two primary CAMs (N-CAM, neural cell adhesion molecule; L-CAM, liver cell adhesion molecule) and one secondary CAM (Ng-CAM, neuron–glia cell adhesion molecule) have been characterized. These molecules have different binding mechanisms and structures and are expressed in a definite spatiotemporal order in development and histogenesis. This order is accompanied by changes in their prevalence, polarity, and chemical structure at the cell surface (so-called local cell surface modulation). CAMs appear to play a key role in the segregation of epithelia, the attachment of tissue sheets, epithelial–mesenchymal transformation, and the control of morphogenetic movement.

The triumph of structural approaches in molecular biology, immunology, and virology tempts the question: Will there be a molecular histology within the next decade? Whatever the answer, a simple consideration of the origins of tissues and their relation to animal form indicates that there must first be a molecular embryology. It is clear, for example, that cytodifferentiation is not equivalent to pattern formation in embryogenesis (Raff and Kaufman, 1983; Slack, 1983) and that the crucial events of mosaic development or of embryonic induction long precede histogenesis but nonetheless strongly determine tissue type. A molecular explanation of cytodifferentiation and of the operation of cell contact molecules is thus necessary but not sufficient. To understand cells in contact in differentiating tissues, it is of singular importance to analyze the fundamentals of development.

At the risk of some oversimplification, I will state in the form of four linked questions the central problems of development that bear upon this task:

1. How does a one-dimensional genetic code specify a three-dimensional animal? (The developmental genetic problem.)
2. How does the answer to this problem conform to the finding that very rapid changes in animal form can occur within and among classes in relatively short evolutionary times? (The evolutionary problem.)
3. What is the relationship of this answer to cytodifferentiation? (The tissue-complexity problem.)
4. How is growth regulated after the establishment of pattern and form? (The growth control problem.)

A consideration of these problems (particularly the first two) in modern terms reminds us that von Baer has prevailed over Haeckel: Ontogeny does not recapitulate phylogeny but, instead, a collection of embryonic cells becomes more and more specialized by means of the mechanisms of heterochrony (Gould, 1977). The modern view is to see these mechanisms (neoteny, recapitulation, hypermorphosis, paedomorphosis, etc.) in terms of evolutionary selection acting on the results of primary processes of development. It focuses particularly on mechanisms by which the epigenetic sequences of such processes are regulated in development. These processes (cell-division cycle control, cell movement, cell adhesion, cell differentiation, and cell death) also operate in adult tissues and thus an understanding of how they operate in early development bears directly upon the question of the molecular basis of histogenesis.

All of these processes are enormously complex at the molecular level. It is not commonly understood that there is a mechanochemical aspect to their regulation. In mosaic development, for example, one must understand the mechanochemical regulation of the unequal partition of cytoplasm in early division. In regulative development, one must understand the mechanochemical regulation of the morphogenetic movements that bring cells of different histories together and in turn bring about embryonic induction. In both kinds of development, defined epigenetic sequences (sequences of events that must occur for subsequent events to happen) can be observed at the cellular level. The culmination of these events is embryonic induction—control of gene expression in terms of the position or place of a cell, a form of milieu-dependent differentiation.

These considerations allow us to restate the developmental genetic problem: What kind of gene products can regulate mechanochemical events to yield epigenetic sequences that result in pattern? And what mechanisms regulate the expression of these gene products? A serious attempt to answer these questions requires a review of several proposed models and signals for such gene products. But before turning briefly to that task, it is reasonable to conjecture that molecules acting at the cell surface to bring cells together would be excellent candidates. These include cell adhesion molecules (CAMs), substrate adhesion molecules (SAMs), and cell junctional molecules (CJMs). Without slighting the importance of the other types of cell contact molecules, the main purpose of this chapter is to consider the evidence that CAMs are prime candidates for the requisite regulatory role.

MODELS AND HYPOTHESES FOR MORPHOGENESIS

It is evident that the basic arrangement in higher vertebrates (a tube within a tube) emerged from the evolutionary selection of major mechanical processes that had taken place during ontogeny. These include the concerted foldings of epithelial sheets, the conversion of epithelia to mesenchyme, and the reciprocal conversion back to epithelia. After the body plan is laid down, histogenesis occurring locally brings about increasing cytodifferentiation and cellular complexity, a major scale change in cellular movements, a supervention of additional refined mechanisms of cell signaling (such as those seen in the nervous system), and incremental additions and contributions of tissue derivatives based on SAMs and CJMs (e.g., hard tissues, desmosomes, etc.). Any model of this embryogenetic and histogenetic sequence must take all of these processes into account.

This is not the place to review models pertaining to morphogenesis in any detail (see Slack, 1983, for a deeper analysis). But it is useful to exclude certain views as being simply incompatible with the requirements and conditions we have discussed above. Structural models of the kind so valuable for the explanation of virus assembly presuppose a large ensemble of evolutionarily maintained and pairwise complementary gene products at the cell surface to account for ordered patterns. An example is Sperry's hypothesis (1963) for the development of the retinotectal map. Such models require exquisite control systems, and confront difficulties both in explaining the plasticity of development and in attempting to account for the evolutionary problem: A mutation in the "wrong" cell marker might require large numbers of simultaneous complementary mutations for recovery of structure.

More dynamic models fall into two classes: equilibrium models and kinetically constrained models. An example of an equilibrium model is Steinberg's (1970) sorting-out hypothesis for histotypy, which treats cells in analogy to molecules in a fluid droplet undergoing phase separations. Such phenomenological models can explain certain events, but they lack molecular detail and are not easily reconciled with the largely irreversible character of large-scale cellular events during histogenesis and development.

Kinetic models have received attention in a series of mathematical formulations based on reaction diffusion ideas that were first clearly proposed for morphogenesis by Turing (1952). They embody several concepts: autocatalytic processes combined with diffusion and local fluctuation; various flow processes as well as short-range positive interactions and long-range inhibitions are also invoked. Such formal models can account for much of what is observed in pattern formation, but unfortunately they are posed in very general terms and usually lack an adequate or detailed description of the participating molecules, of their genetic specification, and of their hierarchical control in tissues through various cellular processes. Formal kinetic models tend to "squash" the description to one level of variables. Nonetheless, it is my surmise that, when all these factors are finally taken into consideration, some revised version of kinetic models will account for many of the early epigenetic sequences during embryogenesis.

Given their various inadequacies, why consider any models at this point? The main reason is that a critical consideration of the limits of various competing models prevents the attribution of wrong properties to molecules that might serve as candidates for regulators of primary processes of morphogenesis. For example, given the frailties of the structural model, it is not reasonable to expect that a description of cell adhesion molecules will account completely for pattern formation. What is needed is not only a chemical account of such molecules but also a hierarchical description of their interaction at the cell surface and their regulation by cellular mechanisms.

CELL SURFACE MODULATION

Various mechanisms that link both molecular and cellular states to kinetic models are embodied in the notions of global and local cell surface modulation (Edelman, 1976, 1983, 1984a, 1985). Global cell surface modulation is a form of alteration in the interaction with the cytoskeleton of a variety of different cell surface receptors as manifested by a decrease in their mobility in the plane of the membrane. It can be induced by *local* cross-linkage of the receptors at a small region of the cell surface; it is thus a propagated event, possibly occurring by means of modificaton of key molecules in the cell cortex via phosphorylation or other enzymatic processes. Global modulation can block expression of already received mitogenic signals and of transforming signals, but it is probably not in itself a direct signal path for control of the primary process of cell division. Nonetheless, alteration in cytoskeletal elements can greatly alter the control of cell division in both positive and negative directions (Edelman, 1976). Global modulation is a prime candidate for the role of linking cell states and cell interactions, based on locale, to such fundamental primary processes as cell-division cycle control, morphogenetic movement, and changes in cell shape, and it may even regulate the expression of adhesion molecules. Despite a convincing body of *in vitro* data on global modulation, however, its direct function *in vivo* has not been established.

In contrast to global modulation, local cell surface modulation refers to the change in number, position, or chemical state of a specific cell surface receptor (e.g., an adhesion molecule) in such a fashion as to change its function, binding strength, or interaction with other molecules (Figure 1). Local modulation can thus serve to change direct cellular interactions as well as the interaction of a cell with molecules released from surrounding cells. An increasing body of evidence indicates that local cell surface modulation of CAMs occurs both *in vitro* and *in vivo* (Edelman, 1983; Edelman, 1984a,b); examples of prevalence modulation, polarity modulation, and chemical modulation (see Figure 1) of CAMs are reviewed later. The idea of local modulation centers around the specificity of a given gene product at the cell surface rather than a change in the whole cell surface. Nonetheless, it also encompasses the possibility that two or more defined cell surface receptors on the same cell may mutually interact to alter their function. Although local modulation can regulate primary processes such as cell movement through its influence on cell adhesion, it does not *in general* have the regulatory potential for determining overall cell state that is a demonstrated property of global modulation.

Local modulation bears directly upon the function of gene products capable of altering mechanochemical epigenetic sequences in regulative development. One of the main themes of this chapter is that CAMs represent a key set of gene products subject to local modulation in development. A review of the properties of these molecules and of their relation to the topological requirements of early and later development will provide background for the hypothesis that CAMs act as regulators of inductive sequences by controlling sequences of morphogenetic movements.

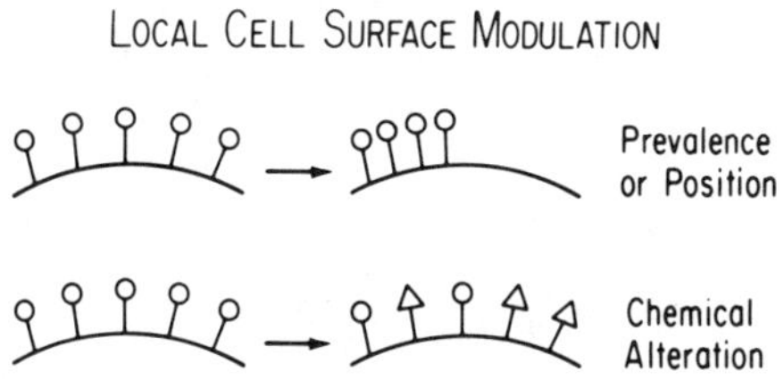

LOCAL CELL SURFACE MODULATION

Figure 1. *Schematic representation of local cell surface modulation.* Various elements represent a specific glycoprotein (for example, N-CAM) on the cell surface. The upper sequence shows modulation by alteration of both the prevalence of a particular molecule and its distribution on the cell surface. The lower sequence shows modulations by chemical modification resulting in the appearance of new or related forms (*triangles*) of the molecule with altered activities. Local modulation is distinct from global modulation, which refers to alterations in the whole membrane that affect a variety of different receptors independent of their specificity. (Modified from Edelman, 1985.)

CELL ADHESION MOLECULES

The strategy used in searching for CAMs was based on two kinds of assays: an immunological detection assay based on the recognition of CAMs present on single cells as they collided and interacted over short times (Brackenbury et al., 1977); and a series of confirmatory perturbation assays (Rutishauser et al., 1978; Buskirk et al., 1980; Fraser et al., 1984) in which antibodies known to be directed against particular CAMs could be shown to disrupt tissue patterns by binding to cells in developing structures.

Detection assays will be considered first (Figure 2). Cells in a tissue were dissociated by digestion of their surface protein molecules with trypsin. After a recovery period, during which these cells were allowed to resynthesize their surface proteins, they were stirred together to allow collision and adhesion over short times in arrangements that depended on the particular assay. In one detection assay, for example, a portion of the cells was fixed in a layer and another labeled portion of cells was thrown onto this layer; the cells that adhered to the layer could then be counted. In another assay, (more rapid but without facilities for watching

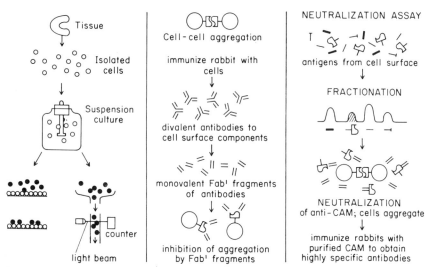

Figure 2. *Immunologically based adhesion assays.* Chick retinal cells are dissociated by light trypsin digestion and allowed to resynthesize their surface molecules in suspension culture (*left panel*). Adhesion is assayed by counting labeled cells bound to a layer of fixed cells or by shaking cells together and measuring the disappearance of single cells in an automatic counter. Antibodies to cell surface components are assessed for the ability to block cell–cell adhesion following conversion to Fab' fragments (*center panel*). Cell surface antigens are fractionated, and fractions that neutralize inhibition by these antibody fragments (*right panel*) are used to reimmunize rabbits and thus obtain antibodies to CAMs of higher specificity. (From Edelman, 1985.)

the interaction of individual cells), the disappearance of single cells into aggregates was determined in an automatic cell counter (Brackenbury et al., 1977).

How may one detect the molecules responsible for the actual ligation of cell surfaces? Taking a cue from earlier work of Gerisch and Malchow (1976), who had used antibodies to perturb adhesion in slime molds, we decided to search for antibodies to CAMs that would specifically block adhesion. Rabbits were immunized with chick brains and retinas and the antibodies were scanned for their ability to block adhesion after they were cleaved into univalent Fab' fragments. (Uncleaved antibodies are divalent and, instead of blocking adhesion, might actually bind two cells together by their CAMs.) After specific adhesion-blocking antibodies were found, a dilemma had to be faced: The antibodies in the rabbit sera were mixed with those directed against other cell surface molecules and could not be used as specific probes to identify CAMs or isolate them. In order to reduce the heterogeneity of the antibody population, a neutralization assay was therefore devised (Brackenbury et al., 1977). In this assay, different fractions of surface protein antigens were tested for their ability to compete with cells for binding to the anti-CAM antibody. If such fractions contained CAMs by this criteria, they were then used to reimmunize rabbits. By iteration of this procedure, antigenic fractions of high specificity were obtained that elicited highly specific anti-CAM antibodies in the rabbits. Once a means had been found for identifying such antigenic CAM fractions, modern techniques of monoclonal antibody production could be used to make very specific antibodies to CAMs in different animal species.

These various antibodies allowed scans for the distribution of CAMs in tissues and provided a means for perturbing the development of tissue patterns. Using anti-N-CAM antibodies labeled by a fluorescent marker, it was found that this molecule was present on all neurons in the central and peripheral nervous systems. Anti-N-CAM antibodies could disrupt the orderly development of neural tissues such as those of the retina (Buskirk et al., 1980) as they grew in tissue culture. By similar approaches (Bertolotti et al., 1980; Gallin et al., 1983; Grumet and Edelman, 1984; Grumet et al., 1984a), several other adhesion molecules (L-CAM from liver and Ng-CAM, a molecule on neurons that binds them to glia) were identified; as we shall see later, knowledge of these other molecules greatly facilitated analyses of the role of adhesion in early neural histogenesis.

Tissue perturbation assays provide additional means for verifying the relationship between binding and distribution of a particular CAM. It is clear, however, that perturbation assays alone will not give deep insights into the function of adhesion molecules. Such insights require a variety of more fundamental approaches, including an analysis of the chemical structure, specificity, and binding mechanism of each particular CAM in very early development, and a study of CAM function and genetic control in later stages of development and histogenesis.

STRUCTURE AND SPECIFICITY OF DIFFERENT CAMS

By means of classical chromatographic fractionation techniques and affinity chromatography, both N-CAM and L-CAM were purified to the point at which a chemical analysis of their structures could be carried out (Thiery et al., 1977; Hoffman et al., 1982; Gallin et al., 1983). Both molecules are glycoproteins, but N-CAM is unusual in that it contains extraordinarily large amounts of sialic acid, a complex, negatively charged sugar that is present in an unusual polymerized form (Rothbard et al., 1982; Finne et al., 1983). Although sialic acid is found attached in much smaller amounts on other cell surface proteins, it has not previously been seen as polysialic acid on proteins in the vertebrate species.

Various cleavages with proteolytic enzymes have been used to construct topographic or linear maps of N-CAM (Cunningham et al., 1983) and L-CAM (Gallin et al. 1983; Figure 3). These maps indicate that the molecules are structurally different. The larger polypeptide chain of N-CAM (which, exclusive of sugar, has a maximal molecular weight of 160 kD) consists of three domains linked by stretches of polypeptide chain susceptible to attack by proteolytic enzymes. The amino-terminal domain projects away from the cell and contains the binding region. A middle domain contains the great bulk of the sialic acid, and the carboxy-terminal domain is associated with the cell membrane. Chemical evidence suggested that a portion of

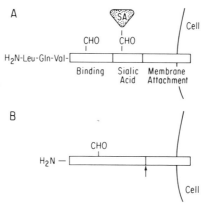

Figure 3. *Linear structures of N-CAM and L-CAM. A:* Three structural and functional regions of N-CAM deduced from studies of the intact molecule and a series of fragments. The amino-terminal region includes a specific binding domain and carbohydrate (CHO) but little, if any, sialic acid; the neighboring region is very rich in sialic acid (SA), present mainly as polysialic acid; the carboxy-terminal region is associated with the plasma membrane. *B:* Linear structure of L-CAM obtained by comparing the intact molecule (with a molecular weight of 124 kD) released by detergent extraction with a fragment (molecular weight, 81 kD) released by trypsin (*arrow*). In both N-CAM and L-CAM the carbohydrate is attached at several sites. (From Edelman, 1985.)

this membrane-associated domain is actually inserted into the lipid bilayer of the membrane (Hoffman et al., 1982; Cunningham et al., 1983). This was originally inferred from the requirements for extracting N-CAM from cell membranes, as well as from the fact that the intact chain could become readily associated with artificial lipid vesicles; it has been strongly confirmed by analyses of the carboxy-terminal region in cells with monoclonal antibodies (Gennarini et al., 1984).

Lipid vesicles have been found to be extremely useful in following the binding behavior of the N-CAM. For example, lipid vesicles can be attached to cells; if the cells were first treated with anti-N-CAM Fab' and washed, it was found that the vesicles containing only N-CAM and lipid would not bind (Cunningham et al., 1983). This suggested that the binding mechanism is homophilic: N-CAM on the membrane of one cell binds to N-CAM on the membrane of the opposed cell. The idea that there may be functional and structural domains in N-CAM has recently received support from preliminary experiments in which N-CAM molecules were visualized by electron microscopy; the picture suggested regions of folded polypeptide separated by bends that might correspond to stretches of polypeptide chains between domains (Edelman et al., 1983b).

The detailed structures of the two other known CAMs are less well worked out, but each is known to differ from the others. L-CAM has a molecular weight of 124 kD (Gallin et al., 1983), but it was first isolated in lower molecular weight forms from embryonic liver (Bertolotti et al., 1980; Figure 3). Unlike N-CAM, it is a "normal" glycoprotein, and it will not mediate cell adhesion unless Ca^{2+} is present. Indeed, unless the L-CAM molecule binds Ca^{2+}, it is susceptible to rapid proteolytic cleavage. It is not definitely known whether L-CAM binding is homophilic. Although L-CAM was isolated originally from liver, it has been found that together with N-CAM it plays a fundamental role in early developing tissues. We will consider the early role of these CAMs at length in another section.

The most recently isolated CAM is Ng-CAM, which mediates the binding of neurons to glial cells (Grumet and Edelman, 1984; Grumet et al., 1984a). Ng-CAM consists of three polypeptides with molecular weights of 200 kD, 135 kD, and 80 kD. At present, it is not known whether they are associated at the cell surface. All of these components are found on neurons but are not detectable on glia; Ng-CAM is therefore probably bound to a structurally different glial adhesion molecule, that is, its binding mechanism is likely to be heterophilic. Despite the fact that their structures and binding specificities differ, it has been shown that N-CAM and Ng-CAM share an antigenic determinant (Grumet et al., 1984b). One monoclonal antibody has been found that reacts with a shared carbohydrate determinant on both molecules but it has not been ruled out that other antibodies may simply recognize a similar peptide fold or very short sequence.

N-CAM, L-CAM, and Ng-CAM do not appear to be cross-specific; that is, in linking cells, no one binds effectively to any of the others. Recent findings (Grumet et al., 1984a) have raised the possibility, however, that N-CAM and Ng-CAM may interact on the surface of the same neuron and possibly between neurons.

CAM Binding

Of all of these molecules, N-CAM is the best studied in terms of its binding properties. One key question dominated studies of the relation of N-CAM structure to function: Does the unusual carbohydrate—particularly the sialic acid—play a role in cell–cell binding? Direct binding has been ruled out; N-CAM molecules from which all of this sugar had been specifically removed by the enzyme neuraminidase have been shown to bind specifically to cells. Nonetheless, two observations (Rothbard et al., 1982) have revealed an unexpected but very important function for this sugar. The first was that N-CAM from embryonic brains had 30 g sialic acid/100 g polypeptide and migrated on electrophoretic gels as a diffuse band with a broad molecular weight distribution. In contrast, N-CAM from adults had only 10 g sialic acid/100 g polypeptide and migrated as two or three sharp bands. Thus, at some time during development, the embryonic (E) form of the molecule must be converted to or exchanged for the adult (A) forms. The second observation was that the A-forms appeared to bind more effectively than the E-form (Hoffman and Edelman, 1983). This bears strongly on mechanisms that might govern the selectivity of cell adhesion in the developing nervous system. Modulations either in prevalence of CAMs at the cell surface or in their individual binding strength through E-to-A conversion would be expected to lead to different interactions among the cells that were subject to other primary processes, such as migration. These altered interactions might in turn lead to the formation of different cellular patterns.

To demonstrate that such modulatory changes in CAMs could actually affect binding, a direct measurement was carried out with artificial lipid vesicles into which N-CAM was inserted; these vesicles were tested for their relative rates of aggregation (Hoffman and Edelman, 1983). Both the E-form and A-forms could be separately inserted into different vesicle fractions (Figure 4). The prediction was that the order of rates would be E–E < E–A < A–A, based on the idea (Edelman, 1983) that the sialic acid in the middle domain (Figure 3) was either altering the shape of the binding domain or was directly repelling the opposing N-CAM molecule from another cell. Vesicles containing the A-form, which has lesser amounts of charged sugar, should aggregate more rapidly. As shown in Figure 4, this prediction was clearly confirmed.

Even more striking was the effect of increasing the amount of N-CAM of a given form in the membrane. A twofold increase led to a greater than 30-fold increase in binding rates (Hoffman and Edelman, 1983). *In vivo*, both E-to-A conversion and surface prevalence changes would be expected to lead to large changes in rates of binding, a necessary condition for any

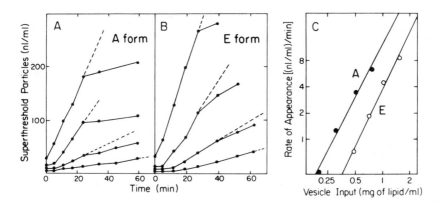

Vesicles (form of N-CAM)	N–CAM/lipid (μg/mg)	k_{agg} (units)[a]
(a) E	17	3.5
A	17	12.2
(b) E	14	1.5
E	19	6.4
E	28	54.0
(c) E(100%)	14	1.4
E(75%) + A(25%)		3.3
E(50%) + A(50%)		6.3
E(25%) + A(75%)		9.3
A(100%)	17	12.2

Figure 4. *Aggregation of reconstituted vesicles containing embryonic (E) or adult (A) forms of N-CAM.* Vesicles were reconstituted from purified N-CAM and lipid, and their aggregation was analyzed in a particle counter. The concentration of superthreshold particles is plotted as a function of time for four concentrations of vesicles. Apparent initial rates of aggregation are calculated from the initial slopes (*broken lines*) of the aggregation curves. A: A-form vesicles (16.3 μg of N-CAM per mg of lipid) aggregated at 0.76, 0.50, 0.30, and 0.20 mg of lipid per ml (curves proceeding from left to right). B: E-form vesicles (16.9 μg of N-CAM per mg of lipid) aggregated at 1.55, 1.03, 0.71, and 0.48 mg of lipid per ml. C: Log-log plot of rate of appearance of superthreshold A-form (*closed circles*) and E-form (*open circles*) particles versus input concentration. Table compares effects on aggregation rates of E- and A-forms (*a*); of alterations in surface amounts of E-form (*b*); and of effects of mixtures of vesicles containing either E- or A-forms (*c*). (From Edelman, 1985.)

kinetically constrained model of pattern formation. It is important to stress that both of these modulatory changes can be graded, and therefore they define a very large number of possible binding states for N-CAM. Although it is an open question as to whether L-CAM will show similar properties, changes in its prevalence during development are at least suggestive.

DYNAMIC SEQUENCES OF CAM EXPRESSION: EMBRYOGENESIS TO ADULT TISSUE

If modulation is a fundamental mechanism in morphogenesis then it must occur in a definite pattern. Studies of the tissue distribution of the known CAMs in fact show that they appear in defined sequences in time and in specific spatial orders. We shall first consider their temporal expression and a list of their tissue distributions. Their potential role in the histogenesis of the nervous system will then be considered in some detail. Finally, we shall attempt to analyze the topological relationships of the CAMs in connection with the establishment of embryonic form.

A summary of CAM distribution in the chick (Thiery et al., 1982; Edelman et al., 1983a) spanning two early epochs of embryonic development through the period of organogenesis and in adult life is presented in Table 1. Several generalizations emerge from this data: (1) Primary CAMs (N-CAM and L-CAM) appear very early in development, before any extensive cell differentiation in organs. (2) The same primary CAMs, each of which appears in more than one germ layer, can be expressed later in the differentiation of particular organs. (3) Certain CAMs are secondary, that is, they are not seen in the earliest stages, are strictly derived from one germ layer, and appear only during histogenesis. Ng-CAM in the nervous system is the best studied example. (4) Despite the addition of secondary CAMs, modulation of primary CAMs continues in later histogenesis.

The known sequences of expression of the primary and secondary CAMs can be summarized in a temporal diagram (Figure 5) that also illustrates three forms of local cell surface modulation. The diagram shows N-CAM and L-CAM (1° set) appearing first together, followed by a sharp divergence of their spatial distribution at neural induction, and the subsequent increases, decreases, or disappearances that are characteristic of each tissue and each CAM. Placodes echo the differential CAM expression; both CAMs are initially present, but subsequently, in placodes destined for neural structures, L-CAM disappears and N-CAM increases. Ng-CAM (2° set, Figure 5) is expressed later, on neurons. The fact that this second neuronal CAM, which appears to be involved in interactions of neurons with glia, is not expressed during early embryogenesis but only at 3 1/2 days in the chick embryo (just before the appearance of glial cells in the central nervous system) suggests that temporal control of CAM gene expression is critical in histogenesis. Recent studies indicate that Ng-CAM

Table 1. Distribution of L-CAM and N-CAM in Three Epochs

0–3-day Embryo	5–13-day Embryo	Adult
L-CAM		
Ectoderm		
Upper layer	Epidermis	Skin: Stratum
Epiblast	Extraembryonic ectoderm	germinativum
Presumptive epidermis		
Placodes		
Mesoderm		
Wolffian duct	Wolffian duct	Epithelium of:
	Ureter	Kidney
	Most meso- and	Oviduct
	metanephric	
	epithelium	
Endoderm		
Endophyll	Epithelium of:	Epithelium of:
Hypoblast	Esophagus	Tongue
Gut primordium and	Proventriculus	Esophagus
buddings	Gizzard	Proventriculus
	Intestine	Gizzard
	Liver	Intestine
	Pancreas	Liver
	Lung	Pancreas
	Thymus	Lung
	Bursa	Thymus
	Thyroid	Thyroid
	Parathyroid	Parathyroid
	Extraembryonic	Bursa
	endoderm	
N-CAM		
Ectoderm		
Upper layer	Nervous system	Nervous system
Epiblast		
Neural plate		
Placodes		
Mesoderm		
Notochord	Striated muscle	[a]
Somites	Adrenal cortex	Cardiac muscle
Dermomyotome	Gonad cortex	Testis
Somato- and	Some mesonephric and	
splanchnopleural	metanephric epithelia	
mesoderm	Somato- and	
Heart	splanchnopleural	
Mesonephric	elements	
primordium	Heart	

Source: Edelman, 1985.

[a] It is not yet known whether adult striated muscle contains N-CAM.

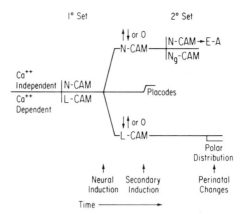

Figure 5. *Schematic diagram showing the temporal sequence of expression of CAMs.* After an initial
differentiation event, N-CAM and L-CAM diverge in cellular distribution and are then
modulated in prevalence (⇅) within various regions of induction or actually disappear
(0) when the mesenchyme appears or cell migration occurs. Note that placodes, which
have both CAMs, echo the events seen for neural induction. Just before the appearance
of glia, a secondary set CAM (Ng-CAM) emerges; unlike the other two CAMs, this CAM
would not be found in the map shown in Figure 10 before 3 1/2 days. In the perinatal
period, a series of epigenetic modulations occurs: E-to-A conversion for N-CAM and
polar redistribution for L-CAM. The diagrammed events are based mainly on work on
the chick. (From Edelman, 1985.)

in the developing central nervous system is expressed mainly on the axons
of fiber tracts or on the somas of cells undergoing migration; that is, its
distribution on the cell shows polarity modulation. Subsequent to the
expression of Ng-CAM and particularly in the perinatal period, E-to-A
conversion occurs in N-CAM, whereas polar redistributions of L-CAM
occur on cells of certain tissues, such as those of the pancreas.

   This dynamic picture indicates that there is a definite schedule of CAM
gene expression during organogenesis. It is noteworthy that the epigenetic
alterations arising from E-to-A conversion of N-CAM must be initiated by
enzymes that are themselves under control of regulatory genes other than
those concerned with CAM expression. Direct confirmation of the precise
times of CAM gene expression and an understanding of their control by
various regulatory genes shall have to await the isolation of cDNA probes
for each of the CAMs.

   As shown in Table 1, the tissue distribution of CAMs in the adult is
concordant with their germ layer origins. In wound healing, and in certain
animals capable of regeneration (such as *Xenopus*), portions of morphogenetic
processes may be reactivated to form tissues, although their sequences
will not be identical to those of original development. Questions concerning
both the distribution and the function of CAMs in the adult arise naturally
from these considerations. Although CAMs appear in adult tissues in
distributions that follow from those seen in embryonic life, in most cases

they constitute a smaller proportion of the tissues, appear only in certain limited tissue locations (Damsky et al., 1982; Thiery et al., 1982; Ogou et al., 1983; Thiery et al., 1984a), and in some tissues (such as the pancreas) they are found distributed on cells in a polar fashion (Edelman et al., 1983a; Gallin et al., 1983). The sparser CAM distribution is perhaps not surprising, for adult tissues contain increased amounts of derivatives of connective tissue, show evidence for cellular interactions with SAMs (molecules that include collagen, laminin, fibronectin, glycosaminoglycans, etc.), and cells of adult tissues have formed specialized junctions of various types by means of cell junctional molecules (CJMs). The central nervous system is an exception: fiber tracts do not have CJMs and SAMs are present in lower amounts than in nonneural tissue. For this reason, it is perhaps not surprising that CAMs predominate; indeed, their behavior in neural histogenesis suggests that they play a central role in both early and late pattern formation.

NEURAL HISTOGENESIS

As indicated in the expression sequence (Figure 5), at least two neuronal CAMs, N-CAM and Ng-CAM, play key roles at those times during which fiber tracts are formed, maps are established, and distinct portions of the nervous system are differentiated. We shall first consider some of the modulatory events that occur at these later times; this allows us to focus on the functions and the specificities of two well-defined CAMs within one tissue during a critical histogenetic period. Inasmuch as the earlier events of neural induction and neurulation involving N-CAM and L-CAM will be connected with an evolutionary hypothesis on the regulatory roles of CAMs in morphogenesis at the early stages of development, we shall defer their discussion until the end of this chapter. Obviously, the natural order of CAM expression in neural tissue is that already considered in the dynamic sequence shown in Figure 5.

Until neural tracts and glial cells appear, N-CAM is the major adhesion molecule in neural tissue (Theiry et al., 1982; Edelman et al., 1983a). It shows definite changes in prevalence during closure of the neural tube and during movement of neural crest cells, as indicated by immunohistochemical staining with anti-N-CAM reagents. In these cells, N-CAM staining disappears at the surface during migration, appearing again at the site of ganglion formation (Thiery et al., 1982; Thiery, 1985).

At 3 1/2 days of development in the chick, Ng-CAM appears on neurons (Hoffman et al., 1982; Grumet and Edelman, 1984; Grumet et al., 1984b; Thiery et al., 1984b). In order to bind to glia selectively in the presence of other neurons, it would be expected that neurons have molecules with specificities different from that of N-CAM. As previously discussed, the major polypeptide components of Ng-CAM differ from those of N-CAM.

Moreover, Ng-CAM, unlike N-CAM, does not alter radically after treatment with neuraminidase (Grumet and Edelman, 1984; Grumet et al., 1984a, b). Ng-CAM mediates the binding of neurons to glial cells (probably astrocytes, although the evidence is not yet complete). It also may be associated with N-CAM at the same cell surface (S. Hoffman, B. A. Cunningham, and G. M. Edelman, unpublished observations), raising the possibility that it also undergoes local modulation. This idea is reinforced by observations (Theiry et al., 1984b) that in the central nervous system Ng-CAM undergoes polarity modulation—its distribution is favored on processes of neurons. Ng-CAM does not appear on glial surfaces, suggesting that neuron–glia interaction involves the heterophilic binding of Ng-CAM to another molecule that so far has not been isolated from glia.

Recent studies on the distribution of Ng-CAM in the developing central and peripheral nervous systems of the chick (Thiery et al., 1984b) indicate that it appears first on postmitotic neurons. It is subsequently expressed in the precursors of motor neurons in the ventral neural tube, and then in optic nerve fibers and ciliary ganglion. At succeeding times, it appears in cranial sensory ganglia, dorsal root ganglia, enteric ganglia, and in the fiber networks of the developing telencephalon. Slightly afterward, it appears in sympathetic ganglia.

At even later stages, during map formation in the retinotectal projection, Ng-CAM appears in fiber bundles of the optic nerve and in descending fibers as they move into the tectum. Most strikingly, it also appears on the bodies of migratory granule cells in the cerebellum at the postmitotic stage (Thiery et al., 1984b). In view of its role in neuron–glia interaction (Grumet and Edleman, 1984; Grumet et al., 1984a) and the known role of glia (Rakic, 1972) in neural migrations of this type, it is tempting to suggest that Ng-CAM is essential for such migrations. Recent experiments in fact suggest that anti-N-CAM antibodies block migration in tissue slices *in vitro;* the fact that antibodies to a mouse determinant called L1 also have been shown to block migration (Schachner et al., 1983) suggests that L1, which has not yet been shown to mediate neuron–glia binding, is Ng-CAM.

The complex but ordered changes in cell surface expression resulting from prevalence modulations of N-CAM and Ng-CAM are later followed by chemical modulation of N-CAM (Figure 6). During tract formation, one still sees evidence of prevalence changes of CAMs, but they are accompanied by perinatal and postnatal variations in the extent of E-to-A conversion of N-CAM at different times in structurally different parts of the brain. Thus, E-to-A conversion is a later event; like changes in prevalence, it changes CAM binding rates (see Figure 4).

A correlated regional and chemical analysis of E-to-A conversion showed strong pattern differences in different gross brain regions (Figure 7; Chuong and Edelman, 1984). This suggests that the rate of conversion, its time of initiation, or its degree in different cell types may differ in these histologically

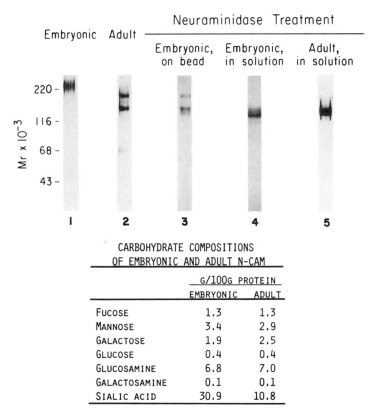

Figure 6. *Electrophoretic patterns and carbohydrate compositions of E- and A-forms of N-CAM.* On SDS-polyacrylamide gels, the E-form shows a diffuse microheterogeneous smear varying from 200 to 250 kD (*lane 1*), whereas the A-form shows two sharp bands at 180 kD and 140 kD (*lane 2*). After treatment with neuraminidase to remove sialic acid completely, the two forms migrate similarly (cf. *lanes 4* and *5*). As shown in the table, the E-form has three times as much sialic acid as the A-form. (From Edelman, 1985.)

different regions. Recent studies show that E- and A-forms of N-CAM have the same attachment sites for carbohydrate, differing mainly in sialic acid content (Crossin et al., 1984). Thus we can conclude with some certainty that E-to-A conversion is an epigenetic event resulting either from enzymatic removal of the sialic acid of surface N-CAM by sialidases, or from turnover of E-forms followed by replacement by A-forms of N-CAM that have lesser amounts of sialic acid. In the latter case, an intracellular enzyme or sialyl transferase responsible for linking sialic acid to N-CAM would be implicated. The significant observation is that during organogenesis, and particularly during the perinatal period, grossly different amounts of E- and A-forms are present in different regions. As indicated by the studies (Hoffman and Edelman, 1983) of binding kinetics reviewed earlier, this would be expected to change the binding efficiencies of various

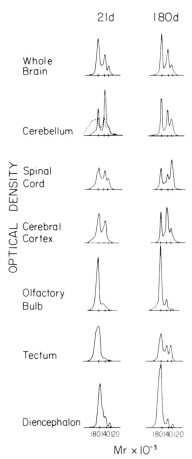

Figure 7. *Differential expression of N-CAM forms by different brain regions of 21- and 180-day-old mice.* Each panel shows a densitometric scan of autoradiographs of immunoblots with anti-N-CAM of brain extracts fractionated by SDS-polyacrylamide gel electrophoresis. Ordinate, optical density; abscissa, apparent molecular weight. The total area under each tracing has been normalized to the same value. The three A-forms (180 kD, 140 kD, and 120 kD) can be seen. Different brain regions are presented from top to bottom in order of decreasing rates of E-to-A conversion. The *dotted line* in cerebellum at 21 days shows the profile from homozygous *staggerer* mice and reveals a delay in E-to-A conversion as compared to the normal profile. (From Edelman, 1985.)

cells in these regions in different ways during histogenesis and thereby alter neural structure during cell differentiation and movement.

While E-to-A conversion is epigenetic, its timing suggested that it might be altered in one or more of the connectional defects seen in mutant animals. The granuloprival mutants (Sidman, 1974; Caviness and Rakic, 1978) provided an opportunity to test this hypothesis. These mutants have major defects in the development of their cerebella; the major symptom is ataxia, which begins to appear in the perinatal period. Of the three extensively investigated mutants, *staggerer* has a connectional defect in neurons, while *reeler* and *weaver* have disorders that also involve the Bergmann glial cells that play a key role in the migration of nerve fibers (Caviness and Rakic, 1978). *Staggerer* is expressed only in homozygous animals; cerebella of these animals show faults in the formation of synapses between parallel fibers and Purkinje cells, which do not appear to have normal tertiary dendritic branches. After failure to make these synapses, the granule cells die in great quantities. The consequence of these anomalies is an ataxic animal with a small and disordered cerebellum, an animal destined to die without further care at about one month after birth.

A defect in N-CAM modulation was hypothesized to be most likely in *staggerer*; this was found to be the case (Edelman and Chuong, 1982). E-to-A conversion was greatly delayed in the cerebellum of the homozygotes but the N-CAM polypeptide appeared to be normal (Figure 7). In contrast, *reeler* and *weaver* mutants had normal schedules of E-to-A conversion. While these findings do not provide an explanation for the cause of disease in *staggerer*, they suggest that one consequence of the genetic defect is a failure in the activity or in the expression of enzymes responsible for E-to-A conversion. It may well be that this leads in turn to a failure to terminate certain key cellular binding events, with a consequent loss of coordination of the various aspects of neural process formation, migration, and synapse formation.

Despite the power and the variety of modulation mechanisms (prevalence, chemical, and polarity modulation) acting on even a single CAM, the histogenetic sequence indicates that there are certain circumstances, such as glial interactions and neurulation, that are likely to require CAMs of different specificity. It is the combination of both specificity and modulation that is important in morphogenetic change. During morphogenesis, the need for CAMs of different specificities appears to be context-dependent, that is, it reflects potentially conflicting interactions among cell types as cytodifferentiation proceeds *pari passu* with cell adhesion and motion.

This account of the role of two CAMs in the histogenesis of a single tissue indicates that intricate modulation events are occurring in many places and involve processes such as fasciculation, neuronal migration, and neuron–glia interaction. It is clear, as shown in Figure 8, that antibodies to N-CAM can perturb such processes, in this case, the retinotectal projection of *Xenopus laevis*. While this confirms the importance of CAMs in such neural mapping, more analytic and chemical approaches are also required for full understanding. Gene expression, transport to the surface, and turnover of CAMs are, for example, involved in a complex web of regulatory biochemical events that remains to be elucidated. Such events bear upon the relation of cytodifferentiation to these primary processes; it is not yet clear whether they resemble the even more fundamental processes connected with embryonic induction and the establishment of animal form.

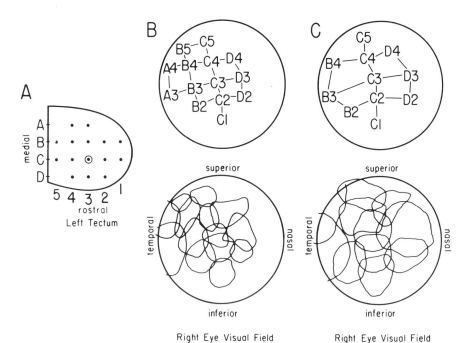

Figure 8. *Perturbation of the retinotectal projection pattern by antibodies to N-CAM.* To map the projection pattern, a metal microelectrode was lowered into the tectum at each of the positions indicated by *dots* on the representative *Xenopus* tectum shown in *A*; the antibody implant was made at the *circled position* near the center of the tectum. The large circles in diagrams *B–C* each represent a 200° range of the field of view of the frog's right eye, with the center of the circle representing the fixation point of the eye. For each electrode position, the center of the responsive region of the frog's visual field is marked in the upper row of circles by the code shown in *A*; the corresponding multiunit receptive fields are shown in the lower row of circles. With implants of preimmune serum, patterns (*B, top*) show the same rectilinear arrangement of receptive fields for the rectilinear set of electrode positions that is seen in normal animals. The size of the receptive fields (*B, bottom*) in the projection shown here was among the largest of those seen in animals with Class 1 projection patterns. With implants of antibodies to N-CAM, patterns (*C, top*) show a distortion of the projection and an enlargement of the receptive fields (*C, bottom*). This distortion is readily seen by comparing the positions of the corresponding receptive fields B3, C3, and D3 in *B* and *C* (see Fraser et al., 1984).

We turn to this aspect of morphogenesis last because its implications are the most far-reaching: The data show a definite early spatial order of CAM expression. A consideration of the topological distribution of primary CAMs prompts a hypothesis relating adhesion to morphogenetic movement and induction. This allows us to place what is known of CAMs in the context of some key problems of development, particularly the developmental genetic problem and the evolutionary problem with which this chapter began.

EARLY EMBRYOGENESIS AND COMPOSITE CAM FATE MAPS

Before considering the detailed changes in their CAM distributions, it may be useful to state two large generalizations that have emerged from observations on the early embryo: (1) CAMs undergo dynamic changes in distribution, sequence, and amount, particularly at regions of both primary and secondary induction; (2) wherever epithelia are converted to mesenchyme, CAMs appear to be lost from the cell surface. A striking example of this phenomenon, already mentioned above, is the prevalence modulation of N-CAM in neural crest cells. When crest cells first appear, they stain for N-CAM (Thiery et al., 1982). As soon as they begin their migration, however, they lose this staining and fibronectin (a SAM) appears on their pathway. (Thiery, 1985). When they reach their destination, and just before they form ganglia, N-CAM reappears on their surface. This classic example of the conversion of a sheet of cells into a mesenchyme consisting of loosely attached mobile cells is clearly correlated with the occurrence of strong CAM modulation.

Antibodies to CAMs may be used to detect whether these molecules are present in very early embryos (Thiery et al., 1982; Edelman et al., 1983a). So far, N-CAM and L-CAM have been found early in the chick; the earliest stage at which they have been detectable in this species is just prior to the formation of the germ layers. During gastrulation, the epiblast and hypoblast stain with fluorescent antibodies to both primary CAMs. Later on, as the primitive streak develops, the middle-layer cells that are migrating do not stain. A remarkable spatial transition then occurs: L-CAM appears on Hensen's node and remains with it for its existence; cells that will become the neural plate lose evidence of L-CAM and stain very strongly with antibodies to N-CAM. Reciprocally, at the border of the ectoderm and neuroectoderm, ectodermal cells stain only with antibodies to L-CAM as do endodermal cells (Figure 9). This emerging pattern of CAM segregation and border formation accompanies the key event of neural induction.

In order to see whether the topology defined by these early events is conserved in later patterns, we employed composite fate maps (Edelman et al., 1983a) constructed by combining classical fate maps of the chick

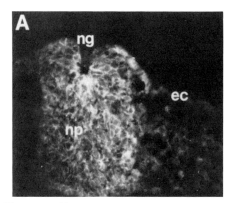
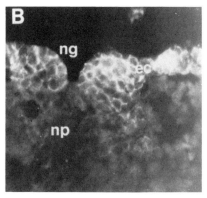

Figure 9. *Change in the distribution of N-CAM and L-CAM during formation of the neural plate and groove (neural induction).* A: Cross section of the neural plate (np) as the neural groove (ng) forms. N-CAM is present in large amounts in the chordamesoderm and neural ectoderm and in small amounts in ectoderm (ec) and adjoining mesoderm. B: L-CAM staining disappears from the neural ectoderm in the neural groove and becomes restricted to the nonneural ectoderm (ec). Just before these events, N-CAM and L-CAM were present in all regions of the blastoderm that give rise to these structures. (From Edelman, 1985.)

embryo with data on identification of tissues that stained with specific anti-CAM antibodies. Fate maps summarize what will become of cells in each embryonic region in terms of the structures to which they will give rise within a defined period of time or epoch. A fate map is only a summary of the events within a fixed time period and a series of such maps is usually necessary to cover several epochs, particularly if one wishes to follow the whole line of development to maturity (Slack, 1983). Fate maps are similar from individual to individual but differ in precision, depending both on the time chosen and the degree to which individual cells in a species give rise to only one exact line of descendants. In the cases considered here, a future four-dimensional distribution (time plus three spatial dimensions) of cells and CAM markers is mapped back onto a two-dimensional surface consisting of the sheet of blastoderm progenitor cells. This results in a topological rather than a topographic map, that is, exact details of structure are sacrificed to reveal connectedness and neighboring relationships. It is these relationships that are particularly important in understanding embryonic induction events.

A composite CAM fate map was derived (Figure 10) by tracing the descendants of blastodermal cells in the chick fate map, as revised by Vakaet (1984; see Figure 8A), for their expression of different kinds of CAM. The first striking feature to be noted in interpreting the map is that the calcium-independent N-CAM regions that will give rise to the neural plate, to the notochord, and to certain parts of the lateral plate mesoderm are completely surrounded by a contiguous, simply connected ring of regions that will express the calcium-dependent L-CAM; the latter regions together comprise the rest of the ectoderm and the endoderm. Second, there is a cephalocaudal coarse gradient of N-CAM antibody staining that is most intense in the region of the neural plate. In underlying regions such as the notochord, the staining is less intense and the pattern is derived from a dynamic sequence—at first, there is no staining for N-CAM, then the notochord (a mesodermal structure) stains intensely, and in later stages the stain disappears. A similar dynamic pattern of N-CAM appearance is seen in the somites just as they segregate into segments. Moreover, the early sequence (L-CAM staining plus N-CAM staining followed by loss of L-CAM staining) is seen in placodes that will form ganglia. Both CAMs are also seen at the apical ectodermal ridge of the limb bud, a key structure necessary for formation of the appendages. A third detailed feature reflected in intersecting boundaries of the map is seen in the mesodermal kidney elements: Both L-CAM and N-CAM appear and disappear in sequences corresponding to the reciprocal embryonic induction of the so-called mesonephric mesenchyme by the tubular structure called the Wolffian duct. L-CAM appears on the Wolffian duct, then N-CAM appears on mesonephric tubules as they organize, and then N-CAM is

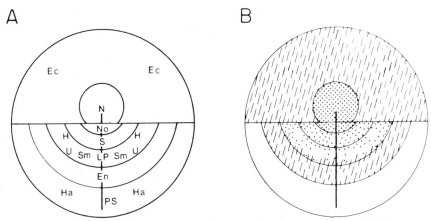

Figure 10. *A: A fate map of the blastodisk* (L. Vakaet, 1984). Areas of cells that will give rise to differentiated tissues are indicated by letters; see below for designations. *B:* Map of cells that will express CAMs. The distribution of N-CAM (*stippled*) or L-CAM (*slashed*) in tissues at 5–14 days (stage 26–40) as determined by immunofluorescence staining is mapped back onto the blastodisk fate map. Cells that will give rise to the urinary tract (U) express both L-CAM and N-CAM. Smooth muscle (Sm) and hemangioblastic (Ha) tissues express neither N-CAM nor L-CAM; areas giving rise to these tissues are blank on this map. Ng-CAM is represented on the neuroectoderm by *small black dots*; it is seen only after three days. The *vertical bar* represents the primitive streak (PS). Ec, intraembryonic and extraembryonic ectoderm; En, endoderm; H, heart; LP, lateral plate (splanchnosomatopleural mesoderm); N, nervous system; No, prechordal and chordamesoderm; S, somite. (From Edelman, 1985.)

replaced by L-CAM as these tubules extend. A fourth feature of the map is that there are regions of the lateral plate mesoderm that will give rise to smooth muscle and other regions containing the hemangioblastic precursors that do not stain at all with either CAM. This raises the possibility that at least one other primary CAM mediates the adhesive interactions of these areas in early development.

Although it is not possible to say exactly how many CAMs will be discovered, it appears likely that the number required for early embryogenesis will not be large (Edelman, 1984a). The evidence reviewed so far suggests that at least two primary CAMs of different specificity (N-CAM and L-CAM) are required to form the distinct epithelial sheets that define the topology of primary induction and neural plate formation. What is required to prove this beyond doubt is to show directly that perturbation of these CAMs alters and blocks morphogenetic development. In any case, the most striking global feature of the composite fate map is that after the first transition in primary induction, the topological relationships of the two primary CAMs remain invariant throughout development.

CELL ADHESION, PATTERN FORMATION, AND MORPHOGENESIS

The combined data on the temporal sequences and spatial pattern of CAMs in tissues suggest that the expression of genes for different CAMs in various sequences during embryonic development, and the mechanisms of transport and epigenetic modulation of CAMs at cell surfaces, both appear to be fundamental in morphogenesis. While modulation mechanisms vary at different stages of embryogenesis in different tissues and are highly dynamic, at all stages, the patterns show a definite order of appearance. A hypothesis on the general role of the CAMs in morphogenesis (Edelman, 1984c) has been formulated on the basis of the CAM expression sequence and the evidence derived from the fate map. This so-called regulator hypothesis bears upon both the developmental genetic problem and the evolutionary problem mentioned at the beginning of this chapter, and its key assumptions and conclusions may be summarized briefly.

According to the hypothesis, CAMs act through adhesion as steersmen or regulators for primary processes, particularly morphogenetic movements, and CAMs play a central role in morphogenesis. CAMs exercise their function as regulators by means of local cell surface modulation. CAM expression is thus connected with the control of motion, which is the result of the play between cellular motility, tension in tissue sheets, and adhesion. A key assumption of the regulator hypothesis is that genes for CAMs are expressed in schedules that are prior to and relatively independent of those for particular networks of cytodifferentiation in different organs. This appears to be reflected in the observation that a single kind of primary CAM is expressed early in the composite fate map and remains in structures that span classical map boundaries not only within a germ layer (e.g., somites, heart, kidney) but also across germ layers (all three for L-CAM; ectoderm and mesoderm for N-CAM). According to the hypothesis, the control of CAM structural genes by regulatory genes is responsible for the body plan as seen in fate maps. In the chick, this plan is reflected in a topological order: a simply connected central region of N-CAM surrounded by a contiguous, simply connected ring of cells expressing L-CAM.

The connection between these developmental genetic events and evolutionary events has to do with covariant alterations in CAM regulatory genes and cell movements during evolution. Morphogenetic movements are the result of the inherent motility of cells and of CAM expression as it is coordinated with the expression of substrate adhesion molecules (SAMs) such as fibronectin (Hay, 1982; J.-P. Thiery, 1985). Such movements, which are regulated by CAM modulation, are responsible for bringing cells of different histories together to result in various embryonic inductions, including neural induction. According to the regulator hypothesis, natural selection acts to eliminate inappropriate movements by selecting against organisms that express CAM genes in sequences leading to failure of embryonic induction. On the other hand, any variant combination of movements and covariant timing of CAM gene expression (resulting from variation in regulatory genes) that leads to appropriate inductive sequences will in general be evolutionarily selected. This mechanism allows for great variation in the details of fate maps from species to species, but at the same time would tend to conserve the basic body plan. Small changes in CAM regulatory genes that do not abrogate this principle of selection could nevertheless lead to large changes in form in relatively short evolutionary times. This model is parsimonious—the total number of genes involved need not be large. Because of the wide dynamic range of cell surface modulation effects (Edelman, 1976, 1983; Hoffman and Edelman, 1983) and their temporal permutations (Thiery et al., 1984b), the developmental and evolutionary effects of the variant expression of a rather small number of genes related to cell adhesion could be momentous.

Various predictions and contrary results that would falsify the regulator hypothesis have been reviewed elsewhere (Edelman, 1984a,c). In the present context, one prediction is particularly germane: If the topology of CAM expression in neural induction is not the same in various animals having similar body plans but *different* morphogenetic movements, then the hypothesis must be abandoned. For example, the frog, which has morphogenetic movements in early embryogenesis that differ greatly from those of the chicken, should nonetheless show a composite CAM fate map topology similar to that of the chicken. If so, it also follows that the timing of regulatory gene action for the primary CAMs would differ in the two species.

The regulator hypothesis focuses attention upon a key question concerning both early and late morphogenesis: What regulates the expression of CAM genes themselves and how are the other primary processes controlled? It remains unknown whether the action of CAM regulatory genes is triggered by morphogens or is triggered by feedback from CAM interactions, by spatial asymmetries at boundaries of cells bearing different CAMs, or by global cell surface modulation (Edelman, 1976) induced by the binding of SAMs. Moreover, it is not known whether CAM gene expression and the ensuing cell–cell adhesion events that irreversibly alter cell movements actually feed back directly to affect other differentiation or division events. At the least, differential CAM expression would be expected to have indirect effects that change the proportional contribution of each primary process to morphogenesis.

HISTOGENESIS RECONSIDERED

As this survey has indicated, no single static factor such as CAM specificity can account for histogenesis, and we are still very far from the goal of defining histology in molecular terms. Nonetheless, the study of CAMs clearly reveals the extraordinary dynamics and hierarchical control that must be exercised on molecular expression at the cell surface during embryogenesis and histogenesis.

While CAMs cannot be sufficient for the establishment of form or for detailed tissue patterns, the evidence suggests that they are absolutely necessary. The continuity of their action from early embryogenesis to adult life is particularly persuasive in suggesting their functional significance. Nonetheless, the other molecules concerned with cell–substrate interaction are also present at early and late developmental stages and it is therefore a reasonable conclusion that, while the major determinants of tissue specificity may derive from the selectivity of CAM modulation, the different SAMs and CJMs also play very important parts in morphogenesis. The function of SAMs is related to cell migration, tissue partition, and the development of hard tissues (Hay, 1982). The function of CJMs is cell connection and communication as seen in gap junctions or in the sealing of the surfaces of epithelial sheets (Gilula, 1978; Lane, 1984).

According to the evidence accumulated so far, CAMs, SAMs, and CJMs are completely separate families of molecules with disparate but conjugate functions. No dynamic view of the molecular basis of morphogenesis would be complete without considering the potential interactions of molecules in all of these groups. Interactions between events conditioned by SAMs and CAMs are already revealed clearly in the case of neural crest cells, as mentioned above. SAMs have been implicated in certain embryonic induction events and it is reasonable to surmise that this may also be related to their support of cell migration (Hay, 1982; J.-P. Thiery, 1985) and to a possible role in global cell surface modulation. Cell junctions and junctional molecules can be studied in a variety of nervous systems (Lane,

1984) as well as in the earliest embryos. It remains to be seen how the temporal expression of all three functional molecular families of cell binding molecules is regulated and coordinated in particular tissues during embryogenesis. One intriguing possibility touched on earlier is that such SAMs as fibronectin and laminin may induce *global* cell surface modulation (Edelman, 1976), which may in turn alter CAM prevalence, itself a local modulation event.

Finally, inasmuch as histology leads to a consideration of pathology, it may not be amiss to suggest that, in addition to *staggerer*, various other pathologies in CAM distribution and control are likely to be uncovered. In view of the widespread distribution of CAMs and their fundamental adhesive roles, there is a high probability that they will be implicated in the pathogenesis of a variety of general and tissue-specific diseases. Almost certainly, they must have a role in the metastasis of particular kinds of cancer cells. Recent studies in our laboratory (Brackenbury et al., 1984; Greenberg et al., 1984) have shown, for example, that transformation of embryonal neural cells by tumor viruses such as the Rous sarcoma virus leads to a loss of N-CAM and of adhesivity. This is a clear-cut demonstration that tumor transformation can be related to loss of a defined CAM and concomitantly of cell adhesion. It suggests a likely permissive basis for the invasiveness of tumor cells into their parent tissues and for their metastasis or spread to other tissues; the relationship of this pathological modulation to alterations in fibronectin and alterations in substrates following transformation remains to be defined. In any event, the study of disease may help to clarify several important issues relating cell adhesion to histogenesis.

REFERENCES

Bertolotti, R., U. Rutishauser, and G. M. Edelman (1980) A cell surface molecule involved in aggregation of embryonic liver cells. *Proc. Natl. Acad. Sci. USA* **77**:4831–4835.

Brackenbury, R., J.-P. Thiery, U. Rutishauser, and G. M. Edelman (1977) Adhesion among neural cells of the chick embryo. I. An immunological assay for molecules involved in cell–cell binding. *J. Biol. Chem.* **252**:6835–6840.

Brackenbury, R., M. E. Greenberg, and G. M. Edelman (1984) Phenotypic changes and loss of N-CAM-mediated adhesion in transformed embryonic chicken retinal cells. *J. Cell Biol.* **99**:1944–1954.

Buskirk, D. R., J.-P. Thiery, U. Rutishauser, and G. M. Edelman (1980) Antibodies to a neural cell adhesion molecule disrupt histogenesis in cultured chick retinae. *Nature* **285**:488–489.

Caviness, V. S., Jr., and P. Rakic (1978) Mechanisms of cortical development. A view from mutations in mice. *Annu. Rev. Neurosci.* **1**:297–326.

Chuong, C.-M., and G. M. Edelman (1984) Alterations in neural cell adhesion molecules during development of different regions of the nervous system. *J. Neurosci.* **4**:2354–2368.

Crossin, K. L., G. M. Edelman, and B. A. Cunningham (1984) Mapping of three carbohydrate attachment sites in embryonic and adult forms of the neural cell adhesion molecule (N-CAM). *J. Cell Biol.* **99**:1848–1855.

Cunningham, B. A., S. Hoffman, U. Rutishauser, J. J. Hemperly, and G. M. Edelman (1983) Molecular topography of N-CAM: Surface orientation and the location of sialic acid-rich and binding regions. *Proc. Natl. Acad. Sci. USA* **80**:3116–3120.

Damsky, C. H., J. Richa, K. Knudsen, D. Solter, and C. A. Buck (1982) Identification of a cell–cell adhesion glycoprotein from mammary tumor epithelium. *J. Cell Biol. (Abstr.)* **95**:22.

Edelman, G. M. (1976) Surface modulation in cell recognition and cell growth. *Science* **192**:218–226.

Edelman, G. M. (1983) Cell adhesion molecules. *Science* **219**:450–457.

Edelman, G. M. (1984a) Cell adhesion and marker multiplicity in neural patterning. *Trends Neurosci.* **7**:78–84.

Edelman, G. M. (1984b) Modulation of cell adhesion during induction, histogenesis, and perinatal development of the nervous system. *Annu. Rev. Neurosci.* **7**:339–377.

Edelman, G. M. (1984c) Cell adhesion and morphogenesis: The regulator hypothesis. *Proc. Natl. Acad. Sci. USA* **81**:1460–1464.

Edelman, G. M. (1985) Molecular regulation of neural morphogenesis. In *Molecular Bases of Neural Development*, G. M. Edelman, W. E. Gall, and W. M. Cowan, eds., pp. 35–59, Wiley, New York.

Edelman, G. M., and C.-M. Chuong (1982) Embryonic to adult conversion of neural cell adhesion molecules in normal and *staggerer* mice. *Proc. Natl. Acad. Sci. USA* **79**:7036–7040.

Edelman, G. M., W. J. Gallin, A. Delouvée, B. A. Cunningham, and J.-P. Thiery (1983a) Early epochal maps of two different cell adhesion molecules. *Proc. Natl. Acad. Sci. USA* **80**:4384–4388.

Edelman, G. M., S. Hoffman, C.-M Chuong, J.-P. Thiery, R. Brackenbury, W. J. Gallin, M. Grumet, M. E. Greenberg, J. J. Hemperly, C. Cohen, and B. A. Cunningham (1983b) Structure and modulation of neural cell adhesion molecules in early and late embryogenesis. *Cold Spring Harbor Symp. Quant. Biol.* **48**:515–526.

Finne, J., U. Finne, H. Deagostini-Bazin, and C. Goridis (1983) Occurrence of alpha 2-8 linked polysialosyl units in a neural cell adhesion molecule. *Biochem. Biophys. Res. Commun.* **112**:482–487.

Fraser, S., B. A. Murray, C.-M. Chuong, and G. M. Edelman (1984) Alteration of the retinotectal map in *Xenopus* by antibodies to neural cell adhesion molecules. *Proc. Natl. Acad. Sci. USA* **81**:4222–4226.

Gallin, W. J., G. M. Edelman, and B. A. Cunningham (1983) Characterization of L-CAM, a major cell adhesion molecule from embryonic liver cells. *Proc. Natl. Acad. Sci. USA* **80**:1038–1042.

Gennarini, G., G. Rougon, H. Deagostini-Bazin, M. Hirn, and C. Goridis (1984) Studies on the transmembrane disposition of the neural cell adhesion molecule N-CAM. A monoclonal antibody recognizing a cytoplasmic domain and evidence for the presence of phosphoserine residues. *Eur. J. Biochem.* **142**:57–64.

Gerisch, G., and D. Malchow (1976) Cyclic AMP receptors and the control of cell aggregation in *Dictyostelium*. *Adv. Cyclic Nucleotide Res.* **7**:59–68.

Gilula, N. B. (1978) Structure of intercellular junctions. In *Intercellular Junctions and Synapses: Receptors and Recognition*, Vol. 2, J. Feldman, N. B. Gilula, and J. D. Pitts, eds., pp. 1–19, Chapman and Hall, London.

Gould, S. J. (1977) *Ontogeny and Phylogeny*, Belknap Press, Cambridge, Mass.

Greenberg, M. E., R. Brackenbury, and G. M. Edelman (1984) Alteration of N-CAM expression following neuronal cell transformation by Rous sarcoma virus. *Proc. Natl. Acad. Sci. USA* **81**:969–973.

Grumet, M., and G. M. Edelman (1984) Heterotypic binding between neuronal membrane vesicles and glial cells is mediated by a specific neuron–glial cell adhesion molecule. *J. Cell Biol.* **98**:1746–1756.

Grumet, M., S. Hoffman, C.-M. Chuong, and G. M. Edelman (1984a) Polypeptide components and binding functions of neuron–glia cell adhesion molecules. *Proc. Natl. Acad. Sci. USA* **81**:267–271.

Grumet, M., S. Hoffman, and G. M. Edelman (1984b) Two antigenically related neuronal CAM's of different specificities mediate neuron–neuron and neuron–glia adhesion. *Proc. Natl. Acad. Sci. USA* **81**:267–271.

Hay, E. D. (1982) Collagen and embryonic development. In *Cell Biology of Extracellular Matrix*, E. D. Hay, ed., pp. 379–409, Plenum, New York.

Hoffman, S., and G. M. Edelman (1983) Kinetics of homophilic binding by E and A forms of the neural cell adhesion molecules. *Proc. Natl. Acad. Sci. USA* **80**:5762–5766.

Hoffman, S., B. C. Sorkin, P. C. White, R. Brackenbury, R. Mailhammer, U. Rutishauser, B. A. Cunningham, and G. M. Edelman (1982) Chemical characterization of a neural cell adhesion molecule purified from embryonic brain membranes. *J. Biol. Chem.* **257**:7720–7729.

Lane, N. J. (1984) A comparison of the construction of intercellular junctions in the CNS of vertebrates and invertebrates. *Trends Neurosci.* **7**:95–99.

Ogou, S. I., C. Yoshida-Noro, and M. Takeichi (1983) Calcium-dependent cell–cell adhesion molecules common to hepatocytes and teratocarcinoma stem cells. *J. Cell Biol.* **97**:944–948.

Raff., R. A., and T. C. Kaufman (1983) *Embryos, Genes, and Evolution*, Macmillan, New York.

Rakic, P. (1972) Mode of cell migration to the superficial layers of fetal monkey neocortex. *J. Comp. Neurol.* **145**:61–84.

Rothbard, J. B., R. Brackenbury, B. A. Cunningham, and G. M. Edelman (1982) Differences in the carbohydrate structures of neural cell-adhesion molecules from adult and embryonic chicken brains. *J. Biol. Chem.* **257**:11064–11069.

Rutishauser, U., W. E. Gall, and G. M. Edelman (1978) Adhesion among neural cells of the chick embryo. IV. Role of the cell surface molecule CAM in the formation of neurite bundles in cultures of spinal ganglia. *J. Cell Biol.* **79**:382–393.

Schachner, M., A. Faissner, J. Kruse, J. Lindner, D. H. Meier, F. G. Rathjen, and H. Wernecke (1983) Cell-type specificity and developmental expression of neural cell-surface components involved in cell interactions and of structurally related molecules. *Cold Spring Harbor Symp. Quant. Biol.* **48**:557–568.

Sidman, R. L. (1974) Contact interaction among developing mammalian brain cells. In *Cell Surface in Development*, A. Moscona, ed., pp. 221–253, Wiley, New York.

Slack, J. M. W. (1983) *From Egg to Embryo*, Cambridge Univ. Press, New York.

Sperry, R. W. (1963) Chemoaffinity in the orderly growth of nerve fiber patterns and connections. *Proc. Natl. Acad. Sci. USA* **50**:703–710.

Steinberg, M. S. (1970) Does differential adhesion govern self-assembly processes in histogenesis? Equilibrium configuration and the emergence of a hierarchy among populations of embryonic cells. *J. Exp. Zool.* **173**:395–433.

Thiery, J.-P. (1985) Roles of fibronectin in embryogenesis. In *Fibronectin*, D. Mosher, ed., Academic, New York (in press).

Thiery, J.-P., R. Brackenbury, U. Rutishauser, and G. M. Edelman (1977) Adhesion among neural cells of the chick embryo. II. Purification and characterization of a cell adhesion molecule from neural retina. *J. Biol. Chem.* **252**:6841–6845.

Thiery, J.-P., J.-L. Duband, U. Rutishauser, and G. M. Edelman (1982) Cell adhesion molecules in early chick embryogenesis. *Proc. Natl. Acad. Sci. USA* **79**:6737–6741.

Thiery, J.-P., A. Delouvée, W. J. Gallin, B. A. Cunningham, and G. M. Edelman (1984a) Ontogenetic expression of cell adhesion molecules: L-CAM is found in epithelia derived from the three primary germ layers. *Dev. Biol.* **102**:61–78.

Thiery, J.-P., A. Delouvée, M. Grumet, and G. M. Edelman (1984b) Initial appearance and regional distribution of the neuron–glia cell adhesion molecule (Ng-CAM) in the chick embryo. *J. Cell Biol.* **100**:442–456.

Turing, A. M. (1952) The chemical basis of morphogenesis. *Philos. Trans. R. Soc. Lond. (Biol.)* **237**:37–72.

Vakaet, L. (1984) Early development in birds. In *Chimeras in Developmental Biology,* N. M. Le Douarin and A. McLaren, eds., pp. 71–87, Academic, New York.

Chapter 8

The Role of Cell Adhesion in Morphogenetic Movements During Early Embryogenesis

JEAN-PAUL THIERY
JEAN-LOUP DUBAND
ANNIE DELOUVÉE

ABSTRACT

In young vertebrate embryos, morphogenesis involves epithelium–mesenchyme interconversion, displacement of cell sheets and of individual cells, and regrouping of cells into defined but transitory structures. The movements of cells, as exemplified by gastrulation and the migration of neural crest cells, require specific adhesive interactions between cells and with the extracellular matrix. Before their final localization, both gastrulating cells and crest cells migrate through transitory, fibronectin-rich pathways. This migration depends on the ability of the cells to adhere to fibronectin (FN): either anti-FN antibodies or peptides synthesized to the cell-binding sequence arrest movement. In contrast, neural and liver cell adhesion molecules (N-CAM and L-CAM) are expressed by nonmigratory cells. Both molecules are observed at the blastula stage but are transitorily lost from gastrulating cells, reappearing with the cessation of movement. N-CAM is maintained in the ectoderm, particularly at the level of the neural plate, whereas L-CAM completely disappears from this structure and the neural crest presumptive territory. N-CAM again diminishes at the surface of neural crest cells but is reacquired at the onset of aggregation into ganglion rudiments. Although a detailed molecular mechanism is still inaccessible, the inverse modulation of these adhesive properties may contribute to the patterning of the neural axis and the overall shaping of the embryo.

Morphogenesis and organogenesis are achieved by a highly complex but ordered series of events. The primary processes directly involved in these events are cell divison, migration, adhesion, differentiation, and death. Our knowledge of the molecular bases of these processes has been developing over the last decade but the strategies used by multicellular organisms are not understood. We are still ignorant of the mechanisms by which the

169

primary processes are interconnected, and in higher organisms we do not have a simple model system in which to approach the problem of morphogenesis. The purpose of this chapter is to review recent data obtained on the mechanisms of cell migration and cell adhesion in the vertebrate embryo that may shed light on some aspects of tissue formation.

GASTRULATION AND NEURAL INDUCTION

Topography and Morphogenetic Movement

In the vertebrate embryo, cleavage and subsequent cell proliferation lead to a variable but limited number of cells. With the exception of some mammals, approximately 10,000–50,000 cells are formed prior to gastrulation. Many of these cells are already organized into an epitheliallike structure. It is not possible to give a general description of the movements and reorganizations that take place during the formation of the primary germ layers in the vertebrates. Indeed, not only the shape of the embryos but also the procedures used by the embryos to establish three layers differ considerably, at least in their detail, in the different classes and even within the same class. For example, one can find fish such as the sturgeon (Ballard and Ginsburg, 1980) that gastrulate like most amphibians, whereas others behave similarly to avian embryos (Ballard, 1973). There are even frog eggs (Gasthroteca) that form an embryonic disk even though gastrulation proceeds, at least to some extent, as it does in other amphibians (Del Pino and Elinson, 1983).

Although early morphogenetic processes can be diversified, it should be stressed that once neural induction has occurred, the body plan that emerges is basically the same in all the species. Therefore, it is important to search for common mechanisms in gastrulation and early neurulation.

The fate of cells from the blastula has been determined through the use of various labeling techniques that allow the construction of fate maps (Vogt, 1929; Spratt and Haas, 1965; Rosenquist, 1966; Vakaet, 1970; Keller, 1975, 1976; Fontaine and Le Douarin, 1977; Ballard, 1982). These representations, though difficult to compare directly because of the great diversity in topography, may be simplified in order to see whether or not the relative position of prospective tissues varied throughout evolution. It appears that the presumptive neural plate is always associated with the ectoderm, the latter being sometimes in continuity with the endoderm (Vogt, 1929; Keller, 1975, 1976; Ballard and Ginsburg, 1980; Vakaet, 1984; M. Snow, unpublished observations). Even within the prospective mesoderm, the different tissues are always positioned in the same order. In fact, as a result of these fairly stereotyped movements, superficial cells of the embryo are released from a well-defined site and spread progressively under the upper layer. In all cases, a defined area of the ectoderm that becomes underlined by a precise population of mesodermal cells is rapidly transformed into a neural plate. The origin of the neurogenic placodes and the neural crest can also be traced back to the stage during which the newly induced neural plate segregates from the ectoderm (Rosenquist, 1981). Interestingly, the placodes and the neural crest lie at the boundary between the ectoderm and the neural plate. Also worth mentioning is the rapid regionalization of the neural plate in the anteroposterior axis (Jacobson, 1959; Saxen and Toivonen, 1962). Perhaps one of the most critical issues in trying to understand the formation of the neural plate resides in the mechanisms that generate movements during gastrulation. The relocation of superficial cells into a three-layered embryo involves both active and passive movements. These movements have been particularly well described in fish (Ballard, 1982), reptiles (Nieuwkoop and Sutasurya, 1979), amphibians (Vogt, 1929; Holtfreter, 1943, 1944; Keller, 1975, 1978, 1981), and birds (Vakaet, 1970), but remain to be defined in mammals.

In amphibians, superficial movements become appreciable at the time of the formation of the blastoporal lip. The appearance of the latter in the marginal zone between the vegetal and animal poles may result from pulling by the bottle cells (Holtfreter, 1943). Nevertheless, the triggering of cell invagination may involve other cells. Cell movements occur in both the animal and vegetal poles. The animal pole expands considerably by a thinning out of its multilayered epithelium and spreading of the most superficial cells (Keller, 1980). It continues to expand ventrally while the blastopore forms a ring occupied by the yolk plug. Concomitantly, meso-dermal cells migrate actively along the blastocoelic roof (Nakatsuji, 1975; Kubota and Durston, 1978), while most of the endoderm may be carried inside passively to form a new cavity, the archenteron.

In the avian embryo, gastrulation is somewhat different since the shape of the embryo is not spherical but discoidal. However, it also proceeds by ingression of cells, the blastopore being replaced by a long, growing slit, the primitive streak. Here again, the streak is the site at which cells of the upper layer invaginate. Ingressing cells then actively move under the basal surface of the upper layer (see Figure 1A). In birds, however, the origin of the definitive endoderm is somewhat different from that of amphibians. Indeed, three deep layers form sequentially, each replacing the previous one; they are called the endophyll, the hypoblast, and the endoderm. The endophyll and hypoblast seem to originate from cells of the upper layer shedding into the underlying cavity (Vakaet, 1970; Eyal-Giladi and Koshav, 1976). On the contrary, the endoderm forms as a consequence of the gastrulation process; that is, from cells moving toward the streak and invaginating through it (Vakaet, 1962). The formation of the first two deep layers does not seem to require active cell movements, while the formation of the endoderm and mesoderm involves long-range displacements (Vakaet, 1970; Vakaet et al., 1980).

Displacement of sheets of cells cannot be precisely analyzed since such a movement may result from several distinct cell behaviors, including changes in cell shape, local migration leading to thinning out of the epi-

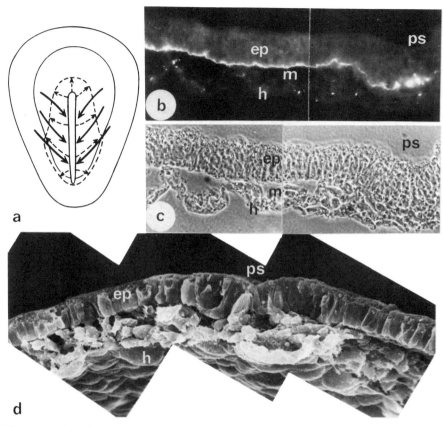

Figure 1. *Gastrulation in the chick embryo. a:* Diagram showing the convergence of epiblastic cells (*arrows*) toward the primitive streak, where they invaginate and subsequently migrate under the epiblast (*dashed arrows*). *b* and *c:* Immunofluorescence staining for FN and phase-contrast micrograph, respectively, of a stage-5 embryo (transverse section; stages according to Vakaet, 1962). The primitive streak (ps) has reached its maximal length and the presumptive mesodermal (m) and endodermal cells move deeper into the groove and expand laterally as a single layer along the epiblast (ep). FN is present at the basal surface of the epiblast. Locally under the primitive streak, this basement membrane is broken off at sites where cells are separating from the epiblast. Note that moving cells and the hypoblast (h) are not surrounded by an FN meshwork. *d:* Scanning electron micrograph of a stage-6 embryo. Cells in the primitive streak send processes to the underlying cavity. Under the epiblast, mesodermal cells occupy the full space and form a multilayered cell population.

thelium, and proliferation. Localized tension in the sheet because of the active migration of marginal cells during epiboly may in part generate local displacement that eventually propagates in the sheet. Unfortunately, so far, changes in cell shape have only been evaluated as a possible mechanism for the formation of the neural plate and of the neural tube. In the urodele embryo, it appears that the transformation of the neural plate from an hemispheric cap to a flat keyhole-shaped structure and later to a tube can be well explained by an intrinsic shrinkage program in the apical domain of the cells and a redistribution of the neural plate cells associated with the notochord (Jacobson, 1978, 1981, and this volume).

Role of Fibronectin in the Migration of Presumptive Mesodermal Cells

Migration of cells from the blastoporal lip or the primitive streak has been studied by classic histological methods, scanning and transmission electron microscopy, and time-lapse microcinematography (Spratt and Haas, 1965; Vakaet, 1970; Nakasutji, 1975; Keller and Schoenwolf, 1977; Kubota and Durston, 1978; Solursh and Revel, 1978; Sanders, 1979). Cells released from the superficial layer emit fine but occasionally long filopodia, allowing anchorage and locomotion. They remain in close apposition while migrating along the superficial layer basal surface. Filopodia interact both with the basal surface and the extracellular matrix (ECM), which is organized into a three-dimensional array of fibrils and interstitial bodies (Trelstad et al., 1967; Sanders, 1979; Nakasutji and Johnson, 1983). The ECM is known to contain large amounts of glycosaminoglycans, including hyaluronic acid (HA) and chondroitin sulfate (CS), fibronectin (FN), and possibly collagens (Johnson, 1977; Sanders, 1979; Vanroelen et al., 1980; Duband and Thiery, 1982a; Mitrani and Farberov, 1982; Sanders, 1982; Harrisson et al., 1984). The basal lamina, an organized thin layer close to the epithelial cells, contains laminin (LN), collagen IV, entactin, and glycosaminoglycans (Sanders, 1979; Leivo et al., 1980; Wu et al., 1983).

The role of FN is important, as indicated by its many different biological properties and primary structure (see Yamada et al., this volume). It consists of a disulfide-bonded dimer of two almost identical polypeptides of approximately 240 kD. Several distinct domains have been localized along the polypeptide chains; FN contains a cell surface binding region and several other sites that interact directly with other ECM components, including collagen and glycosaminoglycans. FN is coded by a single gene containing 48 exons (Hirano et al., 1983), and a result of differential splicing, several messenger RNAs may be expressed (Schwarzbauer et al., 1983; Kornblihtt et al., 1984a, b). At least two forms of fibronectin differ in their ability to form polymers. Immunolocalization of FN during early embryogenesis has revealed its presence in all the regions of active migration (see Figures 1, 2; Duband and Thiery, 1982a, b; Thiery et al., 1982a; Boucaut and Darribère, 1983).

In urodeles, FN appears first in the blastocoelic roof at the early to midblastula stage. FN fibers grow from the center of each cell surface, joining those of the surrounding cells (Boucaut and Darribère, 1983). An intricate network develops subsequently without any obvious orientation (Figure 2A, B). The FN-rich ECM is formed a long time prior to the onset of migration and therefore cannot be considered a signal for the formation

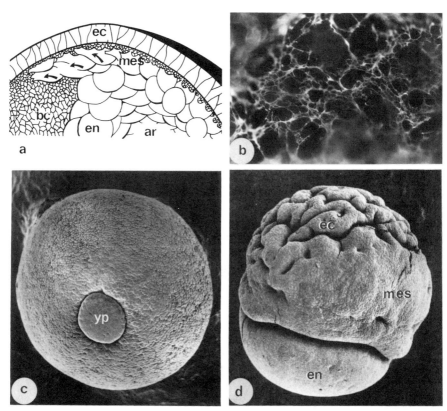

Figure 2. *Gastrulation in* Pleurodeles waltlii *embryos. a:* A schematic representation of part of the blastocoelic roof at the midgastrula stage. Mesodermal cells (mes) migrate along the roof of the ectoderm (ec) in contact with a dense fibrillar matrix. Endodermal cells (en) are internalized together with the mesoderm and form a new cavity, the archenteron (ar). The blastocoelic cavity (bc) is progressively displaced to the ventral aspect of the egg. *b:* A whole-mount preparation of part of the blastocoelic roof ahead of the front of migration. FN-rich fibers form an intricate meshwork without any preferential orientation. *c:* Late gastrula-stage embryo exhibiting a yolk plug (yp) corresponding to the last endodermal cells to invaginate. *d:* Late gastrula-stage embryo was injected with either anti-FN monovalent antibodies or with a cell binding peptide (Arg-Gly-Asp-Ser-Pro-Ala-Ser-Ser-Cys-Pro) at the late blastula stage. Gastrulation did not occur and most of the ectoderm not undergoing normal epiboly developed into highly convoluted structures.

of the blastoporal lip and local ingression of cells along the blastocoelic cavity. However, FN is synthesized by all cells of the blastula. FN mRNAs are maternally derived and are translated at a low frequency during oogenesis and thereafter throughout cleavage (Darribère et al., 1984; Lee et al., 1984). An increase in FN synthesis occurs at the midblastula stage precisely at the time when the meshwork has been seen, by immunofluorescence labeling, to thicken (Darribère et al., 1984). It should be stressed that while FN can be produced by all cells of the blastula, only a limited number of cells of the epiblast are able to retain it at their surface. In particular, endodermal cells do not exhibit FN on their surface and this is consistent with their passive displacement into the embryo (Boucaut and Darribère, 1983).

Studies in amniotes (birds and mammals) are few at the present time. The presence of FN under the upper layer during the migration of the mesoderm has been clearly established both in whole-mount specimens (Critchley et al., 1979) and in sections (Wartiovaara et al., 1979; Duband and Thiery, 1982a; Sanders, 1982; Harrisson et al., 1984).

The appearance of FN under the upper layer has been correlated with primitive streak formation (Duband and Thiery, 1982a). During the ingression of mesodermal cells in the streak, the basement membrane is interrupted locally (Figure 1B, C), thus inducing destabilization of the epithelial structure among cells of the upper layer (see Sugrue and Hay, 1981, 1982). Subsequently, cells send lobopodia into the underlying cavity, separate, and move onto the FN meshwork (Figure 1D).

The way in which cells of the upper layer move toward the streak remains obscure. It may involve a movement of the whole sheet, or local displacement of individual cells or, more simply, cell proliferation. Recent studies support the first hypothesis, since it seems that cells of the upper layer move together with their basement membrane (Sanders, 1984).

The role of FN in the migration of mesodermal cells has been successfully examined in the amphibian embryo through two kinds of perturbation experiments (Boucaut et al., 1984a, b). In the first, the blastocoelic roof was inverted locally, creating a zone completely devoid of ECM, since the latter is never found on the apical side of the epithelium. Mesodermal cells migrated normally under the epiblast except at the modified site, since filopodia could not adhere to the FN-free area. In the second experiment, microinjection of monovalent antibodies to *Ambystoma* FN into the blastocoelic cavity of *Pleurodeles waltlii* blastulae led to a complete inhibition of the normal gastrulation processes. While epiboly of the outer layer still occurred, prospective mesodermal cells did not migrate under the basal surface of the epiblast. In contrast, when the antibodies were introduced slightly later, in mid- or late gastrulae, partial or normal development ensued, particularly with respect to neurulation (Figure 2C, D).

The inhibition of migration by monovalent antibodies (Fab's) has been interpreted as being a consequence of a steric hindrance effect preventing the interactions between mesodermal cell surface receptors and the cell binding domain of FN. However, since Fab' fragments have a diameter of approximately 50 nm, it can be argued that other ECM components, closely associated with FN, could not interact with these cells.

This objection was rejected after a new series of perturbation experiments using FN-derived peptides was performed. The biological properties of large peptides generated after limited proteolysis of FN have been analyzed through the use of *in vitro* cell adhesion assays; it appears that an 11.5-kD fragment contains the cell binding domain (Pierschbacher et al., 1981). A 30-amino-acid peptide, and thereafter 5- to 10-amino-acid peptides, all containing the Arg-Gly-Asp-Ser (RGDS) sequence, were found to prevent the adhesion of cells to an FN-coated substratum (Pierschbacher and Ruoslahti, 1984; Yamada and Kennedy, 1984). A synthetic decapeptide, identical to a highly conserved sequence of fibronectin, and a smaller peptide, both containing the RGDS sequence, inhibited gastrulation in amphibian embryos, whereas a peptide derived from the collagen binding domain or ACTH did not perturb migration (Boucaut et al., 1984b). These results demonstrate unequivocally that a direct interaction between the FN binding domain and the cell surface is an absolute requirement for attachment and migration, whatever other types of interaction may also be necessary for cell locomotion.

It should be stressed that once mesodermal cells have migrated under the epiblast, neural induction cannot be perturbed in the presence of anti-FN antibodies or with the peptides belonging to the FN binding site. Neural induction can also occur when the blastoporal lip remains in contact for four hours with the apical surface of the blastocoelic roof (Duprat and Galandris, 1984) under conditions in which there is no FN-containing matrix (Boucaut et al., 1984a). Therefore neural induction, as opposed to the formation of the primary germ layer, does not require interaction mediated by FN.

THE NEURAL CREST

Definition

The part of neural primordium lying at the boundary between the ectoderm and neural plate gives rise to the neural crest. After a phase of extensive migration through the embryo, neural crest cells differentiate into a great number of cell types, including most elements of the peripheral nervous system, pigments, and connective tissues in the head and the neck (for a review, see Le Douarin, 1982).

One of the most striking characteristics of crest migration is the remarkable precision with which these cells reach their target sites. Numerous experiments using transplantation of ^{3}H-thymidine-labeled neural primordium (Weston, 1963; Noden, 1975) or chick–quail chimeras (for a review, see Le Douarin, 1982) to study the migration pathways and fate of neural crest cells have been undertaken. Additionally, we have carried out a precise study of the mechanism of crest cell migration using markers of cell adhesion in parallel with a monoclonal antibody (NC-1) that recognizes migrating neural crest cells (Vincent et al., 1983; Tucker et al., 1984; Vincent and Thiery, 1984).

Individualization of Neural Crest Cells

Neural crest individualization can be defined as the series of events that lead to the separation of crest cells from the neural epithelium. It is only during or just after individualization that neural crest cells can be identified unambiguously.

The mechanism of crest cell individualization can be summarized as the transition from an epithelial architecture (when crest cells are associated with the neuroectoderm) to a mesenchymal one (when they are in their migratory phase). This mechanism must involve, at least, a local disruption of the basement membrane and the disappearance of cell junctions to allow the release of the cells (see Sugrue and Hay, 1981, 1982; Trinkaus, 1984).

In birds, at cephalic levels, the neural crest individualizes during neural tube closure. Lateral edges of the neural tube bend toward each other and carry the ectoderm above the neural epithelium at the junction of the ectoderm and the neural tube, thus leading to the local fusion of their respective basement membranes (Duband and Thiery, 1982b). Mechanical pressure applied locally during folding, the fusion of the basement membranes, and active proliferation of crest cells all contribute to their detachment from the neural epithelium basement membrane. Furthermore, crest cells are soon in contact with a suitable substrate for motility. Interestingly, in the posterior mesencephalon, crest cells do not release from the neural tube. This is correlated with the absence of both crest cell proliferation and fusion of the basement membranes of the neural tube and the ectoderm (J.-P. Thiery, J.-L. Duband, and A. Delouvée, unpublished observations).

Changes in cell surface properties as well as local synthesis of proteases (Liotta et al. this volume; J. Valinski and N. Le Douarin, unpublished observations) may also be responsible for the egression of crest cells. Finally, cell surface modulation of cell adhesion molecules and the expression of new components (Tucker et al., 1984; Thiery, 1984) may favor epithelium–mesenchyme interconversion.

In the trunk, there is sometimes a great delay between the closure of the neural tube and crest cell individualization. Moreover, no apparent changes in tissue shape are correlated with the emigration of crest cells (Thiery et al., 1982a). However, the basement membrane overlying the neural tube is completely disrupted at the onset of crest cell migration, as assessed by the total disappearance of LN and the appearance of FN around premigrating crest cells (see Figure 4A). As in the head, the presence of a suitable substrate may favor crest emigration, but the triggering of the dissociation of crest cells from the neural tube cannot yet be explained. An "internal clock" is likely to be involved, since *in vitro* crest cell emigration was considerably delayed when neural tubes were explanted several hours before normal crest cell emigration, whereas it occurred within 30 minutes when neural tubes were explanted at the time of their emigration (Newgreen and Gibbins, 1982).

In the trunk, in contrast to the head, crest cell emigration is very progressive; at the beginning, few crest cells escape, and later on the flux of detaching cells greatly amplifies. Concomitantly, the LN meshwork becomes interrupted and completely disappears from the dorsal part of the neural tube.

One could imagine a cascade mechanism in which, for some unknown reason, a few crest cells detached from the neural tube, thus making local breaks in the basement membrane. Such a fragilized basement membrane would allow the further dissociation of many crest cells from the neural tube; only the basement membrane would be destroyed.

In conclusion, we can state that the disruption of the basement membrane bordering the presumptive crest population at the time of its individualization is a general feature common at least to birds and mammals (Nichols, 1981; J.-P. Thiery, J.-L. Duband, and A. Delouvée, unpublished observations). A lot more work will be necessary to understand the mechanism of re-modeling of the basement membrane.

Migration of Crest Cells

Substrate of Migration. At the onset of migration, crest cells encounter an FN-containing ECM (see Figures 3, 4, 5), collagens type I and III (J.-L. Duband, R. Timpl, and J.-P. Thiery, unpublished observations), and high amounts of HA and smaller quantities of CS (Greenberg and Pratt, 1977; Derby, 1978). The relative concentrations of HA and CS change during the time course of migration (Derby, 1978).

Because of its high hydration properties, HA has been thought to create a space for migration. Its synthesis is locally controlled by the tissues adjacent to the migration pathways and by crest cells themselves (Greenberg and Pratt, 1977; Pintar, 1978). However, HA alone does not favor crest cell spreading and motility (Fisher and Solursh, 1979; Newgreen et al., 1982).

In vitro, crest cells do not attach to pure collagen (Newgreen et al., 1982; Rovasio et al., 1983). It should be mentioned that most cell types, except rat hepatocytes (Rubin et al., 1981), are incapable of adhering directly to collagen. Collagens, at least collagen type I, are not involved in crest cell movement, but rather constitute a framework of fibers which maintains the tissue shape. Conversely, FN alone, associated with collagen, or de-posited by fibroblasts, greatly promotes the spreading and movement of crest cells (Rovasio et al., 1983).

It has been shown that the origin and the local concentration of FN can influence the behavior (anchorage motility) of chick heart fibroblasts (Couchman et al., 1982). The anchorage of a nonmigrating cell is correlated with the synthesis by the cell of a dense fibrillar meshwork of FN. Con-

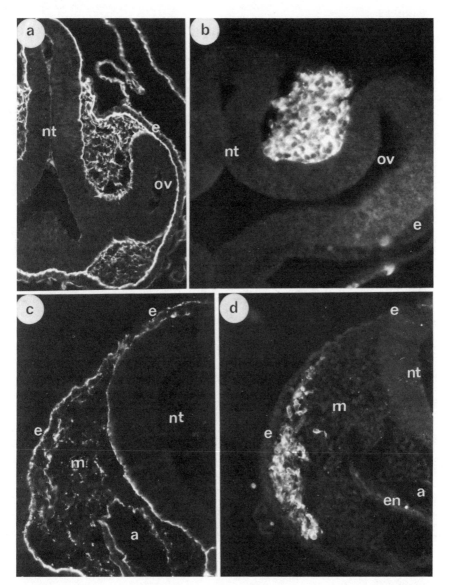

Figure 3. *Migration of neural crest cells at cephalic levels of chick embryos. a* and *b:* Immunofluorescence staining for FN and NC-1, respectively, at the prosencephalon level in a 12-somite-old embryo. Because of the presence of the optic vesicle (ov), crest cells cannot move laterally and ventrally and accumulate in an FN-rich area where they contribute to connective tissues around the eyes. *c* and *d:* Immunofluorescence staining for FN and NC-1 in the mesencephalon of a 12-somite-old-embryo. Crest cells move exclusively between the ectoderm (e) and the loose cephalic mesenchyme (m) as a dense mass surrounded by FN. Note the presence of a cell-free space devoid of FN ahead of crest cells. a, aorta; en, endoderm; nt, neural tube.

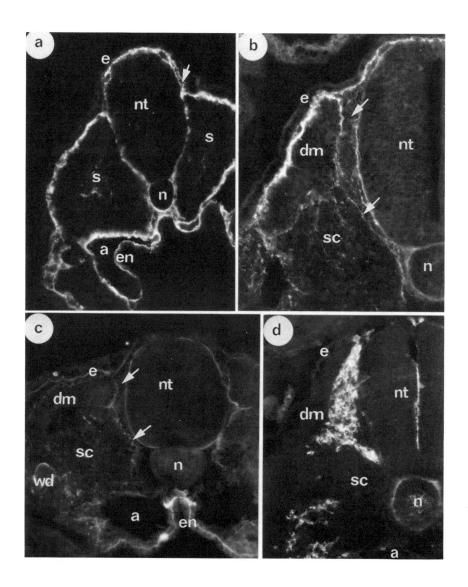

versely, exogenous FN as small plaques at the cell surface greatly enhances cell motility. Interestingly, most crest cells lack the ability to synthesize this protein (Newgreen and Thiery, 1980).

Laminin is present in the basement membranes of epithelia (Figure 4C) but cannot be considered a substrate of crest cell migration since we have never seen it around migrating crest cells. Furthermore, *in vitro* studies have revealed that the crest cell adhesion to LN is low, compared to FN, during the migration process (Rovasio et al., 1983).

The three-dimensional structure of the ECM in crest cell pathways does not appear to influence the direction of crest cell migration (Tosney, 1978, 1982). With the possible exception of the amphibian embryo, the meshwork of the ECM does not seem to exhibit any particular orientation (Löfberg et al., 1980).

Pathways of Migration.   The pathways of migration have been studied extensively by using FN staining in combination with the quail nucleolar marker, acetylcholinesterase, or NC-1, to reveal the cells (see Duband and Thiery, 1982b; Thiery et al., 1982a; Cochard and Coltey, 1983; Le Douarin et al., 1984; Vincent and Thiery, 1984). Since many of the pathways and sites of final localization of crest cells do not exist at the onset of migration, crest cells are not preprogrammed to move to any particular area. Heterotopic grafts of premigratory crest cells have demonstrated that the paths of migration and the sites of arrest are determined by the level of the neuroaxis to which the cells were transplanted (Le Douarin and Teillet, 1974; Noden, 1975).

Even though the particular routes of migration pathways vary, depending on the level of the embryo through which they are passing, their structure is well defined: They are cell-free spaces filled with FN, collagens type I and III, HA, and CS, which are limited by one of two basal laminae of

Figure 4. *Migration of neural crest cells in the trunk of chick embryos. a:* Immunofluorescence for FN at the somite-15 level of an 18-somite-stage embryo. The embryo is composed only of epithelia bounded by FN-rich ECM: neural tube (nt), ectoderm (e), somite (s), notochord (n), and endoderm (en). Crest cells begin to dissociate from the dorsal aspect of the neural tube (*arrow*) and subsequently move in a restricted area defined by the basement membranes of the neural tube and the ectoderm. *b:* Immunofluorescence for FN at the somite-15 level of a 32-somite-stage embryo. The crest cell population (*arrows*) expands lateral to the neural tube and is maintained in an FN-rich area limited by the neural tube, the dermomyotome (dm), and the sclerotome (sc). *c:* Immunofluorescence for LN at the somite-18 level of a 30-somite-stage embryo. LN is found exclusively at the basal surface of the epithelia, and in the medioventral side of the dissociating sclerotome. Crest cells (*arrows*) are located in a space limited by LN but are not enmeshed in LN. *d:* Immunofluorescence for NC-1 at the somite-15 level of a 35-somite-stage embryo. Crest cells are compacted in a limited area defined by the neural tube, der-momyotome, and sclerotome (cf. *b* and *c*). Only a few crest cells are visible ventral to the sclerotome near the aorta (a). They do not originate from the same migration as crest cells that are accumulating on the dorsal side (see Figure 5). wd, wolffian duct.

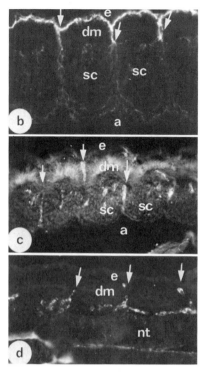

Figure 5. *Migration of crest cells at trunk levels. a:* This scheme represents the trunk of an
embryo viewed from its dorsal side and illustrates the relative position of crest cells
(*diagonal lines*) in relation to the somites (s) and the neural tube (nt). The level of the
sections shown in *b* and *c* is indicated (b,c). *b* and *c:* Immunofluorescence staining for
FN and NC-1 along somites 15–18 in a 29-somite-stage embryo (sagittal section). The
division between two consecutive sclerotomes (sc) is indicated (*arrows*). Crest cells are
distributed as narrow fluxes along the intersomitic arteries. Crest cells "among" sclerotomes
correspond to those that move close to the neural tube. *d:* Immunofluorescence staining
for NC-1 in a 30-somite-stage embryo (longitudinal section). Crest cells are clearly distributed
both between the neural tube and the dermomyotome (dm) and between two consecutive
somites. Intersomitic boundaries are marked by *arrows*. a, aorta; e, ectoderm.

epithelia, except in the digestive tract, where crest cells migrate among
the delaminated splanchnopleural mesoderm (Tucker and Thiery, 1984;
Thiery et al., 1985). In this respect, it is interesting to note that at the
onset of crest cell migration epithelial structures predominate in the embryo.
This is particularly obvious in the trunk, where the cessation of crest cell
movement correlates with the disruption of epithelia into dense mesenchyme.

The chemical composition of the ECM is in part responsible for the
direction of crest cell migration. For example, crest cells do not invade the
notochordal area and the space between the ectoderm and dermomyotome

(see Figure 4D), areas rich in FN and particularly in CS (Derby, 1978). Thus, the presence of FN in cellular spaces limited by basal surfaces is necessary for crest cell migration but not sufficient.

On the other hand, the pattern of the pathways of migration is greatly influenced by the morphogenesis of the embryo. In the head, the presence of local thickening of the neural tube (optic vesicle) or of the ectodermal placodes (particularly the otic placode), prevent ventral migration of crest cells, which locally accumulate and then migrate more rostrally or caudally (see Figure 3A, B). At other cephalic levels (especially in the mesencephalon), crest cells utilize an acellular space under the ectoderm without invading the loose mesenchyme (Figure 3C, D). This pathway leads them to the ventral areas of the head, where they will provide the mesectodermal tissues of the face.

In the trunk, the migration of crest cells is mainly guided by the metamerized structures of the somites. Crest cells use two main pathways, depending on their respective location to the somite: between the somite and the neural tube, and between two consecutive somites (Figures 4, 5). Those using the first pathway are rapidly arrested by the expanding sclerotome, a mesenchymal structure that derives from the dissociating somite (Figure 4B). However, crest cells do not appear to invade most parts of the sclerotome although the latter synthesizes FN. This is probably due to several factors: the cell density of the sclerotome, the three-dimensional organization of the matrix around the sclerotomal cells, the presence of the remaining LN-rich basal lamina along the medial side of the sclerotome (see Figure 4D), and the migratory properties of crest cells themselves. As a consequence of this obstacle, as well as of the increase of the crest population, crest cells aggregate and differentiate into sensory ganglia.

The intersomitic pathways lead crest cells to the aorta (Figure 5), along which they localize and give rise to the sympathetic cells. Crest cells using this route are rapidly separated from crest cells taking the midsomite pathway by the sclerotome that fills the intersomitic space and the perinotochordal area. A third pathway along the basement membrane of the myotome allows crest cells to reach the ventral areas of the embryo where they become the Schwann cells of growing nerves (see Figure 7A, B).

In more rostral regions (somites 1–5), the development of the somites also influences crest cell migration but in a different way: The onset of crest cell migration is delayed in relation to the dissociation of the somite. As a consequence, they do not migrate ventrally; instead, they move laterally between the ectoderm and the dermomyotome, and between the sclerotome and the dermomyotome (Thiery et al., 1982a; Vincent and Thiery, 1984).

Role of FN in the Behavior of Crest Cells.   The importance and the role of FN in crest cell displacement have been approached in both *in vitro* and *in vivo* experiments (Rovasio et al., 1983; Boucaut et al., 1984b).

In addition to the adhesive interactions between crest cells and FN, FN appears to significantly influence the motile properties of these cells. Crest cells explanted onto a pure FN substrate rapidly initiate migration and assume a flattened shape. Conversely, collagen substrates fail to support motility. Very few crest cells can emigrate from a neural tube explanted onto collagen type I. Crest cells also migrate poorly away from neural tubes covered by a collagen gel (Newgreen et al., 1982; Rovasio et al., 1983).

The importance of an FN-containing substrate was further demonstrated by experiments in which crest cells were placed in an FN-depleted medium. When cells were confronted with alternating stripes of FN and glass, they migrated exclusively on the FN stripes (Figure 6). When LN stripes were used instead, few crest cells could move. In normal serum containing some FN, crest cells migrated preferentially in regions containing FN derived from the serum, although some cells tended to aggregate on the LN stripes (Rovasio et al., 1983).

The speed of locomotion of isolated crest cells on different substrata was also measured by time-lapse microcinematography (Rovasio et al., 1983). Crest cells on FN move two to three times faster than on collagen or LN but, surprisingly, the effective distance is unaltered as compared to that on LN or collagen. It should be noted that crest cells, unlike fibroblasts,

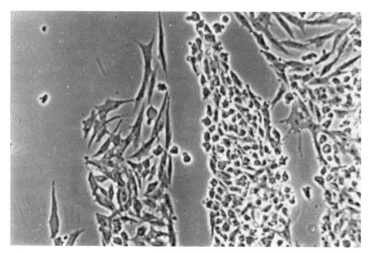

Figure 6. In vitro *migration of neural crest cells on an inhomogeneous FN substrate.* Neural crest cells migrate exclusively in the region of the substratum covered with plasma FN. These pathways resemble the *in vivo* pathways that either become wider or completely obstructed. In all cases, crest cells occupy the full space available in the FN-rich area. Aside from the pioneers, most crest cells remained at high density at all stages. Intense cell proliferation, narrow pathways, and the intrinsic ability to migrate allow the crest cells to direct their migration.

do not exhibit any polarity and thus frequently change their direction of movement.

The directionality of movement can be understood by observing a crest population on FN stripes, which can serve as a model for what happens *in vivo*. It appears that directionality of movement is due to population pressure in the limited suitable area for migration, which prevents backward migration.

The absolute requirement of FN for crest cell movement was proven by perturbation experiments very similar to those described for amphibian gastrulation. Antibodies to FN or small peptides from the binding domain both inhibit crest cell adhesion or movement *in vitro*. When these inhibitors are removed, crest cells reacquire a normal morphology and resume their migration (Rovasio et al., 1983; Boucaut et al., 1984b). When injected in the mesencephalic crest pathways, antibodies or peptides retard the forward stream and provoke crest cell accumulation in the neural lumen. Interestingly, it seems that crest cells blocked in their migration tend to aggregate or use unusual pathways of migration, where the concentration of inhibitor is reduced (i.e., ventrally within the mesenchyme at mesencephalic levels) (Boucaut et al., 1984b; T. J. Poole, K. M. Yamada, and J.-P. Thiery, unpublished observations).

Aggregation of Crest Cells into Ganglia

When the substrate of migration is varied *in vitro* (collagen, serum albumin, denatured FN) crest cells form two- and three-dimensional clusters of closely juxtaposed cells (Thiery et al., 1985). These clusters organize themselves first into epitheliallike structures containing only a few gap junctions, as in the case of dorsal root ganglion rudiment (Aoyama et al., 1984; Thiery et al., 1985). It is therefore unlikely that the formation of ganglion anlagen results from the appearance of differentiated junctions. *In vitro*, the aggregation of crest cells seems a consequence of the lack of space and substratum for migration and/or of the appearance of cell adhesion molecules.

In vivo, different mechanisms may apply for the aggregation of crest cells into sensory and autonomic ganglia (Duband et al., 1984). As already mentioned, crest cells give rise to the dorsal root ganglia, and several cranial sensory ganglia accumulate in a pouch. Physical barriers and the lack of a favorable substrate may be responsible for the cessation of movement. In this area, the ECM is greatly modified: FN, collagens type I and III, and HA disappear while the amount of CS increases. These crest cells reach a very high density before expressing the neural cell adhesion molecule (N-CAM) at their surface (Figure 7C). Thus, N-CAM expression is probably only a secondary event in the further compaction of the sensory ganglion rudiment.

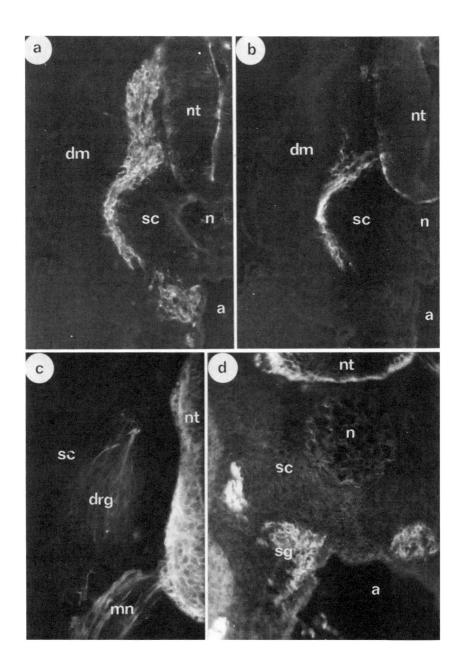

186

In contrast, the sympathetic rudiments form in the FN-rich area of the aorta, even though the mesentery would provide a more ventral pathway. Furthermore, sympathetic precursors express N-CAM just as they reach the aorta and rapidly collect into small clusters (Figure 7D). Neuronal markers such as tyrosine hydroxylase appear in those cells while they are still actively dividing (Rothman et al., 1978). A similar situation is observed in the ontogeny of the enteric plexuses. Finally, the ciliary ganglion, a parasympathetic ganglion associated with the eye, derives from crest precursors expressing N-CAM prior to their arrest. In this case too, there is no obvious physical barrier or lack of substratum that could explain their precise final localization. It remains to be shown that N-CAM is involved in the aggregation and in the maintenance of the aggregated state of the precursors of both neurons and supportive cells of the peripheral nervous system.

Interestingly, the neuron–glia cell adhesion molecule (Ng-CAM; Figure 7B) is not expressed in both ganglion rudiments marked with NC-1 immunolabeling (Figure 7A). Ng-CAM, known to mediate specific interaction between neurons and glia (see Edelman, this volume), is first found at the stage at which motor nerves are emerging from the ventral neural tube. Ng-CAM appears subsequently at the surface of neurons of the peripheral nervous sytem following their birth date (Thiery et al., 1984b). In contrast to N-CAM, Ng-CAM is not likely to be involved in the formation of ganglion rudiments; rather it is expressed when neuron–glia interactions develop.

CELL ADHESION MOLECULES DURING THE ESTABLISHMENT AND PATTERNING OF THE NEUROECTODERM

During gastrulation, whole sheets of cells are rapidly displaced. The mechanism of adhesion between cells in a sheet is now amenable to molecular analysis (see other chapters in this volume). First, well-defined junctions

---

Figure 7. *Appearance of CAMs among the cells of developing peripheral ganglia. a and b:* Immunostaining for NC-1 and Ng-CAM, respectively, in a stage-20 embryo (transverse section; stages according to Hamburger and Hamilton, 1951). Many crest cells are aggregated into sensory ganglion along the neural tube (nt) and sympathetic ganglion next to the aorta (a), while some of them are distributed all along the nerves. *b:* As assessed by the presence of Ng-CAM, numerous nerve fibers have emerged from the ventral part of the neural tube and progressed in the sclerotome (sc). *c:* Immunostaining for N-CAM at the midtrunk level of an embryo at stage 23. The primordium of the sensory ganglion (drg) is stained faintly as compared to the motor nerves (mn) and the motoneurons in the neural tube. *d:* Immunostaining for N-CAM at the midtrunk level of an embryo at stage 21. Crest cells are aggregated near the aorta to form the primordium of the sympathetic ganglion (sg) and exhibit an intense staining for N-CAM (compare with the sensory ganglion in *a–c*). dm, dermomyotome; n, notochord.

such as desmosomes may contribute to the maintenance of an epithelial sheet (Burnside, 1971; Revel and Brown, 1975). Second, cell surface molecules may provide an additional or an independent mechanism for binding between cells (Edelman, 1983).

Two mechanisms for cell–cell adhesion have been described. The first requires calcium and depends primarily on the liver cell adhesion molecule (L-CAM), isolated originally from chick embryo hepatocytes (Gallin et al., 1983). A very similar, if not identical, molecule was found in mammals (Hyafil et al., 1981; Damsky et al., 1983; Yoshido-Noro et al., 1984). L-CAM, a 124-kD glycoprotein, is detected at an early stage of development in the chick embryo (Edelman et al., 1983) and in the fertilized egg in mammals (F. Hyafil, unpublished observations). Its molecular weight remains identical in the different tissues that contain L-CAM throughout embryo-genesis and during adulthood (Thiery et al., 1984a).

The second mechanism does not depend on calcium and involves another cell surface glycoprotein, N-CAM, which was first described in neural tissues (Thiery et al., 1977). The apparent molecular weight of N-CAM decreases during maturation of the nervous system as a result of an important reduction in its carbohydrate content from 30 to 10% in weight; the transition from the embryonic to the adult form of N-CAM corresponds mostly to the loss of sialic acid (Hoffman et al., 1982; Rothbard et al., 1982). The structure of N-CAM, now under current investigation, contains several distinct domains. The binding region is located in the amino-terminal domain, while most of the carbohydrate, organized into several chains of polysialic acid, is localized in the central domain (for more detail see Edelman, this volume).

N-CAM and L-CAM molecules are expressed in a variety of tissues at specific times in development. N-CAM can be detected during early blas-toderm formation in the chick (Edelman et al., 1983). During neural in-duction, L-CAM disappears progressively from the neural plate while N-CAM is enriched in this area. Later, L-CAM accumulates in the nonneural ectoderm and in the endoderm. During organogenesis, L-CAM is maintained in the skin but disappears from invaginated placodes. It is also expressed at the surface of all other developing epithelia, including the kidney tubules and endodermal-derived tissues such as the thyroid, the thymus, the liver, the pancreas, the bursa of Fabricius, and all the endodermal elements of the digestive tract (Thiery et al., 1984a).

N-CAM is found transiently in many tissues of the mesoderm; it increases in amount in the myotomes and in the kidney rudiments. In the ectoderm, neurogenic placodes are well labeled for N-CAM as well as other ectodermal regions, especially in the head (Thiery et al., 1982b).

Special attention has been paid to the peripheral nervous system. It was shown by immunofluorescent labeling that L-CAM, N-CAM, and Ng-CAM are all expressed at different stages of development of the neural crest (Figure 8): L-CAM is likely to be lost permanently from crest cell

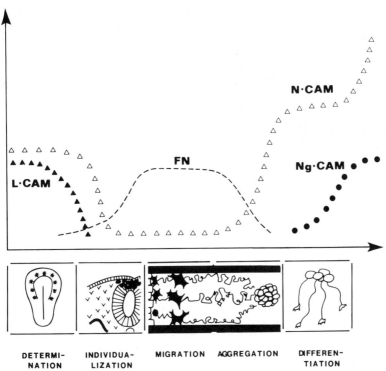

DETERMI-   INDIVIDUA-   MIGRATION   AGGREGATION   DIFFEREN-
NATION   LIZATION   TIATION

Figure 8. *Diagram summarizing neural crest appearance, migration, and differentiation as related to the expression of adhesive properties.* Prior to neural induction, presumptive crest cells express both L-CAM and N-CAM. At the time of their separation from the neural tube, crest cells first lose the adhesion mediated by L-CAM and then progressively that mediated by N-CAM to acquire much greater adhesivity to FN during their migration. When crest cells aggregate into ganglion primordium, the ratio of adhesion to FN versus N-CAM is inverted. The aggregation phase is soon followed by differentiation into neurons characterized by the appearance of Ng-CAM and neurofilaments.

ancestors during early neurogenesis; N-CAM is only reduced at the surface of crest cells during the migratory phase and reappears early during aggregation into ganglion rudiment; Ng-CAM, in contrast, is found on neurons as soon as they leave the cell cycle (Thiery et al., 1984b). Interestingly, adhesive properties of crest cells to fibronectin increase transiently during their migratory phase. This behavior may result from the *de novo* and transient expression of FN receptors.

As in the process of neural induction, the highly dynamic interactive processes occuring during neural crest cell migration have not yet received a detailed molecular analysis. It will be particularly challenging to decipher the mechanisms that control the expression of these different adhesive properties. This analysis will conceivably bring some new insights in one of the most fundamental processes of embryogenesis: epithelium–mes-

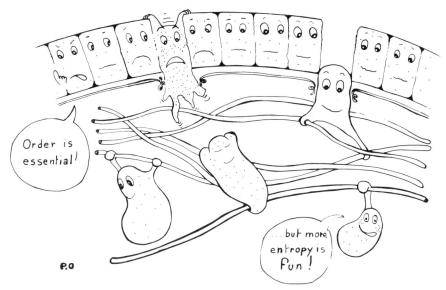

Figure 9. *Epithelium–mesenchyme interconversion.* At specific sites, cells detach from the epithelial layer and migrate within the fibrillar structures of the extracellular matrix. Cell–substrate adhesion can be transient (*hanging cells*) or permanent (*sleeping cells*). In other regions of the epithelium, cells that have not acquired the ability to migrate will not detach from the epithelium, even though the basal lamina has been destroyed locally (see Sugrue and Hay, 1981, 1982, for the behavior of these cells). The transition between epithelium and mesenchyme involves important changes in the adhesive properties of the cells. (Drawn by Bruno Péault.)

enchyme transition (Figure 9). Local anisotropy in cell sheets may define physical boundaries, lines of shearing, which in turn may contribute to the modulation of cell–cell adhesive interactions. The presence of two distinct adhesion mechanisms may facilitate local remodeling in cell sheets while preserving the integrity of the epithelia. What remains a total mystery so far is the intrinsic ability of cells to acquire a new shape, to individualize, and to migrate.

ACKNOWLEDGMENTS

Work from this laboratory was supported by the Centre National Recherche Scientifique, the Institut National de la Santé et de la Recherche Médicale, the Ministère de l'Industrie et de la Recherche, the Ligue Nationale Francaise contre le Cancer, and the Fondation pour la Recherche Médicale. Leslie Blair and Françoise Dieterlen reviewed the manuscript, Monique Denoyelle and Chantal Debain provided excellent technical assistance. Typing and illustration were done by Lydie Obert, Stephane Ouzounoff, and Sophie Tissot.

REFERENCES

Aoyama, H., A. Delouvée, and J.-P. Thiery (1984) Mechanism of neural crest cell aggregation. *Cell Differ.* (in press).

Ballard, W. W. (1973) Morphogenetic movements in *Salmo gaidneri Richardson*. *J. Exp. Zool.* **184**:27–48.

Ballard, W. W. (1982) Morphogenetic movements and fate maps of the Cypriniform teleost, *Catostomus commersoni*. *J. Exp. Zool.* **219**:301–321.

Ballard, W. W., and A. S. Ginsburg (1980) Morphogenetic movements in Acipenserid embryos. *J. Exp. Zool.* **213**:69–103.

Boucaut, J. C., and T. Darribère (1983) Fibronectin in early amphibian embryos. *Cell Tissue Res.* **234**:135–145.

Boucaut, J. C., T. Darribère, H. Boulekbache, and J.-P. Thiery (1984a) Antibodies to fibronectin prevent gastrulation but do not perturb neurulation in gastrulated amphibian embryos. *Nature* **307**:364–367.

Boucaut, J. C., T. Darribère, R. J. Poole, H. Aoyama, K. M. Yamada, and J.-P. Thiery (1984b) Biologically active synthetic peptides as probes of embryonic development: A competitive inhibitor of fibronectin function inhibits gastrulation in amphibian embryos and neural crest cell migration in avian embryos. *J. Cell Biol.* **99**:1822–1830.

Burnside, B. (1971) Microtubules and microfilaments in newt neurulation. *Dev. Biol.* **26**:416–444.

Cochard, P., and P. Coltey (1983) Cholinergic traits in the neural crest: Acetylcholinesterase in crest cells of the chick embryo. *Dev. Biol.* **98**:221–238.

Cohen, A. M., and E. D. Hay (1971) Secretion of collagen by embryonic neuroepithelium at the time of spinal cord somite interaction. *Dev. Biol.* **26**:578–605.

Couchman, J. R., D. A. Rees, M. R. Green, and C. G. Smith (1982) Fibronectin has a dual role in locomotion and anchorage of primary chick fibroblasts and can promote entry into the division cycle. *J. Cell Biol.* **93**:402–410.

Critchley, D. R., M. A. England, J. Wakely, and R. O. Hynes (1979) Distribution of fibronectin in the ectoderm of gastrulating chick embryo. *Nature* **280**:498–500.

Damsky, C. H., J. Richa, D. Solter, K. Knudsen, and C. A. Buck (1983) Identification and purification of a cell surface glycoprotein mediating intercellular adhesion in embryonic and adult tissue. *Cell* **34**:455–466.

Darribère, T., D. Boucher, J. C. Lacroix, and J. C. Boucaut (1984) Fibronectin synthesis during oogenesis and early development of the amphibian *Pleurodeles waltlii*. *Cell Differ.* **14**:171–177.

Del Pino, E., and R. P. Elinson (1983) A novel development pattern for frogs: Gastrulation produces an embryonic disk. *Nature* **306**:589–591.

Derby, M. A. (1978) Analysis of glycosaminoglycans within the extracellular environments encountered by migrating neural crest cells. *Dev. Biol.* **66**:321–336.

Duband, J. L., J.-P. Thiery (1982a) Appearance and distribution of fibronectin during chick embryo gastrulation and neurulation. *Dev. Biol.* **94**:337–350.

Duband, J. L., and J.-P. Thiery (1982b) Distribution of fibronectin in the early phase of avian cephalic neural crest cell migration. *Dev. Biol.* **93**:308–323.

Duband, J. L., G. C. Tucker, T. J. Poole, M. Vincent, H. Aoyama, and J.-P. Thiery (1984) How do the migratory and adhesive properties of the neural crest govern ganglia formation in the avian peripheral nervous system? *J. Cell. Biochem.* **27**:189–203.

Duprat, A. M., and L. Galandris (1984) Extracellular matrix and neural determination during amphibian gastrulation. *Cell Differ.* **14**:105–112.

Edelman, G. M. (1983) Cell adhesion molecules. *Science* **219**:450–457.

Edelman, G. M., W. Gallin, A. Delouvée, B. A. Cunningham, and J.-P. Thiery (1983) Early epochal maps of two different cell adhesion molecules. *Proc. Natl. Acad. Sci. USA* **80**:4384–4388.

Eyal-Giladi, H., and M. Koshav (1976) From cleavage to primitive streak formation. A complementary normal table and a new look at the first stages of the development of the chick. I. General morphology. *Dev. Biol.* **49**:321–337.

Fisher, M., and M. Solursh (1979) The influence of the substratum on the mesenchyme spreading *in vitro. Exp. Cell Res.* **123**:1–14.

Fontaine, J., and N. M. Le Douarin (1977) Analysis of endoderm formation in the avian blastoderm by the use of quail–chick chimeras. The problem of the neurectodermal origin of the cells of the APUD series. *J. Embryol. Exp. Morphol.* **41**:209–222.

Gallin, W. J., G. M. Edelman, and B. A. Cunningham (1983) Characterization of L-CAM, a major cell adhesion molecule from embryonic liver cells. *Proc. Natl. Acad. Sci. USA* **80**:1038–1042.

Greenberg, J. H., and R. M. Pratt (1977) Glycosaminoglycan and glycoprotein synthesis by cranial neural crest cells *in vitro. Cell Differ.* **6**:119–132.

Hamburger, V., and H. L. Hamilton (1951) A series of normal stages in the development of the chick embryo. *J. Morphol.* **88**:49–92.

Harrisson, F., C. H. Vanroelen, J. M. Foidart, and L. Vakaet (1984) Expression of different regional patterns of fibronectin immunoreactivity during mesoblast formation in the chick blastoderm. *Dev. Biol.* **101**:373–381.

Hirano, H., Y. Yamada, M. Sullivan, B. De Crombrugghe, I. Pastan, and K. M. Yamada (1983) Isolation of genomic DNA clones spanning the entire fibronectin gene. *Proc. Natl. Acad. Sci. USA* **80**:46–50.

Hoffman, S., B. C. Sorkin, P. C. White, R. Brackenbury, R. Mailhammer, U. Rutishauser, B. A. Cunningham, and G. M. Edelman (1982) Chemical characterization of a neural cell adhesion molecule purified from embryonic brain membranes. *J. Biol. Chem.* **257**:7720–7729.

Holtfreter, J. (1943) A study of the mechanisms of gastrulation. Part I. *J. Exp. Zool.* **94**:261–318.

Holtfreter, J. (1944) A study of the mechanics of gastrulation. Part II. *J. Exp. Zool.* **95**:171–212.

Hyafil, F., C. Babinet, and F. Jacob (1981) Cell–cell interactions in early embryogenesis: A molecular approach to the role of calcium. *Cell* **26**:447–454.

Jacobson, A. G. (1978) Some forces that shape the nervous system. *ZOON* **6**:13–21.

Jacobson, A. G. (1981) Morphogenesis of the neural plate and tube. In *Morphogenesis and Pattern Formation,* T. G. Connelly, ed., pp. 233–263, Raven, New York.

Jacobson, C. O. (1959) The localization of the presumptive cerebral region in the neural plate of the Axolotl larva. *J. Embryol. Exp. Morphol.* **7**:1–21.

Johnson, K. E. (1977) Extracellular matrix synthesis in blastula and gastrula stages of normal and hybrid frog embryos. III. Characterization of galactose- and glucosamine-labeled materials. *J. Cell Sci.* **25**:335–354.

Keller, R. E. (1975) Vital dye mapping of the gastrula and neurula of *Xenopus laevis*. I. Prospective areas and morphogenetic movements of the superficial layer. *Dev. Biol.* **42**:222–241.

Keller, R. E. (1976) Vital dye mapping of the gastrula and neurula of *Xenopus laevis*. II. Prospective areas and morphogenetic movements of the deep layer. *Dev. Biol.* **51**:118–137.

Keller, R. E. (1978) Time-lapse cinematographic analysis of superficial cell behaviour during and prior to gastrulation in *Xenopus laevis*. *J. Morphol.* **157**:223–248.

Keller, R. E. (1980) The cellular basis of epiboly: An SEM study of deep-cell rearrangement during gastrulation in *Xenopus laevis*. *J. Embryol. Exp. Morphol.* **60**:201–234.

Keller, R. E. (1981) An experimental analysis of the role of bottle cells and the deep marginal zone in gastrulation of *Xenopus laevis*. *J. Exp. Zool.* **216**:81–101.

Keller, R. E., and G. C. Schoenwolf (1977) A SEM study of cellular morphology, contact, and arrangement, as related to gastrulation in *Xenopus laevis*. *Wilhelm Roux' Arch.* **182**:165–186.

Kornblihtt, A. R., K. Vibe-Pedersen, and F. E. Baralle (1984a) Human fibronectin: Molecular cloning evidence for two mRNA species differing by an internal segment coding for a structural domain. *EMBO J.* **3**:221–226.

Kornblihtt, A. R., K. Vibe-Pedersen, and F. E. Baralle (1984b) Human fibronectin: Cell-specific alternative mRNA splicing generates polypeptide chains differing in the number of internal repeats. *Nucleic Acids Res.* **12**:5853–5868.

Kubota, H. Y., and A. J. Durston (1978) Cinematographical study of cell migration in the opened gastrula of *Ambystoma mexicanum*. *J. Embryol. Exp. Morphol.* **44**:71–80.

Le Douarin, N. M. (1982) *The Neural Crest*, Cambridge Univ. Press, England.

Le Douarin, N. M., and M. A. Teillet (1974) Experimental analysis of the migration and differentiation of neuroblast of the autonomic nervous system and of neurectodermal mesenchymal derivatives using a biological cell marking technique. *Dev. Biol.* **41**:162–184.

Le Douarin, N. M., P. Cochard, M. Vincent, J. L. Duband, G. C. Tucker, M. A. Teillet, and J.-P. Thiery (1984) Nuclear cytoplasmic and membrane markers to follow neural crest cell migration. A comparative study. In *The Role of Extracellular Matrix in Development*, R. L. Treslstad, ed., pp. 373–398, Alan R. Liss, New York.

Lee, G., R. Hynes, and M. Kirschner (1984) Temporal and spatial regulation of fibronectin in early *Xenopus* development. *Cell* **36**:729–740.

Leivo, I., A. Vaheri, R. Timpl, and J. Wartiovaara (1980) Appearance and distribution of collagens and laminin in the early mouse embryo. *Dev. Biol.* **76**:100–114.

Löfberg, J., K. Ahlfors, and C. Fällstrom (1980) Neural crest cell migration in relation to extracellular matrix organization in the embryonic axolotl trunk. *Dev. Biol.* **75**:148–167.

Mitrani, E., and A. Farberov (1982) Fibronectin expression during the process leading to axis formation in the chick embryo. *Dev. Biol.* **91**:197–210.

Nakatsuji, N. (1975) Studies on the gastrulation of amphibian embryos: Light and electron microscopic observation of a urodele *Cynops pyrrhogaster*. *J. Embryol. Exp. Morphol.* **34**:669–685.

Nakatsuji, N., and K. E. Johnson (1983) Comparative study of extracellular fibrils of the ectodermal layer in gastrulae of five amphibian species. *J. Cell Sci.* **59**:61–70.

Newgreen, D. F., and I. L. Gibbins (1982) Factors controlling the time of onset of the migration of neural crest cells in the fowl embryo. *Cell Tissue Res.* **224**:145–160.

Newgreen, D. F., and J.-P. Thiery (1980) Fibronectin in early avian embryos: Synthesis and distribution along the migration pathways of neural crest cells. *Cell Tissue Res.* **211**:269–291.

Newgreen, D. F., I. L. Gibbins, J. Sauter, B. Wallenfels, and R. Wütz (1982) Ultrastructural and tissue-culture studies on the role of fibronectin, collagen, and glycosaminoglycans in the migration of neural crest cells in the fowl embryo. *Cell Tissue Res.* **221**:521–549.

Nichols, D. M. (1981) Neural crest formation in the head of the mouse embryo as observed using a new histological technique. *J. Embryol. Exp. Morphol.* **64**:105–120.

Nieuwkoop, P. D., and L. A. Sutasurya (1979) *Primordial Germ Cells in the Chordates,* Cambridge Univ. Press, Cambridge.

Noden, D. M. (1975) An analysis of the migratory behavior of avian cephalic neural crest cells. *Dev. Biol.* **42**:106–130.

Pierschbacher, M. D., and E. Ruoslahti (1984) Cell attachment activity of fibronectin can be duplicated by small synthetic fragments of the molecule. *Nature* **309**:30–33.

Pierschbacher, M. D., E. G. Hayman, and E. Ruoslahti (1981) Location of the cell attachment site in fibronectin with monoclonal antibodies and proteolytic fragments of the molecule. *Cell* **26**:259–267.

Pintar, J. E. (1978) Distribution and synthesis of glycosaminoglycans during quail neural crest morphogenesis. *Dev. Biol.* **67**:444–464.

Revel, J.-P., and S. S. Brown (1975) Cell junctions in development with particular reference to the neural tube. *Cold Spring Harbor Symp. Quant. Biol.* **40**:443–455.

Rosenquist, G. C. (1966) A radioautographic study of labelled grafts in the chick blastoderm. In *Development of Primitive Streak Stages to Stage 12,* Vol. XXXVIII, pp. 71–110, Publication 625, Contributions to Embryology, Carnegie Institution of Washington.

Rosenquist, G. C. (1981) Epiblast origin and early migration of neural crest cells in the chick embryo. *Dev. Biol.* **87**:201–211.

Rothbard, J. B., R. B. Brackenbury, B. A. Cunningham, and G. M. Edelman (1982) Differences in the carbohydrate structures of neural cell-adhesion molecules from adult and embryonic chicken brains. *J. Biol. Chem.* **257**:11064–11069.

Rothman, T., M. D. Gershon, and M. Holtzer (1978) The relationship of cell division to the acquisition of adrenergic characteristics by developing sympathetic ganglion cell precursors. *Dev. Biol.* **65**:322–341.

Rovasio, R. A., A. Delouvée, K. M. Yamada, R. Timpl, and J.-P. Thiery (1983) Neural crest cell migration: Requirement for exogenous fibronectin and high cell density. *J. Cell Biol.* **96**:462–473.

Rubin, K., M. Höök, B. Obrink, and R. Timpl (1981) Substrate adhesion of rat hepatocytes: Mechanisms of attachment to collagen substrates. *Cell* **24**:463–470.

Sanders, E. J. (1979) Development of the basal lamina and extracellular material in the chick embryo. *Cell Tissue Res.* **198**:527–537.

Sanders, E. J. (1982) Ultrastructural immunocytochemical localization of fibronectin in the early chick embryo. *J. Embryol. Exp. Morphol.* **71**:155–170.

Sanders, E. J. (1984) Labelling of basement membrane constituents in the living chick embryo during gastrulation. *J. Embryol. Exp. Morphol.* **79**:113–123.

Saxen, L., and S. Toivonen (1962) *Primary Embryonic Induction*, Academic, New York.

Schwarzbauer, J. E., J. W. Tamkun, I. R. Lemischka, and R. O. Hynes (1983) Three different fibronectin mRNAs arise by alternative splicing within the coding region. *Cell* **35**:421–431.

Solursh, M., and J.-P. Revel (1978) A scanning electron microscope study of the cell shape and cell appendages in the primitive streak region of the rat and chick embryo. *Differentiation* **11**:185–190.

Spratt, N. T., and M. Haas (1965) Germ layer formation and the role of the primitive streak in the chick. *J. Exp. Zool.* **158**:9–38.

Sugrue, S. P., and E. D. Hay (1981) Response of basal epithelial cell surface and cytoskeleton to solubilized extracellular matrix molecules. *J. Cell Biol.* **91**:45–54.

Sugrue, S. P., and E. D. Hay (1982) Interaction of embryonic corneal epithelium with exogenous collagen, laminin, and fibronectin: Role of endogenous protein synthesis. *Dev. Biol.* **92**:97–106.

Thiery, J.-P. (1984) Mechanism of cell migration in the vertebrate embryo. *Cell Differ.* **15**:1–15.

Thiery, J.-P., R. Brackenbury, U. Rutishauser, and G. M. Edelman (1977) Adhesion among neural cells of the chick embryo. II. Purification and characterization of a cell adhesion molecule from neural retina. *J. Biol. Chem.* **252**:6841–6845.

Thiery, J.-P., J.-L. Duband, and A. Delouvée (1982a) Pathways and mechanism of avian truck neural crest cell migration and localization. *Dev. Biol.* **93**:324–343.

Thiery, J.-P., J.-L. Duband, U. Rutishauser, and G. M. Edelman (1982b) Cell adhesion molecules in early chicken embryogenesis. *Proc. Natl. Acad. Sci. USA* **79**:6737–6741.

Thiery, J.-P., A. Delouvée, W. Gallin, B. A. Cunningham, and G. M. Edelman (1984a) Ontogenetic expression of cell adhesion molecules: L-CAM is found in epithelia derived from the three primary germ layers. *Dev. Biol.* **102**:61–78.

Thiery, J.-P., A. Delouvée, M. Grumet, and G. M. Edelman (1984b) Initial appearance and regional distribution of the neuron–glia cell adhesion molecule (Ng-CAM) in the chick embryo. *J. Cell Biol.* **100**:442–456.

Thiery, J.-P., G. C. Tucker, and H. Aoyama (1985) Gangliogenesis in the avian embryo: Migration and adhesion properties of neural crest cells. In *Molecular Bases of Neural Development*, G. M. Edelman, W. E. Gall, and W. M. Cowan, eds., pp. 181–211, Wiley, New York.

Tosney, K. W. (1978) The early migration of neural crest cells in the trunk region of the avian embryo. An electron microscopic study. *Dev. Biol.* **62**:317–333.

Tosney, K. W. (1982) The segregation and early migration of cranial neural crest cells in the chick embryo. *Dev. Biol.* **16**:78–106.

Trelstad, R. L., E. D. Hay, and J.-P. Revel (1967) Cell contact during early morphogenesis in the chick embryo. *Dev. Biol.* **16**:78–106.

Trinkaus, J. P. (1984) *Cells Into Organs: The Forces That Shape The Embryo*, Prentice-Hall, Englewood Cliffs, N.J.

Tucker, G. C., H. Aoyama, M. Lipinski, T. Tursz, and J.-P. Thiery (1984) Identical reactivity of monoclonal antibodies HNK-1 and NC-1: Conservations in vertebrates on cells derived from the neural primordium and on some leukocytes. *Cell Differ.* **14**:223–230.

Tucker, G. C., M. Vincent, and J.-P. Thiery (1984) Mechanisms of neural crest cell migration in the developing gut. *Dev. Biol.* (submitted).

Vakaet, L. (1962) Some new data concerning the formation of the definitive endoblast in the chick embryo. *J. Embryol. Exp. Morphol.* **10**:38–57.

Vakaet, L. (1970) Cinematographic investigations of gastrulation in the chick blastoderm. *Arch. Biol.* **81**:387–426.

Vakaet, L. (1984) Early development of birds. In *Chimeras in Developmental Biology*, N. M. Le Douarin and A. McLaren, eds., pp. 71–87, Academic, New York.

Vakaet, L., C. Vanroelen, and L. Andries (1980) An embryological model of non-malignant invasion or ingression. In *Cell Movement and Neoplasia*, M. De Brabander et al., eds., pp. 65–75, Pergamon, Oxford.

Vanroelen, C., L. Vakaet, and L. Andries (1980) Localization and characterization of acid mucopolysaccharides in the early chick blastoderm. *J. Embryol. Exp. Morphol.* **56**:169–178.

Vincent, M., and J.-P. Thiery (1984) A cell surface marker for neural crest and placodal cells: Further evolution in peripheral and central nervous system. *Dev. Biol.* **103**:468–481.

Vincent, M., J. L. Duband, and J.-P. Thiery (1983) A cell surface determinant expressed early on migrating avian neural crest cells. *Dev. Brain Res.* **9**:235–238.

Vogt, W. (1929) Gestaltunganalyse am Amphibienkeim mit örtlicher Vitalförbung. II. Teil Gastrulation und Mesodermbildung bei Urodelen und Anuren. *Wilhelm Roux' Arch. Entwicklungsmech. Org.* **120**:384–706.

Wartiovaara, J., I. Leivo, and A. Vaheri (1979) Expression of the cell surface-associated glycoprotein, fibronectin, in the early mouse embryo. *Dev. Biol.* **69**:247–257.

Weston, J. A. (1963) A radioautographic analysis of the migration and localization of trunk crest cells in the chick. *Dev. Biol.* **6**:279–310.

Wu, T. C., Y. J. Wan, A. E. Chung, and I. Damjanov (1983) Immunohistochemical localization of entactin and laminin in mouse embryos and fetuses. *Dev. Biol.* **100**:496–505.

Yamada, K. M., and D. W. Kennedy (1984) Dualistic nature of adhesive protein function: Fibronectin and its biologically active peptide fragments can auto-inhibit fibronectin function. *J. Cell Biol.* **99**:29–36.

Yoshido-Noro, C., N. Susuki, and M. Takeichi (1984) Molecular nature of the calcium-dependent cell–cell adhesion system in mouse teratocarcinoma and embryonic cells studied with a monoclonal antibody. *Dev. Biol.* **101**:19–27.

Chapter 9

Structures of Cell Adhesion Molecules

BRUCE A. CUNNINGHAM

ABSTRACT

Structural features of the cell adhesion molecules N-CAM, L-CAM, and Ng-CAM have been determined by chemical and molecular biological techniques. The neural cell adhesion molecule, N-CAM, is distinguished by a large amount of polysialic acid that is covalently attached to one or more of the three asparagine-linked oligosaccharides in the center of the linear structure. This sialic acid can influence N-CAM binding even though it is not in the amino-terminal binding domain, and a decrease in sialic acid is the primary, if not exclusive, structural change in the conversion of N-CAM from an embryonic to an adult form. Some of the asparagine-linked oligosaccharides are also sulfated. The two polypeptides of chicken N-CAM (130 kD and 160 kD) are apparently specified by separate mRNAs derived from a single gene. They differ from each other in their carboxy-terminal regions, which include the portion of the molecule that associates with the membrane. Some amino acids in this portion of the molecule can be phosphorylated or acylated with fatty acids, consistent with the notion that N-CAM is an integral membrane protein and providing additional sites for possible regulation of N-CAM expression and activity. The neuron–glia cell adhesion molecule, Ng-CAM, is also found on neurons, but it differs chemically and functionally from N-CAM with three polypeptides (80 kD, 135 kD, and 200 kD), at least one (135 kD) of which contains asparagine-linked oligosaccharides. Two of the components (80 kD and 200 kD) can be phosphorylated in vitro. *Immunological studies suggest that the two smaller components may be derived from the larger one and that N-CAM and Ng-CAM share at least one antigenic determinant.*

The liver cell adhesion molecule, L-CAM, is distinct from both neuronal CAMs. It contains a single polypeptide (110 kD) that has three complex and one high-mannose asparagine-linked oligosaccharides, but little sialic acid. The molecule is specified by a single mRNA in all of the tissues in which it is detected, but there may be two or three L-CAM genes. Phosphoserine and phosphothreonine are detected in L-CAM but not in the portion of the molecule (fragment Ft1, 81 kD) released from membranes by trypsin, consistent with a variety of data indicating that L-CAM is an integral membrane protein. Both L-CAM-mediated adhesion and the integrity of the L-CAM molecule are dependent on calcium, providing a distinct basis for the regulation of L-CAM activity.

The structural features of the CAMs include a variety of properties that might be involved in the regulation of the expression and activity of each CAM independently or coordinately during embryological development.

The formation of tissues and organs during development is dependent on the binding of cells to other cells and to extracellular substrates. These adhesive interactions have been studied with increasing intensity over the last decade and hold considerable promise for detailed analyses by current immunological, biochemical, and molecular biological techniques.

Our studies have focused on cell adhesion molecules (CAMs) originally detected in two tissues, brain and liver (Edelman, this volume). These include the neural cell adhesion molecule, N-CAM, which mediates calcium-independent adhesion between neuronal cells (Brackenbury et al., 1981) by a homophilic (N-CAM to N-CAM) mechanism (Rutishauser et al., 1982; Edelman, 1983). N-CAM was initially detected in neural retina (Brackenbury et al., 1977; Thiery et al., 1977; Rutishauser et al., 1978), but it is prevalent in brain and throughout the central and peripheral nervous systems (Thiery et al., 1982; Chuong and Edelman, 1984). It is also found on muscle, where it appears to mediate nerve–muscle interactions *in vitro* by the same homophilic mechanism (Grumet et al., 1982; Rutishauser et al., 1983).

We have also examined the liver cell adhesion molecule, L-CAM, and the neuron–glia cell adhesion molecule, Ng-CAM. L-CAM was detected on the basis of its role in calcium-dependent adhesion between hepatocytes (Bertolotti et al., 1980, Gallin et al., 1983), but it has since been detected in a variety of tissues (Edelman et al., 1983a; Thiery et al., 1984). Ng-CAM has been found only on neuronal cells and mediates calcium-independent adhesion between neurons and glia, apparently by a heterophilic mechanism inasmuch as Ng-CAM has not been detected on glia (Grumet and Edelman, 1984; Grumet et al., 1984a). All of these molecules were initially isolated on the basis of their ability to neutralize antibodies that block cell adhesion. Subsequently, *in vitro* and *in vivo* assays have been used to verify their roles in cell adhesion.

N-CAM and L-CAM are expressed on very early embryonic cells and thus have been designated primary CAMs (Edelman, 1984a,b). Ng-CAM appears during more refined development of the nervous system and has a more restricted distribution than either N-CAM or L-CAM (Thiery et al., 1985); it is therefore designated a secondary CAM. All three glycoproteins persist in adult tissues, but they do not necessarily maintain the same levels, distribution, or chemical form throughout development; in fact, the available evidence indicates that all of these factors can vary, and their alteration may dramatically influence developmental events. For example, N-CAM undergoes a transition from an embryonic (E) to an adult (A) form that involves a graded decrease in its sialic acid content (Edelman and Chuong, 1982; Rothbard et al., 1982) and a concomitant increase in the rate of N-CAM to N-CAM binding (Hoffman and Edelman, 1983). Moreover, N-CAM binding is even more dependent on concentration, so changes in the amount of N-CAM may have dramatic consequences for cell–cell interactions.

Our studies have been carried out primarily on CAMs from embryonic chickens, but all of the molecules have counterparts in a variety of other species. N-CAM has been detected in humans (McClain and Edelman, 1982), rats (Jorgensen et al., 1980), mice (Chuong et al., 1982; Hirn et al., 1983), and frogs (Fraser et al., 1984), as well as sharks, salamanders, newts, turtles, snakes, and lizards (Hoffman et al., 1984). L-CAM shares nearly all of its features with a molecule (uvomorulin) associated with the compaction of mouse blastomeres and the aggregation of teratocarcinoma cells (Hyafil et al., 1980; Yoshida and Takeichi, 1982). A similar protein has also been detected in human mammary carcinoma cells (Damsky et al., 1983). Ng-CAM has been detected in mice and shares most of its features with an antigen (L1) defined by a cell migration assay (Rathjen and Schachner, 1984).

All three CAMs are large cell surface glycoproteins and as such share a number of structural features. They are distinct chemical species, however, in accord with their different activities, cellular distributions, and antigenic properties. To identify features that might regulate CAM expression or activity, we have determined the chemical features of each molecule and have produced cDNA probes (Murray et al., 1984; Gallin et al., 1985) to be used for characterizing the amino acid sequences and regulation of these CAMs.

N-CAM

The most distinctive feature of N-CAM is apparent in its behavior during SDS-polyacrylamide gel electrophoresis (Figure 1A). N-CAM from embryonic chicken brain migrates as a diffuse zone (150–250 kD) (Hoffman et al., 1982) that stains blue with the dye Stains-all, a property of polyanions (Green and Pastewka, 1975). The diffuse migration appears to reflect molecular polydispersity in that material from each part of the diffuse zone migrates to its same position if cut from the gel and rerun. The polydispersity and polyanionic characteristics are both due to large amounts of sialic acid present in an unusual polymeric form (Finne, 1982; Hoffman et al., 1982; Finne et al., 1983). Treatment of N-CAM with neuraminidase reduces the pattern on SDS-polyacrylamide gels to two components (140 kD and 170 kD) that stain red with Stains-all, typical of most proteins. Peptide maps (Figure 1B) indicate that these two glycoproteins are very similar. Treatment of N-CAM with endoglycosidase F to remove all asparagine-linked oligosaccharides gives two polypeptides of 130 kD and 160 kD.

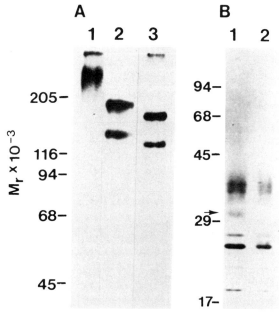

Figure 1. *Molecular properties of N-CAM as detected on SDS-polyacrylamide gels. A:* Electrophoresis on 7.5% polyacrylamide gels of N-CAM (*lane 1*), neuraminidase-treated N-CAM (*lane 2*), and endoglycosidase F-treated N-CAM (*lane 3*) from embryonic chicken brain. *B:* One-dimensional maps of V8 protease peptides from the 170-kD (*lane 1*) and 140-kD (*lane 2*) components obtained after neuraminidase treatment. (Adapted from Hoffman et al., 1982.)

N-CAM from adult tissue (A-form) migrates on SDS-polyacrylamide gels like embryonic (E-form) N-CAM that has been treated with neuraminidase. A variety of data indicate that the primary, if not the only, difference between the E- and A-forms of the molecule is the lower amount of polysialic acid present in the A-form (Rothbard et al., 1982; Rougon et al., 1982). Even the A-form, however, contains significantly more sialic acid than is seen in most glycoproteins.

Nucleic Acid Studies

cDNA was prepared from poly(A)$^+$ RNA obtained from immune-enriched polysomes and was cloned in the plasmid vectors pBR322 and pUC8 (Murray et al., 1984). Synthesis of cDNA for making the pBR322 clones was primed with oligothymidine to obtain sequences at the 3' ends of mRNAs, whereas those for pUC8 were primed with random oligodeoxy-nucleotides from calf thymus DNA.

Bacterial colonies containing recombinant pBR322 plasmids were screened by comparing their hybridization to ^{32}P-labeled cDNA made from enriched mRNA with that made from unenriched poly(A)$^+$ RNA. One colony that reacted preferentially with the enriched mRNA was purified and designated pEC001. RNA selected by hybridization to pEC001 directed the synthesis of the same set of N-CAM polypeptides detected in the translation of immune-enriched mRNA, whereas RNA selected by pBR322 itself or a control probe (p13) (Milner et al., 1983) failed to do so, indicating that pEC001 contains DNA sequences complementary to N-CAM mRNA. We obtained an additional cDNA clone by excising the 640 base-pair insert from pEC001 and using it to screen the pUC8 bacterial colonies. One strain reproducibly hybridized with the pEC001 insert and was designated pEC020. Excision of the inserted DNA followed by agarose gel electrophoresis indicated that the pEC020 insert contained 150–200 base pairs.

To estimate the number of N-CAM mRNAs, poly(A)$^+$ RNA from 9-day-old embryonic chicken brain and 14-day-old embryonic liver was separated by electrophoresis on agarose gels and transferred to nitrocellulose filters. Hybridization was carried out with labeled inserts from both pEC001 and pEC020 (Figure 2A). Both probes hybridized to two large discrete RNA

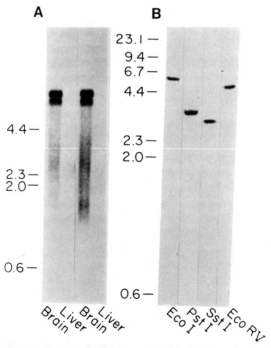

Figure 2. *Blot hybridization analyses of chicken N-CAM mRNA and genomic DNA. A: RNA transfer analysis of poly(A)$^+$ RNA isolated from 10-day-old embryonic chicken brain or liver probed with ^{32}P-labeled pEC001 (left) or pEC020 (right). B: Southern blot analysis of chicken genomic DNA digested with the indicated restriction endonucleases and probed with ^{32}P-labeled pEC001. Sizes (in kilobases) and migration positions of reference DNA fragments are indicated to the left of each panel.*

species in brain; variable amounts of polydisperse, faster-migrating material, assumed to be degradation products of the larger species, were also seen. No components were detected in the liver RNA, consistent with earlier observations that N-CAM is not seen in liver.

The size of the two RNA species (6–7 kb) is sufficient to code for the two polypeptide chains of N-CAM, suggesting that the polypeptides are probably derived from separate mRNAs, rather than the smaller polypeptide being generated from the larger by proteolysis. Earlier data (Cunningham et al., 1983), as well as pulse-chase experiments (Lyles et al., 1984a), support this conclusion.

To estimate the number of N-CAM genes, adult chicken liver DNA was digested with four restriction enzymes that did not cleave the pEC001 insert (Eco RI, Pst I, Sst I, and Eco RV). After electrophoretic separation, the digests were tested for the presence of sequences homologous to the pEC001 insert (Figure 2B). Each digest gave only one hybridizing fragment. The separated Eco RI and Pst I digests were also tested with the pEC020 insert and each gave one hybridizing fragment. These results suggest that there may be only one N-CAM gene and that the two mRNAs might arise by differential splicing. Such differential splicing of a single gene has been described recently for fibronectin and gives rise to a number of fibronectin mRNAs (Tamkun et al., 1984).

These results have recently been confirmed in our laboratory using additional cDNA probes made in the expression vector λgt11. These probes were detected with antibodies to N-CAM and thus contain N-CAM coding sequences. The results were identical to those obtained with the plasmid vectors, verifying that pEC001 and pEC020 are authentic N-CAM probes and supporting the conclusion that N-CAM polypeptides are specified by multiple mRNAs derived from a single gene.

Protein Chemical Studies

Protein chemical data are in good agreement with the conclusions from the studies with the cDNA probes and have allowed us to formulate a working model of the N-CAM molecule (Figure 3). The two N-CAM polypeptides are very similar to each other, having the same amino acid sequences for at least the first 14 residues at the amino terminus and giving similar peptide maps (Cunningham et al., 1983). A variety of data suggest that the 130-kD component probably differs from the 160-kD component in the carboxy-terminal third of the polypeptide, either lacking a segment at the carboxyl terminus of the larger species or having a deletion in this portion of the sequence. For example, an amino-terminal fragment with a molecular weight of 65 kD is common to both polypeptides. This fragment, Fr1, is obtained by incubation of N-CAM in solution at 37°C, causing the molecule to be degraded, apparently by protease activity that is intrinsic to N-CAM or that is a contaminant in N-CAM preparations

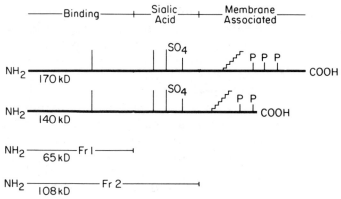

Figure 3. *Molecular model of the N-CAM.* The figure shows the two glycopeptides (170 kD and 140 kD) obtained after neuraminidase treatment, fragment Fr1 produced by spontaneous cleavage of N-CAM in solution, and fragment Fr2 released from cell membranes by V8 protease. *Vertical lines* denote asparagine-linked oligosaccharides. At least one of the oligosaccharides in the central region contains polysialic acid. The *stairstep symbol* denotes covalently bound fatty acid, P indicates phosphate moieties on phosphothreonine and phosphoserine residues, and SO_4 indicates sulfate detected on some asparagine-linked oligosaccharides.

(Hoffman et al., 1982). Fr1 can be isolated from other products on the same monoclonal antibody column used to purify N-CAM. The amino-terminal sequence of Fr1 is the same as that of each of the intact polypeptides (Cunningham et al., 1983), and antibodies that recognize Fr1 recognize both N-CAM polypeptides. Fr1 also appears to include the N-CAM binding region, but it lacks the bulk of the sialic acid. A second fragment, Fr2, can be generated by treating N-CAM with V8 protease. This fragment includes all of Fr1 plus the bulk of the sialic acid; it has a polypeptide segment of 108 kD with the same amino-terminal sequence as the N-CAM polypeptides. This material reacts with the monoclonal antibody anti-N-CAM No. 2, which does not recognize Fr1 and is assumed to be directed against peptide determinants near the sites where the bulk of the carbohydrate is attached. This antibody also recognizes both N-CAM polypeptides (Crossin et al., 1984).

These and other data suggest that the two N-CAM polypeptides are nearly identical except in the carboxy-terminal region, where it is assumed that the protein interacts with the membrane. Whether the two have different cytoplasmic domains or traverse the membrane a different number of times remains to be established. Both N-CAM polypeptides can be phosphorylated *in vitro*, suggesting that both have cytoplasmic domains (Sorkin et al., 1984), and detailed analyses (see below) are consistent with the notion that they differ from each other in this region. Other immunological studies (Gennarini et al., 1984) support this view.

In considering possible mechanisms for modulation of N-CAM activity, transcriptional control of N-CAM levels could be the most effective, because

N-CAM binding appears to be highly dependent on its density on the cell surface (Hoffman and Edelman, 1983). The ability to differentially synthesize two messages offers another possible refinement of control. In some adult tissues, for example retina and muscle, there appears to be a preferential expression of the N-CAM polypeptide with a molecular weight of 140 kD (Hoffman et al., 1982; Rutishauser et al., 1983).

The way in which the N-CAM polypeptides are incorporated into the functional molecule is not clear. Its solubility properties and ability to aggregate, characteristics presumably reflecting its properties both as a membrane protein and as a cell adhesion molecule, have masked details of the molecular organization of N-CAM that might be revealed by physicochemical techniques. The available data indicate that the molecule is multimeric (Hoffman et al., 1982), and this notion is supported by electron microscopy of rotary-shadowed samples (Edelman et al., 1983b). Regular figures that appear as two or more arms protruding from a central hub are seen; the three-armed structures bear some resemblance to the "triskelions" of clathrin (Ungewickell and Branton, 1981), but their detailed morphology is different from that of clathrin, and immunological studies (B. A. Cunningham, unpublished observations) indicate that N-CAM is distinct from clathrin. It is not clear if some N-CAM units contain only 130-kD polypeptides and others only 160-kD polypeptides or whether each form has both. One attractive model (Edelman et al., 1983b) derived from the observed forms is that the central hub includes the region(s) associated with the membrane while the more distal end contains the amino-terminal binding region. The zone between could contain the sialic acid-rich region, the negative charge separating the arms from each other.

Oligosaccharides

The most distinguishing feature of N-CAM is its high content of sialic acid (30 g/100 g protein) (Hoffman et al., 1982), most of which is present in $\alpha2-8$ linked polymers at least five residues long (Finne, 1982; Finne et al., 1983). N-CAM, fragments of N-CAM containing the polysialic acid, and the asparagine-linked oligosaccharides released by endoglycosidase F all are detectable on SDS-polyacrylamide gels as material staining blue with Stains-all, reflecting the polyanionic character of this sialic acid. To date, polymeric forms of sialic acid have been detected only in bacterial polysaccharides (e.g., colominic acid) (Troy et al., 1975) and in an as yet uncharacterized material in fish eggs (Inoue and Iwasaki, 1980).

Sialic acid appears to play a role in the modulation of N-CAM binding, and a decrease in sialic acid is the main, if not exclusive, feature of E-to-A conversion. The sialic acid does not appear to be within the N-CAM binding region (Cunningham et al., 1983), but its removal enhances the rate of N-CAM binding fourfold (Hoffman and Edelman, 1983). Whether this effect is steric, ionic, or conformational is as yet unclear; considering the magnitude of the change in mass and charge, each could contribute to the process. Removal of sialic acid significantly affects the electrophoretic properties of N-CAM; the pI, for example, is shifted an entire unit (Hoffman et al., 1982). As indicated earlier (Figure 1A), sialic acid causes N-CAM to migrate on SDS-polyacrylamide gels as a broad zone that reflects the polydispersity of the molecule. The sialic acid is released from N-CAM by either acid or neuraminidase but at rates lower than those seen for other glycoproteins such as fetuin (Hoffman et al., 1982). Some sialic acid is released by boiling or exposing N-CAM to low pH; under these conditions the material released appears to be oligomers of sialic acid, not free sialic acid (Hoffman et al., 1982; Crossin et al., 1984).

All of the N-CAM sialic acid can be released by endoglycosidase F (Crossin et al., 1984) and no sialic acid is added to N-CAM synthesized in the presence of tunicamycin (Cunningham et al., 1983), indicating that it is attached to asparagine-linked oligosaccharides. Essentially all of it is contained within fragment Fr2, with the majority in the segment represented by the difference between Fr2 and Fr1 (Figure 3) that is in the central region of the polypeptide (Cunningham et al., 1983; Crossin et al., 1984). This region can be isolated from both embryonic and adult N-CAM as a CNBr fragment, the polypeptide portion of which has a molecular weight of 35 kD and reacts with monoclonal antibody anti-N-CAM No. 2. When isolated from embryonic tissue this component migrated on SDS-poly-acrylamide gels as a broad zone of 42–60 kD that stained blue with Stains-all, reflecting the fact that it contains the bulk of the E-form sialic acid. The CNBr fragment from adult N-CAM migrated as a component of 45 kD that did not stain blue with Stains-all.

Treatment of the oligosaccharide-containing CNBr fragments from embryonic and adult N-CAM with neuraminidase reduced each to a component of 42 kD; treatment with endoglycosidase F reduced each to a species of 35 kD. Titration of each neuraminidase-treated CNBr fragment with endoglycosidase F indicated that each had three attachment sites for asparagine-linked oligosaccharides (Crossin et al., 1984).

Other recent studies in our laboratory have detected an additional oligosaccharide-containing CNBr fragment in N-CAM. This peptide can be isolated from purified Fr1 and contains little sialic acid; the oligosaccharide is asparagine-linked, as evidenced by the fact that the mobility of the fragment on SDS-polyacrylamide gels is altered slightly by treatment with neuraminidase and significantly by exposure to endoglycosidase F. This oligosaccharide and the additional three in Fr2 are sufficient to account for the total saccharide estimated to be present in N-CAM (Hoffman et al., 1982).

Taken together, the data indicate that the E- and A-forms of N-CAM have the same amino acid sequences, the same sites for oligosaccharide attachment, and the same number of asparagine-linked oligosaccharides. The only difference appears to be the amount and form of the sialic acid attached to the asparagine-linked oligosaccharides. This difference could arise either because of the increased expression of sialidase or the decreased activity of sialyltransferase during development and maturation. Increased sialidase activity during development has been observed, but pulse-chase experiments (Friedlander et al., 1985) suggest the E-to-A conversion of N-CAM involves decreased sialyltransferase activity.

Other Posttranslational Modifications

In addition to the negatively charged sialic acid residues, we have recently found that N-CAM can be sulfated when brain tissue is incubated with $^{35}SO_4$ (Figure 4A). The ^{35}S label is removed by endoglycosidase F but not by neuraminidase, indicating that it is on asparagine-linked oligosaccharides but not on sialic acid (Sorkin et al., 1984). Whether the sulfate plays any role in E-to-A conversion or in modulating activity remains to be established. Recent studies by others (Lyles et al., 1984b) suggest that in primary cultures of rat neurons, N-CAM may also be sulfated on tyrosyl residues.

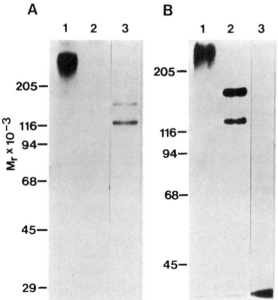

Figure 4. *Phosphorylation and sulfation of N-CAM. A:* N-CAM was isolated by immune precipitation from brain tissue labeled in culture with $^{35}SO_4$. Equal aliquots of this material were mock-treated (*lane 1*), endoglycosidase F-treated (*lane 2*), or neuraminidase-treated (*lane 3*) and were resolved on a 7.5–15% polyacrylamide gradient gel. Incorporated ^{35}S was detected by fluorography. *B:* Brain tissue was labeled in culture with $^{32}PO_4$ and isolated N-CAM was mock-treated (*lane 1*), or neuraminidase-treated (*lane 2*). In addition, N-CAM fragments were released from labeled membranes by V8 protease digestion (*lane 3*). The samples were resolved on a 7.5% polyacrylamide gel and incorporated ^{32}P was detected by autoradiography.

Both N-CAM polypeptides were phosphorylated when brain tissue was cultured with $^{32}PO_4$ (Figure 4B) (Sorkin et al., 1984). Less than 2% of the label was removed by endoglycosidase F, indicating that most, if not all, is on amino acids. Significantly more ^{32}P, relative to [^3H]leucine, was found in the larger component. Phosphothreonine and phosphoserine were detected in both polypeptides following enzymatic digestion and acid hydrolysis, although the ratio of the phosphoamino acids differed significantly. Phosphorylation may thus reflect a difference between the two N-CAM components.

To localize the probable sites of phosphorylation, membranes from cells labeled with $^{32}PO_4$ or [^3H]leucine were each treated with V8 protease under the conditions that generate Fr2. A number of released components, including Fr2, were detected in the ^3H-leucine-labeled material, but none of these were seen in the ^{32}P-labeled material (Figure 4B). These results indicate that the $^{32}PO_4$ is incorporated into a portion of the molecule that is carboxyl terminal to Fr2 (see Figure 3). Transmembrane proteins are often phosphorylated on the portion of the molecule on the cytoplasmic side of the membrane. By analogy, our results suggest that the two N-CAM polypeptides span the membrane and may differ in the size of their cytoplasmic domains or in the number of times they span the lipid bilayer. Enzymes that phosphorylate external segments of membrane proteins have been detected (Kubler et al., 1982), however, so any definitive interpretation of N-CAM phosphorylation will have to await more detailed analyses.

Preliminary results indicate that the membrane-associated region of N-CAM can also be acylated. Cultures of embryonic brain tissue incorporated [^3H]palmitate into N-CAM, and the label was not released by endoglycosidase F. Both polypeptides were labeled. Treatment of membranes with V8 protease gave results comparable to those obtained with $^{32}PO_4$; that is, no palmitate was detected in Fr2, indicating that the fatty acid is in the same general area as the phosphoamino acids. Current evidence suggests that fatty acid acylation helps anchor proteins in the lipid bilayer (Schlesinger and Malfer, 1982).

The major differences between the two N-CAM polypeptides are probably contained within the carboxy-terminal third of the molecule that includes the region associated with the membranes. Our earlier data indicated that N-CAM is an integral membrane protein with at least one segment that penetrates the lipid bilayer and one or more cytoplasmic domains. The phosphorylation and fatty acid acylation seen in this portion of the polypeptides are consistent with this conclusion.

L-CAM

L-CAM presents a less complex structure for analysis (Figure 5). It has a single polypeptide chain that appears to be invariant among the different

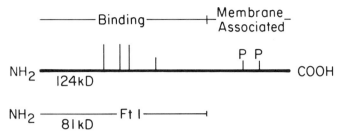

Figure 5. *Molecular model of L-CAM.* The figure shows the intact polypeptide and the fragment Ft1 released from membranes by trypsin. *Vertical bars* denote high-mannose (*short bar*) and complex (*long bars*) asparagine-linked oligosaccharides. L-CAM has phosphoserine and phosphothreonine residues (P) in that portion of the molecule not included in Ft1.

tissues in which it is found (Thiery et al., 1984). It has asparagine-linked oligosaccharides (Cunningham et al., 1984), but they lack the unusual polysialic acid seen in N-CAM. Moreover, there is no obvious E-to-A conversion in L-CAM. The apparent simplicity of this molecule, coupled with its early appearance in embryos and its widespread distribution, make it an excellent candidate for detailed structural analysis and a paradigm for molecules involved in calcium-dependent adhesion. It is, however, an integral membrane protein and available in much smaller amounts than N-CAM, so such analyses are not without constraint.

We turned directly to the use of the λgt11 expression system (Young and Davis, 1983) to produce L-CAM cDNA so that clones could be screened with specific antibodies. Polyclonal antibodies to the trypsin-released fragment Ft1 (see Figure 5) (Gallin et al., 1983) were used to screen for clones that included L-CAM coding segments. cDNA synthesized from poly(A)+ RNA from 9- and 14-day-old embryonic livers was primed with a combination of oligonucleotides that included oligothymidine, random calf thymus oligodeoxynucleotides, and a mixture of synthetic oligodeoxynucleotides prepared on the basis of the partial amino acid sequences of some fragments purified after CNBr cleavage of Ft1.

Three antibody-positive clones were selected and purified. One bacteriophage, designated L-301, has been characterized and used in initial studies of L-CAM mRNA and genomic DNA (Gallin et al., 1985). A lysogen of L-301 was prepared in the bacterial strain Y1089 and the β-galactosidase fusion protein induced in a suspension culture. Immunoblots of the cleared lysate with anti-L-CAM detected a single fusion protein (125 kD); detection was blocked by the addition of Ft1. To further demonstrate that L-301 contained a DNA fragment coding for part of the L-CAM molecule, the induced lysate was blotted onto nitrocellulose filters and used to absorb anti-L-CAM antibody. The bound antibody was eluted and specifically recognized both the L-301 fusion protein and purified L-CAM fragment Ft1. To prepare large amounts of the L-301 insert, it was subcloned in the bacterial plasmid pBR322. The recombinant plasmid was designated pEC301; it hybridized to the insert from L-301 and gave antibody-positive plaques when recloned in λgt11.

The pEC301 insert was used to detect L-CAM mRNA in extracts of chicken liver (Figure 6A). The insert hybridized to a 4-kb species in the poly(A)$^+$ RNA from 11-day-old embryonic chicken liver but gave no detectable component in embryonic brain, consistent with earlier observations that L-CAM concentrations are highest in liver whereas the molecule is not detectable in brain (Gallin et al., 1983).

The mRNA is large enough to code for the L-CAM polypeptide. On membranes, L-CAM is detected as a 124-kD glycopeptide, the polypeptide portion of which has a molecular weight of 110 kD. Recent evidence indicates that L-CAM, like uvomorulin, is synthesized as a larger precursor of 135 kD (Peyrieras et al., 1983; B.C. Sorkin and B.A. Cunningham, unpublished observations). This precursor differs from L-CAM both in its

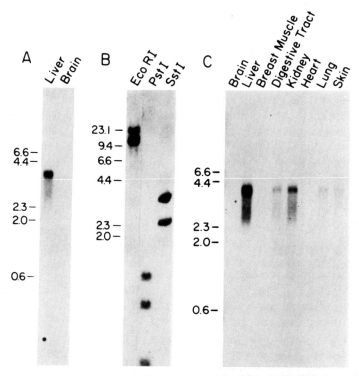

Figure 6. *Blot hybridization analyses of chicken L-CAM mRNA and genomic DNA. A: RNA transfer analysis of poly(A)$^+$ RNA isolated from 10-day-old embryonic chicken liver or brain probed with ^{32}P-labeled pEC301. B: Southern blot analysis of chicken genomic DNA digested with the indicated restriction endonucleases and probed with ^{32}P-labeled pEC301. C: RNA transfer analysis of equal amounts of poly(A)$^+$ RNA isolated from the indicated organs of 10-day-old embryonic chicken, probed with ^{32}P-labeled pEC301.*

carbohydrates and in having a larger polypeptide chain (124 kD). While the mRNA for L-CAM is large enough to specify a polypeptide of this size, the L-CAM coding sequence accounts for a larger proportion of its mRNA than does the coding sequence of N-CAM for its mRNA; we suspect that this is due to a more extensive 3' noncoding segment in N-CAM mRNA.

While the analyses of L-CAM mRNA were in accord with the simple pattern described by immunological studies of the L-CAM protein, assessing the number of L-CAM genes proved to be complex (Gallin et al., 1985). Pooled DNA from 14-day-old embryonic chicken livers was digested with restriction enzymes that did not cleave the pEC301 insert; the products were separated on agarose gels, transferred to nitrocellulose, and hybridized with the pEC301 insert (Figure 6B). Three bands of approximately equal intensity were detected in the Pst I digest, while two bands, one more intense than the other, were detected in both the Eco RI and Sst I digests.

Two simple interpretations could account for these results: There are three L-CAM genes, including possible pseudogenes, or there are introns containing restriction sites within the region coding for pEC301. To test the latter possibility, the pEC301 insert was digested with Hae III. Three fragments were separated, nick-translated, and used to probe replicate blots of the same digests of liver DNA that had been hybridized with the intact pEC301 insert. Two of the Hae III fragments were of about equal size (90–100 base pairs) and gave identical patterns, both hybridizing to the most intense band of the Eco RI and Sst I fragments seen by the intact pEC301 insert, and both hybridizing to the largest and smallest bands seen in the Pst I digest by the intact probe. Surprisingly, the smallest Hae III fragment gave the same pattern as the intact probes; because it is so small (20 base pairs), these results might be artifactual. The data indicate that there may be at least two copies of the L-CAM gene. Because the DNA used for Southern blot analyses was obtained from multiple, outbred animals, there could be allelic genes; the possibility of two closely related functioning genes or of pseudogenes must also be considered.

While L-CAM is most prominent in liver, it is detected in a variety of tissues throughout development and in a number of adult organs, primarily on epithelial cells (Thiery et al., 1984). Immunoblotting studies indicated that in all of these tissues the glycoprotein was the same size, gave the same proteolytic fragments, and reacted with the same antibodies. The amounts of L-CAM and the proportions of the fragments varied among these tissues. To test whether different mRNAs might be involved in the regulation of L-CAM expression in different tissues, poly(A)$^+$ RNA was isolated from a number of organs from 14-day-old chicken embryos and analyzed by Northern blot analyses with the pEC301 insert (Figure 6C). One mRNA component of the same size as that seen in liver was detected in digestive tract, kidney, lung, and skin but was absent in heart and breast muscle (Gallin et al., 1985). The amounts of mRNA parallel the amount of L-CAM protein in each tissue.

L-CAM lacks the unusual amount and form of sialic acid seen in N-CAM, but it is glycosylated at multiple sites (Cunningham et al., 1984). Endoglycosidase F titrations give five identifiable L-CAM species, suggesting that there are four oligosaccharides. Endoglycosidase H appears to be able to remove one of these units, so we have concluded that the molecule has one high-mannose and three complex oligosaccharides. The Ft1 fragment gave similar results with these enzymes, indicating that it contains all of the asparagine-linked oligosaccharides.

The Ft1 fragment also has the same amino-terminal sequence as intact L-CAM, suggesting that, like N-CAM, the amino terminus of L-CAM extends away from the cell surface (Cunningham et al., 1984). L-CAM can also be phosphorylated on serine and threonine residues, but no phosphate was detected in Ft1. This result suggests that L-CAM has a cytoplasmic domain. Other studies on uvomorulin, the murine equivalent of L-CAM, however, have raised the possibility that it is not an integral membrane protein (Peyrieras et al., 1983). Uvomorulin apparently can be extracted by EDTA solutions and its solubility and electrophoretic mobility in various detergents suggest that it lacks a hydrophobic segment that could integrate into the membrane. In contrast, we cannot extract intact L-CAM with EDTA solutions, although we do extract L-CAM fragments. In addition, our experiments with L-CAM in detergents comparable to the experiments with uvomorulin have given different results: We detect L-CAM in both the detergent and aqueous phases when solutions of Triton X-114 are allowed to undergo phase separation, and intact L-CAM has a more significant change in its electrophoretic mobility in anionic and cationic detergents than does Ft1. Both results are consistent with L-CAM being an integral membrane protein.

Although there is no obvious chemical alteration in L-CAM comparable to E-to-A conversion in N-CAM, a number of possible mechanisms remain for regulating L-CAM activity, including modulation of its prevalence and distribution on the cell surface. A distinguishing feature of L-CAM-mediated adhesion is its requirement for calcium. L-CAM itself also appears to bind calcium in that the molecule is more sensitive to proteolysis in calcium-free solution; fragment Ft1 is degraded and only small amounts of material with a maximum molecular weight on 43 kD appear on SDS-polyacrylamide gels (Gallin et al., 1983). Analyses of the role of calcium in L-CAM-mediated adhesion and the agents that regulate it may be a crucial factor in understanding how L-CAM is involved in development.

Ng-CAM

Ng-CAM represents a different class of cell adhesion molecules; it appears late in development, relative to N-CAM and L-CAM, and its binding is apparently heterophilic (Grumet et al., 1984a; Thiery et al., 1985). Compared to N-CAM and L-CAM, little is known chemically about Ng-CAM. It is

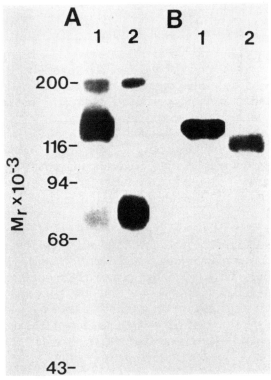

Figure 7. *Molecular components of Ng-CAM. A:* The three components of Ng-CAM were detected by immunoblotting with polyclonal anti-Ng-CAM (*lane 1*); ^{32}P-labeled Ng-CAM was immunoprecipitated with polyclonal anti-Ng-CAM and visualized by autoradiography (*lane 2*). *B:* Ng-CAM before (*lane 1*) or after (*lane 2*) digestion with neuraminidase as detected by immunoblotting with a monoclonal anti-Ng-CAM.

detected as a predominant component with a molecular weight of 135 kD by Coomassie blue staining after SDS-polyacrylamide gel electrophoresis and on immunoblots (Figure 7). Species at 200 kD and 80 kD are also consistently observed in Ng-CAM preparations although in variable amounts (Grumet et al., 1984a). No resemblance between L-CAM and Ng-CAM has been detected chemically or antigenically, and their tissue distributions are markedly different (Thiery et al., 1985). On the other hand, antigenic cross reactions have been observed with some monoclonal antibodies that recognize N-CAM (Grumet et al., 1984a). In at least one of these cases, the common determinant appears to be carbohydrate.

The 80-kD and 135-kD species of Ng-CAM appear to be antigenically related to the larger component but not to each other (Grumet et al., 1984b). Moreover, the largest and smallest components, but not the 135-kD component, can be phosphorylated when brain tissue is incubated in culture with [32]PO$_4$ (Figure 7A). We conclude from these results that the 200-kD component gives rise to the two smaller species, but there is as yet no compelling evidence for this hypothesis. The results also raise the possibility that the 135-kD species is not an integral membrane protein, although detergents are required for extraction of Ng-CAM from membranes, and the 135-kD component can be incorporated into lipid vesicles (Grumet et al., 1984b).

Although Ng-CAM polypeptides migrate as discrete bands, the molecule is glycosylated (Grumet et al., 1984a). When the 135-kD species is treated with endoglycosidase F, it is reduced to a species of 115 kD, and treatment of the same material with neuraminidase (Figure 7B) reduces the molecular weight to 127 kD. The difference between the molecular weights of the neuraminidase- and endoglycosidase F-treated materials suggests that the molecule has four asparagine-linked oligosaccharides of an average weight of 3.5 kD, but the change in molecular weight with neuraminidase is too large to be due solely to sialic acid residues from four typical asparagine-linked complex oligosaccharides, suggesting that the oligosaccharides may have multiple sialic acids attached. It is even possible that Ng-CAM contains some truncated form of the polysialic acid seen in embryonic N-CAM, reminiscent of what is present in adult N-CAM. Such interpretation must await detailed analytical data on purified components, because the effects of sialic acid and other factors on the migration of glycoproteins on SDS-polyacrylamide gels could be complex.

CONCLUSION

The results described here provide an overall view of the cell adhesion molecules N-CAM, Ng-CAM, and L-CAM, particularly with regard to those features that may be involved in the regulation of CAM expression and activity. All three are large cell surface glycoproteins that are probably integrated into the lipid bilayer and contain a cytoplasmic domain that can be phosphorylated on serine and threonine residues. The extracellular portions of each contain multiple asparagine-linked oligosaccharides and are subject to limited proteolysis at discrete sites. Despite their overall similarities, each molecule is a distinct entity with features that distinguish its expression and activity. Among these features are the polysialic acid of N-CAM, the multiple polypeptides of Ng-CAM, and the dependence of L-CAM on calcium. Molecular biological studies have shown that N-CAM and L-CAM are each specified by only a few genes, with possibly only a single gene for N-CAM in the chicken. Transcriptional control and differential splicing could provide critical bases for regulation of CAM expression.

It should now be possible to localize the sites involved in the homophilic binding of N-CAM and L-CAM and to describe the detailed mechanisms of the cell–cell interactions mediated by these molecules. The identification of the putative ligand for Ng-CAM on glial cells should make it possible to pursue similar analyses of the heterophilic binding mediated by Ng-CAM. We should soon know the complete amino acid sequence of these CAMs and the structures of the genes for each. In addition, we are developing DNA probes in other species that will allow us to map the genes within the chromosomes of mice and humans. Moreover, the DNA probes open the way for more detailed analyses of CAM function by techniques such as *in situ* hybridization and transfection of CAM genes.

REFERENCES

Bertolotti, R., U. Rutishauser, and G. M. Edelman (1980) A cell surface molecule involved in aggregation of embryonic liver cells. *Proc. Natl. Acad. Sci. USA* **77**:4831–4835.

Brackenbury, R., J.-P. Thiery, U. Rutishauser, and G. M. Edelman (1977) Adhesion among neural cells of the chick embryo. I. An immunological assay for molecules involved in cell–cell binding. *J. Biol. Chem.* **252**:6835–6840.

Brackenbury, R., U. Rutishauser, and G. M. Edelman (1981) Distinct calcium-independent and calcium-dependent adhesion systems of chicken embryo cells. *Proc. Natl. Acad. Sci. USA* **78**:387–391.

Chuong, C.-M., and G. M. Edelman (1984) Alterations in neural cell adhesion molecules during development of different regions of the nervous system. *J. Neurosci.* **4**:2354–2368.

Chuong, C.-M., D. A. McClain, P. Streit, and G. M. Edelman (1982) Neural cell adhesion molecules in rodent brains isolated by monoclonal antibodies with cross-species reactivity. *Proc. Natl. Acad. Sci. USA* **79**:4234–4238.

Crossin, K. L., G. M. Edelman, and B. A. Cunningham (1984) Mapping of three carbohydrate attachment sites in embryonic and adult forms of the neural cell adhesion molecule (N-CAM). *J. Cell Biol.* **99**:1848–1855.

Cunningham, B. A., S. Hoffman, U. Rutishauser, J. J. Hemperly, and G. M. Edelman (1983) Molecular topography of N-CAM: Surface orientation and the location of sialic acid-rich and binding regions. *Proc. Natl. Acad. Sci. USA* **80**:3116–3120.

Cunningham, B. A., Y. Leutzinger, W. J. Gallin, B. C. Sorkin, and G. M. Edelman (1984) Linear organization of the liver cell adhesion molecule L-CAM. *Proc. Natl. Acad. Sci. USA* **81**:5787–5791.

Damsky, C. H., J. Richa, D. Solter, K. Knudsen, and C. A. Buck (1983) Identification and purification of a cell surface glycoprotein mediating intercellular adhesion in embryonic and adult tissue. *Cell* **34**:455–466.

Edelman, G. M. (1983) Cell adhesion molecules. *Science* **219**:450–457.

Edelman, G. M. (1984a) Modulation of cell adhesion during induction, histogenesis, and perinatal development of the nervous system. *Annu. Rev. Neurosci.* **7**:339–377.

Edelman, G. M. (1984b) Cell adhesion and morphogenesis: The regulator hypothesis. *Proc. Natl. Acad. Sci. USA* **81**:1460–1464.

Edelman, G. M., and C.-M. Chuong (1982) Embryonic to adult conversion of neural cell adhesion molecules in normal and *staggerer* mice. *Proc. Natl. Acad. Sci. USA* **79**:7036–7040.

Edelman, G. M., W. J. Gallin, A. Delouvée, B. A. Cunningham, and J.-P. Thiery (1983a) Early epochal maps of two different cell adhesion molecules. *Proc. Natl. Acad. Sci. USA* **80**:4384–4388.

Edelman, G. M., S. Hoffman, C.-M. Chuong, J.-P. Thiery, R. Brackenbury, W. J. Gallin, M. Grumet, M. E. Greenberg, J. J. Hemperly, C. Cohen, and B. A. Cunningham (1983b) Structure and modulation of neural cell adhesion molecules in early and late embryogenesis. *Cold Spring Harbor Symp. Quant. Biol.* **48**:515–526.

Finne, J. (1982) Occurrence of unique polysialosyl carbohydrate units in glycoproteins of developing brain. *J. Biol. Chem.* **257**:11966–11970.

Finne, J., H. Deagostini-Bazin, and C. Goridis (1983) Occurrence of α2-8 linked polysialosyl units in a neural cell adhesion molecule. *Biochem. Biophys. Res. Commun.* **112**:482–487.

Fraser, S. E., B. A. Murray, C.-M. Chuong, and G. M. Edelman (1984) Alteration of the retinotectal map in *Xenopus* by antibodies to neural cell adhesion molecules. *Proc. Natl. Acad. Sci. USA* **81**:4222–4226.

Friedlander, D. R., R. Brackenbury, and G. M. Edelman (1985) Conversion of embryonic to adult forms of N-CAM: *In vitro* results from *de novo* synthesis of adult forms. *J. Cell Biol.* **101**:412–419.

Gallin, W. J., G. M. Edelman, and B. A. Cunningham (1983) Characterization of L-CAM, a major cell adhesion molecule from embryonic liver cells. *Proc. Natl. Acad. Sci. USA* **80**:1038–1042.

Gallin, W. J., E. A. Prediger, G. M. Edelman, and B. A. Cunningham (1985) Isolation of a cDNA clone for the liver cell adhesion molecule (L-CAM). *Proc. Natl. Acad. Sci. USA* **82**:2809–2813.

Gennarini, G., G. Rougon, H. Deagostini-Bazin, M. Hirn, and C. Goridis (1984) Studies on the transmembrane disposition of the neural cell adhesion molecule N-CAM. A monoclonal antibody recognizing a cytoplasmic domain and evidence for the presence of phosphoserine residues. *Eur. J. Biochem.* **142**:57–64.

Green, M. R., and J. V. Pastewka (1975) Identification of sialic acid-rich glycoproteins on polyacrylamide gels. *Anal. Biochem.* **65**:66–72.

Grumet, M., and G. M. Edelman (1984) Heterotypic binding between neuronal membrane vesicles and glial cells is mediated by a specific neuron–glial cell adhesion molecule. *J. Cell Biol.* **98**:1746–1756.

Grumet, M., U. Rutishauser, and G. M. Edelman (1982) N-CAM mediates adhesion between embryonic nerve and muscle cells *in vitro*. *Nature* **295**:693–695.

Grumet, M., S. Hoffman, and G. M. Edelman (1984a) Two antigenically related neuronal CAMs of different specificities mediate neuron–neuron and neuron–glia adhesion. *Proc. Natl. Acad. Sci. USA* **81**:267–271.

Grumet, M., S. Hoffman, C.-M. Chuong, and G. M. Edelman (1984b) Polypeptide components and binding functions of neuron–glia adhesion molecules. *Proc. Natl. Acad. Sci. USA* **81**:7989–7993.

Hirn, M., M. S. Ghandour, H. Deagostini-Bazin, and C. Goridis (1983) Molecular heterogeneity and structural evolution during cerebellar ontogeny detected by monoclonal antibody of the mouse cell surface antigen BSP-2. *Brain Res.* **265**:87–100.

Hoffman, S., and G. M. Edelman (1983) Kinetics of homophilic binding by embryonic and adult forms of the neural cell adhesion molecule. *Proc. Natl. Acad. Sci. USA* **80**:5762–5766.

Hoffman, S., B. C. Sorkin, P. C. White, R. Brackenbury, R. Mailhammer, U. Rutishauser, B. A. Cunningham, and G. M. Edelman (1982) Chemical characterization of a neural cell adhesion molecule purified from embryonic brain membranes. *J. Biol. Chem.* **257**:7720–7729.

Hoffman, S., C.-M. Chuong, and G. M. Edelman (1984) Evolutionary conservation of key structures and binding functions of neural cell adhesion molecules. *Proc. Natl. Acad. Sci. USA* **81**:6881–6885.

Hyafil, F., D. Morello, C. Babinet, and F. Jacob (1980) A cell surface glycoprotein involved in the compaction of embryonal carcinoma cells and cleavage-stage embryos. *Cell* **21**:927–934.

Inoue, S., and M. Iwasaki (1980) Sialoglycoproteins from the eggs of Pacific herring. *Eur. J. Biochem.* **111**:131–135.

Jorgensen, O. S., A. Delouvée, J.-P. Thiery, and G. M. Edelman (1980) The nervous system-specific protein D2 is involved in adhesion among neurites from cultured rat ganglia. *FEBS Lett.* **111**:39–42.

Kubler, D., W. Pyerin, and V. Kinzel (1982) Protein kinase activity and substrates at the surface of intact HeLa cells. *J. Biol. Chem.* **257**:322–329.

Lyles, J. M., B. Norrild, and E. Bock (1984a) Biosynthesis of the D2-cell adhesion molecule: Pulse-chase studies in cultured fetal rat neuronal cells. *J. Cell Biol.* **98**:2077–2081.

Lyles, J. M., D. Linneman, and E. Bock (1984b) Biosynthesis of the D2-cell adhesion molecule: Post-translational modifications, intracellular transport, and developmental changes. *J. Cell Biol.* **99**:2082–2091.

McClain, D. A., and G. M. Edelman (1982) A neural cell adhesion molecule from human brain. *Proc. Natl. Acad. Sci. USA* **79**:6380–6384.

Milner, R. J., M. D. Brow, D. W. Cleveland, T. M. Shinnick, and J. G. Sutcliffe (1983) Glyceraldehyde 3-phosphate dehydrogenase protein and mRNA are both expressed differentially in adult chickens but not chick embryos. *Nucleic Acids Res.* **11**:3301–3315.

Murray, B. A., J. J. Hemperly, W. J. Gallin, J. S. MacGregor, G. M. Edelman, and B. A. Cunningham (1984) Isolation of cDNA clones for the chicken neural cell adhesion molecule (N-CAM). *Proc. Natl. Acad. Sci. USA* **81**:5584–5588.

Peyrieras, N., F. Hyafil, D. Louvard, H. L. Ploegh, and F. Jacob (1983) Uvomorulin: A nonintegral membrane protein of early mouse embryo. *Proc. Natl. Acad. Sci. USA* **80**:6274–6277.

Rathjen, F. G., and M. Schachner (1984) Immunocytological and biochemical characterization of a new neuronal cell surface component (L1 antigen) which is involved in cell adhesion. *EMBO J.* **3**:1–10.

Rothbard, J. B., R. Brackenbury, B. A. Cunningham, and G. M. Edelman (1982) Differences in the carbohydrate structures of neural cell adhesion molecules from adult and embryonic chicken brains. *J. Biol. Chem.* **257**:11064–11069.

Rougon, G., H. Deagostini-Bazin, M. Hirn, and C. Goridis (1982) Tissue and developmental stage-specific forms of a neural cell surface antigen linked to differences in glycosylation of a common polypeptide. *EMBO J.* **1**:1239–1244.

Rutishauser, U., J.-P. Thiery, R. Brackenbury, and G. M. Edelman (1978) Adhesion among neural cells of the chick embryo. III. Relationship of the surface molecule CAM to cell adhesion and the development of histotypic patterns. *J. Cell Biol.* **79**:371–381.

Rutishauser, U., S. Hoffman, and G. M. Edelman (1982) Binding properties of a cell adhesion molecule from neural tissue. *Proc. Natl. Acad. Sci. USA* **79**:685–689.

Rutishauser, U., M. Grumet, and G. M. Edelman (1983) N-CAM mediates initial interactions between spinal cord neurons and muscle cells in culture. *J. Cell Biol.* **97**:145–152.

Schlesinger, M. J., and C. Malfer (1982) Cerulenin blocks fatty acid acylation of glycoproteins and inhibits vesicular stomatitis and sindbis virus particle formation. *J. Biol. Chem.* **257**:9887–9890.

Sorkin, B. C., S. Hoffman, G. M. Edelman, and B. A. Cunningham (1984) Sulfation and phosphorylation of the neural cell adhesion molecule (N-CAM). *Science* **225**:1476–1478.

Tamkun, J. W., J. E. Schwarzbauer, and R. O. Hynes (1984) A single rat fibronectin gene generates three different mRNAs by alternative splicing of a complex exon. *Proc. Natl. Acad. Sci. USA* **81**:5140–5144.

Thiery, J.-P., R. Brackenbury, U. Rutishauser, and G. M. Edelman (1977) Adhesion among neural cells of the chick embryo. II. Purification and characterization of a cell adhesion molecule from neural retina. *J. Biol. Chem.* **252**:6841–6845.

Thiery, J.-P., J.-L. Duband, U. Rutishauser, and G. M. Edelman (1982) Cell adhesion molecules in early chick embryogenesis. *Proc. Natl. Acad. Sci. USA* **79**:6737–6741.

Thiery, J.-P., A. Delouvée, W. J. Gallin, B. A. Cunningham, and G. M. Edelman (1984) Ontogenetic expression of cell adhesion molecules: L-CAM is found in epithelia derived from the three primary germ layers. *Dev. Biol.* **102**:61–78.

Thiery, J.-P., A. Delouvée, M. Grumet, and G. M. Edelman (1985) Initial appearance and regional distribution of the neuron–glia cell adhesion molecule (Ng-CAM) in the chick embryo. *J. Cell Biol.* **100**:442–456.

Troy, F. A., I. K. Vijay, and N. Tesche (1975) Role of undecaprenyl phosphate in synthesis of polymer containing sialic acid in *Escherichia coli. J. Biol. Chem.* **250**:156–163.

Ungewickell, E., and D. Branton (1981) Assembly units of clathrin coats. *Nature* **289**:420–422.

Yoshida, C., and M. Takeichi (1982) Teratocarcinoma cell adhesion: Identification of a cell surface protein involved in calcium-dependent cell aggregation. *Cell* **28**:217–224.

Young, R. A., and R. W. Davis (1983) Efficient isolation of genes by using antibody probes. *Proc. Natl. Acad. Sci. USA* **80**:1194–1198.

Chapter 10

Calcium-Dependent Cell–Cell Adhesion System: Its Molecular Nature, Cell Type Specificity, and Morphogenetic Role

MASATOSHI TAKEICHI
CHIKAKO YOSHIDA-NORO
YASUAKI SHIRAYOSHI
KOHEI HATTA

ABSTRACT

Ca^{2+} is an essential ion for cell–cell binding in probably all animal species. Recent studies have shown that Ca^{2+} is required by a particular cell adhesion machinery called the Ca^{2+}-dependent cell–cell adhesion system (CDS). This system appears fundamental to the construction of multicellular animals, since its inhibition in any of a variety of ways causes severe reduction in cell–cell adhesiveness in most kinds of tissues. CDS is cell type-specific, as indicated by observations that epithelial cells (e.g., teratocarcinoma stem cells) do not cross-adhere to fibroblasts via CDS. Consistently, the monoclonal antibody ECCD-1, which specifically blocks CDS-dependent cell–cell adhesion in various epithelial tissues, does not react with fibroblasts, brain cells, and so on. Immunoblot analysis revealed that the major constituent of the ECCD-1-sensitive CDS is a 124-kD protein termed cadherin. In attempts to clarify the morphogenetic role of CDS in early mouse development, we found that the ECCD-1-sensitive CDS is responsible for compaction of embryos and that inhibition of compaction with ECCD-1 leads to suppression of development of the inner cell mass. Thus, CDS is involved in the morphogenetic processes of embryos as a regulator of differentiation. By using various monoclonal antibodies that react with CDS with different specificity in different cell types, we shall be able to analyze the molecular mechanisms of specific or selective cell–cell adhesion that are fundamental processes of animal morphogenesis.

Removal of Ca^{2+} from cell surfaces causes or accelerates disaggregation of whole embryos or tissues of various animal species. Recent studies using mammalian (Takeichi, 1977; Takeichi et al., 1979; Urushihara et al.,

1979) and avian cells (Grunwald et al., 1980; Brackenbury et al., 1981; Magnani et al., 1981) have revealed that Ca^{2+} is involved in a specific cell–cell adhesion machinery, which we call the Ca^{2+}-dependent cell–cell adhesion system (CDS). As suggested by the strong effect on cell–cell adhesion caused by the removal of Ca^{2+} from cell surfaces, CDS seems to play a fundamental role in the construction of multicellular animals.

Since Ca^{2+} is not exclusively required by CDS in cell–cell adhesion, the definition of this system seems necessary. CDS is extremely sensitive to protease but becomes resistant in the presence of Ca^{2+}. In other words, treatment of cells with protease in the presence of Ca^{2+} leaves CDS intact, but cell treatment with protease in the absence of Ca^{2+} inactivates it. For example, cells disaggregated with trypsin solution containing Ca^{2+} are able to reaggregate without a lag period, but cells disaggregated with trypsin solution containing EGTA do not reaggregate until CDS is restored following *de novo* protein synthesis. Thus, in our definition, the term CDS refers to the cell adhesion system sensitive to Ca^{2+} in two distinct aspects: it requires Ca^{2+} for its action on cell–cell adhesion and reacts with Ca^{2+} so as to acquire a protease-resistant property. CDS thus defined has been detected in a variety of cell types (almost all cell types with strong mutual cohesiveness).

Ca^{2+} sensitivity is a good marker for the identification of the molecules involved in CDS. Among the cell surface molecules sensitive to trypsin, those that can be protected by Ca^{2+} against proteolysis are regarded as potential candidates. In fact, Yoshida and Takeichi (1982) succeeded in finding a such molecule. They raised antibodies against whole teratocarcinoma F9 cells (anti-TC-F9), Fab fragments of which are able to inhibit aggregation of these cells via CDS. Anti-TC-F9 reacted with several cell surface proteins; one with a molecular weight of 124 kD (originally reported as 140 kD) was detected in cells treated with trypsin in the presence of Ca^{2+} but not in those trypsinized in the presence of EGTA. Further experiments provided evidence supporting the conclusion that this 124-kD protein is a major component of teratocarcinoma CDS.

An important property of CDS is its cell type specificity. One can experimentally disaggregate cells, leaving CDS intact but destroying all other cell adhesion systems, by treatment with trypsin plus Ca^{2+} (TC-treatment). Takeichi et al. (1981) demonstrated that when TC-treated teratocarcinoma cells were artificially mixed with TC-treated fibroblasts of various origins they adhered preferentially to their own cell types, suggesting that the different types have CDS with distinct functional specificity. Consistently, Fab fragments of anti-TC-F9 did not inhibit the CDS-dependent aggregation of fibroblastic cells, indicating that CDS in teratocarcinoma and fibroblastic cells are also immunologically distinct. These results suggest the possibility that CDS plays a key role in segregating different cell types for the construction of tissues and organs.

For a more precise analysis of the molecular properties and morphogenetic roles of CDS, it was necessary to prepare monospecific antibodies that block the action of this system. Yoshida-Noro et al. (1984) succeeded in producing a monoclonal antibody that inhibits CDS in teratocarcinoma cells; more recently, we (K. Hatta and M. Takeichi, manuscript in preparation) obtained a monoclonal antibody affecting CDS-dependent adhesion of mouse brain cells. In this review, we summarize the results of studies in which these monoclonal antibodies have been used to characterize the molecular properties of CDS and to investigate their possible roles in mouse embryonic development.

MONOCLONAL ANTIBODY RECOGNITION OF TERATOCARCINOMA CDS

Teratocarcinoma stem cells are multipotent tumor cells expressing early embryonic phenotypes. Therefore, they have served as a model system for studying cell surface molecules in early embryonic cells. Yoshida-Noro et al. (1984) attempted to produce monoclonal antibodies recognizing CDS in teratocarcinoma cells; the antigens used for immunizing animals were intact teratocarcinoma cells of mouse origin (F9 line). Lymphocytes of a rat receiving injections of F9 cells were fused with mouse myeloma cells to produce a hybridoma. To screen the hybridoma, culture supernatants were assayed for activity in disrupting cell–cell adhesion in teratocarcinoma cell monolayers. A hybridoma clone picked up after this screening produced antibodies, termed ECCD-1, that exhibited the activity described below.

When ECCD-1 was added to the monolayer culture of teratocarcinoma cells, individual cells tended to contract and separate from each other. The antibody did not induce detachment of cells from the culture substrate. Cells whose mutual contact was inhibited by the antibody continued to proliferate, indicating that the antibody had no toxic effect.

The effect of ECCD-1 on aggregation of disaggregated teratocarcinoma cells was then assayed. The results showed that this antibody specifically inhibits CDS-dependent aggregation. The major target molecule for ECCD-1, revealed by Western blot analysis, was found to be the same as the 124-kD protein detected by Yoshida and Takeichi (1982) in studies using the polyclonal antibody anti-TC-F9. Involvement of this 124-kD protein in teratocarcinoma CDS was thus confirmed. ECCD-1 also reacted with other relatively minor components; we termed the proteins recognized by ECCD-1 cadherin.

ROLE OF CALCIUM IN CDS

Yoshida-Noro et al. (1984) observed an interesting phenomenon in a binding assay of ECCD-1 to cell surfaces. Ca^{2+} is required for binding but most other divalent cations are not, although Mn^{2+} is slightly effective. This suggested that the antigens must react with Ca^{2+} to be recognized by

ECCD-1. The authors present evidence supporting this possibility in the following immunoblot experiment. In a Western blot analysis, antigen blots on a nitrocellulose sheet were incubated in an ECCD-1 solution with and without Ca^{2+}. It was found that the antibodies are able to bind with the cadherin bands only in the presence of Ca^{2+}. Also, once bound to the antigens, ECCD-1 molecules were released from the blots by rinsing with a solution containing EGTA. These results clearly demonstrate that the binding of ECCD-1 to the antigens is a Ca^{2+}-dependent process, which suggests that the conformation of cadherin molecules is modified by reaction with Ca^{2+}, and that only their Ca^{2+}-binding form is recognized by ECCD-1. The direct role of Ca^{2+} in CDS, therefore, is to switch the molecular conformation of cadherin from inactive to active form. The change in protease sensitivity of CDS resulting from reaction with Ca^{2+} probably has the same molecular basis.

CELL TYPE SPECIFICITY OF CADHERIN

As mentioned above, CDS has a cell type-specific function. To examine the distribution of cadherin in various cell types, two methods of cell culture were adopted. One was a radioimmunological assay of ECCD-1 binding to cells, and the other was a bioassay of the sensitivity of cell–cell adhesion to ECCD-1. The results are summarized in Table 1. Cadherin appears very early in the development of mouse embryos, probably even

Table 1. Distribution of Cadherin in Mouse Embryonic Tissues and Cell Lines

Cadherin-Positive	Cadherin-Negative
Blastomeres	Mesenchyme
Embryonic ectoderm	(Connective tissues)
Hepatocytes	Brain
Mammary gland	
Epidermis	
Lung alveoli	
Teratocarcinoma stem cells	Visceral endoderm cells
(all cell lines tested)	(PSA-5E)
Mammary tumor cells (MTD-1)	Adrenal tumor cells (Y1)
	Glioma cells (G26-20)
	Neuroblastoma cells
	(Neuro-2a)
	Fibroblastic cells
	(all cell lines tested)

Sources: MTD-1, Enami et al. (1984); PSA-5E, Adamson et al. (1977); Y1, Yasumura et al. (1966); G26-20, Sundarraj et al. (1975); Neuro-2a, Klebe and Ruddle (1969).

at the one-cell stage, and it is a key molecule involved in the compaction of 8- to 16- cell embryos (Shirayoshi et al., 1983). At later developmental stages, this cell adhesion molecule is detected exclusively in epithelial cells in a variety of tissues. All mesenchymal cells were negative in reaction to ECCD-1. For example, hepatocytes in liver, epidermal cells in skin, and epithelial cells in mammary gland were positive, but fibroblastic cells present in these tissues were negative. Cells in brain were also negative. These results suggest that cadherin is an epithelial cell-specific protein in differentiated organs; however, some epithelial cells, including the parietal yolk sac endodermal cell line PSA5-E and adrenal tumor cell line Y1, did not react with ECCD-1. Therefore, cadherin is specific to epithelial cells to a limited extent.

TRYPTIC FRAGMENTS OF CADHERIN

In the immunoblot analysis using ECCD-1, the 124-kD component of cadherin was detected in TC-treated as well as in nontrypsinized F9 cells, but it was not detected in F9 cells treated with trypsin in the absence of Ca^{2+}. The 124-kD component, however, seems to be not entirely protected against tryptic digestion. While it was detected as the only major antigen to ECCD-1 in nontrypsinized cells, another major antigen was detected in TC-treated cells whose molecular weight was about 100 kD. It is possible that this smaller component was produced by tryptic cleavage of the 124-kD component even in the presence of Ca^{2+}. A similar phenomenon was observed when liver cells were used for the detection of antigens to ECCD-1 (Ogou et al., 1983). Cell surface antigens were extracted with detergents from nontrypsinized monolayer cultures of liver cells for immunoblot analysis. From this sample, the 124-kD component was detected as a major antigen, as was found in F9 cells. However, in some cases, a smaller component with a molecular weight of 100 kD was also detected as a second major antigen to ECCD-1, and in the extreme case this smaller component was found to be the only major antigen. It is most likely that this smaller component is a degradation product of the 124-kD component formed by digestion with intrinsic cellular protease in the presence of Ca^{2+}; thus, the 124-kD component could be partially cleaved in the presence of Ca^{2+}. It is not known whether these smaller components of cadherin are active in binding cells.

A product of tryptic cleavage of the 124-kD component formed in the absence of Ca^{2+} has been characterized. By incubating F9 cells with trypsin in the presence of EGTA, a substance that could neutralize the effect of anti-TC-F9 in inhibiting the function of CDS was extracted. This polypeptide had a molecular weight of 34 kD and competed with the 124-kD component in immunoprecipitation with ECCD-1. A fraction with the same activity and a similar size can be extracted from liver cells (Y. Shirayoshi and M.

Takeichi, manuscript in preparation). It should be noted that normal mouse serum shows an activity that neutralizes the CDS-inhibiting effect of anti-TC-F9 (Y. Shirayoshi and M. Takeichi, unpublished observations). This suggests that cadherin molecules (probably their fragments) are released into serum from cell surfaces, possibly as a result of turnover. Interestingly, antibodies raised against mouse whole serum proteins, commercially obtained, can induce disruption of cell–cell adhesion in teratocarcinoma cells, as does ECCD-1 (Y. Shirayoshi and M. Takeichi, unpublished observations). This is consistent with the observation that serum contains substances reactive with the anti-TC-F9. Therefore, antisera against serum proteins can be used as crude cadherin-blocking antibodies.

CADHERIN-RELATED MOLECULES STUDIED IN OTHER LABORATORIES

Intimate adhesion between blastomeres in "compacted" mouse embryos requires Ca^{2+}. Ogou et al. (1982, 1983) showed that mouse early embryos (even at the one-cell stage) have CDS with the same specificity as that seen in teratocarcinoma cells, suggesting the involvement of CDS in compaction. Yoshida-Noro et al. (1984) and Shirayoshi et al. (1983) actually demonstrated that compaction is sensitive to ECCD-1.

Immunological approaches to the identification of molecules involved in the compaction of mouse embryos were originally begun by Jacob's group (Kemler et al., 1977). Their first success was the identification of a tryptic fragment (UMt) of a larger putative molecule, termed uvomorulin, which neutralizes the effect of the compaction-inhibiting antibody (Hyafil et al., 1980; Hyafil et al., 1981). Recent studies by Yoshida-Noro et al. (1984) showed that the 124-kD component recognized by ECCD-1 cross-reacted with a monoclonal antibody raised against UMt. Thus, it turned out that the molecule responsible for compaction sought by Jacob's group was the same as that identified by Yoshida and Takeichi (1982) as the Ca^{2+}-dependent cell–cell adhesion molecule in teratocarcinoma cells.

Epithelial cell–cell adhesion molecules identified in several different laboratories seem to be equivalent to cadherin. L-CAM, the term used for chicken liver cell–cell adhesion molecules identified by Edelman's group, shows the same molecular weight as cadherin, and the tissue distribution of L-CAM and cadherin is quite similar (Gallin et al., 1983). Mammary tumor is one of the targets of ECCD-1, and a cell–cell adhesion molecule of this tumor identified by Damsky et al. (1983) is similar to cadherin in its molecular weight and cell type specificity. A monoclonal antibody prepared by Imhof et al. (1984) affected cell–cell adhesion of canine kidney epithelial cells, recognizing a cell surface molecule similar in molecular weight to cadherin.

These epithelial cell adhesion molecules, detected in different animal species in various laboratories, probably fall into a single molecular class because of their similarity in molecular weight and cell type specificity.

STUDIES ON OTHER CELL TYPE-SPECIFIC CDS

Antibodies that inhibit the function of CDS offer the only effective way of hunting for relevant CDS molecules. No one has yet succeeded in producing antibodies that block CDS in fibroblastic cells; therefore, the molecular nature of fibroblastic CDS is not as clear as is that of cadherin. An early study by Takeichi (1977), however, presented a candidate molecule for fibroblastic CDS. He compared the cell surface proteins in Chinese hamster fibroblast V79 cells treated with trypsin in the presence and absence of Ca^{2+} and detected one protein that is protected by Ca^{2+} from trypsin digestion. This component has a molecular weight a little higher than that of the 124-kD component of cadherin, and as cadherin exhibits the same type of trypsin sensitivity, it could belong to the same group of CDS molecules.

Neural retina has a CDS that is not recognized by ECCD-1. Grunwald et al. (1982) attempted to identify the molecules in chicken neural retina CDS by looking for cell surface components that can be rendered trypsin-resistant with CA^{2+}. They found a protein with such a type of trypsin sensitivity; its molecular weight is 130 kD and it is correlated with the presence of CDS in neural retina cells. Final evidence implicating this molecule in neural retina CDS requires the use of monospecific antibodies. Our recent study using a monoclonal antibody is now close to its goal of identifying the molecules in neural cell CDS, as described below.

A MONOCLONAL ANTIBODY INHIBITING BRAIN CDS

As ECCD-1 does not react with all types of cells, it will be necessary to obtain monoclonal antibodies that recognize CDS in other cells. It should then be possible to study the molecular basis of CDS cell type specificity by comparing molecules with different specificities. Such antibodies would also be helpful in understanding how many types of CDS are present and how these are implicated in the morphogenesis of animals.

To this end, we have been attempting to produce monoclonal antibodies reacting with CDS of various cells types that are not recognized by ECCD-1. Ideally, these antibodies should be as effective in inhibiting the CDS-dependent cell–cell adhesion as ECCD-1. To obtain such antibodies, we have screened hybridomas to detect monoclonal antibodies that induce disruption of cell–cell adhesion of target tissues. With this approach, we

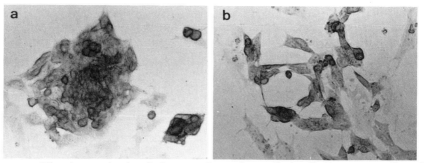

Figure 1. *Effect of a monoclonal antibody on monolayer cultures of brain cells.* Brains were collected from 10.5-day-old mouse fetuses; after dissociation by trypsin treatment, cells were cultured overnight in a medium without (*a*) and with (*b*) this antibody. The cultures were stained with Giemsa solution. *a* and *b* are at the same magnification.

have recently succeeded in obtaining a monoclonal antibody that actively perturbs cell–cell contact in brain cell colonies (Figure 1). A preliminary assay suggests that, in brain cells, this antibody recognizes CDS components that differ from cadherin. Characterization of the target molecules for this antibody is in progress.

MORPHOGENETIC ROLE OF CDS

One approach to understanding the role of CDS in morphogenesis is to examine the effect of disruption of this adhesion system on the development of whole embryos or tissues. Removal of Ca^{2+} is one way to block the function of CDS, but this ion is involved in many processes essential for cellular functions. Thus, antibodies that can specifically block a CDS-dependent process are ideal probes; monoclonal antibodies are particularly useful although ECCD-1 is the only monoclonal now known that actively disrupts the CDS-dependent cell–cell adhesion.

A recent study by Shirayoshi et al. (1983) examined the effect of ECCD-1 on early development of mouse embryos. Embryos cultured *in vitro* in the presence of ECCD-1 cannot undergo compaction, which normally takes place at the late eight-cell stage. Successive cultures of embryos in ECCD-1, however, allows them to display a compacted morphology at the late 16-cell stage. Thereafter, they develop into apparently normal blastocysts even in the continued presence of ECCD-1.

A striking finding from the blastocysts so formed is that they lack an inner cell mass (ICM) and are composed of only trophectoderm cells. The results of this experiment indicate that the intimate cell–cell contact depending upon CDS (cadherin in this CDS) in compacted embryos at the 8- to 16-cell stage is essential for generation of cells with the ICM phenotype.

The mechanism for this CDS-dependent differentiation of blastomeres has yet to be solved, although the possibility has been suggested that CDS-dependent cell–cell contact is responsible for the unequal division of cells to produce multiple cell lineages. CDS is thus essential in the early morphogenetic processes of mouse embryos. After the 16-cell stage, ECCD-1 becomes inaccessible to cadherin-bearing cells that are inside the embryos as a result of the formation of ECCD-1-resistant barriers such as the trophectoderm layer. We therefore cannot study the effect of this antibody on the later developmental processes of mouse embryos. Isolated tissues in organ culture systems may serve this purpose in future studies.

Similar studies are possible using monoclonal antibodies that are able to block cell–cell adhesion via other types of CDS. The monoclonal antibody recognizing brain CDS is a particularly interesting probe for analysis of the morphogenesis of the complex structure of nervous systems.

CDS IN CANCER CELLS

Much attention has been paid to the adhesiveness of cancer cells, since possible abnormalities in cell–cell contact may be a primary cause of abnormal cell proliferation. Abnormality in cell adhesiveness may also be related to metastasis of cancer cells.

A number of studies have been conducted comparing the adhesiveness of normal and transformed fibroblastic cells. One general conclusion reached from them is that transformed cells are less adhesive. It should be pointed out, however, that most studies on cancer cell adhesiveness were performed before we realized that because of the complexity of cell adhesion mechanisms, special precautions are necessary in manipulating cells to assay their adhesiveness. One must therefore be careful when evaluating conclusions drawn from early studies on cell adhesion.

Because multiple cell–cell adhesion mechanisms are present in a cell (Takeichi, 1977; Takeichi et al., 1979; Urushihara et al., 1979), it is important to assay each mechanism individually to measure accurately the adhesive property of any given cell. Comparison of the aggregating property of normal and polyoma-transformed cells (BHK cells) by dissecting the Ca^{2+}-dependent and Ca^{2+}-independent cell–cell adhesion systems was first attempted by Urushihara et al. (1977) and Urushihara and Ikawa (1983). They found that the Ca^{2+}-independent system is significantly less active in transformed cells, whereas the Ca^{2+}-dependent system (CDS) is equally active in normal and transformed cells, suggesting that CDS is not affected by transformation. After this study, we assayed CDS in a variety of normal and transformed cell types and found no correlation between its presence and the malignant transformation. It is important to note that, in malignant cells with a strong mutual cohesiveness, CDS is always highly active. It seems, therefore, that CDS is an essential system for mutual adhesion of tumor cells.

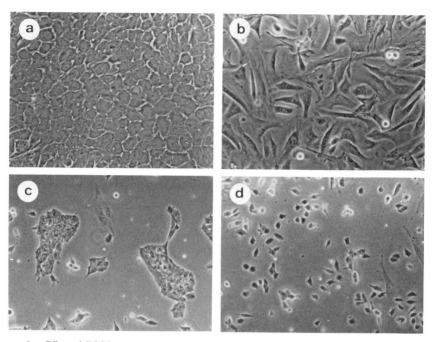

Figure 2. *Effect of ECCD-1 on mammary gland and tumor cells. a* and *b*: Normal mammary gland cells collected from a pregnant C3H mouse. In *a*, freshly collected cells were cultured for two days in normal medium; in *b*, cells were cultured in normal medium for the first day and in normal medium supplemented with ECCD-1 for the second. Note the morphological conversion of cells from epithelial to fibroblastic resulting from the presence of ECCD-1. *c* and *d*: Primary cultures of mammary tumor cells collected from a tumor spontaneously generated in a C3H mouse. In *c*, freshly collected cells were cultured overnight in normal medium; in *d*, cells were cultured overnight in normal medium supplemented with ECCD-1. Note the inhibition of colony formation in *d*. For these experiments, an epithelial cell population was collected from tissues by differential centrifugation after dissociation treatment with 0.1% collagenase in Hanks' solution for one hour at 37°C. *a* and *b*, and *c* and *d*, are at the same magnification, respectively.

With ECCD-1, it is possible to assay accurately the activity of CDS in epithelial cancer cells. We compared the cell–cell adhesion property of normal mammary gland cells and mammary tumor cells genetically produced in C3H and SHN mice (M. Takeichi, unpublished observations). Both normal and tumor cells each formed cohesive monolayer sheets; their cell–cell adhesions were equally disrupted by ECCD-1 (Figure 2). Cells in a mammary tumor line MTD-1 isolated by Enami et al. (1984) were also sensitive to ECCD-1. When grown in collagen gel, these cells formed colonies with a unique dendritic morphology in which they were tightly connected to each other. By adding ECCD-1 to the culture, individual cells acquired the capacity to migrate out from the edges of the colonies (Figure

3). Thus, blocking of CDS caused the release of cells from tumor cell colonies that originally did not permit such migration.

Pitelka et al. (1980) made a precise observation of the cell–cell junctional structure of normal mammary gland and mammary tumor, comparing it with that of the same tumor metastasized into lung, and found no difference in the ultrastructure of cell–cell junctions between the original and metastasized tumors. From this, and the observations described above, it appears that major change in cell–cell adhesiveness is not necessarily required for the metastatic capacity of tumor cells. From this standpoint, it is conceivable that the induction of metastasis of cells is not always due to a permanent change in their adhesive properties but to the temporal breakdown of cell–cell adhesion by some unknown factors. As mentioned above, blocking of CDS results in the release of cells from a tumor. Therefore, if some enzymic factors are capable of digesting the proteins functioning in CDS under physiological conditions, they may be primarily responsible for inducing metastasis of tumor cells.

CONCLUSIONS

Cell–cell adhesion is a complicated system involving multiple classes of molecules. Among them, the Ca^{2+}-dependent cell–cell adhesion molecules

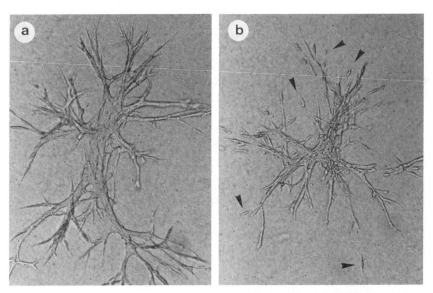

Figure 3. *Effect of ECCD-1 on colony formation of mammary tumor cells (MTD-1 line) in collagen gel.* Clonal cultures were prepared by embedding MTD-1 cells in collagen gel containing medium without (*a*) and with (*b*) ECCD-1. Micrographs were taken after one week in culture. Note the release of many single cells (*arrows*) from a dendritic colony of MTD-1 cells in the presence of ECCD-1. *a* and *b* are at the same magnification.

seem to play the most basic role in the construction of multicellular animals, since blocking of the function of these molecules leads to dissociation of tissues. The importance of Ca^{2+}-dependent cell–cell adhesion molecules is emphasized because they are divided into subclasses with cell type-specific properties. Our experiments showed that cells expressing one type of Ca^{2+}-dependent cell–cell adhesion molecules will segregate from those expressing another type of these molecules when artificially mixed, as seen in combined cultures of teratocarcinoma and fibroblastic cells. This suggests that Ca^{2+}-dependent cell–cell adhesion molecules play a crucial role in segregating different cell types, such as epithelial and mesenchymal cells, during the organization of tissues.

The subclasses of the Ca^{2+}-dependent cell–cell adhesion molecules are similar both in their molecular weight and in their Ca^{2+}-sensitive properties, although they differ in their functional and immunological specificity. It is therefore possible that these molecules belong to the same family and have a basically common structure; the specificity of each subclass could be exerted in a way analogous to that of immunoglobulins. Genetic approaches to the verification of these ideas will be important and attractive subjects in future studies on cell–cell adhesion molecules.

ACKNOWLEDGMENTS

We would like to thank Professor T.S. Okada for his critical comments and encouragement on the projects described in this chapter. We also thank Dr. J. Enami (Dokkyo University) for a mammary tumor cell line and Dr. A. Murakami (Virus Institute, Kyoto University) for spontaneous mammary tumors. This work was supported by research grants from the Ministry of Education, Science, and Culture of Japan.

REFERENCES

Adamson, E. D., M. J. Evans, and G. G. Magrane (1977) Biochemical markers of the progress of differentiation in cloned teratocarcinoma cell lines. *Eur. J. Biochem.* **79**:607–615.

Brackenbury, R., U. Rutishauser, and G.M. Edelman (1981) Distinct calcium-independent and -dependent adhesion systems of chicken embryo cells. *Proc. Natl. Acad. Sci. USA* **78**:387–391.

Damsky, C. H., J. Richa, D. Solter, K. Knudsen, and C. A. Buck (1983) Identification and purification of a cell surface glycoprotein mediating intercellular adhesion in embryonic and adult tissue. *Cell* **34**:455–456.

Enami, J., S. Enami, and M. Koga (1984) Isolation of an insulin-responsive preadipose cell line and a mammary tumor-producing, dome-forming epithelial cell line from a mouse mammary tumor. *Dev. Growth and Differ.* **26**:223–234.

Gallin, W. J., G. M. Edelman, and B.A. Cunningham (1983) Characterization of L-CAM, a major cell adhesion molecule from embryonic liver cells. *Proc. Natl. Acad. Sci. USA* **80**:1038–1042.

Grunwald, G. B., R. L. Geller, and J. Lilien (1980) Enzymatic dissection of embryonic cell adhesive mechanisms. *J. Cell Biol.* **85**:766–776.

Grunwald, G. B., R. S. Pratt, and J. Lilien (1982) Enzymatic dissection of embryonic cell adhesive mechanisms. III. Immunological identification of a component of the calcium-dependent adhesive system of embryonic chick neural retina cells. *J. Cell Sci.* **55**:69–83.

Hyafil, F., D. Morello, C. Babinet, and F. Jacob (1980) A cell surface glycoprotein involved in the compaction of embryonal carcinoma cells and cleavage-stage embryos. *Cell* **21**:927–934.

Hyafil, F., C. Babinet, and F. Jacob (1981) Cell–cell interactions in early embryogenesis: A molecular approach to the role of calcium. *Cell* **26**:447–454.

Imhof, B. A., H. P. Vollmers, S. L. Goodman, and W. Birchmeier (1983) Cell–cell interaction and polarity of epithelial cells: Specific perturbation using a monoclonal antibody. *Cell* **35**:667–675.

Kemler, R., C. Babinet, H. Eisen, and F. Jacob (1977) Surface antigen in early differentiation. *Proc. Natl. Acad. Sci. USA* **74**:4449–4452.

Klebe, R. J., and F. H. Ruddle (1969) Neuroblastoma: Cell culture analysis of a differentiating stem cell system. *J. Cell Biol.* **43**:69a.

Magnani, J. L., W. A. Thomas, and M. S. Steinberg (1981) Two distinct adhesion mechanisms in embryonic neural retina cells. I. A kinetic analysis. *Dev. Biol.* **81**:96–105.

Ogou, S., T. S. Okada, and M. Takeichi (1982) Cleavage-stage mouse embryos share a common cell adhesion system with teratocarcinoma cells. *Dev. Biol.* **92**:521–528.

Ogou, S., C. Yoshida-Noro, and M. Takeichi (1983) Calcium-dependent cell–cell adhesion molecules common to hepatocytes and teratocarcinoma stem cells. *J. Cell Biol.* **97**:944–948.

Pitelka, D. R., S. T. Hamamoto, and B. N. Taggart (1980) Epithelial cell junctions in primary and metastatic mammary tumors of mice. *Cancer Res.* **40**:1588–1599.

Shirayoshi, Y., T. S. Okada, and M. Takeichi (1983) The calcium-dependent cell–cell adhesion system regulates inner cell mass formation and cell surface polarization in early mouse development. *Cell* **35**:631–638.

Sundarraj, N., M. Schachner, and S. E. Pfeiffer (1975) Biochemically differentiated mouse glial lines carrying a nervous system specific cell surface antigen (NS-1). *Proc. Natl. Acad. Sci. USA* **72**:1927–1931.

Takeichi, M. (1977) Functional correlation between cell adhesive properties and some cell surface proteins. *J. Cell Biol.* **75**:464–474.

Takeichi, M., H. S. Ozaki, K. Tokunaga, and T. S. Okada (1979) Experimental manipulation of cell surface to affect cellular recognition mechanisms. *Dev. Biol.* **70**:195–205.

Takeichi, M., T. Atsumi, C. Yoshida, K. Uno, and T. S. Okada (1981) Selective adhesion of embryonal carcinoma cells and differentiated cells by Ca^{2+}-dependent sites. *Dev. Biol.* **87**:340–350.

Urushihara, H., and Y. Ikawa (1983) Modification of the calcium-independent mechanisms of cell adhesion in transformed BHK cells. *Cell Struct. Funct.* **8**:57–65.

Urushihara, H., M. J. Ueda, T. S. Okada, and M. Takeichi (1977) Calcium-dependent and -independent adhesion of normal and transformed BHK cells. *Cell Struct. Funct.* **2**:289–296.

Urushihara, H., H. S. Ozaki, and M. Takeichi (1979) Immunological detection of cell surface components related with aggregation of Chinese hamster and chick embryonic cells. *Dev. Biol.* **70**:206–216.

Yasumura, Y., V. Buonassisi, and G. Sato (1966) Clonal analysis of differentiated function in animal cell cultures. I. Possible correlated maintenance of differentiated function and the diploid karyotype. *Cancer Res.* **26**:529–535.

Yoshida, C., and M. Takeichi (1982) Teratocarcinoma cell adhesion: Identification of a cell surface protein involved in calcium-dependent cell aggregation. *Cell* **28**:217–224.

Yoshida-Noro, C., N. Suzuki, and M. Takeichi (1984) Molecular nature of the calcium-dependent cell–cell adhesion system in mouse teratocarcinoma and embryonic cells studied with a monoclonal antibody. *Dev. Biol.* **101**:19–27.

Chapter 11

Characterization of Cell-CAM 120/80 and the Role of Surface Membrane Adhesion Glycoproteins In Early Events in Mouse Embryo Morphogenesis

CAROLINE H. DAMSKY
JEAN RICHA
MARGARET WHEELOCK
IVAN DAMJANOV
CLAYTON A. BUCK

ABSTRACT

A cell–cell adhesion molecule of cultured mammary epithelial cells has been characterized. This molecule, cell-CAM 120/80, exists as a 120-kD glycoprotein at the cell surface and is found in conditioned medium as a soluble 80-kD glycopeptide. Its further proteolytic degradation by trypsin is inhibited in the presence of Ca^{2+}. The molecule is localized to the cell–cell borders of a wide variety of mature epithelial tissues. An antibody against the purified 80-kD glycopeptide of cell-CAM 120/80, anti-GP80, and an antiserum recognizing a distinct group of 120–160-kD cell–substratum adhesion glycoproteins, anti-GP140, were used to probe early adhesive events in mouse embryo morphogenesis. Anti-GP80 inhibits both compaction of the 8-16-cell embryo and cell–cell adhesion of isolated inner cell masses, while anti-GP140 inhibits blastocyst attachment in vitro. *Thus cell-CAM 120/80 and the 120–160-kD cell–substratum adhesion glycoproteins first identified in differentiated cells are expressed and required for the earliest morphogenetic events in mouse development.*

Adhesive interactions among cells, and between cells and extracellular matrices, are central to morphogenesis. Significant adhesive events begin as early as the eight-cell stage in the mouse, at which time the embryo undergoes the process of compaction (Figure 1). This process transforms the appearance of the embryo from a grapelike cluster of eight distinct

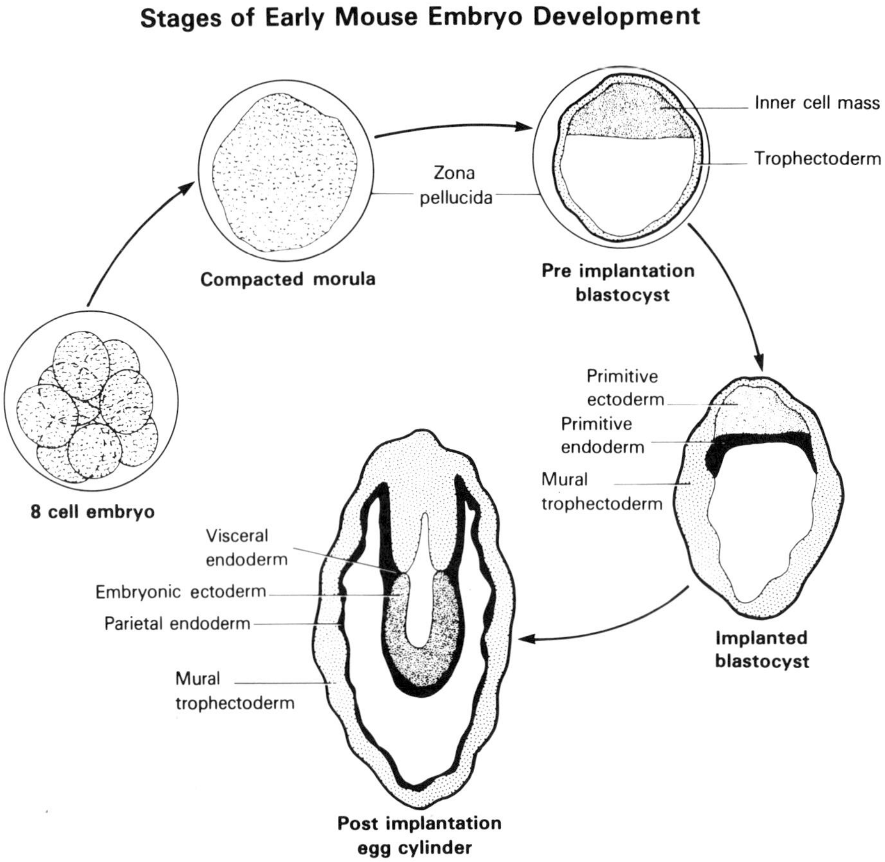

Stages of Early Mouse Embryo Development

Figure 1. *Early morphogenetic events in the mouse embryo.* (After Adamson and Gardner, 1979.)

cells into a tight ball in which cellular boundaries become difficult to discern. This process lays the foundation for the segregation of two distinct cell populations at the blastocyst stage. Cell–matrix adhesion events also come into play early in mouse embryo development. The blastocyst must attach to and invade the uterine wall, and early cell migratory events involving specific cell–matrix interactions are essential for establishing the principal cell layers of both the extraembryonic membranes and the embryo. These events in early mouse development have been thoroughly described and studied both *in vivo* and *in vitro* (for reviews, see Adamson and Gardner, 1979; Hogan et al., 1983; Johnson and Pratt, 1983); such studies have contributed to a greater understanding of the relationship between adhesion, cell shape, and cytoplasmic organization.

Our laboratory has come to appreciate the relationship between adhesion and cell morphology and its importance to morphogenesis as a result of our efforts to identify cell surface molecules that mediate cell–cell and cell–

matrix events. We began these studies using mammary tumor epithelial monolayers grown *in vitro* as a model system. These cells exhibit both cell–cell and cell–matrix interactions and show a polarized distribution of membrane specializations such as microvilli, junctional complexes, and attachment plaques.

Several years ago, efforts to identify cell surface molecules involved in cellular adhesion in mammary epithelial cells led our group to prepare broad spectrum antisera against serum-free medium (SFM) conditioned by several murine and human mammary carcinoma cell lines. When tested on a murine mammary tumor epithelial (MMTE) cell line, two of these antisera had dramatic effects on cellular adhesions. Anti-SFM I, made against SFM conditioned by the target MMTE cell line, caused MMTE cells to round up and detach from the substratum and also disrupted their cell–cell adhesion. Anti-SFM II, made against SFM conditioned by MCF-7, a human mammary carcinoma cell line, had no significant effect on cell–substratum adhesion but caused a disruption of cell–cell interactions in MMTE (Figure 2; Damsky et al., 1981). A closer look at the effects of anti-SFM II on cell–cell adhesion illustrates the close relationship between adhesion, cell shape, and cytoskeletal organization. Transmission electron microscopy of cell monolayers cut parallel to the substratum at the level of junctional complexes shows that, in the control cultures, the surfaces of neighboring cells follow one another's contours closely over considerable distances, and that small bundles of microfilaments form a striking geometric network in the subsurface cytoplasm, interacting frequently with the surface membrane (arrows in Figure 3A). Following treatment with anti-SFM II, the cells become more elongated, the surfaces of neighboring cells are no longer closely apposed, and microfilaments are now arranged in longitudinal bundles (arrows in Figure 3B). Thus, treatment of cells with an antiserum that disrupts cell–cell adhesion also profoundly affects both cell morphology and cytoskeletal organization (Damsky et al., 1981).

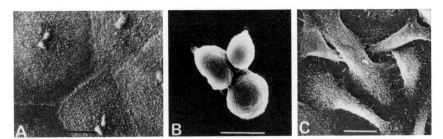

Figure 2. *Scanning electron micrographs showing the effects of adhesion-disrupting antisera on the morphology of mouse mammary tumor epithelia (MMTE). A:* Portions of several neighboring MMTE cells. The cell–cell boundaries of these flat cells are tightly apposed. *B:* MMTE cells after treatment with anti-SFM I. The cells have become rounded and are about to detach. *C:* MMTE cells after treatment with anti-SFM II. Cells remain firmly attached to the substratum but their morphology is fibroblastic. Calibration bars = 10 μm.

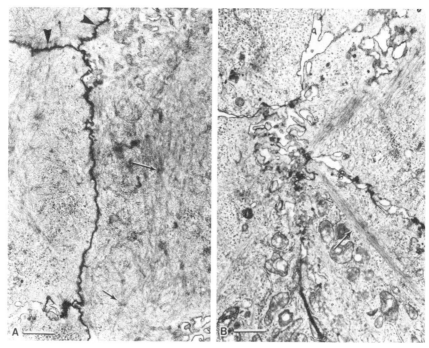

Figure 3. *Effects of anti-SFM II on the ultrastructure of cell surface interactions and cytoskeletal organization. A*: Control; portions of several cells sectioned parallel to the substratum at the level of the junctional complex. Neighboring cell surfaces are closely apposed. Subsurface microfilaments are arranged as an anastomosing network of slender bundles (vertices marked with *arrows*) that interact frequently with the cell surface (*arrowheads*). *B*: MMTE cells treated with anti-SFM II and prepared as above. Neighboring cell surfaces do not follow one another and microfilaments are arranged in longitudinal bundles. Calibration bars = 1 μm.

The effects of anti-SFM I on cell–cell and cell–substratum adhesion can be inhibited independently by different fractions purified from extracts of MMTE cells (Figure 4), indicating that the two types of adhesive interactions are mediated by distinct populations of molecules. The surface antigens recognized by the cell–substratum adhesion-disrupting antibodies in anti-SFM I consisted of a restricted group of glycoproteins of 120–160 kD (Damsky et al., 1981). These glycoproteins are similar to a group of three glycoproteins of 120–160 kD purified previously from hamster fibroblasts and recognized by the polyclonal anti-GP140 serum (Knudsen et al., 1981; Damsky et al., 1982). These cell–substratum adhesion glycoproteins are discussed in detail by Buck et al. (this volume).

We focus first on the identification and characterization of the cell–cell adhesion molecule, which we call cell-CAM 120/80, recognized by anti-SFM II. We then return to the early mouse embryo and describe studies using specific antisera against cell-CAM 120/80 and the cell–substratum adhesion glycoproteins GP120–160 to examine the role of these adhesion molecules in some of the early morphogenetic events illustrated in Figure 1.

IDENTIFICATION AND CHARACTERIZATION OF CELL-CAM 120/80

Material capable of inhibiting the anti-SFM II-induced disruption of cell–cell adhesion of MMTE cells was purified from concentrated SFM conditioned by MCF-7 cells. Following gel filtration and ion exchange chromatography, material with anti-SFM II blocking activity was subjected to sequential lectin affinity chromatography. No activity was bound by wheat germ agglutinin. However, all of the activity was bound by concanavalin A (Con A). The Con A$^+$ material was purified further by anti-SFM II affinity chromatography. In order to increase the specificity of this heterogeneous antiserum, it was adsorbed to and eluted from the surface of intact MCF-7 cells as described previously (Damsky et al., 1983). This absorbed anti-SFM II antiserum, which retained its ability to disrupt cell–cell adhesion of MMTE cells, was then immobilized and used to bind the cell–cell adhesion-related material in the Con A$^+$ fraction. Analysis of the material eluted from the anti-SFM II column by SDS-polyacrylamide gel electrophoresis revealed two bands at about 80 kD and 55 kD. To determine which band was related to the anti-SFM II blocking activity, the gel lane was sliced and eluates from each slice tested in the antibody blocking

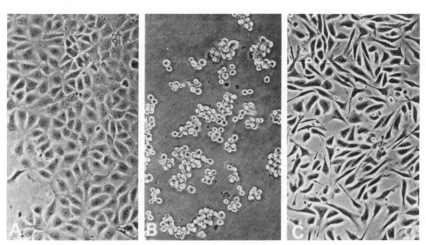

Figure 4. *Selective blocking of cell–substratum adhesion-disrupting antibodies in anti-SFM I by a complex of 120–160-kD glycoproteins purified from MMTE cells. A: Control MMTE monolayer. B: MMTE cells detached by anti-SFM I. C: MMTE cells treated with a mixture of anti-SFM I and purified 120–160-kD glycoproteins. Cells are adherent and spread but are fibroblastic in morphology, showing that anti-SFM I contains both cell–cell and cell–substratum adhesion-disrupting antibodies and that the cell–substratum adhesion-disrupting activity can be selectively inhibited.*

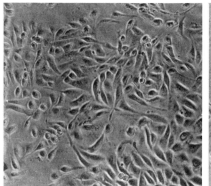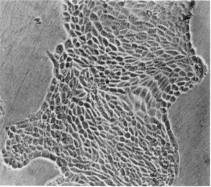

Figure 5. *Blocking of cell–cell adhesion-disrupting activity of anti-SFM II by purified GP80. Left:* MMTE treated with anti-SFM II. *Right:* MMTE treated with a mixture of anti-SFM II and GP80.

assay. All of the anti-SFM II blocking activity (Figure 5) was associated with a glycopeptide in the 80-kD region of the gel.

Antiserum specific for this 80-kD glycopeptide was raised in a rabbit immunized with material eluting from the 80-kD region of the gel. This antiserum, designated anti-GP80, disrupted cell–cell adhesion in MMTE cells at a 20-fold higher titer than anti-SFM II. The specificity of the anti-GP80 was evaluated by Western blotting, as shown in Figure 6. Anti-GP80 bound a single band at 80 kD in unfractionated MCF-conditioned SFM (Figure 6A,C). When exposed to NP40 extracts of MCF-7 cells (Figure 6B), bands at 120 kD and 92 kD were recognized (Figure 6D). NP40 extracts of the JAR human gestational choriocarcinoma cell line contained one reactive band at 120 kD (Figure 6E). A 90-kD anti-GP80 binding peptide could also be detected in JAR extracts when protease inhibitors were omitted during detergent extraction, suggesting that the 90-kD species was a breakdown product of the 120-kD form. No material was recognized by anti-GP80 in NP40 extracts of WI-38 human fibroblasts (Figure 6F).

The relationship between the 120-kD form of cell-CAM recognized by anti-GP80 in NP40 extracts of epithelioid cells and the 80-kD glycopeptide purified from conditioned SFM was examined further by trypsinizing extensively NP40 extracts of JAR or MCF-7 cells in the presence and absence of Ca^{2+}. In the presence of Ca^{2+}, trypsinization of NP40 extracts of either JAR or MCF-7 converted all anti-GP80 binding material to a molecular weight of 80 kD (Figure 6H). In the absence of Ca^{2+}, proteolysis destroyed all GP80 binding activity (Figure 6I). Thus, cell-CAM 120/80 joins the family of cell adhesion molecules protected from complete proteolytic breakdown by Ca^{2+}.

The fact that the 80-kD glycopeptide is found naturally in conditioned medium and can also be produced by trypsinization of the 120-kD form in the presence of Ca^{2+}, suggests that proteolysis may play a role in the turnover of cell-CAM 120/80 and perhaps in the regulation of its function as an adhesive molecule. This idea is supported by the extreme protease sensitivity of cell-CAM 120/80. An elaborate mixture of protease inhibitors is a necessary component of the detergent extraction buffer in order to maintain the molecule in its 120-kD form.

Having established that the 120-kD molecule was the likely cell surface-associated form of cell-CAM, we then asked how tightly associated it was with the cell membrane. JAR choriocarcinoma cells were treated in the presence of protease inhibitors, with 1 M urea, hypertonic and hypotonic salt solutions, and EDTA in attempts to extract cell-CAM from the cell in its 120-kD form in the absence of detergent. Extracts were tested both for

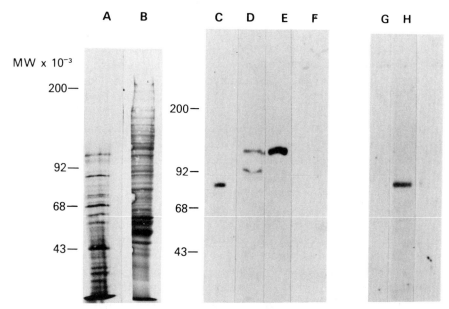

Figure 6. *Evaluation of the specificity of anti-GP80 antiserum.* Immunoblotting experiments with protein A-purified anti-GP80 IgG and ^{125}I-labeled protein A. *Lane A*: Autoradiogram of concentrated, unfractionated SFM produced by ^{35}S-methionine-labeled MCF-7 cells. *Lane B*: Autoradiogram of an NP40 extract of MCF-7 cells labeled with [^{35}S]methionine. *Lane C*: Western blot analysis of unlabeled, unfractionated MCF-7 SFM reveals a single band at about 80 kD. *Lane D*: NP40 extract of MCF-7 cells; bands at 120 kD and 90 kD. *Lane E*: NP40 extract of JAR choriocarcinoma cells; a single band at 120 kD. *Lane F*: NP40 extract of WI-38 human fibroblasts. Anti-GP80 detects nothing. *Lane G*: MCF-7 NP40 extract incubated with anti-GP80 plus purified GP80. All anti-GP80 binding activity is neutralized. *Lane H*: MCF-7 NP40 extract trypsinized in the presence of calcium. All material recognized by anti-GP80 migrates to 80 kD. *Lane I*: MCF-7 NP40 extract trypsinized as in *Lane H* but with 2 mM EDTA instead of calcium. Anti-GP80 binding is not detected. Results similar to those in *Lanes G–I* were obtained when JAR NP40 extracts were used instead of MCF-7 (not shown). (From Damsky et al., 1983.)

their anti-GP80 blocking activity on MMTE cells and by immunoblotting. None of the treatments described extracted the 120-kD form of cell-CAM. In some experiments, hypotonic shock and treatment with 5 mM EDTA released a small fraction of the total anti-GP80 blocking activity, but analysis of these extracts by Western blotting showed the activity to be associated with the 80-kD fragment and thus probably arising by proteolysis during extraction. The 120-kD form of cell-CAM was, however, efficiently extracted with NP40. These results suggest that cell-CAM 120 is an integral cell surface molecule.

EXPRESSION OF CELL-CAM 120/80 IN MAMMALIAN CELLS AND TISSUES

Using anti-GP80, cell-CAM 120/80 was found to be located in the cell–cell boundaries of a wide variety of cultured epithelial cells and adult epithelial tissues. Figure 7 shows examples of the distribution of cell-CAM 120/80 on cultured A431 human epidermoid cells (Figure 7A,B) and in the small intestine of a 17-day-old mouse embryo (Figure 7C,D). Staining was restricted largely to cell–cell boundaries and was reduced or absent from the free surfaces of cells. Table 1 summarizes the results of screening a variety of cultured human and mouse cells, as well as normal and tumor tissues from a variety of organs. In general, cell-CAM was found on all the epithelial cell lines and tissues (both simple and striated) examined and in some, but not all, tumors or tumor cell lines of epithelial origin. Cell-CAM was not found in fibroblasts, parietal endodermlike differentiated teratocarcinoma cell lines (F9-AC-C1-9, PYS-2), in muscle, connective tissue, endothelium, or peripheral nerve. Cell-CAM was absent from some types of liver tumors and from the cultured hepatoblastoma cell line HepG2. Whether cell-CAM's absence in these cases represents a real lack of expression or can be explained by enhanced proteolytic activity (or other such secondary cause) in these epithelial tissues remains to be determined.

Although cell-CAM is present in most, if not all, differentiated epithelia, antisera against it are not able to disrupt cell–cell interactions in most of these cells and tissues. In fact, MMTE cells with their sensitivity to anti-SFM II and anti-GP80 are the exception. Epithelial cells in differentiated tissue are held together by several kinds of adhesive interactions. These include zonulae occludentes (tight junctions), zonulae adherentes (intermediate junctions), and desmosomes. Cell-CAM 120/80 is probably not a component of the zonulae adherentes or desmosomes since it is absent from heart muscle cells, which contain elaborate examples of these junctions in their intercalated disks. Cell-CAM 120/80 is also not likely to be associated with tight junctions because as shown below in studies with early mouse embryos, the appearance of cell-CAM 120/80 precedes the appearance of tight junctions. Thus we must conclude that cell-CAM 120/80 is distributed diffusely along cell–cell borders of epithelia rather than being associated

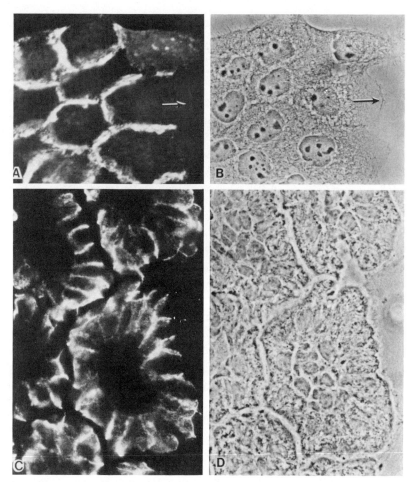

Figure 7. *Localization of cell-CAM 120/80. A*: A431 human epidermoid carcinoma cells. Anti-GP80 staining is restricted to the cell–cell boundaries of an island of cells. *Arrow* indicates free border of island where no staining is seen. *B*: Phase image of same field. *C and D*: Small intestine of 17-day-old mouse embryo. Anti-GP80 staining is restricted to cell–cell borders of epithelial cells (C). The connective tissue in the center of the cross-sections of villi is not stained. Phase image of same field (D).

with a particular junctional specialization.[†] Further, those epithelial cells whose specialized junctions are not fully developed or have been rendered less effective by long-term growth in culture or by malignant transformation are probably held together primarily by cell-CAM 120/80 and are particularly sensitive to anti-SFM II and anti-GP80. We would also predict that cell-CAM 120/80 is involved in an early stage (ontogenetically) in the process of establishing the stable, multiple adhesive strategy of mature epithelia.

[†] Note added in proof: Recent evidence (Boller et al., 1985, *J. Cell Biol.* **100**:327) indicates association of uvomorulin with intermediate junctions in mature intestinal epithelia.

Table 1. Presence of Cell-CAM in Selected Cells and in Normal and Tumor Tissue[a]

Cell/Tissue	Presence of Cell/CAM	Cell/Tissue	Presence of Cell/CAM
Mouse origin		*Human origin*	
ECC (F9)	+ A	Teratocarcinoma: 2102E	+ A,B
F9 retinoic acid	− A	JAR, choriocarcinoma	+ A,B
PYS-2	− A	Fibroblasts	− A,B
Fibroblasts	− A	MCF-7 mammary carcinoma	+ A,B
MMTE	+ A,B	Liver tissue: Hepatocytes	+ B
		Bile duct	+ B
Kidney (adult)	+ A,B	Liver hepatocarcinoma tissue	− B
Uterus (7-D pregnant)	+ B	Liver hepatocarcinoma carcinoma cells	− B
Intestine (17-D embryonic)	+ B	Liver hepatoblastoma HepG2	− B
		Lung alveolar tissue	+ B
		Lung adenocarcinoma	+ B
		Breast adenocarcinoma	+ B
		Kidney	+ B
		Skin: Epidermis	+ A,B
		Dermis	− A,B
		Squamous cell carcinoma	+ B

[a] + A, − A: The presence or absence of cell-CAM 120/80 as detected by the ability of tissue or cells to absorb MMTE cell–cell adhesion-disrupting activity from anti-SFM II or anti-GP80 by the procedure described in Damsky et al., 1983. + B, − B: The presence or absence of cell-CAM 120/80 as observed by immunofluorescence of frozen tissue sections or fixed cultured cells. Immunofluorescence of human tissue and tumor specimens was performed by Dr. Ivan Damjanov, Department of Pathology, Hahnemann Medical College, Philadelphia, PA.

These suggestions are supported by the studies on early mouse embryos described in the next section. The role of cell-CAM 120/80 in adult epithelia is not clear but this molecule may be important in maintaining tissue architecture and epithelial identity during the natural turnover and renewal processes of mature epithelial tissues.

CELL-CAM 120/80 EXPRESSION AND FUNCTION IN EARLY DEVELOPMENT

Having produced antisera that could specifically disrupt cell–cell interaction in MMTE and having identified the molecule on the cell surface with which these antisera reacted to trigger the changes in morphology in these cells, we were then in a position to determine whether the antigen was expressed and functional in embryonic tissue.

As stated earlier, the early mouse embryo undergoes several well-defined morphogenetic events that can be studied *in vitro*. Compaction of the 8–16-cell embryo has been particularly well studied. This process results in a cell–cell contact-induced polarization of eight-cell blastomeres such that their free apical surfaces become rich in microvilli and Con A receptors. These surfaces are relatively nonadhesive. The basolateral surfaces, on the other hand, become smooth and more adhesive, promoting a flattening of neighboring blastomeres upon one another (for a review, see Johnson and Pratt, 1983). This asymmetry is thought to provide the basis for segregating two distinct populations of cells at the next division. Those with a free apical surface will become trophoblast, while those surrounded on all sides by other cells form the inner cell mass (ICM). Compaction coincides with the establishment of intercellular communication via gap junctions (Lo and Gilula, 1979) and precedes the formation of tight junctions required for blastocoel formation. Anti-GP80 inhibits compaction of eight-cell embryos and is effective in disrupting compacted embryos at least until the 16–32-cell stage (Figure 8). The inhibition of compaction is reversible (Figure 9) and does not interfere with cell division (Damsky et al., 1983). Embryos prevented from compacting by anti-GP80, however, are not able to form tight junctions or undergo blastulation. As a result, the segregation of the ICM and trophoblast as distinct cell populations that occurs during blastulation is not apparent.

Many cells in the decompacted embryo are able to accumulate fluid (Figure 9B), a property displayed by trophoblast cells. Whether all cells in the anti-GP80-treated embryos can do this has not been investigated carefully, but fluid accumulation would suggest that interference with cell–cell interactions by anti-GP80 inhibits development of asymmetry in the blastomeres, rendering them all identical as they continue to divide in the presence of antiserum. If anti-GP80 is added after the 32-cell stage, compaction is not disrupted and normal blastulation occurs (Damsky et al., 1983). Thus, there is a short period of time during which anti-GP80 can affect cell–cell adhesion when applied to intact embryos. The antiserum loses its effectiveness at a time when tight junctions are forming in the embryo.

If normal blastulation is permitted to occur, a layer of trophoblast cells surrounds the ICM and forms a fluid-filled cavity with the compact ICM attached eccentrically to a portion of the trophoblast layer. Eventually, a layer of primitive endoderm cells is segregated from the surface of the ICM exposed to the blastocoelic cavity. The remaining ICM cells (epiblast) become the embryonic ectoderm (Figure 1). Anti-GP80 has no effect when added to intact blastocysts, presumably because well-developed tight junctions between trophoblast cells prevent access of the antiserum to the

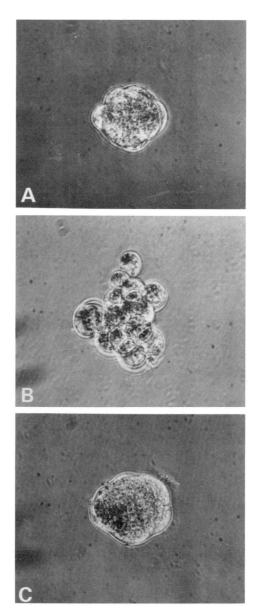

Figure 8. *Effect of anti-GP80 on precompaction mouse embryos.* A: Embryo has been cultured for 24 hours following removal from mouse; compaction has proceeded normally. B: Embryo has been cultured 24 hours in anti-SFM II or anti-GP80; compaction has been prevented. Cell division, however, is not affected. C: Embryo has ben incubated with a mixture containing anti-SFM II or anti-GP80 and about 0.1 μg or purified GP80 eluted from SDS-polyacrylamide gel slices; compaction has proceeded normally. Embryos treated this way are indistinguishable from controls. ×110. (From Damsky et al., 1983.)

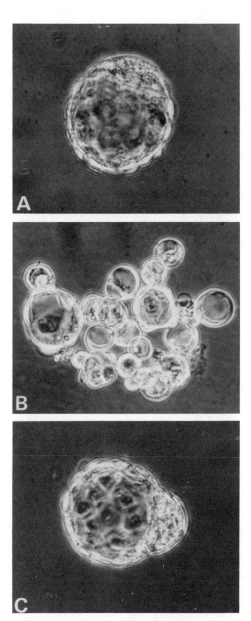

Figure 9. *Reversibility of the effects of anti-GP80 or anti-SFM II on precompaction mouse embyros.* A: Embryo has been cultured for 48 hours. The blastocyst has formed normally. B: Embryo cultured in anti-SFM II or anti-GP80 for 48 hours. Cell division has proceeded (cf. Figure 6B), but compaction and blastocyst formation have been blocked. Some cells are vacuolated, indicating that they are accumulating fluid. C: Embryo cultured in anti-SFM II or anti-GP80 for 24 hours and in normal medium for an additional 24 hours. Compaction and blastocyst formation have proceeded normally. This embryo is indistinguishable from the control. ×110. (From Damsky et al., 1983.)

interior of the embryo. However, following removal of the trophoblast by
immunosurgery (Solter and Knowles, 1975), the isolated ICM will continue
to develop *in vitro*, forming a discrete endoderm layer completely surrounding the compact group of remaining ICM cells (Hogan and Tilly,
1978). If anti-GP80 is added to the isolated ICM prior to the segregation
of a discrete edoderm layer (Figure 10A), it is able to disrupt cell–cell
adhesion (Figure 10B). In fact, the disrupted ICM looks very much like a
decompacted 8–16-cell embryo. If anti-GP80 is added after the endoderm
layer has formed, it is no longer effective in disrupting cell–cell interactions

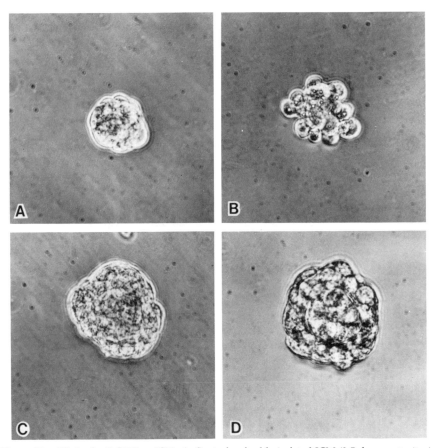

Figure 10.   *Effects of anti-GP80 on ICM. A:* Control; a freshly isolated ICM (3.5 days postcoitus).
Cells appear as a tight cluster of homogeneous cells. *B:* Isolated ICM treated with anti-
GP80 for six hours. ICM is disrupted into a loose aggregate. *C:* Isolated control ICM
cultured for 48 hours *in vitro.* A discrete cell layer has segregated from the rest of the
ICM. If anti-GP80 is added at this time, there is no effect. *D:* Freshly isolated ICM treated
with anti-GP80 for six hours, as in *B,* and then cultured for an additional 40 hours in
normal medium. A discrete layer similar to that found in the control embryo in *C* has
formed, showing that ICM disruption induced by anti-GP80 is reversible.

(Figure 10C). Decompaction of isolated ICM by anti-GP80 is reversible in a majority of embryos. Reversibility appears to depend on the morphology of the cells of the decompacted ICM. Some cells in some of the decompacted ICM appear to accumulate fluid in a manner analogous to cells of the 8–16-cell embryos following prolonged decompaction by exposure to anti-GP80. In decompacted ICM that show no fluid accumulation, reversal is complete and a normal-appearing endoderm layer is formed around the ICM (Figure 10D).

With respect to the decompacted morphology and inhibition of segregation of a new cell layer (the endoderm), the effect of anti-GP80 on isolated ICM is analogous to its effect on 8–16-cell embryos. This may reflect a similarity between the segregation of endoderm and presumptive ectoderm from the ICM, and segregation of trophoblast and ICM cells from the eight-cell embryo. The observation by Gardner (1982) that primitive endoderm cells of the late blastula stage appear to have more microvilli than ectoderm cells supports this idea. In the blastocyst, the outer layer of ICM cells faces two different environments. Some of the cells are exposed to the blastocoelic cavity, while the rest are completely surrounded by other ICM cells. As with the segregation of ICM and trophoblast cells, asymmetry in the cell–cell contacts of the outer layer of ICM cells might be required for a subsequent polarization of the surface membrane and cytoplasmic components in this cell layer. The subsequent cell division in this layer would then result in the segregation of endodermal cells.

These early adhesive events all occur without cell translocation and during a time when all cells involved in the segregation events are in constant contact with one another. Cell-CAM's role, therefore, may not be that of a specific cell recognition molecule per se. Instead, cell-CAM may play a permissive role, providing a particular pattern of cell–cell contacts that is required for subsequent polarization and segregation of cytoplasmic and surface components. Cell-CAM 120/80 is present in the embryo at least as early as the four-cell stage and perhaps earlier (J. Richa and D. Solter, unpublished observations), while compaction and the subsequent asymmetric allocation of cytoplasm, which segregates trophoblast and ICM cells, does not occur until the 8–16-cell stage. Therefore compaction, and probably by analogy, endoderm segregation, are complex events in which cell-CAM 120/80 plays a crucial and necessary role, but not a solitary one.

EXPRESSION OF CELL–MATRIX ADHESION GLYCOPROTEINS IN EARLY DEVELOPMENT

Cell–matrix adhesion events also play important roles in early devlopment. In this section, we discuss (1) the attachment and spreading of the blastocyst *in vitro* (and by analogy, its attachment and implantation *in vivo*); and (2) the migration of parietal endoderm cells out from the primitive endoderm to colonize the roof of the blastocoelic cavity. Anti-SFM I and anti-GP140, which disrupt cell–substratum adhesion of cultured fibroblasts and epithelial cells, were tested for their effects on blastocyst attachment to the substratum *in vitro*. Anti-SFM I, made against SFM conditioned by MMTE cells (Damsky et al., 1981), has a broad specificity, while anti-GP140 is much more specific, being directed against material highly enriched for the 120–160-kD cell–substratum adhesion glycoproteins purified from hamster fibroblasts (Knudsen et al., 1981). Both antisera inhibit blastocyst attachment (Figure 11B) and can remove attached and spread blastocysts from the substratum. The effects of these antisera can be inhibited by preincubation with 120–160-kD cell–substratum adhesion glycoproteins purified from either BHK fibroblasts or MMTE cells (Figure 11C), but not by purified GP80. These results show that the trophoblast cells of the blastocyst express the GP120–160 substratum adhesion glycoproteins *in vitro* at a time when they would be preparing to attach and implant in the uterine wall *in vivo*. Anti-GP140 has no effect on the adhesive behavior of 8–16-cell embryos or isolated ICM prior to segregation of an endoderm layer, suggesting that GP120–160 substratum adhesion glycoproteins play no role in the early cell–cell adhesion events described above.

The importance of cell–matrix interactions is again clearly evident as a subpopulation of cells derived from the lateral margins of the primitive endoderm prepares to migrate out from its original location to encircle the entire blastocoelic cavity underneath the mural trophoectoderm (Figure 1). These migrating cells form the parietal endoderm of the yolk sac. They

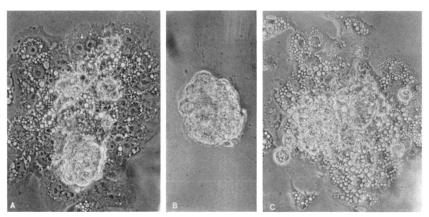

Figure 11. *Inhibition of blastocyst attachment in vitro by anti-GP140 raised against cell–substratum adhesion glycoproteins isolated from BHK fibroblasts. A*: Control; attached and spread blastocyst. *B*: Blastocyst exposed to anti-GP140; attachment is inhibited. *C*: Blastocyst exposed to a mixture of anti-GP140 and 120–160-kD glycoproteins purified from MMTE NP40 extracts as described in Damsky et al. (1981). The effects of anti-GP140 have been inhibited and blastocyst attachment and spreading proceeds normally. ×70.

migrate on a thin layer of fibronectin-containing extracellular matrix that is produced by the trophoectoderm (Semoff et al., 1982). As they populate the roof of the blastocoelic cavity, these parietal endoderm cells secrete large quantities of collagen type IV, laminin, and other basement membrane molecules, thereby producing the very thick Reichert's membrane. In addition to becoming migratory and able to secrete large quantities of basement membrane molecules, these cells become fibroblastic in shape, lose their adhesiveness for one another, and begin to produce the vimentin type of intermediate filament protein instead of exclusively the cytokeratin types (Lane et al., 1983). Thus, a highly complex differentiation process has occurred in these cells. The endoderm cells remaining over the ICM (presumptive embryonic ectoderm) form the extraembryonic visceral endoderm. These cell remain closely associated with one another, maintain an epithelioid morphology, and produce the cytokeratin type of intermediate filament protein only.

The expression or involvement of either cell-CAM 120/80 or the 120–160-kD cell–substratum adhesion glycoproteins during the migration of parietal endoderm is unknown at this time. However, data are available on the expression of these adhesion molecules on F9 embryonal carcinoma stem cells and their parietal endodermlike derivatives, PYS-2 and F9-AC-C1-9. If F9 monolayers are treated with retinoic acid, they become less tightly packed, more spread out, and behave as individual cells rather than as clusters or aggregates. These cells resemble parietal endoderm on the basis of several biochemical markers, in particular, their ability to secrete collagen type IV and laminin (for a review, see Hogan et al., 1983). PYS-2 cells, derived from the same testicular tumor as F9, also resemble parietal endoderm. These cell lines have been used extensively as *in vitro* models for the differentiation of parietal endoderm (Hogan et al., 1983). We have determined by antibody absorption studies that F9 cells express cell-CAM 120/80. However, neither PYS-2 nor F9-AC-C1-9 cells are able to remove the cell–cell adhesion-disrupting activity from anti-GP80. In addition, detergent extracts of these cells are not able to block the decompaction of 8–16-cell mouse embryos induced by anti-GP80, suggesting that neither of these parietal endodermlike cell types expresses cell-CAM 120/80. This is consistent with their morphology in culture, where they appear fibroblastic. PYS-2 and F9-AC-C1-9 cells do express the 120–160-kD cell–substratum adhesion glycoproteins, however. This conclusion is based not only on absorption studies, but on the ability of anti-GP140 to detach them from the substratum. Purified GP120–160 glycoproteins are able to block the adhesion-disrupting effects of anti-GP140. It is tempting to speculate on the expression of adhesion antigens in primitive and parietal endoderm *in vivo*, based on the experiments with teratocarcinoma cells in culture. We would predict, from these experiments, that the cell-CAM 120/80 cell–cell adhesion molecule would be present on the primitive endoderm but absent on the migrating parietal endoderm, while the GP120–160 cell–substratum adhesion glycoproteins would be present on both. These speculations will have to be confirmed by localization studies, in embryos, using anti-GP80 and antibodies specific for anti-GP120–160.

The results of our studies thus far on the effects of adhesion-disrupting antibodies on morphogenetic events in the early mouse embryo demonstrate the presence of two sets of important adhesion molecules: cell-CAM 120/80, which is critical to compaction and segregation of the primitive endoderm; and the GP120–160 cell–substratum adhesion glycoproteins, which are likely to be required for blastocyst attachment and parietal endoderm migration.

RELATIONSHIP OF CELL-CAM 120/80 TO OTHER CELL–CELL ADHESION MOLECULES

In the past several years, a number of cell–cell adhesion molecules or "activities" have been identified. Takeichi and his colleagues (Takeichi, 1977; Takeichi et al., 1981) have developed the concept that these adhesion activities can be grouped according to their dependence on Ca^{2+} for protection from complete proteolytic degradation. As illustrated by the experiment in Figure 6, cell-CAM 120/80 falls into the calcium-dependent category. Its molecular weight (120 kD), the molecular weight of its major trypsin-stable fragment (80 kD), its ability to bind Con A, but not to WGA, and its tissue distribution, are properties shared by cadherin (Ogou et al., 1983; Yoshida-Noro et al., 1984), uvomorulin (Hyafil et al., 1981; Peyrieras et al., 1983), and a 123-kD surface glycoprotein recently reported by Vestweber and Kemler (1984), all identified in mouse embryonal carcinoma cells, and by L-CAM from embryonic chick liver (Gallin et al., 1983; Thiery et al., 1984). The biological activity of anti-GP80 on the early mouse embryo is, in general, similar to that of anti-cadherin, anti-uvomorulin, and GP123, although there are some interesting differences in detail (discussed below). The most logical interpretation of these data is that cell-CAM 120/80, uvomorulin, cadherin, and GP123 are the same or very similar molecules. Since cell-CAM 120/80 has been isolated from differentiated cells (mammary epithelium), and cadherin, uvomorulin, and GP123 from embryonal stem cells, it will be of interest to determine whether there are significant differences in the polypeptide or carbohydrate structures that might suggest that differential expression of a multigene family or posttranslational modifications in this adhesion molecule accompany differentiation.

The effects of early adhesive events in the mouse embryo have been documented for uvomorulin (Hyafil et al., 1981), cadherin (Shirayoshi et al., 1983), cell-CAM 120 (Damsky et al., 1983), and GP123 (Vestweber and Kemler, 1984). Antibodies to all inhibit compaction but some differences exist. Anti-GP80 disrupts compaction and prevents subsequent formation of tight junctions, and therefore blastulation and ICM formation. An anti-cadherin monoclonal antibody interferes with compaction, but permits junction formation and blastulation. The blastocysts attach and spread, but contain no ICM cells. Because of this, Shirayoshi et al. (1983) have been able to determine that cadherin may have a specific effect on the segregation of the ICM rather than on the process of blastulation as a whole. The differences in the actions of these two antibodies may result from the polyclonal nature of the monospecific anti-GP80. Presumably several determinants on cell-CAM 120/80 are bound by anti-GP80, and this may result in a more extensive disruption of cell–cell interactions that interfere more completely with the degree of contact required for junction formation, and therefore blastulation as well as the segregation of ICM.

Studies by Shur (1983) on cell–cell adhesion in mouse embryonal carcinoma stem cells have raised the interesting possibility that cell surface galactosyl transferase plays a role in the cell–cell adhesion of these cells. According to this idea, neighboring embryonal carcinoma cells are held together because their surface galactosyl transferase stably binds polyvalent lactosaminoglycans (90 kD), which are excellent substrates for this enzyme. Since the activated sugar, UDP galactose, is virtually absent from the external medium, the enzymatic transfer of galactose to the lactosaminoglycan, which would liberate the lactosaminoglycan from the enzyme, cannot go to completion. Several experiments support this hypothesis. For instance, addition of UDP-galactose reduces cell–cell adhesion presumably by driving the reaction to completion and liberating the polyvalent lactosaminoglycan substrate. Embryonal carcinoma cells adhere well to substrates derivatized with lactosaminoglycan. Embryonal carcinoma cell–cell adhesion can be perturbed by reagents that selectively interfere with galactosyl transferase activity. Two other properties of this system are similar to the cell-CAM 120/80-mediated adhesion system. Galactosyl transferase is protected from proteolytic degradation by Ca^{2+}, and the lactosaminoglycan substrate, like cell-CAM 120/80, uvomorulin and cadherin, is absent from parietal endodermlike PYS cells. Data such as these raise the possibility that the calcium-dependent cell–cell adhesion system that involves cell-CAM 120/80 (or uvomorulin or cadherin) is related to the galactosyl transferase/lactosaminoglycan complex described above. On the other hand, it may be that both systems are functioning during cell–cell adhesion and that interfering with either substantially reduces cell–cell adhesion in embryonal carcinoma cells.

Yet another adhesion system has been described for embryonal carcinoma cells by Grabel et al. (1983). This involves a calcium-independent cell surface fucan/mannan-specific lectin which, along with a calcium-dependent adhesion mechanism (presumably uvomorulin/cadherin/cell-CAM), contributes to the total adhesion strategy of embryonal carcinoma cells. Those workers see a partial inhibition of embryonal carcinoma cell adhesion if either mechanism is inhibited and complete blockage if both mechanisms are inhibited. Since the events surrounding embryonal carcinoma cell adhesion and mouse embryo compaction are complex, it is conceivable that the lectin, the lactosaminoglycan-galactosyltransferase complex, and cell-CAM 120/180 are all important.

SUMMARY

Cell-CAM 120/80 is a 120-kD, calcium-dependent, surface membrane-associated, cell–cell adhesion molecule. It was first identified and isolated in its soluble 80-kD form from serum-free medium conditioned by human mammary carcinoma cells. It is present on a wide variety of normal and tumor epithelial tissues but is by no means restricted to mature tissue. It is clear that an identical or very similar molecule is expressed very early in development and that it is involved in the earliest cell adhesion event in the embryo (compaction). Whether it is the only cell–cell adhesion mechanism involved at that time is not clear, but it is possible to specifically block this mechanism and inhibit compaction. Our results and those of Shirayoshi et al. (1983) imply that cell-CAM 120/80 (cadherin) is required for segregation of ICM from trophoblast, and endoderm from ICM. Equally interesting are the events not affected by treatment of early embryos with decompacting antibodies. Cleavage proceeds on schedule. Cells treated with anti-GP80 accumulate fluid at a time when they would be doing so during blastulation, even if they cannot form the tight junctions necessary to produce a functioning trophoblast. Substrate adhesion events are independent of cell–cell adhesion since blastocyst attachment and trophoblast spreading take place in embryos decompacted by anti-cadherin, and antibodies against substrate adhesion glycoproteins (anti-GP140) have no effect on compaction or endoderm segregation. Thus, although the effects of antibodies against cell–cell adhesion molecules are dramatic, they are discrete, affecting neither the embryo's biological clock nor the functioning of other adhesion mechanisms.

FUTURE GOALS

Some major questions remain concerning the function and regulation of cell-CAM 120/80. As yet we have no information on the molecule(s) with which cell-CAM 120/80 interacts. Does it function as an adhesive molecule by binding to another cell-CAM 120/80 molecule on an adjacent cell or does it have a distinct receptor? Is cell-CAM 120/80 a transmembrane molecule capable of interacting directly with components on the inner face of the plasma membrane? If not, with what integral (trans) membrane component does it associate? Cell-CAM 120/80 is found on a wide variety of adult tissues and in the early embryo. Are the molecules from these different sources identical or are they representatives of a gene family of closely related glycopeptides whose expression is regulated during development? Cell-CAM 120/80 is present on all cells of the eight-cell embryo but is absent from many tissues in the adult. The time at which particular groups of cells stop expressing cell-CAM 120/80 is not yet known in the mouse embyro. However, based on experiments with F9 and parietal endodermlike teratocarcinoma cells, we predict that the primitive and visceral endoderm express cell-CAM 120/80, while the parietal endoderm does not. However, these cells probably do express the 120–160-kD cell–substratum adhesion glycoproteins. Studies on L-CAM (Thiery et al., 1984) and N-CAM (Edelman et al., 1983) suggest that cell adhesion molecules can disappear and reappear on particular groups of cells at different times throughout development. This indicates that the regulation of cell adhesion molecules during development is complex indeed.

Although we are a long way from the final goal of explaining morphogenesis, we have succeeded in identifying a cell–cell adhesion molecule and a cell–substratum adhesion complex that are expressed very early in embryonic development. We have shown that specific antibodies against these molecules can be powerful tools in dissecting their roles in particular adhesion events in early development. Additional studies that include probes against other membrane- and matrix-associated adhesion molecules should contribute strongly to an increased understanding of morphogenesis.

ACKNOWLEDGMENTS

The authors are grateful to Dr. Davor Solter for his helpful discussions and continued interest in this work, to Dr. Peter Andrews for critical reading of the manuscript, and to Mrs. Helen Schorr for preparation of the manuscript. Work originating in the authors' laboratory is supported by CA-32311, CA-27909, CA-27932, CA-10815, and HD-12487 from the National Institutes of Health, and PCM-81-18801 from the National Science Foundation. M. W. is the recipient of NRSA CA-07572.

REFERENCES

Adamson, E. D., and R. L. Gardner (1979) Control of early development. *Br. Med. Bull.* **35**:113–119.

Damsky, C. H., K. A. Knudsen, R. J. Dorio, and C. A. Buck (1981) Manipulation of cell–cell and cell–substratum interactions in mouse mammary tumor epithelial cells using broad spectrum antisera. *J. Cell. Biol.* **89**:173–184.

Damsky, C. H., K. A. Knudsen, and C. A. Buck (1982) Integral membrane glycoproteins related to cell–substratum adhesion in mammalian cells. *J. Cell Biochem.* **18**:1–13.

Damsky, C. H., J. Richa, D. Solter, K. Knudsen, and C. A. Buck (1983) Identification and purification of a cell surface glycoprotein mediating intercellular adhesion in embryonic and adult tissue. *Cell* **34**:455–466.

Edelman, G. M., W. J. Gallin, A. Delouvée, B. A. Cunningham, and J.-P. Thiery (1983) Early epochal maps of two different cell adhesion molecules. *Proc. Natl. Acad. Sci. USA* **80**:4384–4388.

Gallin, W. J., G. M. Edelman, and B. A. Cunningham (1983) Characterization of L-CAM, a major cell adhesion molecule from embryonic liver cells. *Proc. Natl. Acad. Sci. USA* **80**:1038–1042.

Gardner, R. L. (1982) Investigation of cell lineages and differentiation in the extraembryonic endoderm of the mouse. *J. Embryol. Exp. Morphol.* **68**:175–198.

Grabel, L. B., M. S. Singer, E. R. Martin, and S. D. Rosen (1983) Teratocarcinoma stem cell adhesion: The role of divalent cations and a cell surface lectin. *J. Cell Biol.* **96**:1532–1537.

Hogan, B. L. M., and R. Tilly (1978). *In vitro* development of inner cell masses isolated immunosurgically from mouse blastocysts. I. Inner cell masses from 3.5-day p.c. blastocysts incubated for 24 hours before immunosurgery. *J. Embryol. Exp. Morphol* **76**:6071–6078.

Hogan, B. L. M., D. P. Barlow, and R. Tilly (1983) F9 teratocarcinoma cells as a model for the differentiation of parietal and visceral endoderm in the mouse embryo. *Cancer Surveys* **2**:115–140.

Hyafil, F., C. Babinet, and F. Jacob (1981) Cell–cell interactions in early embryogenesis: A molecular approach to the role of calcium. *Cell* **26**:447–454.

Johnson, M. H., and H. P. M. Pratt (1983) Cytoplasmic localizations and cell interactions in the formation of the mouse blastocyst. In *Time, Space and Pattern in Embryonic Development*, W. R. Jeffrey and R. A. Raff, eds., pp. 287–312, Alan R. Liss, New York.

Knudsen, K. A., P. E. Rao, C. H. Damksy, and C. A. Buck (1981) Membrane glycoproteins involved in cell–substratum adhesion. *Proc. Natl. Acad. Sci. USA* **76**:6071–6078.

Lane, E. B., B. L. M. Hogan, M. Kurkinen, and J. Garrels (1983) Coexpression of vimentin and cytokeratins in parietal endoderm cells of the early mouse embryo. *Nature* **303**:701–703.

Lo, C., and N. B. Gilula (1979) Gap functional communication in the preimplantation mouse embryo. *Cell* **18**:399–409.

Ogou, S., C. Yoshida-Noro, and M. Takeichi (1983) Calcium-dependent cell–cell adhesion molecules common to hepatocytes and teratocarcinoma stem cells. *J. Cell Biol.* **97**:944–948.

Peyrieras, N., F. Hyafil, D. Louvard, H. Ploegh, and F. Jacob (1983) Uvomorulin: A nonintegral membrane protein of early mouse embryos. *Proc. Natl. Acad. Sci. USA* **80**:6274–6277.

Semoff, S., B. L. M. Hogan, and C. R. Hopkins (1982) Localization of fibronectin, laminin/entactin and entactin in Reichert's membrane by immunoelectron microscopy. *EMBO J.* **1**:1171–1175.

Shirayoshi, Y., T. S. Okada, and M. Takeichi (1983) The calcium-dependent cell–cell adhesion system regulates inner cell mass formation and cell surface polarization in early mouse development. *Cell* **35**:631–638.

Shur, B. D. (1983) Embryonal carcinoma cell adhesion: The role of surface galactosyltransferase and its 90 kD lactosaminoglycan substrate. *Dev. Biol.* **99**:360–372.

Solter, D., and B. Knowles (1975) Immunosurgery of mouse blastocyst. *Proc. Natl. Acad. Sci. USA* **72**:5099–5102.

Takeichi, M. (1977) Functional correlation between cell adhesive properties and cell surface properties and some cell surface proteins. *J. Cell Biol.* **75**:464–474.

Takeichi, M., J. Atsumi, K. U. Yoshida, and T. S. Okada (1981) Selective adhesion of embryonal carcinoma cells and differentiated cell by Ca^{2+}-dependent sites. *Dev. Biol.* **87**:340–350.

Thiery, J.-P., A. Delouvée, W. J. Gallin, B. A. Cunningham, and G. M. Edelman (1984) Ontogenic expression of cell adhesion molecules: L-CAM is found in epithelia derived from the three primary germ layers. *Dev. Biol.* **102**:61–78.

Vestweber, D., and R. Kemler (1984) Rabbit antiserum against a purified surface glycoprotein decompacts mouse preimplantation embryos and reacts with specific adult tissues. *Exp. Cell Res.* **152**:169–178.

Yoshida-Noro, C., N. Suzuki, and M. Takeichi (1984) Molecular nature of the calcium-dependent cell–cell adhesion system in mouse teratocarcinoma and embryonic cells studied with a monoclonal antibody. *Dev. Biol.* **101**:19–27.

Chapter 12

Functional and Structural Aspects of the Cell Surface in Mammalian Nervous System Development

MELITTA SCHACHNER
ANDREAS FAISSNER
GÜNTHER FISCHER
GERHARD KEILHAUER
JAN KRUSE
VOLKER KÜNEMUND
JÜRGEN LINDNER
HEIDE WERNECKE

ABSTRACT

Cell surface glycoprotein L1 is a molecular entity distinct from the neural cell adhesion molecule N-CAM but is also involved in a Ca^{2+}-independent adhesion mechanism among neural cells. During development of the mouse central nervous system, L1 appears after and is coexpressed with N-CAM on all postmitotic neuronal cell types investigated so far. During reaggregation of cerebellar cells, L1 acts synergistically with N-CAM. It plays a role in the migration of granule cell neurons in postnatal cerebellar cortex and in the fasciculation of neurites, depending on the culture substrate. L1 appears to be involved in neuron–neuron adhesion but not in neuron–astrocyte or astrocyte–astrocyte adhesion. Structural similarities between the two molecules are detected by the monoclonal antibody L2, which recognizes a functionally active carbohydrate epitope. The L2 epitope appears on all major neural cell types; however, not all N-CAM or L1 antigen-positive cells carry the L2 epitope. The observation that not all N-CAM molecules express the L2 epitope points to a molecular diversity that may have functional implications for modulating adhesion during development. Since the L2 epitope is also present on myelin-associated glycoprotein (MAG) and other as yet unidentified neural cell surface components, we would like to speculate that the L2 epitope-expressing group of molecules forms a family of functionally important and structurally related glycoproteins.

Questions about number, structural diversity, and functional mechanisms of cell surface molecules involved in cell–cell interactions during formation of the nervous system have tantalized developmental neurobiologists at least since Roger Sperry interpreted his experiments on the specification of retinotectal contacts. That these contacts are based, at least to some degree, on recognition between cell surface molecules on neighboring cells is now generally accepted. How many of these molecules are needed to construct the adult mammalian nervous system remains to be elucidated. Furthermore, we have yet to understand the functional mechanisms that underlie the molecules' action at the cellular and molecular levels and whether these molecules are members of a family with a common basic molecular architecture, specialized for a unique functional task by variation of particular, strategically important sites. It seems unlikely that as many distinct molecules specify the highly ordered connectivity between the retina and the tectum as there are cells in the retina, each of which would carry a unique molecular marker and to which another set of markers would be complementary. It seems equally unlikely that only one or two cell surface molecules are able to specify such a complex organ as the brain by simple variation, in time and space, of the molecules' appearance and variation in the strength of affinity, stabilization within the membrane, and deadhesion mechanisms. In this chapter we address the question of the functional and structural heterogeneity of cell surface molecules involved in recognition among neural cells.

To investigate this question, an immunological approach has been used. The advantage of the immunological approach lies in the possibility of isolating molecules by immunoaffinity procedures, studying their structural properties, and at the same time using the antibodies as inhibitors of function, even at the submolecular level of antigenic sites or epitopes. The unique specificity of monoclonal antibodies also allows the recognition of structural similarities between molecules prior to a complete elucidation of each molecule's structure at the level of the amino acid or nucleotide sequence and carbohydrate moiety. Thus, the immunological approach, by using antibodies marked with visual tags, allows not only the dissection of structrual and functional properties, but also the demonstration of the cellular and subcellular localization of antigens in histological sections at different developmental stages.

To study the development of the nervous system in structural and functional terms, we have found it advantageous to investigate the cerebellar cortex of the mouse. This part of the central nervous system displays several features characteristic of the formation of the nervous system, including cell proliferation, migration, aggregation, cytodifferentiation, cell death, synapse formation, and finally, elimination of synapses (Cowan, 1982). The cerebellum contains only five neuronal cell types, which are organized in units of simple geometric arrays repeated throughout the cortex. In mammals, some major events in neuron proliferation and migration, synaptogenesis, and synapse elimination occur postnatally, and thus become more amenable to experimental analysis. Populations and subpopulations of several neuronal and glial cell types have been identified immunocytochemically, isolated, and maintained in culture (Schachner, 1982), so that the cellular and molecular processes of cell interactions can be analyzed and manipulated *in vitro* through control of the cellular and molecular conditions of the culture environment. In the mouse, a number of mutants are available that show selective cell death based on developmental abnormalities in cell–cell interactions (see Weber and Schachner, 1984).

L1 CELL SURFACE GLYCOPROTEIN

Structure

Monoclonal antibody to the L1 cell surface antigen was obtained by immunization with a glycoprotein fraction from the cerebella of 8- to 10-day-old inbred C57BL/6J mice (Rathjen and Schachner, 1984). The glycoprotein fraction was prepared from detergent-solubilized, enriched plasma membranes and obtained by affinity chromatography on lens lentil lectin. This purification procedure was chosen to enrich for particular glycoproteins of the surface membrane and to dissect the molecular heterogeneities previously recognized by polyclonal antibodies such as NS-4 (Goridis et al., 1978; Rohrer and Schachner, 1980).

L1 monoclonal antibody fulfilled certain criteria that made it interesting for further study. It reacted with the cell surface of neurons identified on live cerebellar cells in culture and in histological sections of early postnatal cerebellar cortex, predominantly with postmitotic granule cell neurons about to migrate. It has been suggested that migration of granule cell neurons proceeds by surface contact guidance on radial Bergmann glial processes (Rakic, 1982). To investigate whether the appearance of L1 antigen on neurons before migration was a functional prerequisite for migration or a fortuitous event in differentiation, it seemed necessary to probe for the antigen's involvement in cell–cell adhesion and in migration. As a first step toward probing the functional properties of the L1 antigen, L1 had to be isolated and polyclonal antibodies prepared against it in the hope that functionally important molecular sites also would be recognized. This strategy appeared appropriate, since monoclonal antibodies do not necessarily detect the antigenic sites that are implicated in function.

Preparative amounts of L2 antigen were obtained through immunoaffinity chromatography by use of a monoclonal L1 affinity column (Rathjen and Schachner, 1984). Enriched membrane fractions from adult mouse brain were solubilized by treatment with detergent, passed over the affinity column, and eluted at a high pH. Two prominent bands in the regions of

140 kD and 200 kD were revealed by SDS-polyacrylamide gel electrophoresis. The presence of carbohydrates associated with these two polypeptide bands could be demonstrated by biosynthetic labeling with radioactive fucose or glucosamine. L1 antigen isolated from the C1300 mouse neuroblastoma clone N2A consisted of a similar set of two bands in the 140-kD and 200-kD range, but with slightly higher molecular weights than for the antigens isolated from cerebellar cells. These differences in molecular weight are due to differences in carbohydrate composition, as could be shown by biosynthetic labeling of polypeptides with [^{35}S]methionine in the presence of tunicamycin (A. Faissner, D. Kubler, G. Keilhauer, V. Kinzel, and M. Schachner, unpublished observations). Interestingly, the 200-kD band in cultured neuroblastoma and cerebellar cells was the most prominent, thus showing a different disposition when compared to L1 antigen purified from adult or ealy postnatal mouse brain. Also, the relative proportion of the two protein bands varied with different preparations. Supplementation of buffers with such protein inhibitors as leupeptin, pepstatin, phenyl-methylsulphonylfluoride, and alpha-2 macroglobulin and the calcium-chelating compounds EGTA and EDTA did not influence the relative proportions of the two bands.

During postnatal cerebellar development from day 0 onward, the apparent molecular weight of L1's two glycoprotein bands did not change. Apparently, L1 antigen was also unchanged in four neurological mouse mutants with known cerebellar abnormalities—*reeler*, *weaver*, *staggerer*, and *Purkinje cell degeneration* (Faissner et al., 1984b).

L1 antigen is a molecular entity distinct from antigens related or similar to N-CAM (neural cell adhesion molecule; see Edelman, this volume), BSP-2, and D2, as could be shown by immunospot binding and Western blot analysis (Faissner et al., 1984a). Polyclonal as well as monoclonal antibodies against these molecules reacted with the molecular species used for immunization but not with L1 and vice versa. Polyclonal L1 antibodies specifically recognized the two glycoprotein bands of L1 antigen immunopurified by the monoclonal antibody column and also gave an identical staining pattern in histological sections and on cultured cells, as determined by immunofluorescence double-labeling procedures on adult and early postnatal mouse cerebellar cortex.

Function

Aggregation of Neural Cells.   The first indication that L1 was involved in cell–cell interactions came from the observation that Fab fragments of the polyclonal L1 antibody were able to inhibit short-term reaggregation of single neural cells in suspension. Cerebellar cells expressing L1 antigen, as detected by indirect immunofluorescence (Rathjen and Schachner, 1984; Faissner et al., 1984a), were prevented from reaggregating in the presence of Fab fragments of polyclonal, but not of monoclonal, antibodies. Since monoclonal antibodies were shown to bind to the cell surface of cerebellar cells, inhibition by polyclonal antibodies was taken as evidence for specific, functional, site-dependent interference with aggregation. Like cerebellar cells, the N2A clone of the mouse neuroblastoma C1300 showed L1 antigen-dependent aggregation. Also, inhibition of Ca^{2+}-independent aggregation was seen with Fab fragments of polyclonal, but not of monoclonal, L1 antibodies.

Since in addition to L1, cerebellar cells also express N-CAM, it seemed important to investigate whether the two biochemically distinct cell adhesion molecules are also functionally distinct. Early postnatal cerebellar, rather than C1300 neuroblastoma, cells were used, because they are derived from normal brain tissue at a developmental stage during which important events in cell–cell interactions take place. Furthermore, C1300 neuroblastoma cells express a biochemically altered form of L1 antigen.

In single-cell suspensions obtained from eight-day-old mouse cerebellum with the bacterial protease dispase, more than 95% of the cells were N-CAM/BSP-2 antigen-positive and approximately 70% were L1 antigen-positive, as shown by indirect immunofluorescence. Inhibition of aggregation depended on the concentration of Fab fragments from polyclonal L1 and N-CAM/BSP-2 antibodies (Figure 1; Table 1). When the two antibodies were applied simultaneously in equal amounts, the inhibitory effect on aggregation was more than additive at antibody concentrations up to 0.4

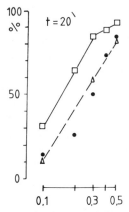

Figure 1. *Inhibition of aggregation of eight-day-old cerebellar cells as a function of concentration of Fab fragments from polyclonal L1 and N-CAM/BSP-2 antibodies alone and in combination. L1, closed circles; N-CAM, triangles; L1 and N-CAM, squares.* Concentrations of Fab fragments are plotted on a logarithmic scale and represent final concentrations. When used in combination, L1 and N-CAM antibodies are present in equal amounts. The percentage of inhibition = [aggregation (control) − aggregation (+ Fab)/aggregation (control)] × 100. Aggregation is expressed as percentage decrease in particle number and varied between 55 and 70% after 20 minutes. Percentages are mean values from duplicates in three independent experiments. For SDs, see Table 1. (Modified from Faissner et al., 1984a.)

Table 1. Inhibition of Aggregation (%) of Eight-Day-Old Cerebellar Cells in the Presence of Fab Fragments of Polyclonal L1 and N-CAM/BSP-2 Antibodies Alone and in Combination [a]

Concentration of Fab Fragments (mg/ml)	L1	N-CAM	L1 and N-CAM
0.1	15.0 ± 11.5	10.5 ± 10.1	31.5 ± 12.1
0.2	26.4 ± 11.0		64.0 ± 11.2
0.3	50.9 ± 17.8	58.6 ± 14.8	84.6 ± 9.3
0.4	73.7 ± 7.9		88.1 ± 6.5
0.5	84.7 ± 11.6	83.1 ± 12.5	92.5 ± 6.8

[a] Experimental procedures and calculations are given in Figure 1 legend. The numbers are mean values of duplicates from three independent experiments ± SDs. (From Faissner et al., 1984a.)

mg/ml, compared with equal concentrations of each antibody alone. Saturation levels were reached at concentrations of about 0.75 mg/ml for the combined antibodies at 20 minutes of aggregation. These experiments show that two structurally and functionally distinct adhesion mechanisms exist among early postnatal cerebellar cells and that they act in a more than additive manner. Synergistic effects are potent means for the modulation of cell surface affinities by a limited number of cell surface molecules. How the two cell adhesion molecules interact with each other to effect this synergism remains to be elucidated. It is possible, for instance, that they enhance their individual activities via topographical proximity within the plasma membrane surface by cooperative effects on stabilization of the cytoskeleton or by ligand–receptor relationships between L1 and N-CAM/BSP-2. However, the synergistic effects observed in live cells are difficult to interpret, and further elucidation of the molecular mechanisms involved will require dissection of the assay system's complexity into its individual components. It should be noted here that, in addition to the two Ca^{2+}-independent adhesion systems, another Ca^{2+}-dependent adhesion mechanism, whose molecular basis is as yet undefined, has been found on early postnatal cerebellar cells (Fischer and Schachner, 1982).

Migration of Granule Cell Neurons.   To gain insight into the involvement of the L1 antigen in granule cell migration, we have used an *in vitro* assay system that displays migratory behavior in small tissue explants during several days of suspension culture (Lindner et al., 1983). Migration of ^{3}H-thymidine pulse-labeled granule cells was observed in cerebellar folium explants from 10-day-old C57BL/6J mice by following the translocation of autoradiographically identifiable cell bodies from the external granular layer through the molecular layer into the internal granular layer over a period of three days (Table 2). The temporal sequence of passage from one layer to the next was similar when a thymidine pulse was given *in vivo* or *in vitro* and migration allowed to occur *in vitro* in serum-free conditions. Neuronal cell loss was not observed in the *in vitro* culture conditions during the limited three-day time period. Furthermore, Bergmann glia were oriented in a radial fashion, as could be seen by immunolabeling with antiserum to glial fibrillary acidic protein. Expression of L1 antigen was also observed as it is *in vivo*, except that the entire width of the external granular layer was strongly L1 antigen-positive after explant cultivation for three days *in vitro*, whereas only the internal part of the external granular layer carried detectable L1 antigen levels *in vivo*. Migration *in vitro* tended to occur with a temporal profile similar to that *in vivo*, with a slight retardation in entry of granule cells into the molecular layer and an acceleration of passage through the molecular layer (Table 2).

Granule cell migration was inhibited when Fab fragments of polyclonal and monoclonal L1 antibodies were added to explant cultures after the thymidine pulse (Table 2). In contrast to the aggregation experiments, Fab fragments of monoclonal antibodies interfered with the migration of cells to an extent comparable to that seen with polyclonal antibodies. The retardation in granule cell translocation was less pronounced when concentrations of Fab fragments were reduced and fragments were added to explants one day after the thymidine pulse (J. Lindner et al., unpublished observations). The incubation of explants with Fab fragments of polyclonal L1 antibody for 15 hours before the thymidine pulse resulted in a migration pattern similar to that observed when antibodies were applied directly after the thymidine pulse. Retardation in granule cell migration was not due to death of labeled cells in the presence of antibody, because staining of the explants with propidium iodide on the third day of culture did not show significant cell death in the external granular or molecular layers. In all cases, the number of thymidine-labeled cells was not detectably affected by L1 antibodies. This was confirmed by counting the absolute number of [^{3}H]thymidine-labeled cells per unit area of a segment comprising all three layers in the histological section.

To verify the specificity of the influence of the L1 antibody on granule cell migration, Fab fragments of nonimmunized rabbits and rabbits immunized with mouse liver membranes were assayed. Antibodies to mouse liver membranes bound to the cell surface not only of neurons, but also of glia and fibroblastlike cells in monolayer cultures. The two antibodies did not detectably interfere with cell migration. Fab fragments of antibodies to N-CAM/BSP-2 also interfered slightly with migration, but to a considerably lesser extent than did L1 antibodies (J. Lindner et al., unpublished results). Furthermore, inhibition of migration in the presence of both antibodies, L1 and N-CAM/BSP-2, was merely additive. Migration also was not modified in the presence of choleragenoid (Willinger and Schachner, 1980) or tetanus toxin.

Table 2. Relative Distribution (%) of ^{3}H-Thymidine Pulse-Labeled Cells in Cerebellar Folia of 10-Day-Old C57BL/6J Mice after Three Days *In Vivo* and *In Vitro*, with or without Antibodies[a]

After thymidine pulse *in vitro* or *in vivo*	95 ± 1.4 EGL 2 ± 0.9 ML 3 ± 0.6 IGL n = 1738 (12)
Pulse label *in vivo*, maintained *in vivo*	14 ± 13.3 6 ± 1.3 80 ± 12.0 n = 805 (2)
Pulse label *in vivo*, maintained *in vitro*	6 ± 3.8 38 ± 6.2 56 ± 2.4 n = 968 (1)
Pulse label *in vitro*, maintained *in vitro*	10 ± 1.6 37 ± 2.3 53 ± 2.3 n = 5353 (8)
Fab fragments of polyclonal L1 antibody (labeled *in vitro*)	33 ± 3.6 33 ± 3.3 34 ± 0.6 n = 3773 (8)
Fab fragments of monoclonal L1 antibody (labeled *in vitro*)	21 ± 4.7 47 ± 4.1 32 ± 4.1 n = 2416 (4)
Fab fragments of anti-liver membrane antibody (labeled *in vitro*)	16 ± 0.8 30 ± 0.7 54 ± 1.5 n = 2384 (3)
Fab fragments of IgG from nonimmune sera (labeled *in vitro*)	16 ± 4.5 33 ± 3.8 51 ± 0.7 n = 1416 (2)

Source: Modified from Lindner et al., 1983.

The observation that migration of granule cells can be modified *in vitro* by antibodies to L1 antigen suggests an involvement of one or both components of L1 in cell–cell interactions during or before migration. Possible explanations for this include the modification of adhesive forces among granule cells before migration or interference with Bergmann glia–granule cell neuron interaction by L1 antibodies. It is remarkable that not all small neurons labeled by [³H]thymidine in the external granular layer are prevented from translocation by L1 antibodies. This may be because stellate and basket cells, which assume their final position in the molecular layer, display a different, L1 antigen-independent, translocation mechanism. Furthermore, it is not known whether all granule cell neurons move from the external to the internal granular layer by interaction with Bergmann glial processes. Indeed, the ability of small neurons to move within the cerebellar cortex in the absence of normal cytoarchitecture has been documented in the case of several neurological mutations of the mouse. It is noteworthy that, although only polyclonal antibodies inhibit adhesion among neural cells, both mono- and polyclonal antibodies affect cell migration. It is conceivable that migration, involving a more complex interplay between cells, is more easily perturbed than is short-term reaggregation of monodisperse cells.

To study the cellular mechanisms underlying granule cell migration, we decided to investigate which cerebellar cell types are involved in L1 antigen-dependent adhesion.

Cell Type-Specific Adhesion. To elucidate which cerebellar cell types use the L1-specific adhesion mechanism, it was necessary to isolate pure populations of glia and neurons and measure adhesive forces between these cells in a quantitative manner. The two major populations of the cerebellar cortex are: (1) late-born, small neurons—the granule, stellate, and basket

---

Table 2. (*Continued*)

[a] After ³H-thymidine pulse labeling of cerebellar cells from 10-day-old C57BL/6J mice *in vivo* or *in vitro*, explants were maintained in suspension culture for three days with or without addition of antibodies (1 mg/ml⁻¹) after termination of the thymidine pulse. For *in vivo* labeling, animals were injected with 10 μCi per g body weight [³H]thymidine (6.7 Ci mmol⁻¹). Autoradiographs were prepared from 10-μm-thick sections. Labeled cell bodies having more than eight grains were counted in fields of standard size. For simplicity, labeled cell bodies in the Purkinje cell layer were included in the counts for IGL. Data represent mean values of percentages ± SDs. *n* represents the total number of cells counted. Values in parentheses indicate the number of independent experiments. First, second, and third rows of numbers represent percentages of labeled cells in EGL, ML, and IGL, respectively. EGL, external granular layer; ML, molecular layer; IGL, internal granular layer.

cells, of which granule cells make up approximately 90%; and (2) astrocytes, which contain mostly Bergmann glial cells and constitute the major glial cell population. Granule cell neurons and Bergmann glial processes engage in a cell surface interaction that may underlie postnatal migration of granule cells, so care was taken to isolate these populations at developmental stages when neuron–glia interactions are known to occur. We succeeded in isolating

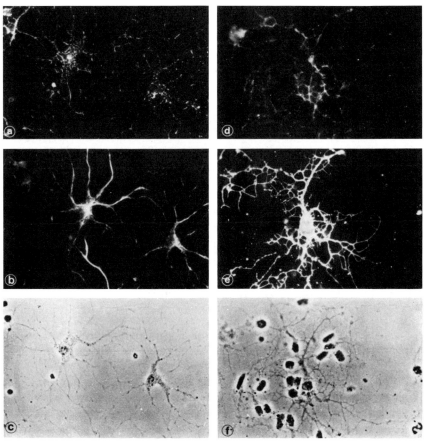

Figure 2. *Double immunofluorescence labeling of monolayer cultures from five-day-old C57BL/6J mouse cerebellum after two days* in vitro *with polyclonal N-CAM/BSP-2, vimentin, and O4 antibodies.* N-CAM antibodies (*a*) react with vimentin-positive glial cells (*b*). *c*: Phase contrast micrograph of field shown in fluorescence images *a* and *b*. N-CAM antibodies (*d*) react with O4 antigen-positive oligodendrocytes (*e*). *f*: Phase contrast micrograph of field shown in fluorescence images *d* and *e*. Cell culture and double immunofluorescence procedures were carried out as described by Schnitzer and Schachner (1981). The tendency for N-CAM to be expressed more on less differentiated glia was seen for astrocytes and oligodendrocytes: more vimentin- and O4 antigen-positive cells and fewer glial fibrillary acidic protein- and O1 antigen-positive cells were also N-CAM-positive. × 400. (Modified from Keilhauer et al., 1985.)

Table 3. Inhibition of Adhesion (%) Between Enriched Populations of Neurons and Astrocytes from Early Postnatal Mouse Cerebellum in the Presence of Fab Fragments from L1, L2, and N-CAM/BSP-2 Antibodies Alone and in Combination[a,b]

Antibody	Fab Concentration (mg/ml)	Neuron[+] to Neuron[++]	Neuron[+] to Astrocyte[++]	Astrocyte[+] to Astrocyte[++]
None	0.0	0 ± 3	0 ± 2	0 ± 2
Poly liver	1.0	2 ± 4	−3 ± 2	1 ± 1
Poly L1	1.0	27 ± 4	0 ± 3	−1 ± 2
Poly N-CAM	1.0	40 ± 7	29 ± 7	24 ± 2
	2.0	41 ± 3	32 ± 3	23 ± 4
Poly L1 + Poly N-CAM	0.5 + 0.5	36 ± 4	nd	nd
	1.0 + 0.5	36 ± 2	nd	nd
Mono L2	1.0	1 ± 5	25 ± 10	15 ± 3
Mono L2 + Poly N-CAM	1.0 + 1.0	56 ± 3	34 ± 6	26 ± 3

Source: Modified from Keilhauer et al., 1985.

[a] Poly, polyclonal antibody; mono, monoclonal antibody; nd, not done; [+], probe cells; [++], target cells.

[b] Astrocyte- and neuron-enriched probe and target cells were obtained from single-cell suspensions of three- to eight-day-old C57BL/6J mouse cerebella by a combination of centrifugation through Percoll and complement-dependent immunocytolysis, using antibodies directed against the surface of neurons, oligodendrocytes, and fibroblastlike cells (Keilhauer et al., 1985). Monolayers were used as target cells after three days in culture with more than 99% of all cells expressing N-CAM and L1 antigen in the case of neurons, and N-CAM in the case of astrocytes. The L2 epitope was detected on approximately 50% of all neurons and astrocytes. Single-cell suspensions of probe cells were obtained by a combination of trypsin–EDTA treatment for 15 minutes at room temperature (Rathjen and Schachner, 1984). To stain the probe cells vitally, fluorescein diacetate was included during the trypsinization step. Probe and target cells were treated with Fab fragments of antibodies for 20 minutes on ice prior to the adhesion tests. Probe cells were added to target cells in Costar wells and incubated for 30 minutes at room temperature in a reciprocal water shaker at 40 cycles per minute. Unbound probe cells were removed from the monolayer target cells by gentle washings. Aggregation among probe cells was negligible in the presence and absence of Fab fragments. Adhering cells were scored with an inverted fluorescence microscope. Adhesion of probe cells in the absence of Fab fragments amounts to 80% ± 15 and 50% ± 15 binding of the total input cells for neurons and astrocytes, respectively. The percentage of inhibition = [adhesion (control) − adhesion (+ Fab)/adhesion (control)] × 100. Numbers are mean values from several experiments ± SD.

pure populations of astrocytes and small neurons, monitored adhesion molecule expression (Figure 2), and determined adhesion of single-cell suspensions of probe cells to monolayer target cells in the presence and absence of Fab fragments (Table 3).

Fab fragments of polyclonal L1 antibody interfered with neuron–neuron but not with neuron–astrocyte or astrocyte–astrocyte adhesion, even when cerebellar astrocytes isolated from cerebella of different postnatal ages or maintained *in vitro* for varying time periods were used as target cells. As controls, Fab fragments of antibodies to liver membranes that reacted intensely with the cell surfaces of both astrocytes and neurons did not inhibit the attachment of probe to target cells in any combination. Fab fragments of polyclonal N-CAM/BSP-2 antibodies inhibited neuron–neuron adhesion by approximately 40%, and neuron–astrocyte, astrocyte–astrocyte, and astrocyte–neuron adhesion by about 25 to 30%. An equimolar mixture of polyclonal L1 and N-CAM/BSP-2 antibodies resulted in additive inhibition values, when compared to the individual antibodies at identical final concentrations. It is noteworthy that, in contrast to the aggregation assay, adhesion between probe and target cells did not obey synergistic mechanisms. It is likely that the less than additive effects in the adhesion system are due to an interdependence between adhesion molecules, to the lower temperatures used to prevent self-aggregation among probe cells, and to the anchoring of one cellular partner, the target cells, at the substrate surface. Under these conditions, lateral mobility of adhesion molecules within the membrane and cooperative effects among them may therefore have been reduced.

From these experiments, it seems evident that distinct adhesion mechanisms operate between neurons and astrocytes in the developing mouse cerebellar cortex at the time of postnatal cell migration. The L1 antigen seems to be involved only in neuron–neuron interaction, whereas N-CAM/BSP-2 mediates not only adhesion of neurons to neurons, but also of neurons to astrocytes and of astrocytes to astrocytes. The observation that L1 antigen appears to specialize in neuron–neuron interaction raises fundamental questions regarding the involvement of L1 in migration of granule cell neurons. Since L1 antibodies interfere with migration even before cells leave the external granular layer, it is possible that a prerequisite for successful migration is the interaction of granule cell neurons or axons with each other. The fact that only the postmitotic granule cells preparing to migrate, and not the proliferating ones, express L1 antigen points to the possibility that L1 promotes the sorting out of more differentiated neurons from less differentiated ones. Our findings also exclude N-CAM/BSP-2 on astrocytes as a candidate for the neuronal L1 "receptor" in this system. Furthermore, it seems unlikely that L1 is functionally similar to another adhesion molecule characterized in chicken brain—Ng-CAM, which promotes neuron–glia interaction (Grumet and Edelman, 1984). Certain structural similarities between L1 and Ng-CAM, however, cannot be overlooked and should be investigated further.

Neurite Fasciculation. L1 antigen is not only present on the cell bodies of postmitotic, premigratory granule cells in the external granular layer and the cell bodies of postmigratory granule cells in the internal granular layer for the short time period of several days, but is also very strongly present on axons of granule cells that fasciculate as parallel fibers in the molecular layer. To investigate whether L1 antigen also mediates fasciculation among L1 antigen-positive neurites, several attempts were made to interfere with the pattern of fasciculation of neurites in culture in the presence of Fab fragments of L1 antibodies. After unsuccessful attempts to modify fasciculation of L1 antigen-positive neurites in cultures of dorsal root and sympathetic ganglia and in long-term reaggregate cultures of early postnatal cerebellum (Lindner et al., 1985), an *in vitro* assay system from early postnatal mouse cerebellar cortex was developed (Fischer et al., 1985) in which L1 antibody showed an effect (Figures 3, 4). Not only was the pattern of neurite outgrowth and the movement of small neuronal cell bodies away from the explant different from that of the control culture to which no antibody was added, but also the speed of neurite outgrowth was increased (Fischer et al., 1985). Several other antibodies that were tested as Fab fragments also had an impact on the outgrowth pattern from the explant, but this effect was quantitatively and qualitatively different from the one observed with

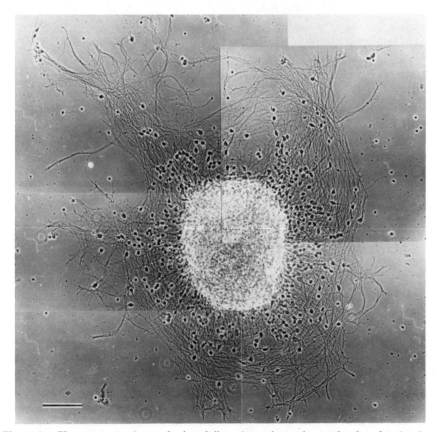

Figure 3. *Phase contrast micrograph of cerebellar microexplant cultures after four days* in vitro. Cerebella from six-day-old C57BL/6J mice, from which meninges and deep cerebellar nuclei had been removed, were forced through a nylon mesh to obtain explants approximately 100–300 μm in diameter. Microexplants were seeded onto poly-D-lysine coated glass cover slips in serum-free, hormone-supplemented medium (Fischer, 1982). Four days after seeding, the microexplants were fixed with paraformaldehyde for microscopic examination. Calibration bar = 100 μm. (Modified from Fischer et al., 1985.)

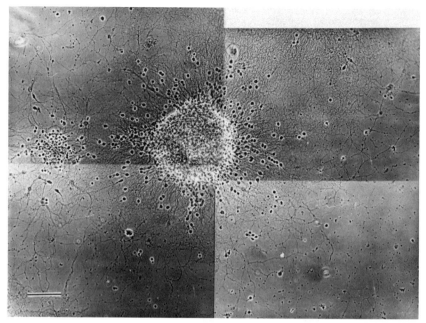

Figure 4. *Phase contrast micrograph of cerebellar microexplant cultures after four days* in vitro *in the presence of Fab fragments of monoclonal L1 antibody.* Microexplant cultures were prepared as described in the Figure 3 legend. One day after seeding Fab fragments, 0.3 mg/ml of monoclonal L1 antibody were added to the culture. Three days later cultures were fixed with paraformaldehyde for microscopic examination. Calibration bar = 100 μm. (Modified from Fischer et al., 1985.)

L1 antibody (V. Künemund and M. Schachner, unpublished observations). As can be seen in Figure 5, the interference of L1 antibodies with fasciculation depended on the substrate on which neurite outgrowth occurred. Although little effect of L2 antibody was seen when neurites extended on a carpet of glial fibrillary acidic protein-positive astrocytes, a distinct effect on neurite fasciculation was observed in regions without cellular substrate (Figure 5). This observation further supports the notion that L1 antigen is not as apparently involved in neuron–glia interaction as in neuron–neuron, or better, neurite–neurite interaction. Our observations also support previous evidence that interaction among neurites depends critically on competing forces exerted by other cells or acellular substrates that are contacted by the outgrowing neurites.

In view of the fact that growth cones are also strongly L1 antigen-positive, it should be pointed out here that L1 does not appear to be involved in synapse formation among cerebellar neurons (Lindner et al., 1985) or between nerve and muscle (Mehrke et al., 1984). Also, the synaptic and electrical activity of cultured neurons are not affected by L1 antibodies (Kettenmann et al., 1983).

THE L2 EPITOPE AND STRUCTURAL SIMILARITIES AMONG ADHESION MOLECULES

Monoclonal L2 antibodies were obtained from hybridoma clones resulting from immunization and fusion as described for L1. These antibodies were shown, by immunospot binding and Western blot analysis, to react with

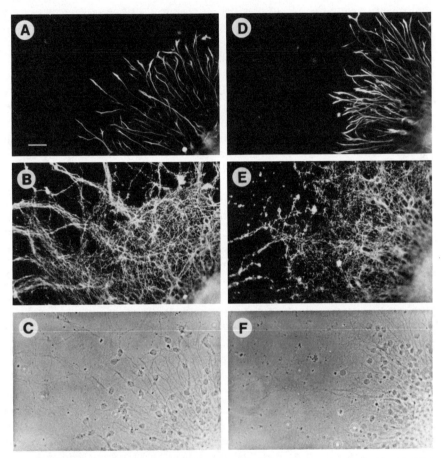

Figure 5. *Double immunofluorescence labeling of microexplant cultures maintained with and without Fab fragments of monoclonal L1 antibodies for three days, using tetanus toxin receptors and glial fibrillary acidic protein as markers.* Microexplant cultures were prepared as described in the Figure 3 legend. One day after seeding, culture medium was replaced with new medium containing no antibody (A–C) or Fab fragments of monoclonal L1 antibodies (D–F) at concentrations of 0.3 mg/ml. Three days later double immunofluorescence labeling procedures were carried out as described by Schnitzer and Schachner (1981). Staining for glial fibrillary acidic protein is shown in A and D, for tetanus toxin receptors in B and E. Phase contrast micrographs of fields shown in fluorescence images A–B, and D–E are shown in C and F, respectively. Calibration bar = 20 μm. (Modified from Fischer et al., 1985.)

both adhesion molecules L1 and N-CAM/BSP-2 and also with the myelin-associated glycoprotein (MAG) (Kruse et al., 1984). In addition, the L2 antibodies recognize, in Western blots, other as yet unidentified cell surface glycoproteins from adult mouse brain. Western blot analysis showed that the monoclonal antibody HNK-1 directed against a subpopulation of human lymphocytes, including natural killer cells, also reacted with L1, N-CAM/BSP-2, MAG, and other unidentified glycoproteins in a pattern similar to that observed for the L2 antibodies. The shared antigenic site or epitope recognized by the L2/HNK-1 antibody was found in the carbohydrate and not in the protein moiety of the L2-positive glycoproteins.

To determine which cell types express the L2 epitope, immunocyto-chemical studies were carried out on cultured cells by using double im-munolabeling techniques with established cell type-specific neural markers. In monolayer cultures of early postnatal cerebellar cells, the L2 and HNK-1 antibodies reacted with a subpopulation of tetanus toxin-, N-CAM-, and L1 antigen-positive neurons, glial fibrillary acidic protein- and vimentin-positive astrocytes, and O4 and O1 antigen-positive oligodendrocytes, but not with fibronectin-positive fibroblastlike cells. The two antibodies were able to block each other in binding. With time in culture, the percentage of N-CAM/BSP-2-positive neurons that were also L2/HNK-1-positive de-creased from approximately 50% after one day to approximately 15% after seven days in culture.

To investigate the paradoxical finding that only some neurons that are recognized by poly- and monoclonal N-CAM or L1 antibodies express the L2 epitope, sequential immunoprecipitation of iodinated N-CAM/BSP-2, immunopurified from adult mouse brain, was performed with antibodies to N-CAM/BSP-2 and L2 (Kruse et al., 1984). These experiments clearly showed that not all N-CAM/BSP-2 molecules of both the adult and embryonic forms of N-CAM express the L2 epitope. Since the 180-kD and 140-kD bands of N-CAM/BSP-2 are heterogeneous with respect to expression of the L2 epitope, the most likely explanation for the absence of L2 on N-CAM/BSP-2-bearing neurons is a heterogeneity among these molecules that is distinct from the molecular heterogeneity represented by the adult and embryonic forms. On the other hand, some neurons may express only the 120-kD component. A similar heterogeneity may also exist for L1, since in this case too, not all L1-positive neurons were found to express the L2 epitope.

These experiments show that two cell adhesion molecules share a common carbohydrate moiety also present on MAG. However, we cannot determine from them whether the two antibodies recognize the same epitope or whether two distinct epitopes closely associated with each other on the carbohydrate chains are present, so that the antibodies compete with each other for binding. In any case, the L2/HNK-1 epitopes are expressed on the same select group of molecules. It has been suggested that MAG mediates the interaction of the periaxonal membrane of myelin with the axon. It is possible, therefore, that the common moiety is not only a sign for fortuitous structural homologies between neural cell surface constituents, but also indicates that the L2-expressing molecules perform similar functional tasks. To probe for the role of the L2/HNK-1 carbohydrate moiety in cell interactions, we used the adhesion system described under "Cell Type-Specific Adhesion," above. Fab fragments of the monoclonal L2 antibody did not reduce neuron–neuron binding over control values, but decreased neuron–astrocyte and astrocyte–astrocyte adhesion by approximately 20% (Table 3). Interestingly, polyclonal N-CAM/BSP-2 antibodies produced a more than additive inhibition in the presence of monoclonal L2 antibody, when compared to the individual antibodies in neuron–neuron, but not in neuron–astrocyte or astrocyte–astrocyte, adhesion.

These experiments show that the L2 epitope carries an adhesive function which manifests itself in neuron–astrocyte, astrocyte–astrocyte and, in conjunction with N-CAM/BSP-2, neuron–neuron adhesion. Modulation of affinities among neural cells have previously been shown to be affected by modifying the carbohydrate chains, as in the adult and embryonic forms of N-CAM/BSP-2 (Hoffman and Edelman, 1983; Sadoul et al., 1983). Removal of more than half of the polysialic acid residues from the embryonic form resulted in the stronger adhesivity that is characteristic of the adult form. Our experiments show that the L2 epitope is another functionally important carbohydrate moiety. Its modulatory action on N-CAM adhesion is more than additive in neuron–neuron interaction, in which the L2 epitope or a sterically neighboring carbohydrate domain by itself does not play a significant role. The fact that the L2 epitope is present on N-CAM/BSP-2, L1, MAG, and other as yet unidentified cell surface glycoproteins, prevents a more detailed analysis of its action in molecular terms at present. Nevertheless, the presence of the L2 epitope on subpopulations of all major neural cell types at particular developmental stages (Wernecke et al., 1985) would make this particular carbohydrate portion an additional candidate for modulating adhesive forces among neural cells. The family of neural cell surface glycoproteins that share the L2 carbohydrate epitope could form the structural roster upon which affinity-modulating carbohydrate chains would change during development. The functional significance of this heterogeneity in the intact developing nervous system remains to be elucidated.

CONCLUSION

Two cell surface molecules that are implicated in cell–cell interactions during development in the mammalian nervous system have now been characterized. The first recognized is N-CAM and the second is the L1 antigen. Yet other neural cell surface glycoproteins are on the verge of being characterized in functional terms by using the shared L2 carbohydrate epitope as a leader, excluding certain cell surface glycoproteins such as the Thy-1, M2, and M6 antigens, but including others, among them MAG. The functional importance of the L2/HNK-1 epitope itself in Ca^{2+}-independent adhesion would seem to guarantee that the protein backbone carrying this carbohydrate moiety is part of the adhesion system. The family of L2/HNK-1 cell surface glycoproteins may comprise a limited set of structurally changing molecular architecture upon which a set of developmentally regulated carbohydrate chains might add further possibilities for permutations. In addition, a selected combination of adhesion molecules within the surface membrane and at topographically distinct sites of the cell (e.g., axon, dendrite, cell body) might further contribute to a differential equilibrium between adhesive and deadhesive forces among the cell surfaces of neighboring cells.

That the L2/HNK-1 eptitope may be used outside the nervous system is not unlikely, since it is expressed on a subpopulation of lymphocytes, including natural killer cells. Whether the adhesive properties of cells in the nervous or immune systems are affected in human gammopathies accompanied by demyelinating neuropathy or by antibodies to the carbohydrate moiety of MAG and other as yet unidentified glycoproteins (see Nobile-Orazio et al., 1984) remain open questions.

REFERENCES

Cowan, W. M. (1982) A synoptic view of the development of the vertebrate central nervous system. *Life Sci. Res. Rep.* **24**:7–24.

Faissner, A., J. Kruse, C. Goridis, E. Bock, and M. Schachner (1984a) The neural cell adhesion molecule L1 is distinct from the N-CAM related groups of surface antigens BSP-2 and D2. *EMBO J.* **3**:733–737.

Faissner, A., J. Kruse, J. Nieke, and M. Schachner (1984b) Expression of neural cell adhesion molecule L1 during development, in neurological mutants and in the peripheral nervous system. *Dev. Brain. Res.* **15**:69–82.

Fischer, G. (1982) Cultivation of mouse cerebellar cells in serum-free, hormonally defined media: Survival of neurons. *Neurosci. Lett.* **28**:325–329.

Fischer, G., and M. Schachner (1982) *In vitro* aggregation of dissociated mouse cerebellar cells. I. Demonstration of different aggregation mechanisms. *Exp. Cell Res.* **39**:285–296.

Fischer, G., V. Künemund, and M. Schachner (1985) Neurite fasciculation in cerebellar microexplant cultures is affected by antibodies to the cell surface glycoprotein L1. *J. Neurosci.* (in press).

Goridis, C., J. A. Joher, M. Hirsch, and M. Schachner (1978) Cell surface proteins of cultured brain cells and their recognition by anti-cerebellum (anti-NS-4) antiserum. *J. Neurochem.* **31**:531–539.

Grumet, M., and G. M. Edelman (1984) Heterotypic binding between neuronal membrane vesicles and glial cells is mediated by a specific cell adhesion molecule. *J. Cell Biol.* **98**:1746–1756.

Hoffman, S., and G. M. Edelman (1983) Kinetics of homophilic binding by embryonic and adult forms of the neural cell adhesion molecule. *Proc. Natl. Acad. Sci. USA* **80**:5762–5766.

Keilhauer, G., A. Faissner, and M. Schachner (1985) Differential inhibition of neuron-neuron, neuron-astrocyte, and astrocyte-astrocyte adhesion by L1, L2, and N-CAM antibodies. *Nature* **316**: 728–730.

Kettenmann, H., M. Wienrich, and M. Schachner (1983) Antibody L1 ejected from a micropipette identifies single neurons without altering electrical activity. *Neurosci. Lett.* **41**:85–90.

Kruse, J., R. Mailhammer, H. Wernecke, A. Faissner, I. Sommer, C. Goridis, and M. Schachner (1984) Neural cell adhesion molecules and myelin-associated glycoprotein share a common, developmentally early carbohydrate moiety recognized by monoclonal antibodies L2 and HNK-1. *Nature* **311**:153–155.

Lindner, J., F. G. Rathjen, and M. Schachner (1983) L1 mono- and polyclonal antibodies modify cell migration in early postnatal mouse cerebellum. *Nature* **305**:427–430.

Lindner, J., P. M. Orkand, and M. Schachner (1985) Histotypic pattern formation in cerebellar reaggregate cultures in the presence of antibodies to L1 cell surface antigen. *Neurosci. Lett.* **55**:145–149.

Mehrke, G., H. Jockusch, A. Faissner, and M. Schachner (1984) Synapse formation and synaptic activity in nerve–muscle cocultures are not inhibited by antibodies to neural cell adhesion molecule L1. *Neurosci. Lett.* **44**:235–239.

Nobile-Orazio, E., A. P. Hays, N. Latov, G. Perman, J. Golier, M. E. Shy, and L. Freddo (1984) Specificity of mouse and human monoclonal antibodies to myelin-associated glycoprotein. *Neurology* **34**:1336–1342.

Rakic, P. (1982) The role of neuronal–glial cell interaction during brain development. *Life Sci. Res. Rep.* **20**:25–38.

Rathjen, F. G., and M. Schachner (1984) Immunocytological and biochemical characterization of a new neuronal cell surface component (L1 antigen) which is involved in cell adhesion. *EMBO J.* **3**:1–10.

Rohrer, H., and M. Schachner (1980) Surface proteins of cultured mouse cerebellar cells. *J. Neurochem.* **35**:792–803.

Sadoul, R. M. Hirn, H. Deagostini-Bazin, G. Rougon, and C. Goridis (1983) Adult and embryonic mouse neural cell adhesion molecules have different binding properties. *Nature* **304**:347–349.

Schachner, M. (1982) Immunological analysis of cellular heterogeneity in the cerebellum. In *Neuroimmunology*, J. C. Brockes, ed., pp. 215–250, Plenum, New York.

Schnitzer, J., and M. Schachner (1981) Expression of Thy-1, H-2, and NS-4 cell surface antigens and tetanus toxin receptors in early post-natal and adult mouse cerebellum. *J. Neuroimmunol.* **1**:429–456.

Weber, A., and M. Schachner (1984) Maintenance of immunocytologically identified Purkinje cells from mouse cerebellum in monolayer culture. *Brain Res.* **311**:119–130.

Wernecke, H., J. Lindner, and M. Schachner (1985) Cell type-specificity and developmental expression of the L2/HNK-1 epitopes in mouse cerebellum. *J. Neuroimmunol.* **9**:115–130.

Willinger, M., and M. Schachner (1980) GM1 ganglioside as marker for neuronal differentiation in mouse cerebellum. *Dev. Biol.* **74**:101–117.

Chapter 13

Adhesion Reactions of Rat Hepatocytes: Cell Surface Molecules Related to Cell–Cell and Cell–Collagen Adhesion

BJÖRN ÖBRINK
CARIN OCKLIND
PER ODIN
KRISTOFER RUBIN

ABSTRACT

We are using rat hepatocytes as a model system for studies on the molecular mechanisms of cell–cell and cell–matrix adhesion reactions. Through an immunological approach, we have identified three cell surface molecules, cell-CAM 105 (CAM for cell adhesion molecule) and CDP (calcium-dependent protein), both of which are involved in cell–cell adhesion, and collagen-CAM, which is involved in cell attachment to collagen. Cell-CAM 105 is an integral membrane glycoprotein with an apparent molecular weight of 105 kD; CDP is a peripheral membrane protein with an apparent molecular weight of 70 kD that can be released from the plasma membrane by chelation of Ca^{2+}; collagen-CAM is an integral membrane glycoprotein with an apparent molecular weight of 120–150 kD. Specific antibodies against collagen-CAM inhibit attachment to collagen but do not inhibit cell–cell adhesion, whereas specific antibodies against cell–CAM 105 or CDP inhibit cell–cell adhesion but not attachment to collagen. Antibodies against cell-CAM 105 or CDP also prevent colony formation of hepatocytes in culture, which suggests that the adhesive interactions of cell-CAM 105 and CDP are important for the formation of intercellular junctions. Cell-CAM 105 and CDP exhibit immunological and structural similarities, but both proteins occur as distinct components on the cell surface, with different sensitivities to treatment with trypsin or EGTA. Furthermore, immunofluorescence microscopy of frozen tissue sections indicates that the two proteins occupy different areas of the cell surface, cell-CAM 105 being localized specifically to the bile canalicular regions, whereas CDP has a more random distribution. Thus, our results indicate that different molecular forms of basically the same cell adhesion molecule might occur on the cell surface, and that they might have different functions in the adhesive reactions of cells. A change in the balance between various forms of cell adhesion molecules might allow cells to modulate their adhesive interactions.

Interactions at the cell surface are of utmost importance in the morphogenetic events of embryonic development and in the maintenance of the differentiation, structure, and function of tissues (Moscona, 1976; Moscona et al., 1983; Ekblom, 1984). These interactions include the binding of hormones and growth factors to their receptors, cell–cell recognition and adhesion, junction formation, and recognition and attachment to extracellular matrix components. Hormone–receptor interactions are fairly well understood at the molecular level, but little is known about the mechanisms whereby cells interact with each other or with the surrounding matrix. As a first step in increasing our knowledge of cell–cell and cell–matrix interactions, we must identify and characterize the cell surface molecules that take part in them. In order to do so, we must first define what type of cellular interaction we want to study, since both cell–cell and cell–matrix interactions include a variety of processes. In this chapter we discuss only the interactions that are of importance for morphogenetic events, leaving, for example, cellular interactions in the immune system out of the discussion.

It is obvious that interactions can occur between dissimilar cell types (i.e., heterotypic interactions) and between cells of the same type (i.e., homotypic interactions). Examples of the former are epithelial–mesenchymal interactions giving rise to embryonic induction, for example, in kidney development or in the formation of exocrine glands (Bernfield and Wessells, 1970; Ekblom, 1984). An example of homotypic interaction is the condensation and aggregation of mesenchymal cells during the formation of kidney tubules (Ekblom, 1984). Cell–cell interactions are also involved in the regulation of motility and proliferation, which is exemplified by such well-known phenomena as contact inhibition of movement and of growth. Contact inhibition of movement is closely correlated with the formation of specialized cell–cell contact areas that triggers a reorganization of the actin-filament system (Heaysman and Pegrum, 1973), and it is easy to envisage that such a reorganization can affect the motility pattern of cells. The nature of contact inhibition of growth is less clear, but in recent years there have been reports indicating that the formation of specific cell–cell contacts is involved in this phenomenon (Lieberman et al., 1982). The formation of such contacts might trigger signals leading to an arrest of the cell cycle. Another example of cell–cell interactions is the formation of intercellular junctions, which in itself includes several different types of interactions, since in vertebrate cells at least four different types of morphologically identifiable junctions are known (McNutt and Weinstein, 1973).

Cell–matrix interactions are important not only for anchoring cells in tissues, but also seem to be a prerequisite for both motility and growth of normal cells. Cellular locomotion *in vitro*, which involves a specific organization and participation of the cytoskeleton (Geiger, 1983), can only occur when cells are attached to a solid substrate containing proper matrix molecules. With regard to growth, it is well known that normal cells *in vitro* have to be both attached and spread on a solid surface to undergo mitosis (Folkman and Moscona, 1978; Ben-Ze'ev et al., 1980). Spreading involves a specific organization of the cytoskeleton, which probably is triggered by the formation of cell–matrix contact areas in the plasma membrane.

Like cell–cell interactions, cell–matrix interactions also make up a heterogeneous group, and it has been demonstrated that vertebrate cells are able to recognize and bind directly to collagens, fibronectin, and laminin (Rubin et al., 1984), which are protein constituents of matrix structures. In a few cases, cell surface molecules that are involved in attachment to laminin (Lesot et al., 1983; Malinoff and Wicha, 1983; Rao et al., 1983) and collagen (Koehler et al., 1980; Ocklind et al., 1980; Chiang and Kang, 1982; Mollenhauer and von der Mark, 1983) have been identified. Some interesting differences between laminin-binding and collagen-binding molecules have been observed. The laminin-binding molecules, which have been identified in different cell types, all have the same molecular weight (67–69 kD) and have a high affinity for laminin. In contrast, the collagen-binding molecules, which also have been found in different cell types, exhibit molecular weights varying from 31 to 150 kD and seem to have a low affinity for collagen.

It is advantageous to have a simple and direct assay to use in the identification and purification of the cell surface molecules that are involved in these interactions. This requirement makes complex phenomenological events such as contact inhibition of movement less suited to serve as a basis for such an assay. Therefore, studies of this kind have been based largely on the various adhesion reactions that cells undergo, in the knowledge that cell recognition and adhesion in itself is a primary process in development. However, some cellular interactions leading to complex phenomena probably are mediated by direct molecular bindings, which may be strong enough to be detected in cell adhesion assays.

In recent years, cell surface molecules involved in cell adhesion reactions have been discovered (Hausman and Moscona, 1976; Müller and Gerisch, 1978; Urushihara and Takeichi, 1980; Nielsen et al., 1981; Marchase et al., 1982; Ocklind and Öbrink, 1982; Damsky et al., 1983; Edelman, 1983; Gallin et al., 1983; Imhof et al., 1983; Ogou et al., 1983; Peyriéras et al., 1983; Siu et al., 1983). Such molecules have been named cognins, adhesins, adherins, aggregation factors, or simply cell adhesion molecules. However, it must be stressed that several different methods have been used to determine cell adhesion in these studies. The most popular approaches have been to determine (1) the early phases of cell aggregation or of binding to a solid surface; (2) the late phases of cell aggregation primarily measured as the increase in size of aggregates; and (3) the dissociation of cells in contact with other cells, or of cells from a solid surface (Marchase et al., 1976; Grinnell, 1982; Öbrink, 1982). Clearly, these reactions involve processes that might be of completely different natures and accordingly might involve different molecular mechanisms. This must be borne in mind when different cell adhesion molecules are compared.

A particularly useful approach has been to combine adhesion assays with immunological methods, taking advantage of the ability of antibodies directed against cell surface antigens to inhibit cell adhesion. In this way, cell adhesion molecules have been identified in neural cells (Edelman, 1983), hepatocytes (Ocklind and Öbrink, 1982; Ogou et al., 1983; Gallin et al., 1983), epithelial cells (Damsky et al., 1983; Imhof et al., 1983), teratocarcinoma cells (Ogou et al., 1983; Peyriéras et al., 1983), fibroblasts (Urushihara and Takeichi, 1980; Oesch and Birchmeier, 1982), and cellular slime molds (Müller and Gerisch, 1978; Siu et al., 1983), among others (see chapters in this volume).

HEPATOCYTES AS A MODEL SYSTEM

Our approach to the understanding of cellular recognition phenomena at the molecular level has been to concentrate on one cell type and to study both its cell–cell and cell–matrix interactions. We have found that normal rat hepatocytes make an excellent model system for several reasons. Rat liver is available in large amounts, it is easy to prepare pure hepatocytes in large numbers, and well-characterized membrane fractions can be prepared, facilitating the isolation and purification of cell adhesion molecules for biochemical analysis. After partial hepatectomy, the remaining liver rapidly regenerates (Higgins and Anderson, 1931); regenerating hepatocytes, which have several characteristics in common with embryonic hepatocytes (Bissell et al., 1973), proliferate, and in a short time, together with other types of liver cells, actively form new physiologically functioning liver tissue. Thus regenerating liver provides an excellent system for studies of cellular interactions involved in morphogenesis. Furthermore, several hepatocyte-derived cell lines, both malignant and nonmalignant, are available, and procedures for the isolation and cultivation of embryonic rat hepatocytes are well established (Leffert et al., 1979).

Hepatocytes occur in the liver in a highly organized pattern, exhibiting matrix interactions, homotypic cell–cell interactions, and possibly heterotypic cell–cell interactions (Figure 1). The extracellular matrix of hepatocytes is found in the space of Disse between the endothelial cells lining the sinusoids and the sinusoidal face of the hepatocytes. In ruminants, a classic basement membrane is seen; other species such as man and rat lack a true basement membrane and instead have collagen fibers and reticular fibers at this location (Ito and Shibasaki, 1968; Grubb and Jones, 1970). Collagens type I, III, and IV, and fibronectin have been observed in the space of Disse by immunohistochemical techniques (McGee, 1982). The hepatocytes make extensive contacts with each other; all four types of junctions seen in

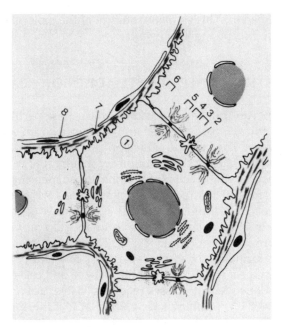

Figure 1. *Schematic presentation of the organization of hepatocytes in the mammalian liver.* The hepatocytes (1) are arranged in plates and three major regions of their surface can be distinguished: the sinusoidal surface facing the space of Disse (7); the contiguous face, where the intercellular junctions (3-6) are found; and the membrane lining the bile canaliculi (2). The bile canaliculi are sealed by tight junctions (3); beyond the tight junctions, intermediate junctions (4), demosomes (5), and gap junctions (6) are situated. The relative positions of the junctions on the contiguous face have been emphasized and are not drawn to exact scale. Gap junctions may also occur at other locations. The sinusoids are lined by endothelial cells (8). The extracellular matrix of the hepatocytes is deposited between the endothelial cells and the hepatocytes, in the space of Disse (7). It consists of reticular fibers and collagen fibers; collagens type I, III, and IV and fibronectin have been localized to this area by immunohistochemical techniques.

epithelia—zonula occludens (tight junction), zonula adherens (intermediate junction), macula adherens (desmosome), and macula communicans (gap junction)—have been observed by electron microscopy. The tight junctions seal the bile canaliculi, where the primary bile is secreted, from the rest of the hepatocyte surface.

The formation of junctions and of bile canaliculi is a complicated morphogenetic event that can be studied *in vitro* because isolated hepatocytes reform junctions and bile canaliculi after approximately 24 hours in culture (Wanson et al., 1977; Miettinen et al., 1978). Bile secretion *in vitro* can easily be visualized with fluorescein, which is taken up by hepatocytes and specifically secreted into the bile canaliculi (Barth and Schwarz, 1982). Since the formation of intact bile canaliculi is a prerequisite for bile secretion, the simple functional test utilizing fluorescein secretion can be used to monitor their formation. This simplifies studies of the molecular mechanisms involved in these morphogenetic events.

CHARACTERISTICS OF HEPATOCYTE ADHESION REACTIONS

We isolate hepatocytes from young rats by a collagenase-perfusion procedure (Öbrink, 1982), and the isolated cells are then used to investigate various adhesion processes. The cell–cell adhesion process we study is the reaggregation that occurs during the first hour of incubation; it is measured as the disappearance of single cells from a shaken suspension (Ocklind and Öbrink, 1982). The cell–matrix adhesion processes we study are the attachments that occur during the first 30–60 minutes of incubation; it is measured as the binding of suspended cells to a surface of matrix molecules coated on culture dishes (Öbrink, 1982). These processes represent the first measurable formation of stable cell–cell or cell–matrix bonds.

In our initial studies, we found that both cell–cell and cell–matrix adhesion are temperature-dependent processes requiring viable, actively metabolizing cells (Öbrink et al., 1977; Rubin et al., 1978). Rapid adhesion of both kinds takes place at 37°C, but no cell–cell adhesion is observed at 27°C or lower; the temperature has to be lowered to 8°C to block completely attachment to collagen or asialoglycoproteins. However, significant adhesion of hepatocytes to fibronectin occurs even at 4°C.

When rat hepatocytes are incubated in a simple balanced salt solution at 37°C, 60–80% of the cells adhere to each other and form aggregates within 60 minutes. The role of divalent cations in hepatocyte intercellular adhesion has not been definitely clarified. We have found evidence that Ca^{2+} is required (Rubin et al., 1977), but others have presented data indicating that Ca^{2+} might inhibit, while Mg^{2+} stimulates, adhesion (Schmell et al., 1982). It seems likely that both are required. Rat hepatocytes show a remarkable selectivity in their initial cell–cell adhesion and do not adhere significantly to other cell types. Initially, we found that this selectivity was species-specific and that rat hepatocytes did not adhere to chicken hepatocytes (Öbrink et al., 1977). Later, Sieber and Roseman (1981) and Albanese et al. (1982) demonstrated that rat hepatocytes also adhere selectively to each other when challenged with rat Leydig's cells, rat preadipocytes, or rat cardiac muscle cells. It has been speculated that the galactoprotein receptor, discovered by Morell et al. (1968), that binds asialoglycoproteins might be involved in rat hepatocyte intercellular adhesion. This receptor is one of the best characterized on the surface of mammalian hepatocytes. Asialoglycoproteins are very rapidly bound to the galactoprotein receptor and taken up by the cells. However, we demonstrated that asialoglycoproteins do not in any way affect the aggregation of rat hepatocytes, indicating that the galactoprotein receptor has nothing to do with their intercellular adhesion (Öbrink and Ocklind, 1978). This was later confirmed by Kuhlenschmidt et al. (1982).

In our analyses of cell–substrate adhesion, we again found that hepatocytes bind in a selective manner (Höök et al., 1977; Rubin et al., 1978). They do not bind to nontreated culture dishes or to dishes coated with albumin or a variety of other plasma proteins. However, they attach efficiently to dishes coated with a variety of components including collagens, fibronectin, laminin (Johansson et al., 1981), asialoglycoproteins, concanavalin A, insulin, EGF, polycationized ferritin, and antibodies directed against the cell surface. It seems as if they are able to bind to almost any substance for which they have receptors or acceptors. Further analysis revealed that attachment to all these substances occurs by different mechanisms. Binding to asialoglycoproteins, for example, is mediated by the galactoprotein receptor, which does not mediate hepatocyte attachment to collagen, fibronectin, or laminin. Separate molecular mechanisms seem to operate for attachment to each of the latter proteins, and we have clearly demonstrated that fibronectin is neither required nor involved in adhesion of hepatocytes to collagen (Rubin et al., 1978, 1979, 1981a).

We investigated the attachment of rat hepatocytes to collagen in detail and found that these cells are able to bind equally efficiently to collagen types I, II, III, IV, and V (Rubin et al., 1981a). Divalent cations are required, Mg^{2+} being more efficient in promoting binding than Ca^{2+} (Rubin et al., 1978). The cells are able to bind to both denatured and native collagen, in both the monomeric and fibrillar forms, although much more efficiently to the native protein. Furthermore, separated collagen α-chains, all the cyanogen bromide peptides of the α1(I)-chain, and most of the cyanogen bromide peptides of the α2-chain can serve as efficient attachment substrates (Rubin et al., 1981a). Thus we have concluded that the cell binding sites occur in multiple copies along the collagen molecule. A low but significant binding of rat hepatocytes to substrates made of synthetic collagenlike peptides with the structures $(Gly-Pro-X)_n$ and $(Gly-X-Pro)_n$ has also been observed (Rubin et al., 1981a).

The ability of rat hepatocytes to adhere to all types of collagens, fibronectin, and laminin distinguishes them from some other cell types. Some epithelial cells (Kleinman et al., 1981) and adult cardiac myocytes (Borg et al., 1984), for example, are able to attach efficiently only to collagen type IV and laminin and not to other types of collagen or to fibronectin. Neonatal cardiac myocytes, on the other hand, resemble hepatocytes in that they can bind to all types of collagen, fibronectin, and laminin (Borg et al., 1984). Embryonic fibroblasts attach to both fibronectin and laminin (Couchman et al., 1983). Some cells of fibroblastic origin are able to attach directly to collagen (Grinnell and Minter, 1978), whereas other fibroblastic cells seem to need the cooperation of fibronectin in order to bind to collagen (Kleinman et al., 1981). Thus hepatocytes, neonatal cardiac myocytes, and perhaps embryonic fibroblasts are pluripotent in their mode of attachment to extracellular matrix components, whereas terminally differentiated cardiac myocytes and epithelial cells seem to have a more restricted attachment behavior.

An observation we made is that the spreading of hepatocytes occurs only after attachment to collagen, fibronectin, or laminin (Rubin et al., 1981b). Cell spreading, which involves organization of the cytoskeleton, is an important process in the regulation of morphogenetic events. Collagen, fibronectin, and laminin probably represent components to which the cells are attached *in vivo*, which has led to the speculation that spreading of hepatocytes is a physiological response triggered primarily by binding to extracellular matrix components. Hepatocyte spreading occurs on both native and denatured collagen. Interestingly, we found that antibodies against fibronectin that inhibit spreading on fibronectin also inhibit spreading on denatured collagen, whereas the spreading reaction on native collagen was unaffected by these antibodies (Rubin et al., 1981b). It has been demonstrated that rat hepatocytes secrete fibronectin (Voss et al., 1979); thus endogenously produced fibronectin seems to be involved in hepatocyte spreading on denatured collagen but not on native collagen. This is in contrast to the initial binding reaction of hepatocytes to native and denatured collagen, both of which occur without the involvement of fibronectin (Rubin et al., 1981a).

IDENTIFICATION OF CELL-CAM 105

To identify adhesion-related cell surface components, we adopted the immunological approach (Öbrink and Ocklind, 1978; Ocklind and Öbrink, 1982). Fab fragments of rabbit antibodies produced against liver plasma membranes or isolated hepatocytes effectively inhibited hepatocyte aggregation (Figure 2). Solubilized plasma membrane components neutralized this inhibition, and we used this neutralizing ability to follow the purification of molecules that were involved in the adhesion reaction that was blocked by the original antibodies. To facilitate both purification and determination of the neutralizing activity, we purified papain-solubilized fragments of the molecule. The purified components were used to produce more specific antibodies, which again were found to inhibit intercellular adhesion of hepatocytes (Figure 2). The new antibodies were then used to identify the intact molecule in detergent-solubilized plasma membranes. In this manner we identified a membrane protein that we named cell-CAM 105 (CAM for cell adhesion molecule), the blocking of which inhibited hepatocyte aggregation (Ocklind and Öbrink, 1982).

Cell-CAM 105 is a glycoprotein with an apparent molecular weight of 105 kD in SDS-polyacrylamide gel electrophoresis. It cannot be solubilized from plasma membranes by high or low salt concentrations or by chelation of divalent cations; detergents are required. It is thus an integral membrane protein. We have purified cell-CAM 105 by using immunoaffinity chromatography. Cell-CAM 105 contains a large amount of carbohydrate, among which N-acetylglucosamine and sialic acid are found. One characteristic

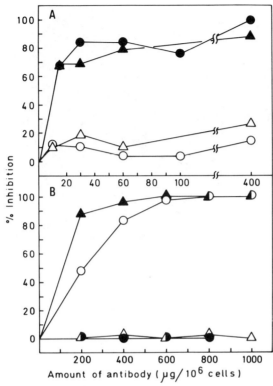

Figure 2. *Effects of various antibodies on intercellular (A) and cell-collagen (B) adhesion of rat hepatocytes.* The cells were preincubated for 30 minutes at 4°C in the presence of the indicated amounts of antibodies. Intercellular adhesion was determined as the decrease in single cells as a result of aggregation after incubation on a gyratory shaker at 37°C for 40 minutes. Cell attachment to collagen substrates was determined by measuring the number of cells that had attached after incubation at 37°C for 30 minutes. *A:* Aggregation was determined in the presence of the indicated amounts of Fab fragments of anti-plasma membrane antibodies (*filled triangles*), anti-cell-CAM (*filled circles*), anti-collagen-CAM (*open circles*), and preimmune serum (*open triangles*). *B:* Attachment to collagen was determined in the presence of the indicated amounts of IgG molecules of anti-plasma membrane antibodies (*filled triangles*), anti-cell-CAM (*filled circles*), anti-collagen-CAM (*open circles*), and preimmune serum (*open triangles*). The percentages of inhibition of adhesion reactions are given on the ordinates. (From Ocklind et al., 1980.)

feature of this protein is that a constant autodegradation is going on, even in the purest preparations. The major degradation product is a peptide with an apparent molecular weight of 70 kD.

Specific antibodies have been used in immunofluorescence microscopy to study the tissue distribution and the cell surface localization of cell-CAM 105 (Ocklind et al., 1983). We found the protein in liver, and also in the simple epithelia of the small intestine, the large intestine, the gastric mucosa, the kidney tubules, and the glandular epithelium of the parotid gland. It was not seen in a number of other tissues, including stratified

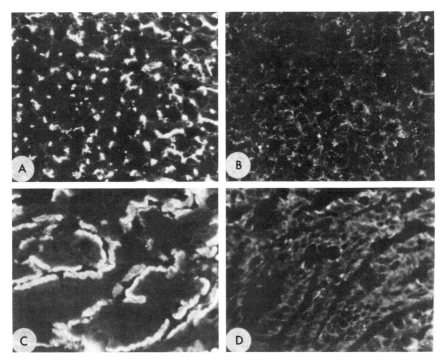

Figure 3. *Immunofluorescence microscopy on frozen sections of rat liver and small intestine.* Cryostat sections were acetone-fixed and sequentially incubated with antiserum diluted 1:10 and FITC-labeled goat anti-rabbit IgG (Nordic Immunological Laboratories) diluted 1:40. *A*: Liver incubated with anti-cell-CAM. *B*: Liver incubated with anti-CDP. *C*: Small intestine incubated with anti-cell-CAM. *D*: Small intestine incubated with anti-CDP.

epithelia, interstitial connective tissue, skeletal muscle, heart muscle, smooth muscle, or neuronal tissue. In the liver, it was seen exclusively along the bile canaliculi (Figure 3). However, on isolated hepatocytes it was seen all over the cell surface, although it was not randomly located on all the cells. Thus a significant reorganization of this protein occurs when the cells are dissociated. In the simple epithelia, cell-CAM 105 is specifically located at the apical cell surface (Figure 3), which is in agreement with the localization to the bile canalicular region in the liver. We have speculated that cell-CAM 105 might be a component of the junctional complex that is found in the bile canalicular region or at the apical surface, since it is involved in intercellular adhesion. However, a localization on the microvilli of the bile canalicular and apical surfaces cannot yet be ruled out.

IDENTIFICATION OF COLLAGEN-CAM

We used a similar immunological approach to identify cell surface molecules related to hepatocyte attachment to collagen. Both intact IgG molecules and Fab fragments of antibodies against liver plasma membranes effectively inhibited attachment to dishes coated with collagen but not attachment to fibronectin-coated dishes (Rubin et al., 1979). Detergent-solubilized membrane components that neutralized antibody-mediated inhibition of attachment to collagen were partially purified by sequential chromatography on lentil lectin-Sepharose and collagen-Sepharose gels (Ocklind et al., 1980). The purified components were used to raise rabbit antibodies, which were found to effectively inhibit adhesion to collagen (Figure 2) but not to fibronectin (Rubin et al., 1984). These antibodies, denoted anti-collagen-CAM, were not completely monospecific, but the major components immunoprecipitated from detergent-solubilized membranes had apparent molecular weights of about 110–150 kD in SDS-polyacrylamide gel electrophoresis (Ocklind et al., 1980). Further analysis by preparative SDS-polyacrylamide gel electrophoresis showed that the components, which were able to neutralize the anti-collagen-CAM-mediated inhibition of attachment to collagen, migrated as components with apparent molecular weights of 120–150 kD. These components could not be solubilized from the membranes by low or high salt concentrations or by chelation of divalent cations. Thus collagen-CAM is an integral membrane glycoprotein(s) with apparent molecular weight of 120–150 kD.

DIFFERENT MECHANISMS OPERATE IN CELL–CELL AND CELL–COLLAGEN ADHESION

One important question that we now could address with our adhesion-blocking antibodies was whether the same or different molecular mechanisms operate in the early phases of cell–cell and cell–matrix adhesion. It has been argued that cell–cell adhesion is essentially the same phenomenon as cell–substrate adhesion. These arguments have been based largely on immunohistochemical and ultrastructural studies, which have revealed striking similarities between certain cell–cell and cell–matrix contact areas (Rees et al., 1978; Chen and Singer, 1982). Furthermore, fibronectin, the first well-characterized adhesive molecule to be described, has been attributed a role in both cell–substrate adhesion and cell–cell adhesion (Yamada and Olden, 1978).

We investigated the effects of anti-cell-CAM and anti-collagen-CAM on both cell–cell and cell–collagen adhesion of hepatocytes. The results were clear-cut and demonstrated a remarkable specificity in the inhibitory effects of the two types of antibodies: Anti-cell-CAM inhibited cell–cell adhesion but not cell–collagen adhesion; anti-collagen-CAM inhibited cell–collagen adhesion but not cell–cell adhesion (Figure 2). These results demonstrate that cell-CAM 105 and collagen-CAM are immunologically different molecules and that different molecular reactions at the cell surface take part in the early phases of cell–cell and cell–collagen adhesion of rat hepatocytes.

A CALCIUM-DEPENDENT PROTEIN (CDP) IS INVOLVED
IN HEPATOCYTE CELL–CELL ADHESION

We discovered that isolated liver plasma membranes contain a protein with an apparent molecular weight around 70 kD that can be released from the membranes by chelation of Ca^{2+} with EGTA and can bind to the membranes once again upon addition of Ca^{2+} (Öbrink et al., 1976). We named this protein CDP (calcium-dependent protein). Given the calcium-dependence of intercellular adhesion in some cellular systems (Takeichi et al., 1979; Brackenbury et al., 1981; Thomas et al., 1981; McClay and Marchase, 1982) and the known calcium-dependence of some of the intercellular junctions (Sedar and Forte, 1964), we considered it interesting to investigate whether CDP might be involved in the cell adhesion of hepatocytes. We purified CDP, produced rabbit antibodies against it, and tested the effect of the antibodies on hepatocyte adhesion (Ocklind et al., 1984). We found that anti-CDP mimicked anti-cell-CAM and effectively inhibited cell–cell, but not cell–collagen, adhesion of rat hepatocytes (Figure 4).

A simple explanation of the inhibition exerted by anti-CDP would be that it is a degradation product of cell-CAM 105, formed during the isolation

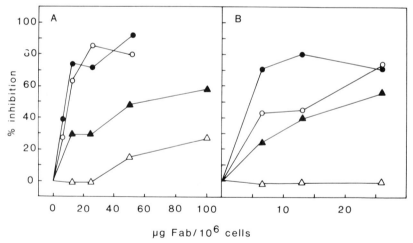

μg Fab/10⁶ cells

Figure 4. *Effects of anti-cell-CAM and anti-CDP on the aggregation of EGTA-treated (A) and trypsin-treated (B) hepatocytes.* Before the aggregation experiments, cells were either treated with 1 mM EGTA for 10 minutes or with 0.1 μg/ml of crystalline trypsin for 15 minutes. The cells were then washed and the aggregation was performed in a balanced salt solution containing calcium. Nontreated cells served as controls. The aggregation and the effect of the antibodies were determined as described in Figure 2. *A*: Nontreated cells + anti-cell-CAM Fab (*filled circles*); EGTA-treated cells + anti-cell-CAM Fab (*open circles*); nontreated cells + anti-CDP Fab (*filled triangles*); EGTA-treated cells + anti-CDP Fab (*open triangles*). *B*: Nontreated cells + anti-cell-CAM Fab (*filled circles*); trypsin-treated cells + anti-cell-CAM Fab (*open circles*); nontreated cells + anti-CDP Fab (*filled triangles*); trypsin-treated cells + anti-CDP Fab (*open triangles*). (From Ocklind et al., 1984.)

of the protein, and that the anti-CDP antibodies recognized and blocked cell-CAM 105 on the cell surface. This would be in agreement with our finding that the major degradation product of cell-CAM 105 has an apparent molecular weight of 70 kD. However, the situation is not as simple as that. We could demonstrate that both CDP and cell-CAM 105 occur as separate and distinct molecules on the hepatocyte surface (Ocklind et al., 1984). Furthermore, treatment of hepatocytes with either trypsin or EGTA selectively made their aggregation insensitive to blocking by anti-CDP Fab, whereas anti-cell-CAM Fab inhibited adhesion of trypsin- and EGTA-treated cells in the same way it inhibited nontreated cells (Figure 4). The change in responsiveness to anti-CDP resulting from EGTA treatment was correlated with a release of CDP from the cell surface to the medium. Cell-CAM 105, on the other hand, was not released or degraded by EGTA treatment. Thus we have concluded that both cell-CAM 105 and CDP take part in the intercellular adhesion of rat hepatocytes as separate components.

In further analyses, we found, however, that anti-CDP antibodies recognized, in addition to the 70-kD CDP, a 105-kD protein that was also recognized by anti-cell-CAM; that is, they seemed to recognize cell-CAM 105. The binding of anti-CDP to the 105-kD protein was much weaker than its binding to the 70-kD protein or than the binding of anti-cell-CAM to the 105-kD protein. The reverse situation was not true; that is, anti-cell-CAM antibodies did not recognize CDP, which is not unreasonable since CDP is a smaller molecule than cell-CAM 105. On the basis of this immunological cross-reactivity we have suggested that CDP and cell-CAM 105 are closely related (Ocklind et al., 1984). Further support for this suggestion was obtained by peptide mapping after partial proteolysis by staphylococcal V8 protease of purified CDP and cell-CAM 105. As shown in Figure 5, the two proteins yielded very similar peptide patterns.

It might be argued that if anti-CDP recognizes cell-CAM 105, anti-CDP antibodies should still be able to inhibit aggregation of EGTA-treated or trypsin-treated hepatocytes since anti-cell-CAM antibodies do so. In fact they did, but at much higher concentrations and to a lesser extent than that observed for nontreated cells. This is in good agreement with our observation that anti-CDP did not react with the 105-kD protein as efficiently as did anti-cell-CAM.

Anti-CDP antibodies were used in immunofluorescence microscopy of frozen sections of rat liver and small intestine (Figure 3). In both tissues the staining pattern was different from that observed with anti-cell-CAM. In the liver, anti-CDP staining was more diffuse and did not seem to be specifically localized to the bile canaliculi. In the intestinal mucosa, it was primarily seen at the lateral surfaces of the epithelial cells with no prominent staining of their apical regions. These results indicate that CDP may be localized to other areas of the cell surface than is cell-CAM 105.

To conclude, we have evidence indicating that rat hepatocytes have two distinct proteins, cell-CAM 105 and CDP, involved in their cell–cell

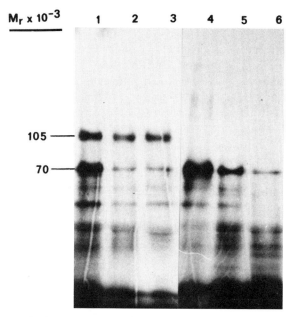

Figure 5. *Peptide mapping of cell-CAM 105 and CDP.* Cell-CAM 105 was purified by immunoaffinity chromatography followed by preparative SDS-polyacrylamide gel electrophoresis
and electroelution from the gel slices. CDP was purified by two-dimensional electrophoresis
(isoelectric focusing in the first dimension and SDS-polyacrylamide gel electrophoresis
in the second) and electroelution. The purified proteins were labeled with ^{125}I and
incubated with staphylococcal V8 protease (Sigma Chemical Co.) for 90 minutes at room
temperature, according to Cleveland et al. (1977). The digested proteins were electrophoresed on 5–15% polyacrylamide slab gels under reducing conditions and autoradio-
graphed. *Lane 1*: Cell-CAM 105 + 250 μg V8 protease. *Lane 2*: Cell-CAM 105 + 100 μg
V8 protease. *Lane 3*: Cell-CAM 105 + 50 μg V8 protease. *Lane 4*: CDP + 50 μg V8
protease. *Lane 5*: CDP + 100 μg V8 protease. *Lane 6*: CDP + 250 μg V8 protease.

adhesion interactions. Cell-CAM 105 and CDP are immunologically and
chemically similar but seem to be localized to different areas of the cell
surface and are perturbed in different ways by EGTA or trypsin.

We do not yet have any detailed information on how cell-CAM or CDP
are related to the adhesion events of hepatocytes. However, they do not
seem to act completely independent of each other during the early phases
of cell–cell adhesion. Thus, simultaneous addition of anti-cell-CAM Fab
and anti-CDP Fab did not result in a simple additive inhibition of hepatocyte
aggregation (Figure 6). Rather, it seemed as if one antibody excluded the
effect of the other. This behavior might indicate that cell-CAM 105 and
CDP interact with each other, either directly or sterically, during this phase
of the adhesion process. Our data indicate that intercellular adhesion of
rat hepatocytes is a very complex process.

EARLY AND LATE EFFECTS OF ANTI-CELL-CAM AND ANTI-CDP

As already pointed out, our assay for cell–cell adhesion is aimed at investigations of adhesive events occurring during the first hour of cellular reaggregation. Cell-CAM 105 and CDP are thus by definition important for the early phases of hepatocyte recognition and adhesion. Homotypic cell–cell adhesion is, however, a complex phenomenon that might involve

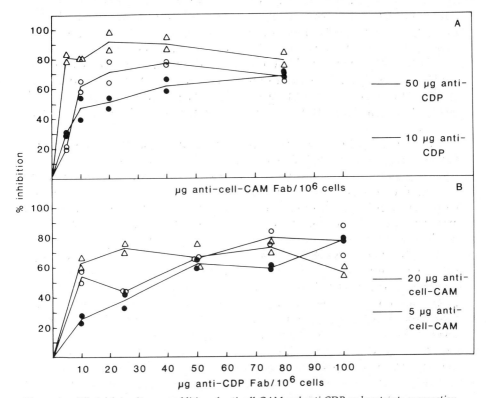

Figure 6. *Effects of simultaneous addition of anti-cell-CAM and anti-CDP on hepatocyte aggregation.* Aggregation experiments performed as described in Figure 2. The values of duplicate incubations are shown. *A:* Addition of various amounts of anti-cell-CAM Fab alone (*filled circles*); various amounts of anti-cell-CAM Fab + a constant amount of 10 μg of anti-CDP Fab/10[6] cells (*open circles*); various amounts of anti-cell-CAM Fab + a constant amount of 50 μg of anti-CDP Fab/10[6] cells (*open triangles*). To the right are shown the degrees of inhibition seen with 10 μg of anti-CDP Fab/10[6] cells alone and 50 μg of anti-CDP Fab/10[6] cells alone. *B:* Addition of various amounts of anti-CDP Fab alone (*filled circles*); various amounts of anti-CDP Fab + a constant amount of 5 μg of anti-cell-CAM Fab/10[6] cells (*open circles*); various amounts of anti-CDP Fab + a constant amount of 20 μg of anti-cell-CAM Fab/10[6] cells (*open triangles*). To the right are shown the degrees of inhibition seen with 5 μg of anti-cell-CAM Fab/10[6] cells alone and 20 μg of anti-cell-CAM Fab/10[6] cells alone. (From Ocklind et al., 1984.)

several different adhesion reactions, including junction formation. There is a clear possibility that cells, even if their early adhesive events are inhibited by blocking of cell-CAM 105 or CDP, might overcome this block and at a later stage form stable cell–cell bonds by some mechanism not involving either cell-CAM 105 or CDP.

In order to study this possibility, we investigated the effects of prolonged incubation in the presence of anti-cell-CAM or anti-CDP antibodies. This could not be done in suspension culture, since rat hepatocytes need to be attached to a collagen or a fibronectin substrate to survive prolonged culture *in vitro*. Furthermore, attachment to a proper matrix substrate may modulate cell–cell adhesion events. It is therefore of great interest to investigate whether intercellular adhesion is influenced by antibodies against cell-CAM 105 or CDP under these conditions. Accordingly, we seeded freshly isolated hepatocytes in dishes coated with native collagen monomers. After a 30-minute attachment period, fresh medium containing either anti-cell-CAM or anti-CDP antibodies was added and incubation was continued at 37°C.

During the first five hours, hepatocytes so cultured go through a spreading reaction and, when seeded densely, a confluent monolayer containing cells in close contact with each other is formed. After 15–20 hours in culture, the cells undergo a reorganization and form islands of epithelioid, well-spread cells in close contact with each other, with cell-free spaces between the islands. This process is due partly to cell detachment and partly seems to be the result of a cellular contraction phenomenon.

No effects of anti-cell-CAM or anti-CDP could be observed by light microscopy during the first five hours of incubation. The cells spread normally and showed close contact with each other even in the presence of the antibodies. However, after nine hours in culture, both anti-cell-CAM and anti-CDP caused a subtle change in the morphology of the cell layer, which became clearly manifest after 19 hours (Figure 7): The cells rounded up and became clearly separated from each other. Islands of aggregated cells, which were seen in the control cultures, were not seen in the presence of anti-cell-CAM or anti-CDP. These effects were more pronounced in the presence of Fab fragments than in the presence of intact IgG molecules or whole antiserum. The same relative difference in effect between Fab fragments and IgG molecules is also observed in the early phases of cell–cell adhesion in suspension, the Fab fragments being more potent in inhibiting aggregation. This difference most likely is due to the monovalency of the Fab fragments and the divalency of the IgG molecules, the latter being able to agglutinate the cells.

Thus both anti-cell-CAM antibodies and anti-CDP antibodies significantly affect the formation of stable cell–cell bonds between hepatocytes, not only in the early phase of reaggregation, but also after prolonged culture on a physiological matrix. In the presence of anti-cell-CAM or anti-CDP Fab

Öbrink, Ocklind, Odin, and Rubin   293

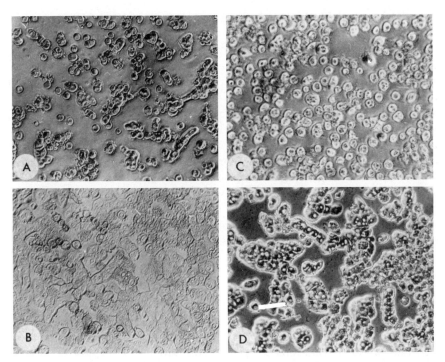

Figure 7. *Effects of anti-cell-CAM and anti-CDP on hepatocytes in culture.* Rat hepatocytes were seeded on collagen-coated dishes. After incubation for 30 minutes at 37°C, the medium containing unattached cells was removed and the dishes were washed. Fresh RPMI medium with or without Fab fragments was added, and incubation of the attached cells was continued at 37°C. After various incubation times the cells were fixed with 2.5 % glutaraldehyde in a balanced salt solution and were photographed in an inverted phase-contrast microscope. A: Hepatocytes incubated in the presence of anti-cell-CAM Fab (1 mg/ml) for 19 hours. B: Hepatocytes incubated in the absence of antibodies for 19 hours. C: Hepatocytes incubated in the presence of anti-CDP Fab (1 mg/ml) for 19 hours. D: Hepatocytes incubated in the absence of antibodies for 19 hours. The cells in A and B were from the same batch and were cultured at the same time. The cells in C and D were prepared and cultured at a subsequent time. Therefore, the cells in A and B should be compared, as should those in C and D. Both anti-cell-CAM and anti-CDP clearly prevented colony formation. No such effects were seen in the presence of immunoglobulins isolated from preimmune sera.

fragments, the collagen-attached hepatocytes probably never formed any stable cell–cell bonds, even though the cells spread normally and were in close contact with each other during the first ten hours in culture. The absence of stable cell–cell bonds then became manifest during the reorganization phase. These results strongly indicate that both cell-CAM 105 and CDP are important for recognition events leading to such morphogenetic events as junction formation.

FUNCTIONAL DOMAINS OF CELL ADHESION MOLECULES

The immunological and chemical similarities between cell-CAM 105 and CDP and their apparently different cell surface localizations and modes of action in cell–cell adhesion lead to interesting speculations. Since the major degradation product of purified cell-CAM 105, seen both as a result of autodegradation and after V8 protease-induced degradation, is a 70-kD protein, there is a clear possibility that this peptide might represent a functional domain of cell-CAM 105, the domain being relatively resistant to proteolytic degradation. The similarities between cell-CAM 105 and CDP suggest that CDP might represent this functional domain and is a processing product of cell-CAM 105, or that the two proteins have a common precursor. If this is true, it would mean that the 70-kD domain might have a different function when it occurs as a free component compared to when it occurs as part of the intact cell-CAM 105. As described earlier, cell-CAM 105 is an integral membrane glycoprotein found in or close to the bile canaliculi, whereas CDP is a calcium-dependent peripheral membrane protein with a more diffuse cell surface localization. Hence different molecular forms of basically the same adhesion molecule might have different functions in cell recognition and adhesion. This, then, may be another type of surface modulation of cell adhesion molecules, which may be important for cell regulation of adhesive interactions. Known examples of surface modulation of cell adhesion molecules include differences in temporal and spatial distribution as well as modulation of the carbohydrate portion of the molecules (Rothbard et al., 1982; Thiery et al., 1982; Edelman, 1983; Edelman et al., 1983).

The hypothesis on functional domains in cell adhesion molecules is supported by recent observations of other cell adhesion molecules. These molecules, like cell-CAM 105, exhibit degradaton products of characteristic sizes. N-CAM of chicken neuronal tissue has an apparent molecular weight of 150–250 kD and a characteristic degradation product with an apparent molecular weight of 65 kD (Hoffman et al., 1982). The corresponding apparent molecular weights of the intact molecules and of their major degradation products are 124 kD, 94 kD, and 81 kD for L-CAM of embryonic chicken liver cells (Gallin et al., 1983); 120 kD, 100 kD, 86 kD and 82 kD for uvomorulin of mouse preimplantation embryos and teratocarcinoma cells (Peyriéras et al., 1983); 120 kD and 80 kD for cell-CAM 120/80 of mammary epithelial cells (Damsky et al., 1983); 124 kD and 104 kD for cadherin of embryonic mouse hepatocytes and teratocarcinoma cells (Ogou et al., 1983); and 130 kD and 40 kD for Arc of MDCK epithelial cells (Imhof et al., 1983). Another feature these molecules have in common with cell-CAM 105 and CDP is that they all have been identified with immunological approaches similar to the ones we used in our identification work. However, the adhesion assays differed from ours. It is thus possible that several cell adhesion molecules contain globular domains that are exposed on the extracellular face of the plasma membrane and are important for the adhesive events of the respective cells.

We should consider the possible relationships between the various cell adhesion molecules mentioned above. Edelman and his coworkers have suggested that only a few different cell adhesion molecules exist in an organism and that they are subject to modulation at the molecular and/or the cellular levels (Edelman, 1983; Gallin et al., 1983). Both they and Ogou et al. (1983) have suggested that cell-CAM 105 may be a degradation product of L-CAM or cadherin. We have not yet proved that cell-CAM 105 is not a degradation product of a larger molecule, although so far we have no indications that this is the case. Cell-CAM 105, however, has a tissue distribution different from that of L-CAM (Edelman et al., 1983), cadherin (Ogou et al., 1983), and cell-CAM 120/80 (Damsky et al., 1983) as shown by immunofluorescence microscopy. The last three proteins are present in stratified epithelia such as epidermis; cell-CAM 105 is not. It is thus possible that cell-CAM 105 is distinct from L-CAM, cadherin, and cell-CAM 120/80. Similarities in the biochemical properties, the tissue distribution, the cell surface localization, and the functional properties of L-CAM, cadherin, cell-CAM 120/80, and uvomorulin suggest, however, that these proteins may represent the same or very similar molecules. It is obvious that direct comparative analyses are required to determine how the various cell adhesion molecules are related.

CONCLUSIONS

Over the last few years, several macromolecules that seem to be involved in cell adhesion reactions have been identified. This has been made possible by combining immunological methods with adhesion assays. Cell-CAM 105, CDP, and collagen-CAM, which we have found in rat hepatocytes, were all discovered by such an approach. The finding of adhesion-related cell surface molecules is an important step in studies aiming to reveal how cells recognize and interact with their proximal environment, and it forms a solid basis for a continued analysis of these phenomena at the molecular level. However, it must be emphasized that at present we have very little information as to how these cell adhesion molecules work. Are they involved in the direct recognition and binding reactions of cells, or do they regulate these phenomena indirectly by perturbing the functional organization of the cell surface? In the continued characterization of cell adhesion molecules it will be most important to analyze such questions. One way of doing so will be to investigate how, and with what other cell surface-associated molecules, the cell adhesion molecules interact. It will also be important, as mentioned above, to compare different cell adhesion molecules and their modes of function, as well as to investigate the relationships and functional implications of the different cell adhesion mechanisms that coexist in several cell types.

ACKNOWLEDGMENTS

This work was supported by grants from the Swedish Medical Research Council (Project nos. 05200, 06349, 06686, and 06833), and Konung Gustaf V: s 80-årsfond.

REFERENCES

Albanese, J., M. S. Kuhlenschmidt, E. Schmell, C. W. Slife, and S. Roseman (1982) Studies on the intercellular adhesion of rat and chicken hepatocytes: Tissue-specific adhesion in mixtures of hepatocytes and heart myocytes. *J. Biol. Chem.* **257**:3165–3170.

Barth, C. A., and L. R. Schwarz (1982) Transcellular transport of fluorescein in hepatocyte monolayers: Evidence for functional polarity of cells in culture. *Proc. Natl. Acad. Sci. USA* **79**:4985–4987.

Ben-Ze'ev, A., S. R. Farmer, and S. Penman (1980) Protein synthesis requires cell-surface contact while nuclear events respond to cell shape in anchorage-dependent fibroblasts. *Cell* **21**:365–372.

Bernfield, M. R., and N. K. Wessells (1970) Intra- and extracellular control of epithelial morphogenesis. *Dev. Biol. (Suppl.)* **4**:195–249.

Bissell, D. M., L. E. Hammacker, and U. A. Meyer (1973) Parenchymal cells from adult rat liver in nonproliferating monolayer culture. I. Functional studies. *J. Cell Biol.* **59**:722–734.

Borg, T. K., K. Rubin, E. Lundgren, K. Borg, and B. Öbrink (1984) Recognition of extracellular matrix components by neonatal and adult cardiac myocytes. *Dev. Biol.* **104**:86–96.

Brackenbury, R., U. Rutishauser, and G. M. Edelman (1981) Distinct calcium-independent and calcium-dependent adhesion systems of chicken embryo cells. *Proc. Natl. Acad. Sci. USA* **78**:387–391.

Chen, W. T., and S. J. Singer (1982) Immunoelectron microscopic studies of the sites of cell–substratum and cell–cell contacts in cultured fibroblasts. *J. Cell Biol.* **95**:205–222.

Chiang, T. M., and A. W. Kang (1982) Isolation and purification of collagen α1(I) receptor from human platelet membrane. *J. Biol. Chem.* **257**:7581–7586.

Cleveland, D. W., S. G. Fischer, M. W. Kirschner, and U. K. Laemmli (1977) Peptide mapping by limited proteolysis in sodium dodecyl sulfate and analysis by gel electrophoresis. *J. Biol. Chem.* **252**:1102–1106.

Couchman, J. R., M. Höök, D. A. Rees, and R. Timpl (1983) Adhesion, growth and matrix production by fibroblasts on laminin substrates. *J. Cell Biol.* **96**:177–183.

Damsky, C. H., J. Richa, D. Solter, K. Knudsen, and C. A. Buck (1983) Identification and purification of a cell surface glycoprotein mediating intercellular adhesion in embryonic and adult tissue. *Cell* **34**:455–466.

Edelman, G. M. (1983) Cell adhesion molecules. *Science* **219**:450–457.

Edelman, G. M., W. J. Gallin, A. Delouvée, B. A. Cunningham, and J.-P. Thiery (1983) Early epochal maps of two different cell adhesion molecules. *Proc. Natl. Acad. Sci. USA* **80**:4384–4388.

Ekblom, P. (1984) Basement membrane proteins and growth factors in kidney differentiation. In *The Role of Extracellular Matrix in Differentiation*, R. L. Trelstad, ed., pp. 173–206, Alan R. Liss, New York.

Folkman, J., and A. Moscona (1978) Role of cell shape in growth control. *Nature* **273**:345–349.

Gallin, W. J., G. M. Edelman, and B. A. Cunningham (1983) Characterization of L-CAM, a major cell adhesion molecule from embryonic liver cells. *Proc. Natl. Acad. Sci. USA* **80**:1038–1042.

Geiger, B. (1983) Membrane–cytoskeleton interaction. *Biochim. Biophys. Acta* **737**:305–341.

Grinnell, F. (1982) Cell–collagen interactions: Overview. *Methods Enzymol.* **82**:499–503.

Grinnell, F., and D. Minter (1978) Attachment and spreading of baby hamster kidney cells to collagen substrata: Effects of cold-insoluble globulin. *Proc. Natl. Acad. Sci. USA* **75**:4408–4412.

Grubb, D. J., and A. L. Jones (1970) Ultrastructure of hepatic sinusoids in sheep. *Anat. Rec.* **170**:75–80.

Hausman, R. E., and A. A. Moscona (1976) Isolation of retina-specific cell-aggregating factor from membranes of embryonic neural retina tissue. *Proc. Natl. Acad. Sci. USA* **73**:3594–3598.

Heaysman, J. E., and S. M. Pegrum (1973) Early contacts between fibroblasts. *Exp. Cell Res.* **78**:71–78.

Higgins, G. M., and R. M. Anderson (1931) Experimental pathology of the liver. I. Restoration of the liver of the white rat following partial surgical removal. *Arch. Pathol.* **12**:186–202.

Hoffman, S., B. C. Sorkin, P. C. White, R. Brackenbury, R. Mailhammer, U. Rutishauser, B. A. Cunningham, and G. M. Edelman (1982) Chemical characterization of a neural cell adhesion molecule purified from embryonic brain membranes. *J. Biol. Chem.* **257**:7720–7729.

Höök, M., K. Rubin, Å. Oldberg, B. Öbrink, and A. Vaheri (1977) Cold-insoluble globulin mediates the adhesion of rat liver cells to plastic petri dishes. *Biochem. Biophys. Res. Commun.* **79**:726–733.

Imhof, B. A., H. P. Vollmers, S. L. Goodman, and W. Birchmeier (1983) Cell–cell interaction and polarity of epithelial cells: Specific perturbation using a monoclonal antibody. *Cell* **35**:667–675.

Ito, T., and S. Shibasaki (1968) Electron microscopic study on the hepatic sinusoidal wall and the fat-storing cells in the normal human liver. *Arch. Histol. Jpn.* **29**:137–192.

Johansson, S., L. Kjellén, M. Höök, and R. Timpl (1981) Substrate adhesion of rat hepatocytes: A comparison of laminin and fibronectin as attachment proteins. *J. Cell Biol.* **90**:260–264.

Kleinman, H. K., R. J. Klebe, and G. M. Martin (1981) Role of collagenous matrices in the adhesion and growth of cells. *J. Cell Biol.* **88**:473–485.

Koehler, J. R., E. D. Nudelman, and S. Hakomori (1980) A collagen-binding protein on the surface of ejaculated rabbit spermatozoa. *J. Cell Biol.* **86**:529–536.

Kuhlenschmidt, M. S., E. Schmell, C. W. Slife, T. B. Kuhlenschmidt, F. Sieber, Y. C. Lee, and S. Roseman (1982) Studies on the intercellular adhesion of rat and chicken hepatocytes: Conditions affecting cell–cell specificity. *J. Biol. Chem.* **257**:3157–3164.

Leffert, H. L., K. S. Koch, T. Moran, and M. Williams (1979) Liver cells. *Methods Enzymol.* **58**:536–544.

Lesot, H., U. Kühl, and K. von der Mark (1983) Isolation of a laminin-binding protein from muscle cell membranes. *EMBO J.* **2**:861–865.

Lieberman, M. A., C. E. Keller-McGandy, T. A. Woolsey, and L. Glaser (1982) Binding of isolated 3T3 surface membranes to growing 3T3 cells and their effect on cell growth. *J. Cell. Biochem.* **20**:81–93.

McClay, D. R., and R. B. Marchase (1982) Calcium-dependent and calcium-independent adhesive mechanisms are present during initial binding events of neural retina cells. *J. Cell. Biochem.* **18**:469–478.

McGee, J. (1982) Liver. In *Collagen in Health and Disease*, J. B. Weiss and M. I. V. Jackson, eds., pp. 414–424, Churchill Livingstone, Edinburgh.

McNutt, N. S., and R. S. Weinstein (1973) Membrane ultrastructure at mammalian intercellular junctions. *Prog. Biophys. Mol. Biol.* **26**:45–101.

Malinoff, H. L., and M. S. Wicha (1983) Isolation of a cell surface receptor protein for laminin from murine fibrosarcoma cells. *J. Cell Biol.* **96**: 1475–1479.

Marchase, R. B., K. Vosbeck, and S. Roth (1976) Intercellular adhesive specificity. *Biochim. Biophys. Acta* **457**:385–416.

Marchase, R. B., L. A. Koro, C. M. Kelly, and D. R. McClay (1982) A possible role for ligatin and the phosphoglycoproteins it binds in calcium-dependent retinal cell adhesion. *J. Cell. Biochem.* **18**:461–468.

Miettinen, A., I. Virtanen, and E. Linder (1978) Cellular actin and junction formation during reaggregation of adult rat hepatocytes into epithelial sheets. *J. Cell Sci.* **31**:341–353.

Mollenhauer, J., and K. von der Mark (1983) Isolation and characterization of a collagen-binding glycoprotein from chondrocyte membranes. *EMBO J.* **2**:45–50.

Morell, A. G., R. A. Irvine, I. Sternlieb, I. H. Scheinberg, and G. Ashwell (1968) Physical and chemical studies on ceruloplasmin. V. Metabolic studies on sialic acid-free ceruloplasmin *in vivo*. *J. Biol. Chem.* **243**:155–159.

Moscona, A. A. (1976) Toward a molecular basis of neuronal recognition. In *Neuronal Recognition*, S. H. Barondes, ed., pp. 205–226, Chapman and Hall, London.

Moscona, A. A., M. Brown, L. Degenstein, L. Fox, and B. M. Soh (1983) Transformation of retinal glia cells into lens phenotype: Expression of MP26, a lens plasma membrane antigen. *Proc. Natl. Acad. Sci. USA* **80**:7239–7243.

Müller, K., and G. Gerisch (1978) A specific glycoprotein as the target site of adhesion blocking Fab in aggregating *Dictyostelium* cells. *Nature* **274**:445–449.

Nielsen, L. D., M. Pitts, S. R. Grady, and E. J. McGuire (1981) Cell–cell adhesion in the embryonic chick. Partial purification of liver adhesion molecules from liver membranes. *Dev. Biol.* **86**:315–326.

Öbrink, B. (1982) Hepatocyte–collagen adhesion. *Methods Enzymol.* **82**:513–529.

Öbrink, B., and C. Ocklind (1978) Cell surface component(s) involved in rat hepatocyte intercellular adhesion. *Biochem. Biophys. Res. Commun.* **85**:837–843.

Öbrink, B., H. Lindström, and N. G. Svennung (1976) Calcium-requirement for a reversible binding of membrane proteins to rat liver plasma membranes. *FEBS Lett.* **70**:28–32.

Öbrink, B., M. S. Kuhlenschmidt, and S. Roseman (1977) Adhesive specificity of juvenile rat and chicken liver cells and membranes. *Proc. Natl. Acad. Sci. USA* **74**:1077–1081.

Ocklind, C., and B. Öbrink (1982) Intercellular adhesion of rat hepatocytes: Identification of a cell surface glycoprotein involved in the initial adhesion process. *J. Biol. Chem.* **257**:6788–6795.

Ocklind, C., K. Rubin, and B. Öbrink (1980) Different cell surface glycoproteins are involved in cell–cell and cell–collagen adhesion of rat hepatocytes. *FEBS Lett.* **121**:47–50.

Ocklind, C., U. Forsum, and B. Öbrink (1983) Cell surface localization and tissue distribution of a hepatocyte cell–cell adhesion glycoprotein (cell-CAM 105). *J. Cell Biol.* **96**:1168–1171.

Ocklind, C., P. Odin, and B. Öbrink (1984) Two different cell adhesion molecules—cell-CAM 105 and a calcium-dependent protein—occur on the surface of rat hepatocytes. *Exp. Cell Res.* **151**:29–45.

Oesch, B., and W. Birchmeier (1982) New surface component of fibroblast's, focal contacts identified by a monoclonal antibody. *Cell* **31**:671–679.

Ogou, S. I., C. Yoshida-Noro, and M. Takeichi (1983) Calcium-dependent cell–cell adhesion molecules common to hepatocytes and teratocarcinoma stem cells. *J. Cell Biol.* **97**:944–948.

Peyriéras, N., F. Hyafil, D. Louvard, H. L. Ploegh, and F. Jacob (1983) Uvomorulin: A nonintegral membrane protein of early mouse embryo. *Proc. Natl. Acad. Sci. USA* **80**:6274–6277.

Rao, N. C., S. H. Barsky, V. P. Terranova, and L. A. Liotta (1983) Isolation of a tumor cell laminin receptor. *Biochem. Biophys. Res. Commun.* **111**:804–808.

Rees, D. A., R. A. Badley, C. W. Lloyd, D. Thom, and C. G. Smith (1978) Glycoproteins in the recognition of substratum by cultured fibroblasts. In *Cell–Cell Recognition*, A. S. G. Curtis, ed., pp. 241–260, Cambridge Univ. Press, Cambridge.

Rothbard, J. B., R. Brackenbury, B. A. Cunningham, and G. M. Edelman (1982) Differences in the carbohydrate structures of neural cell-adhesion molecules from adult and embryonic chicken brains. *J. Biol. Chem.* **257**:11064–11069.

Rubin, K., L. Kjellén, and B. Öbrink (1977) Intercellular adhesion between juvenile liver cells: A method to measure the formation of stable lateral contacts between cells attached to a collagen gel. *Exp. Cell Res.* **109**:413–422.

Rubin, K., Å. Oldberg, M. Höök, and B. Öbrink (1978) Adhesion of rat hepatocytes to collagen. *Exp. Cell Res.* **117**:165–177.

Rubin, K., S. Johansson, I. Pettersson, C. Ocklind, B. Öbrink, and M. Höök (1979) Attachment of rat hepatocytes to collagen and fibronectin: A study using antibodies against cell surface components. *Biochem. Biophys. Res. Commun.* **91**:86–94.

Rubin, K., M. Höök, B. Öbrink, and R. Timpl (1981a) Substrate adhesion of rat hepatocytes: Mechanism of attachment to collagen substrates. *Cell* **24**:463–470.

Rubin, K., S. Johansson, M. Höök, and B. Öbrink (1981b) Substrate adhesion of rat hepatocytes: On the role of fibronectin in cell spreading. *Exp. Cell Res.* **135**:127–135.

Rubin, K., T. K. Borg, R. Holmdahl, L. Klareskog, and B. Öbrink (1984) Interactions of mammalian cells with collagen. *CIBA Found. Symp.* **108**:93–116.

Schmell, E., C. W. Slife, M. S. Kuhlenschmidt, and S. Roseman (1982) Studies on the intercellular adhesion of rat and chicken hepatocytes: Conditions for stimulation by liver plasma membranes. *J. Biol. Chem.* **257**:3171–3176.

Sedar, A. W., and J. G. Forte (1964) Effects of calcium depletion on the junctional complex between oxyntic cells of gastric glands. *J. Cell Biol.* **22**:173–188.

Sieber, F., and S. Roseman (1981) Quantitative analysis of intercellular adhesive specificity in freshly explanted and cultured cells. *J. Cell Biol.* **90**:55–62.

Siu, C. H., B. Des Roches, and T. Y. Lam (1983) Involvement of a cell-surface glycoprotein in the cell-sorting process of *Dictyostelium* discoideum. *Proc. Natl. Acad. Sci. USA* **80**:6596–6600.

Takeichi, M., H. S. Ozaki, K. Tokunaga, and T. S. Okada (1979) Experimental manipulation of cell surface to affect cellular recognition mechanisms. *Dev. Biol.* **70**:195–205.

Thiery, J.-P., J. L. Duband, U. Rutishauser, and G. M. Edelman (1982) Cell adhesion molecules in early chicken embryogenesis. *Proc. Natl. Acad. Sci. USA* **79**:6737–6741.

Thomas, W. A., B. A. Edelman, S. M. Lobel, A. S. Breitbart, and M. S. Steinberg (1981) Two chick embryonic adhesion systems: Molecular vs tissue specificity. *J. Supramol. Struct. Cell Biochem.* **16**:15–27.

Urushihara, H., and M. Takeichi (1980) Cell–cell adhesion molecule: Identification of a glycoprotein relevant to the Ca^{2+}-independent aggregation of chinese hamster fibroblasts. *Cell* **20**:363–371.

Voss, B., S. Allam, J. Rauterberg, K. Ullrich, V. Gieselmann, and K. von Figura (1979) Primary cultures of rat hepatocytes synthesize fibronectin. *Biochem. Biophys. Res. Commun.* **90**:1348–1354.

Wanson, J. C., P. Drochmans, R. Mosselmans, and M. F. Ronveaux (1977) Adult rat hepatocytes in primary monolayer culture: Ultrastructural characteristics of intercellular contacts and cell membrane differentiations. *J. Cell Biol.* **74**:858–877.

Yamada, K. M., and K. Olden (1978) Fibronectins—adhesive glycoproteins of cell surface and blood. *Nature* **275**:179–184.

Section 4

Extracellular Matrix
and Cell–Substrate Adhesion

Chapter 14

Fibronectin: Molecular Approaches to Analyzing Cell Interactions with the Extracellular Matrix

KENNETH M. YAMADA
MARTIN J. HUMPHRIES
TAKAYUKI HASEGAWA
ETSUKO HASEGAWA
KENNETH OLDEN
WEN-TIEN CHEN
STEVEN K. AKIYAMA

ABSTRACT

Fibronectin is a developmentally regulated, multifunctional glycoprotein with roles in cell adhesion, migration, and differentiation. It functions via at least seven specialized intramolecular regions. After proteolytic dissection, structural domains that still interact with collagen, fibrin, heparin, other molecules, and the plasma membrane can be isolated. Various combinations of these functional domains are thought to mediate each of fibronectin's activities. These specialized domains apparently evolved from a simple repetitive gene and protein structure. Each is present in both plasma and cellular fibronectins, which are probably derived from a single gene by differential splicing of precursor mRNA molecules.

The mechanism that binds fibronectin on the plasma membrane is under intensive investigation. Fibronectin is bound by an interaction of moderate affinity, and the interaction of radioactive fibronectin is competitively inhibited by unlabeled fibronectin or its cell-binding domain. In cell adhesion assays, the attachment of cells to fibronectin is also autoinhibited by high concentrations of fibronectin in solution, its cell-binding domain, or even synthetic peptides containing the putative recognition signal (Gly)-Arg-Gly-Asp-Ser. Therefore, fibronectin functions in a dualistic manner, either mediating adhesion or becoming a competitive inhibitor if present at excessive concentrations in solution.

Candidates for a fibronectin receptor include gangliosides and glycoproteins of 47 kD and 140 kD. In particular, monoclonal antibodies against the 140-kD protein complex colocalize with fibronectin in adhesive contacts. They also inhibit fibronectin-mediated adhesion and, in analogy to fibronectin itself, can become positive effectors of cell adhesion if adsorbed onto substrates. Further analyses of fibronectin function should provide insights into the molecular mechanisms of cell surface interactions and morphogenesis.

Fibronectin, together with several other glycoproteins at the cell surface and in the extracellular matrix, can dramatically affect cell migration, morphology, and certain differentiated phenotypes. Central to the action of fibronectin is its activity as a multifunctional cell adhesion protein, interacting specifically with a variety of biologically important extracellular molecules. Dissection of the fibronectin molecule into its component functional domains, combined with biochemical and recombinant DNA analyses, is currently providing insight into its molecular mechanisms of action in a variety of cellular processes.

This review emphasizes key current concepts, with no attempt to present a comprehensive bibliography; we emphasize that each reference presented is often only one of several pertinent publications. A number of comprehensive reviews on fibronectin are now available (e.g., see Ruoslahti et al., 1981; Alitalo and Vaheri, 1982; Aplin and Hughes, 1982; Hynes, 1982; Hynes and Yamada, 1982; Yamada, 1983; Akiyama and Yamada, 1983; Furcht, 1983; Mosher, 1985).

Locations of Fibronectin

Fibronectin exists as a freely soluble plasma protein (termed plasma fibronectin) in vertebrate blood, where it is present at the substantial concentration of 0.3 g/l. A second major form situated on cell surfaces is loosely termed cellular fibronectin. More narrowly, this form is named after the cell type of origin (e.g., "fibroblast cellular fibronectin"). Cellular fibronectins are produced by fibroblasts, endothelial cells, and other cells producing extracellular matrices with a high content of fibronectin *in vitro*. This cellular form is relatively insoluble at physiological pH and becomes organized into striking fibrillar arrays on the cell surface and in the extracellular matrix (Figure 1). Similar fibrillar matrices of fibronectin appear *in vivo* at certain times in development.

Fibrillar extracellular matrices containing fibronectin can be isolated by extracting dense cultures of fibroblasts *in vitro* with detergents (Chen et al., 1978; Hedman et al., 1979). Several experimental findings suggest that these matrices utilize fibronectin rather than collagen as the major structural component and that they may be transient, developmentally regulated structures, specialized for involvement in specific developmental events: (1) Fibronectin, not collagen, is the major constituent of such a matrix (Chen et al., 1978); extensive accumulation of collagen fibrils appears to be a subsequent event. (2) Inhibition studies with anti-fibronectin antibodies suggest that the distribution of collagen *in vitro* depends on the integrity of this fibronectin matrix (McDonald et al., 1982). (3) Analyses of the *in vivo* neural crest migratory pathway reveal large amounts of fibronectin and glycosaminoglycans, but apparently rather limited amounts of fibrillar collagen (Newgreen and Thiery, 1980; Mayer et al., 1981; Weston, 1983).

Figure 1. *Distribution of cellular fibronectin in the extracellular matrix.* Immunofluorescence localization of fibronectin in a confluent culture of chick embryo fibroblasts, as detected by affinity-purified goat anti-fibronectin antibodies labeled with fluorescein isothiocyanate. Note the dense matrix of fibronectin fibrils (*fluorescence*) completely surrounding the cells (*dark regions*). Calibration bar = 25 μm.

Fibronectin in Morphogenesis

Fibronectin-containing matrices are particularly prominent *in vivo* at sites of tissue remodeling and cell migration. Fibronectin often appears at the time of a migratory event and is situated in locations appropriate for a substrate for cell migration. A number of examples now exist: In amphibian and avian gastrulation, fibrils of fibronectin lie along the routes of massive cell movement (Critchley et al., 1979; Duband and Thiery, 1982b; Boucaut et al., 1984); microinjection of anti-fibronectin antibodies inhibits amphibian gastrulation (Boucaut et al., 1984; see also Thiery et al., this volume); it has also been shown that sea urchin primary mesenchyme cells express a fibronectinlike antigen only at the time of ingression (Katow et al., 1982).

Fibronectin is particularly prominent in the migratory pathway of neural crest cells at specific times in development (Newgreen and Thiery, 1980; Mayer et al., 1981; Duband and Thiery, 1982a; Thiery et al., 1982). Isolated neural crest cells selectively utilize such carpets of fibronectin for migration *in vitro* (Rovasio et al., 1983). Amphibian primordial germ cells also migrate for long distances along fibronectin-rich pathways (Heasman et al., 1981). Even the formation of feather buds in avian development appears to involve an accumulation of fibronectin at the site of morphogenetic movement (Mauger et al., 1982).

In the adult organism, fibronectin reappears at sites of tissue remodeling. Soon after wounding, a scaffolding of fibrin and fibronectin appears; collagen appears at a significantly later time. Fibroblastic cells seem to require fibronectin in order to adhere to fibrin, and adhesion is reported to be particularly effective if the fibronectin is covalently cross-linked to fibrin by transglutaminase (Grinnell et al., 1980).

Ephithelial cells may also utilize fibronectin substrates for migration during wound healing. In corneal epithelial wounds, fibronectin appears rapidly at the time of initial cell migration; anti-fibronectin antibodies inhibit corneal wound healing (Nishida et al., 1983). As the healing process continues, fibronectin appears to be supplanted by laminin, collagen type IV, and bullous pemphigoid antigen; it is possible that fibronectin serves as an early migratory substrate and is replaced by other attachment molecules for subsequent migration and stable adhesive interactions (Clark et al., 1982). In addition, there are numerous possible correlations of increased fibronectin levels in tissues with the process of fibrosis in various human diseases; elevations in local fibronectin concentrations may contribute to fibrotic tissue remodeling, although the mechanisms remain obscure (for a review, see Akiyama and Yamada, 1983).

Fibronectin also is found in association with embryonic basement membranes, according to light microscopy, and it may be part of the basal lamina per se. In general, much less fibronectin appears to be present in adult basement membranes (Wartiovaara and Vaheri, 1980; Hynes and Yamada, 1982; Furcht, 1983). The functions of fibronectin at these sites during embryogenesis are still unclear.

Fibronectin and Evolution

By criteria including immunofluorescence and radioimmunoassays with antibodies raised against human plasma fibronectin, a fibronectin-related molecule has been detected in all vertebrates and in most invertebrates examined (e.g., Akiyama and Johnson, 1983). A fibronectinlike molecule can be isolated from sea urchins and marine sponges, although no structural comparisons to fibronectins of higher organisms have been reported. There is no evidence as yet for any fibronectinlike molecules in plants or in bacteria.

The presence of a fibronectin-related molecule in multicellular organisms throughout the animal kingdom underlines its general importance. One obvious function of the molecule is as an adhesive glycoprotein, although this hypothesis requires testing. It is also interesting that such molecules can still be recognized by an antibody to human fibronectin, which suggests that some structural feature(s) of the molecule have been conserved throughout evolution.

THE FUNCTIONS OF FIBRONECTIN

The possible biological functions of fibronectin have been explored by many investigators with preparations purified from cell surfaces and from plasma. Table 1 summarizes the known major biological activities of fibronectin, as identified through a variety of *in vitro* assay systems. Particularly

Table 1. Biological Activities of Fibronectin

Cell–substrate adhesion
    Cell attachment to collagen and gelatin
    Cell attachment to fibrin
    Cell adhesion and spreading on glass and plastic

Cell–cell adhesion

Cell migration
    Stimulation of cell motility
    Haptotaxis or chemotaxis

Cell morphology
    Maintenance of flattened and spread cell shape with
    few cell surface microvilli
    Alignment of confluent fibroblasts into parallel arrays
    Stimulation of actin microfilament bundle organization

Stimulation or inhibition of cytodifferentiation

Stimulation of growth or cytokinesis

Nonimmune opsonic activity for macrophages

interesting are its functions in cell adhesion, cell migration, and embryonic differentiation.

Cell Adhesion

Fibronectin has now become a prototypic cell–substrate adhesion molecule, although it also displays weaker effects on cell–cell adhesion. Its activity is generally thought to be dependent on attachment to a substrate or to other fibronectin molecules to form multimeric complexes. For example, Klebe (1974) discovered that adhesive activity was retained by substrates preincubated with fibronectin; the fibronectin bound to collagenous substrates and remained active even after extensive washing. In contrast, cells exposed to fibronectin-containing serum and then washed showed no evidence of any continued association with fibronectin. An extensive literature now supports the concept of maximal effectiveness only after fibronectin adsorption to a substrate or aggregation to form a multivalent complex. Recent experiments appear to explain this finding as a simple consequence of the low-affinity interaction of fibronectin with its cellular receptor.

    Fibronectin was the first cell–collagen adhesive glycoprotein identified. It is also involved in the adhesion of cells to plastic surfaces *in vitro* and serves as an attachment factor for the culture of certain cells (Barnes and

Sato, 1980). As noted above, fibronectin-mediated adhesion may also be important in the adherence of fibroblasts to fibrin in wound healing. Fibronectin also binds readily to heparin and heparan sulfate, an interaction that may contribute to cell adhesion by strengthening substrate associations with heparan sulfate proteoglycan at the cell surface (Laterra et al., 1983). Alternatively, this binding interaction might be important in the organization of heparan sulfate proteoglycan in the extracellular matrix.

The cellular form of fibronectin also mediates cell aggregation of embryonic cells and aldehyde-fixed erythrocytes (for a review, see Yamada, 1983). This effect appears weaker than that exhibited by classic cell–cell adhesion molecules, and it might be viewed as a form of cell–matrix interaction in which the fibronectin functions as a small fragment of extracellular matrix.

Cell Migration

Purified fibronectin stimulates the migration of a variety of cells *in vitro*, including embryonic neural crest cells. A combination of fibronectin-coated substrate and high cell density *in vitro* mimics the conditions necessary for the rapid, directional rates of cell migration characteristic of neural crest cells *in vivo* (Rovasio et al., 1983).

Fibronectin also stimulates axonal extension by cells of the peripheral nervous system. Another attachment molecule, laminin, is more effective for neurite extension by central nervous system cells (Rogers et al., 1983). Besides promoting increased rates of cell migration, fibronectin can also mediate directional migration in a chemotaxis chamber by a process of haptotaxis or chemotaxis (see Hynes and Yamada, 1982).

Embryonic Differentiation

Purified cellular fibronectin can modulate the differentiation of certain types of cells *in vitro* (Figure 2), although the mechanisms of these metabolic effects are still unknown. Exogenously added cellular fibronectin promotes the adrenergic phenotype of neural crest cells (Sieber-Blum et al., 1981; Loring et al., 1982). Although this effect is not simply the result of an increased plating efficiency of adrenergic cells on a fibronectin substrate, it is not clear whether fibronectin acts by gradually increasing the proportion of a preexisting subpopulation committed to forming adrenergic cells, or is acting more directly to induce catecholamine synthesis in uncommitted cells. Analysis of the mechanisms of this effect could provide insight into the means by which extracellular molecules regulate intracellular events.

Fibronectin can also function as an inhibitory molecule in the processes of myoblast fusion, adipocyte differentiation, and regulation of the chondrogenic phenotype. Myoblasts cultured on collagenous substrates require fibronectin for attachment, a process normally necessary for growth and subsequent successful fusion into myotubes (Chiquet et al., 1979). However, if endogenous or exogenous fibronectin is present in excess, it can delay

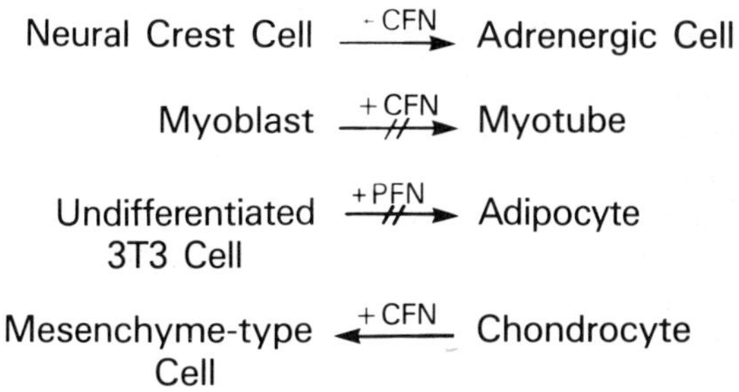

Neural Crest Cell $\xrightarrow{\text{- CFN}}$ Adrenergic Cell

Myoblast $\xrightarrow{\text{+ CFN}}\!\!\!/\!\!/$ Myotube

Undifferentiated $\xrightarrow{\text{+ PFN}}\!\!\!/\!\!/$ Adipocyte
3T3 Cell

Mesenchyme-type $\xleftarrow{\text{+ CFN}}$ Chondrocyte
Cell

Figure 2. *Effects of fibronectin on cytodifferentiation.* Cellular fibronectin (CFN) or plasma fibronectin (PFN) can alter the differentiated phenotype of certain cells as indicated. See text for details and references.

or inhibit the fusion process (Podleski et al., 1979). This inhibitory effect might involve physical interference with membrane fusion events, since cytodifferentiation can apparently proceed even though fusion is prevented (K. Olden, unpublished observations).

Fibronectin also inhibits the differentiation of 3T3 cells into adipocytes (Spiegelman and Ginty, 1983). Culturing cells on plasma fibronectin substrates markedly inhibits the characteristic changes in cell morphology and the induction of such lipogenic enzymes as fatty acid synthetase and glycerophosphate dehydrogenase. Concomitant maintenance of cells in a rounded configuration reverses the inhibitory effects of fibronectin, suggesting that it may act by increasing cell adherence and flattening, followed by effects on actin organization and as yet unknown cytoskeletal effects on nuclear function.

Fibronectin also has dramatic metabolic effects on the expression of the differentiated phenotype of chondrocytes. Cells treated with cellular fibronectin switch from the synthesis of collagen type II characteristic of chondrocytes to the synthesis of collagen type I; in addition, they substantially decrease rates of synthesis of the sulfated proteoglycans characteristic of chondrocytes (Pennypacker et al., 1979; West et al., 1979). In analogy to its effects on 3T3 adipocytes, one possible explanation of this fibronectin-mediated switch in biosynthesis to a less differentiated phenotype is that fibronectin causes flattening and spreading of these cells, thereby inhibiting chondrogenesis (Solursh et al., 1982). Alternatively, fibronectin might have more direct effects on biosynthetic processes. It is of obvious importance to attempt to separate the effects of fibronectin on cell morphology from possible direct effects on differentiation pathways.

THE FUNCTIONAL DOMAINS OF FIBRONECTIN

The biochemical analysis of a complex molecule becomes possible if its activities can be divided into simpler subsets of functions. Fibronectin is now thought to be a multifunctional molecule comprising a series of

Table 2. Interactions of the Fibronectins[a]

Collagens types I–IV and gelatin*
Fibrin and fibrinogen*
Heparin and heparan sulfate*
Factor XIII_a transglutaminase*
Cell surfaces of eukaryotic cells and bacteria*
Hyaluronic acid
Actin
DNA
Gangliosides
$C1_q$ component of complement
Asymmetric acetylcholinesterase
Thrombospondin
Polyamines

[a] The asterisks indicate the most extensively characterized interactions. For specific references, see reviews listed at the beginning of this chapter, for example, Yamada, 1983.

specific binding sites that mediate its functions. The binding activities of fibronectin are listed in Table 2; the first five interactions are the most intensively studied. These interactions appear to be specific, and they can be analyzed by filter-binding assays or affinity chromatography. The former approach permits the estimation of binding constants, for example, $K_D = 10^{-7}$ to 4×10^{-9} M for heparin (Yamada et al., 1980a). The most successful approach, however, has been the isolation of active fragments of the molecule in order to examine functions inherent in isolated domains.

Proteolytic Dissection and Isolation of Domains

A particularly successful approach now applied by a number of laboratories has been to isolate the domains of fibronectin (and of other related molecules such as laminin) that are responsible for binding to specific ligands such as collagen and heparin. The general approach is to cleave fibronectin with a set of proteases under conditions favoring retention of protease-resistant domains (even with the broad-spectrum protease pronase) and then to isolate the domain(s) that bind to an affinity column containing a covalently attached ligand such as collagen.

This general approach has permitted the isolation of biologically active fragments of fibronectin that span the entire molecule (for reviews, see Yamada, 1983; Mosher, 1985). Two unexpected findings in such studies were (1) the importance of performing analyses only under physiological conditions, since artifactual binding occurs in nonphysiological salt solutions (Hayashi and Yamada, 1983), and (2) the occasional generation of proteolytic fragments with greater biological activity than that of the intact molecule itself. In general, however, the isolated domains are believed to reflect subsets of the activities present in the intact molecule.

An example of the relatively clean cleavage of purified human plasma fibronectin into discrete, protease-resistant polypeptide products by trypsin is shown in Figure 3A. The limited number of cleavage products is probably the result of cleavage at relatively flexible regions of the polypeptide chain, because the molecule contains large numbers of potential cleavage sites that are not susceptible to cleavage under these controlled conditions of proteolysis.

Such mixtures of proteolytic fragments are then analyzed by affinity chromatography using various ligands bound by fibronectin. Preparative affinity columns are also used to purify these binding domains (e.g., Hayashi and Yamada, 1983; other studies are described in the reviews listed at the beginning of this chapter). A set of such fragments purified by a combination of affinity and ion exchange chromatography is shown

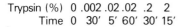

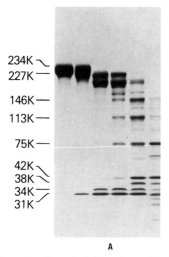

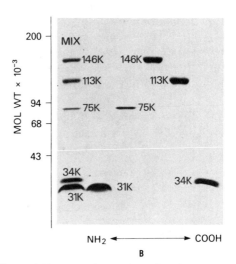

Figure 3. *Proteolytic dissection and isolation of fibronectin domains.* Fibronectin and its fragments were analyzed on SDS-polyacrylamide gels and stained for protein with Coomassie blue (see Hayashi and Yamada, 1983, for details). *A:* Purified human plasma fibronectin consists of a doublet of 234 kD and 227 kD. Exposure to increasing amounts of trypsin and longer periods of incubation result in cleavage of the molecule into a discrete series of protease-resistant domains ranging from 31 to 146 kD in apparent size. *B:* Fragments purified from tryptic digests of fibronectin according to Hayashi and Yamada (1983) on an SDS-polyacrylamide gel stained for protein. Apparent molecular weights of standards are indicated at the left. The first lane, labeled MIX, contains a mixture of each of the purified fragments. Twice the amount of each was analyzed in each of the next five lanes. The fragments are arranged according to the linear sequence of functional domains in the fibronectin molecule, with the amino terminus to the left.

in Figure 3B. These purified domains have been mapped on the molecule through the combined efforts of a number of laboratories; in Figure 3B they have been arranged according to their location in the molecule.

Linear Map of the Functional Domains of Fibronectin

A current summary of the domain structure of fibronectin is depicted in Figure 4. This multifunctional molecule comprises a linear series of at least seven functional domains. The figure shows the plasma fibronectin molecule, which consists of two similar polypeptide subunits linked by disulfide bonds at one end. Both subunits contain the same linear sequence of structural domains, arrayed like a string of beads along the polypeptide chain. Under physiological conditions, the molecule appears to be relatively compact, but not spherical; proteases probably cleave flexible regions of polypeptide separating the domains. In low salt or at extremes of pH, however, the molecule expands to form a relatively linear, V-shaped structure (Erickson and Carrell, 1983; Holly et al., 1984). Ca^{2+} also appears to stabilize the shape of the protein (Holly et al., 1984).

Although the two subunits contain the same overall organization of functional domains, there is a clear difference in domain structure between them toward the carboxy terminus (e.g., see Hayashi and Yamada, 1983). In addition, cellular fibronectin contains more than two subunits, which may also be linked at their carboxy termini; there is also evidence for heterogeneity of the cellular form in this region (Bernard et al., 1984).

The Amino-Terminal Domain.   The domain situated at the amino terminus is cleaved from the intact molecule by a variety of proteases. It binds to a

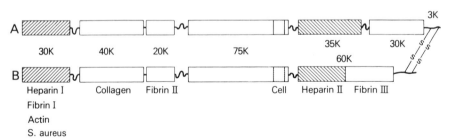

Figure 4. *Current map of the structural and functional domains of human plasma fibronectin.* The amino terminus is at the *left* and the carboxy terminus is at the *right*. The *rectangles* represent the protease-resistant domains described in the text, which are connected by flexible, protease-susceptible lengths of polypeptide chain (*wavy lines*). The two subunits, designated A and B, are very similar in organization. They differ at a region toward the carboxy terminus. Various proteases cleave the domains apart, generating fragments of the sizes indicated (e.g., 30K indicates a fragment with an apparent molecular weight of 30 kD). The binding activities of each domain are listed below the B chain; each activity appears to be present in both subunits. Some domains bind to the same ligand and are numbered with roman numerals, beginning at the amino terminus.

rather surprising number of ligands. Its interaction with heparin is modulated by physiological concentrations of divalent cations; high levels of Ca^{2+} prevent binding (Hayashi and Yamada, 1982). This domain also contains the highest affinity site on fibronectin for binding to fibrin (Sekiguchi et al., 1981; Hayashi and Yamada, 1983). Other materials, including actin (Keski-Oja and Yamada, 1981) and the bacterium *Staphylococcus aureus* (Mosher and Proctor, 1980), are bound by this region of fibronectin. This domain also serves as the primary site for transglutaminase-mediated cross-linking of fibronectin to fibrin or to collagen (Mosher et al., 1980).

The Collagen-Binding Domain.   The next domain is the only binding site on fibronectin for collagen. This region is rich in disulfide bonds, which are essential for collagen-binding activity. The collagen-binding domain binds with highest avidity to denatured collagens; the biological significance of this preference is not known. Nevertheless, it also binds to native type I, type III, and other collagens (Engvall et al., 1978). The collagen-binding domain is the most heavily glycosylated region of fibronectin, containing approximately three N-linked oligosaccharide units per domain. If glycosylation of fibronectin is prevented by tunicamycin, this domain becomes over 12-fold more susceptible to proteolytic degradation (Bernard et al., 1982).

A second, weak binding site for fibrin has been identified adjacent to the collagen-binding site (Seidl and Hörmann, 1983). This fibrin-binding site is destroyed easily by proteases such as trypsin, suggesting that it is not as highly structured as are the other binding regions of fibronectin.

The Cell-Binding or Cell-Recognition Site.   A region of crucial importance for adhesive function is located approximately three-quarters of the length of the molecule from the amino terminus and was initially termed the cell-binding domain (Hahn and Yamada, 1979; Ruoslahti and Hayman, 1979). This region alone can mediate attachment and spreading of cells on inert substrates. The smallest proteolytic fragment retaining full biological activity is one of 75 kD (Hayashi and Yamada, 1983). By using monoclonal antibodies, Pierschbacher et al. (1981) were the first to isolate this functional region in a pepsin-resistant fragment of 11.5 kD; it has rather variable biological activity, perhaps because of its small size.

There is a high degree of homology between amino acid sequences from this region in human, bovine, and rat plasma fibronectins (cf., Pierschbacher et al., 1982; Petersen et al., 1983; Schwarzbauer et al., 1983). Remarkably, cell attachment and competitive inhibition activities are still retained even by synthetic peptides from this region that are only four to six amino acids in length (Pierschbacher and Ruoslahti, 1984; Yamada and Kennedy, 1984). This suggests that cells recognize and bind to fibronectin, rather than vice versa. That is, in contrast to the function of other domains, which appear to be sites that recognize and bind to target molecules, this short peptide sequence would not be expected to form a binding site for some cell surface target molecule. Instead, some receptorlike moiety on the plasma membrane may have the three-dimensional conformation necessary to recognize this small region of primary sequence on fibronectin.

Thus, this region might be more accurately termed the cell-recognition site. It is not yet clear whether all of the binding of fibronectin by cells can be accounted for by this site. For example, binding may be stabilized by a heparin-binding site (Laterra et al., 1983). In addition, it is possible that amino acid sequences adjacent to the peptide recognition sequence might modify its recognition, for example, in terms of cell type specificity or in overall affinity for cells.

The Heparin-Binding Domain. The next domain along the polypeptide backbone contains the binding site with the greatest affinity for heparin. This region is devoid of carbohydrates, yet it is resistant to limited digestion by even the broad-spectrum protease pronase (Hayashi et al., 1980). The binding of fibronectin to heparin via this site is resistant to inhibition by divalent cations.

The Fibrin-Binding Domain. The final domain before the carboxy-terminal disulfide domain binds weakly to fibrin. It also contains one of the two free sulfhydryl groups in fibronectin (e.g., see Hayashi and Yamada, 1983). Most interestingly, it lies close to a region of structural difference between the A and B subunits of human plasma fibronectin, according to proteolytic cleavage analyses (Richter and Hörmann, 1982; Hayashi and Yamada, 1983). To speculate, this site of difference between the subunits might exist because it contains a recognition mechanism between complementary sites on the A and B chains. Such a recognition site might be needed to assemble the dimers of fibronectin from subunits originally synthesized as monomers in the rough endoplasmic reticulum.

The Carboxy-Terminal Domain. This domain, containing the disulfide bonds cross-linking the two chains of plasma fibronectin, is small, and the interchain disulfide bonds are only 16–20 residues from the carboxy terminus. This highly asymmetric location theoretically permits the widest expansion and separation of the amino-terminal ends of the subunits, which may assist the molecule in spanning large distances.

Function of Domains in Cell Adhesion

The linear sequence of modular domains on fibronectin provides the mechanism for a single molecule to mediate several types of adhesive interactions with a variety of molecules. Figure 5 is a schematic summary of studies analyzing the combinations of domains needed for certain biological activities.

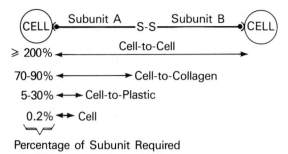

Percentage of Subunit Required

Figure 5. *Schematic summary of the requirements for fibronectin domains in different adhesive activities.* Two cells connected by disulfide-linked subunits of fibronectin are depicted (not drawn to scale). Cellular receptors attach to ball-shaped recognition sites on fibronectin. The numbers at the *left* indicate the percentage of one subunit required for a particular adhesive activity. For example, cell–cell adhesion appears to require two or more subunits, whereas a tiny cell-recognition sequence constituting only 0.2% of a subunit is sufficient for binding to cells. See text for discussion.

Cell–cell interactions appear to require a multivalent form of the molecule, as indicated by a loss of hemagglutinating activity when the disulfide linkage region connecting the subunits is cleaved (Hahn and Yamada, 1979). The actual size of the disulfide-linked aggregate required for agglutination is not yet known with certainty, but it may be larger than a dimer; a multimer may be needed to provide sufficient length to span between cells, because the cell-recognition region is not close to the free amino-terminal end of the molecule.

In contrast, a monomer or a large fragment of a subunit can mediate the attachment of cells to collagen (Hahn and Yamada, 1979). However, further proteolytic cleavage separating cell-recognition and collagen-binding sites leads to a loss of this activity. As discussed above, these fragments still retain the ability to interact with the cell surface or to bind to collagen, respectively.

The existence of multiple domains on fibronectin might permit cooperative interactions between domains. For example, the binding of heparin to fibronectin increaes the strength of its interaction with gelatin (see Yamada, 1983). The presence of two binding sites for heparin and three for fibrin may permit cooperative interactions between homologous binding sites. Moreover, the differences in regulation of the two heparin-binding sites by Ca^{2+} may permit some regulation of these interactions.

THE DIFFERENT FORMS OF FIBRONECTIN

In addition to differences in glycosylation between fibronectins from different sources, functional and probable polypeptide differences exist between the plasma and cellular forms of the molecule. Although quite similar in composition and domain structure, these two major forms of fibronectin differ in some, but not in other, biological activities. Cellular fibronectins have substantially higher specific activities in assays involving cell–cell interactions such as hemagglutination and restoration of a more normal phenotype to transformed cells. Yet both forms have equal activities in mediating the attachment of cells to substrates (Yamada and Kennedy, 1979).

These biological similarities and differences are mirrored in the structures of the proteins. They have identical linear sequences of domains, yet there appear to be at least three sites of differences between the forms, according to proteolytic mapping (Hayashi and Yamada, 1981). Similarly, most antibodies recognize both forms of the molecule, but one monoclonal antibody shows preferential (but not exclusive) binding to cellular fibronectin (Atherton and Hynes, 1981).

Nevertheless, the cellular and plasma forms of fibronectin appear to be derived from a single gene. The broadest supportive evidence to date derives from an evolutionary study of structural comparisons of the two forms of fibronectin isolated from different species (Akiyama and Yamada, 1985a). Two different methods of analysis—immunological comparisons of homology and two-dimensional peptide mapping—each demonstrate that plasma and cellular fibronectins from the same species contain many more primary structural similarities than do fibronectins of the same type from different species. These structural differences between species are quite clear, even in the face of greater similarities in biological activity across species lines than between types of fibronectin (i.e., human and chicken cellular fibronectins are more similar in hemagglutinating activity than are human cellular and plasma fibronectins).

The simplest interpretation is that both plasma and cellular fibronectins are derived from a single gene that has been subject to evolutionary drift during the process of evolution of different species. The products of such a gene would be similar within a species, but different in different species. In contrast, the biological differences between forms would be generated by posttranscriptional events. These data do not, however, rule out the possibility that other minor forms of fibronectin might be produced by additional fibronectin genes.

To assess the total number of fibronectin genes, it is necessary to examine DNA. Multiple hybridizations during the screening to obtain the fibronectin gene did not reveal any evidence of heterogeneity that might suggest more than one gene (Hirano et al., 1983). More direct Southern blot analyses of genomic DNA using human cDNA clones also suggest the existence of only one gene for fibronectin (Kornblihtt et al., 1983), although further screening of other clones under conditions of decreased stringency of hybridization will be useful in ruling out the possibility of other fibronectinlike genes. Nevertheless, the combination of these DNA hybridization experiments with the protein evolutionary data indicates that only one gene is responsible for synthesis of both the plasma and cellular forms of fibronectin.

THE FIBRONECTIN GENE

Isolation and Structure

The fibronectin gene has been isolated as a set of five recombinant DNA clones spanning the entire chicken fibronectin gene sequence (Figure 6). The length of the fibronectin gene—48 kb—ranks it as one of the largest genes identified to date (Hirano et al., 1983). It is also unusually complex in structure in terms of its complement of alternating coding units (exons)— a remarkable 48 or more—and intervening sequences (introns). There must be at least this number of accurate mRNA splicing events to produce each mature fibronectin mRNA molecule.

Although complex, the gene units have a common denominator in terms of size, as determined by electron microscope R-loop analysis. With the exception of those at each end of the gene, individual exons are invariably small; they are relatively uniform in size, averaging 150 base pairs in length (Hirano et al., 1983). In contrast, the noncoding intervening sequences (introns) vary widely in size (Figure 6). At this first level of analysis, no evidence exists for regions of distinctive gene structure that correspond to the functional domain structure of the protein.

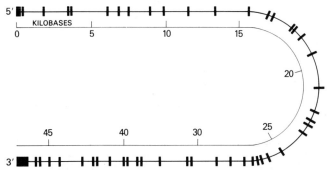

Figure 6. *The fibronectin gene.* The gene encoding chicken cellular fibronectin was isolated in the form of overlapping recombinant DNA clones in a bacteriophage vector. The *vertical bars* indicate coding regions (exons), which are separated by intervening sequences (introns). The first and last exons at the 5' and 3' ends are larger, but the remaining exons are similar in size (though slightly more variable in size than depicted here). In contrast, the introns are variable in length. The gene contains at least 48 exons, according to electron microscopy, and the entire gene is approximately 48 kb in length. This gene appears to be the source of both plasma and cellular fibronectins. See Hirano et al., 1983, and the text for experimental details and discussion.

Evolution of the Fibronectin Gene

How might such a multifunctional cell interaction protein have arisen in evolution? The two most extreme possibilities are depicted in Figure 7. One simple and elegant mechanism postulates that a series of gene units encoding different protein domains joined together. For fibronectin, domains binding to heparin, collagen, cells, and fibrin could, theoretically, have been involved. In this model, the early evolution of the animal kingdom might have included a series of recombinations that linked together gene pieces encoding such domains to form the final fibronectin gene.

At the opposite extreme, the gene could have arisen by an extensive series of gene duplications of one or a few primordial genes; the protein would then include a large number of repetitive protein units. Formation of the specialized functional domains of present-day fibronectin would have required subsequent evolutionary modification of specific polypeptide regions.

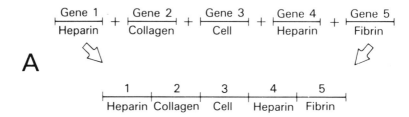

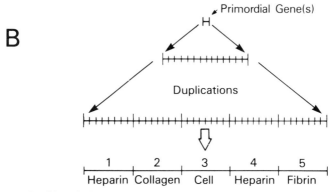

Figure 7. *Possible evolutionary mechanisms for generation of the gene for fibronectin.* Two extreme possibilities are depicted. *A*: The gene is postulated to have been generated by recombinational fusion of a series of five smaller genes encoding protein products binding to the indicated ligands. The final gene would encode a large, multifunctional protein containing all of the binding sites. *B*: Intermediate sets of genes would be generated by duplication of one or more tiny, primordial gene units. The initial protein products would have repeating sequences; evolutionary modification by mutation or recombination would be required to generate the present-day series of binding sites for different ligands.

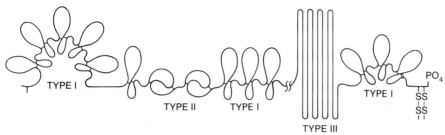

Figure 8. *Schematic representation of the known protein sequences of fibronectin.* The basic organization of the fibronectin molecule is shown in terms of repeating peptide units of homologous amino acid sequences. Type I homologies are located throughout the molecule, and they consist of disulfide-linked loop structures roughly 4.5–5.0 kD in size. Type II units have been identified as yet only in the collagen-binding region and are similar in size to type I units. Type III units are present in the heparin-binding and cell-recognition regions and are twice the size of the other two units. The carboxy terminus contains two interchain disulfide bonds and one phosphate group. (From Petersen et al., 1983; more limited sequencing by Kornblihtt et al., 1983, and Schwarzbauer et al., 1983.)

The structure of the fibronectin gene favors the second hypothesis, that is, a gene evolving by means of a massive series of duplications of one or more tiny primordial genes. Crucial to this conclusion is the sequence of the gene and the protein. Although sequencing is not completed, it is apparent even from current protein and DNA sequences that the gene and protein contain a number of repetitive units. The protein sequence may contain at least three different types of homologous units, with lengths that either match, or are twice the size of, the predicted protein units encoded by the average-size exon (Figure 8; Petersen et al., 1983). This evolutionary mechanism resembles that proposed for the collagen genes, which are thought to have evolved after duplication of a 54-base-pair primordial gene unit (Yamada et al., 1980b). In contrast to the collagens, however, which are relatively simple repeating proteins, fibronectin evolution is likely to have involved extensive modifications to yield such distinctive functional domains.

Derivation of Multiple Fibronectins from One Gene

Although only one fibronectin gene is presently thought to account for the two general forms of fibronectin, there is strong evidence that several classes of mRNAs encode fibronectin (Schwarzbauer et al., 1983; Kornblihtt et al., 1984). The simplest explanation of these findings is that the original precursor mRNA transcript from the fibronectin gene is processed differentially; that is, different splicing mechanisms can apparently generate several different mRNA molecules that would produce slightly differing protein molecules. Although hepatocyte fibronectin mRNAs show evidence of splicing differences within an exon (Schwarzbauer et al., 1983), these differences may not account for those among the major forms of fibronectin; they may instead account for heterogeneity between plasma fibronectin subunits.

In contrast, Kornblihtt et al. (1984) have identified an unusual, charged, putative difference region in amino acid sequences, deduced from comparisons of mRNAs of different cell types, that may arise from differential splicing at introns. This latter type of differential splicing may account for at least some of the differences between the major forms of fibronectin. The intriguing possibility now exists that there is actually a series of fibronectin molecules produced by subtle differences in splicing by different cell types, or even within one cell. This mechanism could generate a panoply of closely related adhesive proteins with differing properties, each of which could theoretically be better suited to a specific developmental event.

THE FIBRONECTIN RECEPTOR

Identification of the molecule(s) on the cell surface that bind to fibronectin has proven to be surprisingly difficult. A major impediment to progress has been the unusually low affinity of the putative receptor, which has prevented its isolation by simple affinity chromatography. In fact, until recently, even the binding characteristics of the receptor were unknown. Fibronectin binding to fibroblasts was examined by measuring the binding of fibronectin-coated beads or of apparent aggregates at 4°C (e.g., see Grinnell et al., 1982), and it was concluded that fibronectin interactions with fibroblastic cells probably require aggregation or substrate adsorption.

Binding Characteristics of the Receptor

Recently, direct binding of soluble fibronectin to fibroblasts was demonstrated in an assay emphasizing physiological binding solutions and high concentrations of fibronectin in order to be able to measure a putative receptor of only moderate affinity (Akiyama and Yamada, 1985b). Binding of human plasma fibronectin was found to be specific and saturable. Figure 9 shows a Scatchard analysis of the binding of tritiated fibronectin to baby hamster kidney (BHK) fibroblasts in suspension culture. There is evidence for a class of receptor with moderate affinity ($K_D = 8 \times 10^{-7}$ M). This affinity is too low for conventional affinity chromatography and seems to explain why a receptor has not been isolated by simply recovering a plasma membrane component that binds to immobilized fibronectin.

The affinity of the putative fibronectin receptor permits half-saturation at the seemingly high concentration of 370 µg/ml fibronectin. This figure should be compared with the concentration of fibronectin in plasma, which is 300 µg/ml. The simplest explanation of this difference is that the affinity

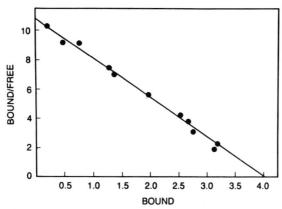

Figure 9. *Scatchard analysis of the binding of [³H]fibronectin to BHK cells in suspension.* Binding data were analyzed by plotting the ratio of the bound concentration of fibronectin to the free concentration (in units of $\mu l/10^7$ cells as a function of the bound concentration of fibronectin (in units of $\mu g/10^7$ cells). The slope indicates an apparent dissociation constant of $K_D = 8.4 \times 10^{-7}M$ and the X-intercept indicates 500,000 sites per cell. See Akiyama and Yamada, 1985b, for experimental details.

is necessarily this low to prevent saturation of the receptor by fibronectin from plasma and lymph; it guarantees that cells can still interact with fibronectin organized as a multivalent complex, even in the presence of normal concentrations of soluble fibronectin (see the section of this chapter on the dualistic activities of fibronectin).

The apparent number of these putative fibronectin receptors per cell is high—500,000 per cell (Akiyama and Yamada, 1985b). This number may limit the possible candidates for the receptor to the more abundant plasma membrane constituents, and it suggests that once the receptor is identified, it should be possible to isolate large enough quantities for structural studies.

Competitive binding studies also suggest that a 75-kD cell-recognition fragment of fibronectin competes more effectively for this receptor function than does intact fibronectin (Akiyama and Yamada, 1985b). This anomalous increase in avidity after cleavage may be reflected in biological studies showing increased activity of fragments of fibronectin in certain assays such as the inhibition of fibronectin-mediated cell attachment.

Although the putative receptor activity is destroyed by trypsinization in the absence of divalent cations, it is resistant to proteolysis in the presence of Ca^{2+} (Akiyama and Yamada, 1985b). Similarly, the capacity of cells to adhere to fibronectin is reportedly sensitive to trypsin only in the absence of Ca^{2+} (Oppenheimer-Marks and Grinnell, 1984). These findings suggest that the fibronectin "receptor" is dependent on a protein component and that it is a divalent cation-stabilized molecule. Its function, however, does not appear to have a strict requirement for divalent cations because cells will bind soluble fibronectin and fibronectin-coated beads even in the presence of EDTA.

A. Normal Cell-Substrate   B. Excess of Adhesion   C. Excess of Active Fragment
         Adhesion   Protein   or Recognition Site

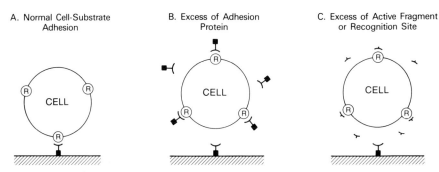

Figure 10. *Dualistic nature of cell adhesion protein function. A:* In normal cell–substrate adhesion mediated by a receptor, the adhesive molecule binds to the substrate and mediates adhesion by binding to cellular receptors (R). *B:* In the presence of an excess of adhesion protein in solution, the putative receptors are fully occupied, interfering with binding of substrate-attached molecules to the receptors. *C:* An excess of active fragments or peptide fragments recognized by the receptor is also inhibitory. A cell adhesion molecule could thus display postive or negative activities, depending on its concentration and relationship to a substrate.

Dualistic Nature of Adhesive Protein Function

If an adhesive protein functions by binding to a specific receptor, it is theoretically possible for an excess of the adhesion molecule to inhibit its own function. Figure 10 presents this argument schematically: Normally, the molecule binds to a substrate and to an available receptor on the cell surface. If all available receptors are saturated with free adhesive protein in solution, no receptors will be available to mediate attachment to substrate-adsorbed proteins.

Fibronectin functions in this dualistic manner (Yamada and Kennedy, 1984). If fibronectin is coated on a substrate and further protein adsorption is blocked by incubation with bovine serum albumin, cells readily attach and spread on the fibronectin. If soluble fibronectin is present during attachment, adhesion is progressively inhibited by increasing quantities of fibronectin in solution (Figure 11; Yamada and Kennedy, 1984). Inhibition is also mediated by the 75-kD cell-recognition fragment of fibronectin; as for inhibition of binding of tritiated fibronectin to cells, the concentration of this fragment required for maximal inhibition (0.5–1.0 mg/ml) is significantly lower than the concentration of intact molecule required, even on a molar basis.

Even short peptides from a highly conserved sequence in the cell-recognition domain containing the specific sequence (Gly)-Arg-Gly-Asp-Ser retain inhibitory activity (half-maximal inhibition at 3×10^{-4} to $10^{-5}M$), whereas a series of other peptides tested as controls were inactive (Figures 11, 12; Yamada and Kennedy, 1984, 1985). Similar inhibition has been found by Pierschbacher and Ruoslahti (1984), although at higher concentrations (Figure 12).

Another theoretical prediction is that multivalent fibronectin should eventually be a more effective competitor for the putative fibronectin receptor. The autoinhibition by fibronectin and its fragments is progressively diminished with increasing amounts of fibronectin adsorbed to the substrate; that is, more mulitvalent fibronectin on the substrate appears to be able to compete more effectively for cellular binding sites. Moreover, inhibition by soluble fibronectin is gradually lost with increased incubation time, as might be expected if the fibronectin in solution is gradually outcompeted by multivalent fibronectin (Yamada and Kennedy, 1984). It appears likely that the adsorbed, multivalent configuration of fibronectin competes more effectively by being able to maintain a very high local concentration of available molecules: When one fibronectin molecule detaches from the receptor, its place is more likely to be taken by an immobilized fibronectin molecule held in close proximity rather than by a soluble fibronectin molecule. In hepatocytes, Johansson and Höök (1984) have recently reported that soluble fibronectin is only a weak and transient inhibitor of its own function; it appears likely that, in that system, adsorbed fibronectin is at an even greater advantage in competing for hepatocyte receptors.

The observation of this dualistic action of fibronectin suggests an explanation for the absence of a high-affinity fibronectin receptor. Such a receptor would become saturated and biologically inactive after exposure to the normal levels of soluble fibronectin in blood. The observed affinity of the receptor is consistent with its efficacy in highly efficient interactions with an adsorbed cell–substrate adhesion protein, but it is low enough to avoid inhibition by fibronectin in body fluids. It is nevertheless possible that pathologically high concentrations of fibronectin, or local concentrations

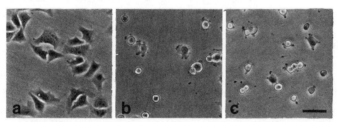

Figure 11. *Autoinhibition of fibronectin function by soluble fibronectin and a biologically active peptide fragment.* All three panels show BHK cells on substrates coated with 3 μg/ml plasma fibronectin. *a*: Control culture, showing fibronectin-mediated adhesion and spreading of cells. *b*: Inhibition of adhesion and spreading by the presence of 10 mg/ml purified plasma fibronectin in solution. *c*: Inhibition of adhesion and spreading by the presence of 0.2 mg/ml synthetic peptide containing the sequence Gly-Arg-Gly-Asp-Ser-Pro-Cys (peptide II in Figure 12, *top*). Calibration bar = 50 μm. See Yamada and Kennedy, 1984, and the text for experimental details and discussion.

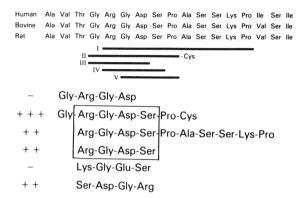

Human	Ala	Val	Thr	Gly	Arg	Gly	Asp	Ser	Pro	Ala	Ser	Ser	Lys	Pro	Ile	Ser	Ile
Bovine	Ala	Val	Thr	Gly	Arg	Gly	Asp	Ser	Pro	Ala	Ser	Ser	Lys	Pro	Val	Ser	Ile
Rat	Ala	Val	Thr	Gly	Arg	Gly	Asp	Ser	Pro	Ala	Ser	Ser	Lys	Pro	Val	Ser	Ile

-          Gly-Arg-Gly-Asp

+ + +   Gly-Arg-Gly-Asp-Ser-Pro-Cys

  + +   Arg-Gly-Asp-Ser-Pro-Ala-Ser-Ser-Lys-Pro

  + +   Arg-Gly-Asp-Ser

   −   Lys-Gly-Glu-Ser

  + +   Ser-Asp-Gly-Arg

Figure 12. *Conserved amino acid sequence from the cell-recognition region of fibronectin and related synthetic peptides (top).* Selected sets of published amino acid sequences are compared; the data were obtained from direct protein sequencing of human plasma fibronectin (Pierschbacher et al., 1982) and bovine plasma fibronectin (Petersen et al., 1983), or deduced from a cDNA sequence of rat hepatocyte fibronectin (Schwarzbauer et al., 1983). Note that the sequences are identical, except for one conservative substitution. This region is unusually hydrophilic, suggesting that it is particularly well exposed on the outer surface of the molecule. Synthetic peptides matching portions of these sequences were synthesized and compared for biological activity. Peptides I, II, and IV were active, but peptides III and V were not; peptide II and a pentapeptide containing the amino acids in peptides III and IV are especially active (Pierschbacher and Ruoslahti, 1984; Yamada and Kennedy, 1984, 1985). Key amino acid residues in the cell recognition region (*bottom*). The boxed residues appear to be essential for activity; although the tetrapeptide sequence (*boxed*) is the minimal sequence needed, the presence of the additional glycine substantially increases biological activity. Substitution of amino acids with the same charge (i.e., lysine for arginine and glutamic for aspartic) does not preserve specificity. Surprisingly, the inverted tetrapeptide sequence remains active, suggesting that only the amino acids and not the polypeptide backbone are being recognized (K. M. Yamada and D. W. Kennedy, unpublished observations).

close to sites of secretion, might exceed optimal levels and cause local inhibition of adhesion.

Candidates for the Fibronectin Receptor

Presently, several candidates for the fibronectin receptor exist. It is conceivable that there might be more than one mechanism by which cells bind to fibronectin, or that several molecules together are required for binding. The most attractive current candidates are 47-kD and 140-kD glycoproteins, and gangliosides.

140-kD Glycoprotein Complex. Monoclonal antibodies initially raised against chicken myoblasts affinity purify a complex of three to four glycoproteins averaging 140 kD in apparent size (Greve and Gottlieb, 1982; Neff et al., 1982). These antibodies block myoblast adhesion to gelatin-coated substrates, which suggests that they may interfere with fibronectin-mediated adhesion. In recent experiments, one of these monoclonal antibodies was found to inhibit fibronectin-mediated adhesion and spreading of normal and transformed chick embryo fibroblasts (Chen et al., 1985a). In analogy to the dualistic nature of fibronectin itself, however, this inhibitory antibody can also become a mediator of cell adhesion if appropriately attached to a substrate.

Immunofluorescence localization of these 140-kD antigens reveals a combination of diffuse labeling of the whole plasma membrane and localized streaks at cell–substrate contacts (Neff et al., 1982; Chen et al., 1985b). Immunoelectron microscopy, using the monoclonal antibody, shows localization of the 140-kD antigens on the plasma membrane at cell–extracellular matrix contact sites (Chen et al., 1985b). A comparison of the localization patterns of fibronectin, 140-kD complex, and several cytoskeletal proteins reveals a pattern of colocalization of the 140-kD complex with fibronectin fibrils and microfilament bundles throughout early spreading events and after transformation-induced alterations (Chen et al., 1985a). It may also be pertinent that the estimated number of 140-kD antigenic sites on cells is very close to the number of estimated fibronectin receptors. These findings strongly suggest an association of fibronectin with the 140-kD membrane protein complex. It will therefore be of considerable importance to determine whether this glycoprotein complex contains the receptor for fibronectin or whether it acts more indirectly by interacting with the receptor or by mediating part of the adhesive process in general.

47-kD Glycoprotein. Aplin et al. (1981) have identified a glycoprotein in photoaffinity cross-linking studies that may be a fibronectin receptor. Crude antibodies against this and possibly other glycoproteins inhibit fibronectin-mediated adhesion (Hughes et al., 1981). Since this glycoprotein is also labeled when cells adhere to a variety of lectins, it is conceivable that it is involved in cell spreading events in general or that it is an unusually exposed molecule on the ventral surface of cells. Studies of the susceptibility of fibronectin receptor function to proteases in the presence and absence of Ca^{2+} are consistent with a role for this glycoprotein as a receptor, although a 120–140-kD protein also may not be eliminated (Tarone et al., 1982; Oppenheimer-Marks and Grinnell, 1984).

Gangliosides. Gangliosides with several sialic acid residues have been implicated as possible fibronectin receptors. Gangliosides are competitive inhibitors in a series of standard assays for fibronectin function (for a review, see Yamada et al., 1983). Moreover, a ganglioside-deficient cell line displays defective organization of fibronectin, according to immunofluorescence studies. Reconstitution of gangliosides in these cells restores a normal fibrillar network of fibronectin on the cell surface (Yamada et al., 1983). The pattern of fibronectin fibrils is matched by the pattern of gangliosides, as determined with fluorescently labeled gangliosides (Spiegel et al., 1985). These results appear to demonstrate that gangliosides can serve as fibronectin receptors or as part of a receptor mechanism. They do not, however, prove that gangliosides are the normal receptor for fibronectin on all cells.

Because of the complexity of current data concerning the fibronectin receptor, it is possible that several molecules could act independently or as part of a receptor complex. Much more investigation is needed to resolve these questions.

PROSPECTS FOR THE FUTURE

Many important questions remain to be explored. For example, little is known about the regulatory mechanisms that determine when and where fibronectin matrices are established. Recombinant DNA probes should complement antibodies in the examination of transcriptional and subsequent levels of control of fibronectin biosynthesis during morphogenetic events.

Once fibronectin is synthesized, it must still be organized into fibrillar matrices. Factors influencing fibrillogenesis, for example, other matrix molecules, remain to be identified. The actual mechanisms of fibril formation remain obscure, but it is possible that new amino acid sequence information will permit the synthesis of peptide probes that can be used to explore this process.

The detailed mechanisms by which fibronectin matrices stimulate cell migration need to be determined. For example, further analyses of the types and amounts of different cytoskeletal proteins associated with cell contacts in migratory and nonmigratory cells on fibronectin compared to other substrates may provide insights into how cell movement is stimulated.

The fibronectin "receptor" remains to be identified and characterized. One goal will be to establish an *in vitro* model system in which purified receptors interact with fibronectin cell-recognition sequences. It may then be possible to reconstruct how interactions with fibronectin lead to alterations in cytoskeletal organization, for example, by changes in affinity for certain motility-related proteins. In addition, the possibility of developmental regulation of receptor numbers or specificity remains to be determined; changes in receptors could have major effects on morphogenesis.

Finally, a number of basic biochemical and functional questions remain. The actual physical mechanisms by which each of the domains interacts with its ligand need elucidation. The physical differences in the forms of fibronectin and their relationships to the known functional differences remain to be elucidated. In fact, the reasons for the existence of different forms of fibronectin need explanation. One exciting approach to this type of problem is the use of highly specific antibodies raised against specific synthetic peptide sequences to attempt to inhibit various functions *in vivo*. A complementary approach is to use these peptides directly as specific competitive inhibitors of the plethora of fibronectin biological activities. It seems likely that within the next few years these approaches to the many remaining questions should yield interesting insights into the functions of fibronectin and of related molecules in adhesion, morphogenesis, and other important biological processes.

ACKNOWLEDGMENTS

We thank Dorothy Kennedy for valuable assistance and the National Cancer Institute for support. E. H. and K. O. were supported by NIH Grant GM-29804; W.-T. C. received support from NSF Grant PCM-83-02882 and NIH Grant HL-31762; S. K. A. was supported by NCI fellowship CA06782.

REFERENCES

Akiyama, S. K., and M. D. Johnson (1983) Fibronectin in evolution: Presence in invertebrates and isolation from *Microciona prolifera. Comp. Biochem. Physiol.* **76B**:687–694.

Akiyama, S. K., and K. M. Yamada (1983) Fibronectin in disease. In *Connective Tissue Diseases*, B. M. Wagner, P. Fleischmajer, and N. Kaufman, eds., pp. 55–96, Williams and Wilkins, Baltimore.

Akiyama, S. K., and K. M. Yamada (1985a) Comparisons of evolutionarily distinct fibronectins: Evidence for the origin of plasma and fibroblast cellular fibronectins from a single gene. *J. Cell. Biochem.* **27**:97–107.

Akiyama, S. K., and K. M. Yamada (1985b) The interaction of plasma fibronectin with fibroblastic cells in suspension. *J. Biol. Chem.* **260**:4492–4500.

Alitalo, K., and A. Vaheri (1982) Pericellular matrix in malignant transformation. *Adv. Cancer Res.* **37**:111–158.

Aplin, J. D., and R. C. Hughes (1982) Complex carbohydrates of the extracellular matrix. Structure, interactions and biological roles. *Biochim. Biophys. Acta* **694**:375–418.

Aplin, J. D., R. C. Hughes, C. L. Jaffe, and N. Sharon (1981) Reversible cross-linking of cellular components of adherent fibroblasts to fibronectin and lectin-coated substrata. *Exp. Cell Res.* **134**:488–494.

Atherton, B. T., and R. O. Hynes (1981) A difference between plasma and cellular fibronectins located with monoclonal antibodies. *Cell* **25**:133–141.

Barnes, D., and G. Sato (1980) Serum-free medium. A review. *Cell* **22**:649–655.

Bernard, B. A., K. M. Yamada, and K. Olden (1982) Carbohydrates selectively protect a specific domain of fibronectin against proteases. *J. Biol. Chem.* **257**:8549–8554.

Bernard, B. A., S. K. Akiyama, S. A. Newton, K. M. Yamada, and K. Olden (1984) Structural and functional comparisons of chicken and human cellular fibronectins. *J. Biol. Chem.* **259**:9899–9905.

Boucaut, J. C., T. Darribere, H. Boulekbache, and J.-P. Thiery (1984) Prevention of gastrulation but not neurulation by antibodies to fibronectin in amphibian embryos. *Nature* **307**:364–366.

Chen, L. B., A. Murray, R. A. Segal, A. Bushnell, and M. L. Walsh (1978) Studies on intercellular LETS glycoprotein matrices. *Cell* **14**:377–391.

Chen, W.-T., E. Hasegawa, T. Hasegawa, C. Weinstock, and K. M. Yamada (1985a) Development of cell surface linkage complexes in cultured fibroblasts. *J. Cell Biol.* **100**:1103–1114.

Chen, W.-T., J. M. Greve, D. I. Gottlieb, and S. J. Singer (1985b) The immunocytochemical localization of 140-kD cell adhesion molecules in cultured chicken fibroblasts, and in chicken smooth muscle and intestinal epithelial tissues. *J. Histochem. Cytochem.* **33**:576–586.

Chiquet, M., E. C. Puri, and D. C. Turner (1979) Fibronectin mediates attachment of chicken myoblasts to a gelatin-coated substratum. *J. Biol. Chem.* **254**:5475–5482.

Clark, R. A. F., J. M. Lanigan, P. DellaPelle, E. Manseau, H. F. Dvorak, and R. B. Colvin (1982) Fibronectin and fibrin provide a provisional matrix for epidermal cell migration during wound reepithelialization. *J. Invest. Dermatol.* **79**:264–269.

Critchley, D. R., M. A. England, J. Wakely, and R. O. Hynes (1979) Distribution of fibronectin in the ectoderm of gastrulating chick embryos. *Nature* **280**:498–500.

Duband, J. L., and J.-P. Thiery (1982a) Distribution of fibronectin in the early phase of avian cephalic neural crest cell migration. *Dev. Biol.* **93**:308–323.

Duband, J. L., and J.-P. Thiery (1982b) Appearance and distribution of fibronectin during chick embryo gastrulation and neurulation. *Dev. Biol.* **94**:337–350.

Engvall, E., E. Ruoslahti, and E. J. Miller (1978) Affinity of fibronectin to collagens of different genetic types and to fibrinogen. *J. Exp. Med.* **147**:1584–1595.

Erickson, H. P., and N. A. Carrell (1983) Fibronectin in extended and compact conformations: Electron microscopy and sedimentation analysis. *J. Biol. Chem.* **257**:14539–14544.

Furcht, L. T. (1983) Structure and function of the adhesive glycoprotein fibronectin. In *Modern Cell Biology*, Vol 1., B. Satir, ed., pp. 53–117, Alan R. Liss, New York.

Greve, J. M., and D. I. Gottlieb (1982) Monoclonal antibodies which alter the morphology of cultured chick myogenic cells. *J. Supramol. Struct.* **18**:221–230.

Grinnell, F., M. Feld, and D. Minter (1980) Fibroblast adhesion to fibrinogen and fibrin substrata: Requirement for cold-insoluble globulin (plasma fibronectin). *Cell* **19**:517–525.

Grinnell, F., B. R. Lang, and T. V. Phan (1982) Binding of plasma fibronectin to the surfaces of BHK cells in suspension at 4°C. *Exp. Cell Res.* **142**:499–504.

Hahn, L.-H. E., and K. M. Yamada (1979) Isolation and biological characterization of active fragments of the adhesive glycoprotein fibronectin. *Cell* **18**:1043–1051.

Hayashi, M., and K. M. Yamada (1981) Differences in domain structures between plasma and cellular fibronectins. *J. Biol. Chem.* **256**:11292–11300.

Hayashi, M., and K. M. Yamada (1982) Divalent cation modulation of fibronectin binding to heparin and to DNA. *J. Biol. Chem.* **257**:5263–5267.

Hayashi, M., and K. M. Yamada (1983) Domain structure of the carboxyl-terminal half of human plasma fibronectin. *J. Biol. Chem.* **258**:3332–3340.

Hayashi, M., D. H. Schlesinger, D. W. Kennedy, and K. M. Yamada (1980) Isolation and characterization of a heparin-binding domain of cellular fibronectin. *J. Biol. Chem.* **255**:10017–10020.

Heasman, J., R. O. Hynes, A. P. Swan, V. Thomas, and C. C. Wylie (1981) Primordial germ cells of *Xenopus* embryos: The role of fibronectin in their adhesion during migration. *Cell* **27**:437–447.

Hedman, K., M. Kurkinen, K. Alitalo, A. Vaheri, S. Johansson, and M. Höök (1979) Isolation of the pericellular matrix of human fibroblast cultures. *J. Cell Biol.* **81**:83–91.

Hirano, H., Y. Yamada, M. Sullivan, B. de Crombrugghe, I. Pastan, and K. M. Yamada (1983) Isolation of genomic DNA clones spanning the entire fibronectin gene. *Proc. Natl. Acad. Sci. USA* **80**:46–50.

Holly, F. J., K. Dolowy, and K. M. Yamada (1984) Comparative surface chemical studies of cellular fibronectin and submaxillary mucin monolayers: Effects of pH, ionic strength, and presence of calcium ions. *J. Colloid Interfacial Sci.* **100**:210–215.

Hughes, R. C., T. D. Butters, and J. D. Aplin (1981) Cell surface molecules involved in fibronectin-mediated adhesion. A study using specific antisera. *Eur. J. Cell Biol.* **26**:198–207.

Hynes, R. O. (1982) Fibronectin and its relation to cell structure and behavior. In *Cell Biology of the Extracellular Matrix*, E. D. Hay, ed., pp. 295–334, Plenum, New York.

Hynes, R. O., and K. M. Yamada (1982) Fibronectins: Multifunctional modular glycoproteins. *J. Cell Biol.* **95**:369–377.

Johansson, S., and M. Höök (1984) Substrate adhesion of rat hepatocytes: On the mechanism of attachment to fibronectin. *J. Cell Biol.* **98**:810–817.

Katow, H., K. M. Yamada, and M. Solursh (1982) Occurrence of fibronectin on the primary mesenchyme cell surface during migration in the sea urchin embryo. *Differentiation* **28**:120–124.

Keski-Oja, J., and K. M. Yamada (1981) Isolation of an actin-binding fragment of fibronectin. *Biochem. J.* **193**:615–620.

Klebe, R. J. (1974) Isolation of a collagen-dependent cell attachment factor. *Nature* **250**:248–251.

Kornblihtt, A. R., K. Vibe-Pedersen, and F. E. Baralle (1983) Isolation and characterization of cDNA clones for human and bovine fibronectins. *Proc. Natl. Acad. Sci. USA* **80**:3218–3222.

Kornblihtt, A. R., K. Vibe-Pedersen, and F. E. Baralle (1984) Human fibronectin: Molecular cloning evidence for two mRNA species differing by an internal segment coding for a structural domain. *EMBO J.* **3**:221–226.

Laterra, J., J. E. Silbert, and L. A. Culp (1983) Cell surface heparan sulfate mediates some adhesive responses to glycosaminoglycan-binding matrices, including fibronectin. *J. Cell Biol.* **96**:112–123.

Loring, J., B. Glimelius, and J. A. Weston (1982) Extracellular matrix materials influence quail neural crest cell differentiation *in vitro. Dev. Biol.* **90**:165–174.

McDonald, J. A., D. G. Kelley, and T. J. Broekelmann (1982) Role of fibronectin in collagen deposition: Fab' to the gelatin-binding domain of fibronectin inhibits both fibronectin and collagen organization in fibroblast extracellular matrix. *J. Cell Biol.* **92**:485–492.

Mauger, A., M. Demarchez, D. Herbage, J.-A. Grimaud, M. Druguet, D. Hartmann, and P. Sengel (1982) Immunofluorescent localization of collagen types I and III, and of fibronectin during feather morphogenesis in the chick embryo. *Dev. Biol.* **94**:93–105.

Mayer, B. W., E. D. Hay, and R. O. Hynes (1981) Immunocytochemical localization of fibronectin in embryonic chick trunk and area vasculosa. *Dev. Biol.* **82**:267–286.

Mosher, D. F., ed. (1985) *Fibronectin*, Academic, New York (in press).

Mosher, D. F., and R. A. Proctor (1980) Binding and factor XIIIa-mediated cross-linking of a 27-kilodalton fragment of fibronectin to *Staphylococcus aureus*. *Science* **209**:927–929.

Mosher, D. F., P. E. Schad, and J. M. Vann (1980) Cross-linking of collagen and fibronectin by factor XIIIa. Localization of participating glutaminyl residues to a tryptic fragment of fibronectin. *J. Biol. Chem.* **255**:1181–1188.

Neff, N. T., C. Lowrey, C. Decker, A. Tovar, C. Damsky, C. Buck, and A. F. Horwitz (1982) A monoclonal antibody detaches embryonic skeletal muscle from extracellular matrices. *J. Cell Biol.* **95**:654–666.

Newgreen, D., and J.-P. Thiery (1980) Fibronectin in early avian embryos: Synthesis and distribution along the migration pathways of neural crest cells. *Cell Tissue Res.* **211**:269–291.

Nishida, T., S. Nakagawa, T. Awata, Y. Ohashi, K. Watanabe, and R. Manabe (1983) Fibronectin promotes epithelial migration of cultured rabbit cornea *in situ*. *J. Cell Biol.* **97**:1653–1663.

Oppenheimer-Marks, N., and F. Grinnell (1984) Calcium ions protect cell–substratum adhesion receptors against proteolysis: Evidence from immunoadsorption and electroblotting studies. *Exp. Cell Res.* **152**:467–475.

Pennypacker, J. P., J. R. Hassell, K. M. Yamada, and R. M. Pratt (1979) The influence of an adhesive cell surface protein on chondrogenic expression *in vitro*. *Exp. Cell Res.* **121**:411–415.

Petersen, T. E., H. C. Thogersen, K. Skorstengaard, K. Vibe-Pedersen, P. Sahl, L. Sottrup-Jensen, and S. Magnusson (1983) Partial primary structure of bovine plasma fibronectin: Three types of internal homology. *Proc. Natl. Acad. Sci. USA* **80**:137–141.

Pierschbacher, M. D., and E. Ruoslahti (1984) The cell attachment activity of fibronectin can be duplicated by small synthetic fragments of the molecule. *Nature* **309**:30–33.

Pierschbacher, M. D., E. G. Hayman, and E. Ruoslahti (1981) Location of the cell-attachment site in fibronectin with monoclonal antibodies and proteolytic fragments of the molecule. *Cell* **26**:259–267.

Pierschbacher, M. D., E. Ruoslahti, J. Sundelin, P. Lind, and P. Peterson (1982) The cell attachment domain of fibronectin. Determination of the primary structure. *J. Biol. Chem.* **257**:9593–9597.

Podleski, T. R., I. Greenberg, J. Schlessinger, and K. M. Yamada (1979) Fibronectin delays the fusion of L6 myoblasts. *Exp. Cell Res.* **123**:104–126.

Richter, H., and H. Hörmann (1982) Early and late cathepsin D-derived fragments of fibronectin containing the C-terminal interchain disulfide cross-link. *Hoppe-Seyler's Z. Physiol. Chem.* **363**:351–364.

Rogers, S. L., P. C. Letourneau, S. L. Palm, J. McCarthy, and L. Furcht (1983) Neurite extension by peripheral and central nervous system neurons in response to substratum-bound fibronectin and laminin. *Dev. Biol.* **98**:212–220.

Rovasio, R. A., A. Delouvée, K. M. Yamada, R. Timpl, and J.-P. Thiery (1983) Neural crest cell migration: Requirements for exogenous fibronectin and high cell density. *J. Cell Biol.* **96**:462–473.

Ruoslahti, E., and E. G. Hayman (1979) Two active sites with different characteristics in fibronectin. *FEBS Lett.* **97**:221–224.

Ruoslahti, E., E. Engvall, and E. G. Hayman (1981) Fibronectin: Current concepts of its structure and functions. *Collagen Relat. Res.* **1**:95–128.

Schwarzbauer, J. E., J. W. Tamkun, I. R. Lemischka, and R. O. Hynes (1983) Three different fibronectin mRNAs arise by alternative splicing within the coding region. *Cell* **35**:421–431.

Seidl, M., and H. Hörmann (1983) Affinity chromatography on immobilized fibrin monomer, IV. Two fibrin-binding peptides of a chymotryptic digest of human plasma fibronectin. *Hoppe-Seyler's Z. Physiol. Chem.* **364**:83–92.

Sekiguchi, K., M. Fukuda, and S. Hakomori (1981) Domain structure of hamster plasma fibronectin. Isolation and characterization of four functionally distinct domains and their unequal distribution between two subunit polypeptides. *J. Biol.Chem.* **256**:6452–6462.

Sieber-Blum, M., F. Sieber, and K. M. Yamada (1981) Cellular fibronectin promotes adrenergic differentiation of quail neural crest cells *in vitro*. *Exp. Cell Res.* **133**:285–295.

Solursh, M., T. F. Linsenmayer, and K. L. Jensen (1982) Chondrogenesis from single limb mesenchyme cells. *Dev. Biol.* **94**:259–264.

Spiegel, S., K. M. Yamada, B. E. Hom, J. Moss, and P. H. Fishman (1985) Fluorescent gangliosides as probes for the retention and organization of fibronectin by ganglioside-deficient mouse cells. *J. Cell Biol.* **100**:721–726.

Spiegelman, B. M., and C. A. Ginty (1983) Fibronectin modulation of cell shape and lipogenic gene expression in 3T3-adipocytes. *Cell* **35**:657–666.

Tarone, G., G. Galetto, M. Prat, and P. M. Comoglio (1982) Cell surface molecules and fibronectin-mediated cell adhesion: Effect of proteolytic digestion of membrane proteins. *J. Cell Biol.* **94**:179–186.

Thiery, J.-P., J. L. Duband, and A. Delouvée (1982) Pathways and mechanisms of avian trunk neural crest cell migration and localization. *Dev. Biol.* **93**:324–343.

Wartiovaara, J., and A. Vaheri (1980) Fibronectin and early mammalian embryogenesis. *Dev. Mamm.* **4**:233–266.

West, C. M., R. Lanza, J. Rosenbloom, M. Lowe, H. Holtzer, and N. Avdalovic (1979) Fibronectin alters the phenotypic properties of cultured chick embryo chondroblasts. *Cell* **17**:491–501.

Weston, J. A. (1983) Regulation of neural crest cell migration and differentiation. In *Cell Interactions and Development: Molecular Mechanisms*, K. M. Yamada, ed., pp. 153–184, Wiley, New York.

Yamada, K. M. (1983) Cell surface interactions with extracellular materials. *Annu. Rev. Biochem.* **52**:761–799.

Yamada, K. M., and D. W. Kennedy (1979) Fibroblast cellular and plasma fibronectins are similar but not identical. *J. Cell Biol.* **80**:492–498.

Yamada, K. M., and D. W. Kennedy (1984) Dualistic nature of adhesive protein function: Fibronectin and its biologically active peptide fragments can auto-inhibit fibronectin function. *J. Cell Biol.* **99**:29–36.

Yamada, K.M., and D. W. Kennedy (1985) Amino acid sequence specificities of an adhesive recognition signal. *J. Cell Biochem.* **28**:99–104.

Yamada, K. M., D. R. Critchley, P. H. Fishman, and J. Moss (1983) Exogenous gangliosides enhance the interactions of fibronectin with ganglioside-deficient cells. *Exp. Cell Res.* **143**:295–302.

Yamada, K. M., D. W. Kennedy, K. Kimata, and R. M. Pratt (1980a) Characterization of fibronectin interactions with glycosaminoglycans and identification of active proteolytic fragments. *J. Biol. Chem.* **255**:6055–6063.

Yamada, Y., V. E. Avvedimento, M. Mudryj, H. Ohkubo, G. Vogeli, M. Irani, I. Pastan, and B. de Crombrugghe (1980b) The collagen gene: Evidence for its evolutionary assembly by amplification of a DNA segment containing an exon of 54 bp. *Cell* **22**:887–892.

Chapter 15

Laminin Receptor

LANCE A. LIOTTA
ULLA M. WEWER
C. NAGESWARA RAO
GAIL BRYANT

ABSTRACT

Laminin is a high molecular weight glycoprotein found in the basement membrane. The basement membrane is a specialized type of extracellular matrix positioned between organ parenchymal cells and the interstitial space. Interactions of these cells with the basement membrane may be regulated by specific cell surface receptors that bind to laminin. The laminin receptor has been identified and isolated from human and murine cells. It has a molecular weight of approximately 70 kD and a binding coefficient of 2 nM; there are 20,000–100,000 receptors per cell. The laminin receptor may be only one of a family of cell surface receptors for extracellular matrix proteins.

The exchange of signals between cells and their environment is largely a cell surface phenomenon and is dependent on the existence of associated transducer elements or "receptors." Protein-signaling molecules such as water-soluble hormones and growth factors bind to specific proteins, receptors, on the surface of the target cells they influence. Such receptors were initially identified through the use of ligands labeled with radioactive iodine. Binding between the ligand and the receptor is characterized in the following terms: It is saturable, displaceable, temperature sensitive, and is inevitably followed by internalization. Combined morphological and biochemical analyses have been most informative in defining the topology of these processes (Goldstein et al., 1979; Pastan and Willingham, 1981; Hopkins, 1983; Schlessinger et al., 1983; Willingham and Pastan, 1983). It is generally agreed that many cells employ a common mechanism for the internalization of receptor–ligand complexes: receptor-mediated endocytosis via specialized regions on the cell membrane known as coated pits (coated on the intracellular surface with clathrin; Pearse, 1976). The subsequent processing of the receptor–ligand complex once inside the cell may take different pathways, although in principle involving the same basic organelle, known as the endosome (receptor-body; the term indicates that ligands entering the cell via specific receptors appear in this organelle). Three major intracellular routing pathways exist: (1) The receptor–ligand complex is carried to the lysosomes and becomes degraded (down regulation); epidermal growth factor follows such a pathway (Carpenter and Cohen, 1976; Das and Fox, 1978; Schlessinger et al., 1978). (2) The receptor–ligand complex becomes dissociated as a result of a low pH in the peripheral compartment of the endosome, after which the ligand is delivered to the lysosome while the receptor becomes recycled back to the cell surface; examples of this type of routing are seen in low-density lipoprotein (Anderson et al., 1977) and asialoglycoprotein (Bridges et al., 1982). (3) Neither the receptor nor the ligand undergo lysosomal influence, but both are retrieved and recycled back to the cell surface; here an example is transferrin (Dautry-Varsat et al., 1983; Klausner et al., 1983).

CELL SURFACE RECEPTORS FOR COMPONENTS OF THE EXTRACELLULAR MATRIX

The biology of the extracellular matrix is a rapidly expanding research area. One type of extracellular matrix is the basement membrane, which underlies organ parenchymal cells and separates them from the underlying stroma. The basement membrane serves a number of important functions, including preservation of tissue architecture, cell attachment, and selective filtration of macromolecules (Kefalides et al., 1979; Farquhar et al., 1982; Vracko, 1982). Basement membranes are composed of a collagen type IV backbone and noncollagenous proteins such as laminin, entactin, nidogen, and proteoglycans (Martinez-Hernandez and Amenta, 1983). Interactions between cells and this matrix influence a variety of normal and pathological conditions. Some of these interactions may be mediated by specific cell surface receptors that bind matrix components. The laminin receptor is the first such receptor to be identified and isolated (Rao et al., 1982b, 1983; Lesot et al., 1983; Malinoff and Wicha, 1983; Terranova et al., 1983). It is conceivable that the laminin receptor may be only one of a class of "matrix receptors" on cell surfaces.

LAMININ RECEPTOR

Laminin is a high molecular weight glycoprotein (800–1000 kD) found in all basement membranes (Chung et al., 1979; Timpl et al., 1979). It has been isolated and characterized from rodent yolk-sac tumors and parietal endoderm cell lines such as the EHS tumor, the L2 tumor, M1536-B3 and

PYS-1 cells, and recently from normal human (authentic) basement membranes of placenta (Chung et al., 1979; Timpl et al., 1979; Hogan, 1980; Risteli and Timpl, 1981; Martinez-Hernandez et al., 1982; Engvall et al., 1983; Wewer et al., 1983). Structurally, it consists of one 400-kD polypeptide chain and three 200-Kd chains linked together by disulfide bonds. Electron microscope visualization of the native molecule after rotary shadowing shows that laminin is cross-shaped, with a long arm (72 nm) and three short arms (36 nm); the thickness of the arms averages 2 nm. At the ends of the rods are seen seven or more globular domains (Engel et al., 1981). Studies on protease-generated fragments of laminin have revealed that the cross-shaped intersection is contained in the 200-kD chains. The intersection of the short arms contains numerous disulfide bonds and is relatively resistant to proteases. The long arm is protease-labile (Ott et al., 1982; Rao et al., 1982a,b).

Identification, Isolation, and Characterization

A number of biological functions of laminin have been elucidated. These include morphogenesis, stimulation of neurite outgrowth, cell migration, and cell differentiation. Laminin plays a key role in the attachment of a number of different cell types to the basement membrane extracellular matrix (Terranova et al., 1980; Hogan, 1981; Couchman et al., 1983; Timpl et al., 1983a,b). Such interactions are important for the structural organization of basement membranes, as they are for the adhesion of the cells to the tissue. Although the detailed cellular events involved in the adhesion process have not been revealed, we can recognize several phases. The initial step is most likely to include an interaction between cell surface receptors and appropriate ligands (e.g., laminin) on the substrate. Subsequent steps may involve changes in cell shape mediated by cytoskeletal reorganization. The laminin receptor was first identified by showing that laminin exhibits saturable and displaceable binding to the surface of some normal and malignant cell lines (Rao et al., 1983; Terranova et al., 1983). The binding domain on the laminin molecule resides on the protease-resistant, cross-shaped intersection of the laminin short arms (Rao et al., 1982b; Terranova et al., 1983).

Three different laboratories have now identified a similar laminin-binding protein. The laminin receptor characterized in our laboratory from human breast carcinoma cells (MCF-7) or mouse melanoma cells (BL-6) has a binding coefficient of 2 nM and a molecular weight of about 70 kD; there are 20,000–100,000 receptors per cell (Figure 1; Rao et al., 1983; Terranova et al., 1983). Primary and metastatic human breast carcinoma cells contain a laminin-binding protein with a similar molecular weight (Barsky et al., 1984a). The laminin receptor isolated by Malinoff and Wicha (1983) from mouse fibrosarcoma cells has a similar binding coefficient, molecular weight, and number of receptors per cell. A laminin-binding protein isolated from

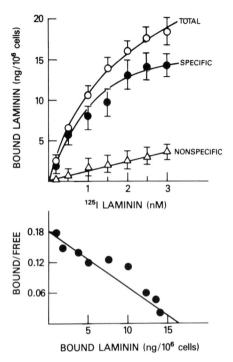

Figure 1. *Laminin-binding assays.* Saturation curve of specific binding to suspended MCF-7
5A9 breast carcinoma cells (*top*). Cells were removed from monolayer cultures using
EDTA (0.2 g/l) in phosphate-buffered saline without Ca^{2+} or Mg^{2+} and suspended (10^5
cells/0.25 ml) in a Tris/saline buffer containing 0.15 M NaCl, 2.5 mM $CaCl_2$, 5 mM $MgCl_2$,
and 1.0% bovine serum albumin and incubated with [^{125}I]laminin (0.1–10 nM) at 25°C
for 40 minutes. The ligand-bound cell fraction was separated by centrifugation. Results
expressed are the mean ± SD of five separate binding assays. Specific binding (*filled
circles*) represents the difference between binding in the absence (*open circles*) and in the
presence (*open triangles*) of 100-fold excess unlabeled laminin. Scatchard plot of
the specific binding data (*bottom*) for the MCF-7 cells was linear (r = 0.95) yielding a K_D
= 1.65 nM and a B_{max} = 17 ng of laminin bound per 10^6 cells (approximately 50,000
binding sites per cell). For receptor binding assays, see Hollenberg and Nexø, 1981.

cultures of rat myoblasts in von der Mark's laboratory also has a molecular
weight in the range of 70 kD (Lesot et al., 1983). To isolate the laminin
receptor, we begin by preparing plasma membranes; the membranes are
then extracted with detergent, and finally, by using laminin affinity chromatography, the receptor is isolated.

Monoclonal Antibodies

We have developed a panel of monoclonal antibodies against the purified
laminin receptor extracted from human breast carcinoma plasma membranes
(Liotta et al., 1985). Two antibodies (LR1 and LR2) were found to differ in

their effects on laminin binding to the receptor. In solid-phase radioimmunoassays, LR1 and LR2 bound with equal titer to the purified receptor. In immunoblot analysis, both LR1 and LR2 recognize a single (approximately 70 kD) component among all the proteins extracted from the membranes of breast carcinoma tissue (Figure 2). The antibodies also bind with equal titer to isolated microsomal membranes or living breast carcinoma cells. No binding to serum components was evident. When added together with labeled ligand, LR1 (but not LR2) produced a dose-dependent inhibition of specific laminin binding to human breast carcinoma cells (Figure 3). Accordingly, LR1 was found to inhibit the attachment of human breast carcinoma cells (MCF-7) and human melanoma cells (A2058) to the basement membrane surfaces of authentic human amnion (Figure 4; S. Togo, U. M. Wewer, C. N. Rao, and L. A. Liotta, manuscript submitted).

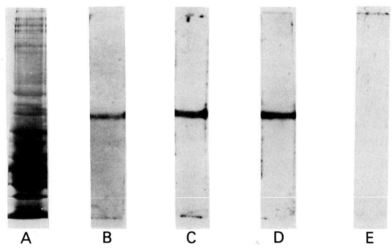

A   **B**   **C**   **D**   **E**

Figure 2. *Monoclonal antibodies against laminin receptor.* Immunoblot of human breast carcinoma plasma membrane extracts with anti-laminin receptor monoclonal antibodies. Plasma membranes from human breast carcinoma were extracted with NP40 and run (reduced) on a 7% SDS-polyacrylamide electrophoresis gel. The gel was blotted onto nitrocellulose (NC) paper. The NC paper was incubated with the anti-laminin receptor antibodies, followed by incubation with rabbit anti-mouse IgM. The bound antibodies were detected autoradiographically by using ^{125}I-labeled protein A. *Lane A*: Total extracted proteins transferred to NC paper. *Lane B*: Iodinated, purified, laminin receptor antigen from human breast carcinoma cells. *Lane C*: LR2 blotted to total protein extract (replicate extract shown in *lane A*). A single component is recognized among all the protein bands. *Lane D*: LR1 blotted to total protein extract (replicate extract shown in *lane A*). A single component is recognized. *Lane E*: Control immunoblot of total extract using a monoclonal antibody directed against a human thymus antigen. No immunoreactivity is present. Control immunoblots of the total extract using rabbit anti-mouse IgM alone, or ^{125}I-labeled protein A alone, were completely devoid of immunoreactivity. Immunoblots of human serum, using the anti-laminin receptor monoclonal antibodies, were not immunoreactive.

Mab INHIBITION OF
SPECIFIC LAMININ BINDING TO
THE MCF-7 LAMININ RECEPTOR

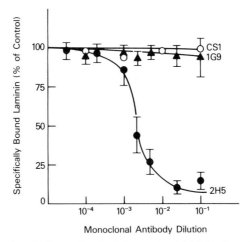

Figure 3. *Monoclonal antibodies against laminin receptor inhibit binding of laminin.* Binding of [^{125}I]laminin to MCF-7 breast carcinoma cells. Results expressed are the mean ± SD of four experiments. Antibody LR1 (2H5, *filled circles*), but not LR2 (lG9, *filled triangles*), inhibited specific binding. Control cell supernatant (CS1, *open circles*) alone was devoid of inhibitory activity.

We conclude that the two classes of antibodies may therefore recognize different structural domains on the receptor molecule. The monoclonal antibodies may prove to be useful tools for: (1) studying the distribution of the receptor in tissues (Hand et al., 1985); (2) identifying fragments of the receptor comprising different functional domains; and (3) investigating whether or not the laminin receptor is internalized following binding of the ligand and determining the route of internalization.

Fate of the Laminin–Receptor Complex: Internalization?

Brown et al. (1983) reported that the fibrosarcoma laminin receptor also has the ability to bind to actin. The indicated stoichiometry was calcuated to be one laminin receptor per three actin monomers at saturation. Moreover, the laminin receptor causes actin to form bundles. Based on these data, Brown et al. (1983) suggest that the laminin receptor may be an integral membrane protein that is capable of connecting the extracellular matrix protein laminin with the cytoskeleton proteins inside the cell. Whether or not the laminin receptor spans the plasma membrane remains to be determined. Clearly, the function of the receptor does not require anchorage to fully polymerized actin, since the receptor is fully functional in suspended cells or isolated plasma membranes. The laminin receptor might alternatively bind indirectly to actin via vinculin or a spectrinlike molecule. Observations by Sugrue and Hay (1981) also suggest that epithelial cells can alter the organization of the cytoskeleton in response to binding exogenous matrix molecules at their cell surfaces.

Very little is known about the events following the binding of laminin to its cell membrane receptor (Figure 5). The results mentioned above might appear to contradict internalization of the laminin–receptor complex. Our preliminary data, however, does favor an internalization process, at least for suspended cells. The laminin-binding process is temperature sensitive, as has been described for other kinds of receptors; after binding has taken place, the ligand is no longer accessible for trypsin. Using immunofluorescence techniques, we have been able to show that the receptor and the ligand are localized inside the cell after incubation at 37°C for one hour (G. Bryant, C. N. Rao, C. Haberen, and L. A. Liotta, manuscript submitted). A hypothesis that is consistent with all these data is as follows: For normal cells, the laminin receptor may be internalized only at the free apical cell surface where soluble laminin (possibly secreted by the same cell) is not anchored to the extracellular matrix. On the basally oriented region of the same cell, however, receptors may be immobilized by the laminin that is "tethered" in the basement membrane.

Most receptors generate a "second messenger" following binding of the ligand. For some cells, this second messenger may be cyclic AMP or

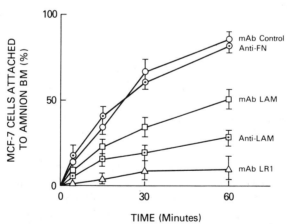

Figure 4. *Monoclonal antibodies to the laminin receptor from human breast carcinoma inhibit attachment of human MCF-7 breast carcinoma cells to the basement membrane surfaces of whole human amnion.* The cultured cells were harvested with EDTA and applied, with the designated antibodies, to the amnion basement membrane suspended in a serum-free RPM1-1640 medium at matched dilutions. mAb control, control IgM; anti-FN, polyclonal anti-fibronectin antibodies; mAb LAM, monoclonal anti-laminin antibodies; Anti-LAM, polyclonal anti-laminin antibodies; mAb LR1, anti-human laminin receptor IgM antibodies. LR1 significantly reduced the rate of attachment and cell spreading (Barsky et al., 1984b; S. Tago, U. M. Wewer, C. N. Rao, and L. A. Liotta, manuscript submitted).

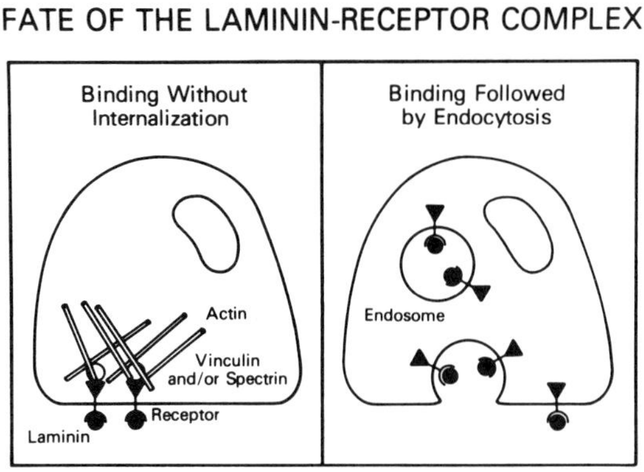

FATE OF THE LAMININ-RECEPTOR COMPLEX

Figure 5. *Schematic representation of two alternative hypotheses concerning the fate of the laminin–receptor complex. A:* The complex may be immobilized inside the cell by cytoskeleton proteins, or it may be immobilized toward the external side of the cell because laminin is integrated as an insoluble part of the basement membrane. *B:* As another possibility, displayed at cell borders where no basement membranes are present (e.g., at apical cell surfaces, in cells in suspension, and in some malignant carcinoma cells) the laminin–receptor complex might be internalized. Data are not yet available to let us make any definite conclusions.

calcium fluxes. We do not as yet know how cells are programmed to respond to the binding of laminin to its receptor. Does laminin alter the pattern of protein synthesis necessary for the attachment process? Is it possible that the response of the cell to laminin binding may depend on the tissue type and its state of differentiation? What is the second messenger for the laminin receptor? How is receptor synthesis regulated? Is the laminin receptor a protein kinase similar to other growth factor receptors? All these and more questions await further studies using appropriate biochemical and molecular genetic approaches.

Application or Implication of the Laminin Receptor in Cancer Research

The number and/or percentage of exposed laminin receptors may be altered in human carcinomas (Liotta et al., 1983; Hand et al., 1985). Breast carcinoma tissue contains a higher number of unoccupied receptors compared to benign breast tissue (Barsky et al., 1984a). This may be concomitant with the disorganized state of basement membranes observed in carcinomas (Albrechsten et al., 1981; Barsky et al., 1983; Haglund et al., 1984). We hypothesize that the laminin receptors of normal epithelium may be polarized at the basal surface and normally saturated with laminin in the basement membrane. In contrast, the laminin receptors on invading carcinoma cells may be distributed over the entire surface of the cell.

Laminin is thought, through its receptor, to play a role in the process of metastasis (Terranova et al., 1982; Varani et al., 1983; Barsky et al.,

1984b). In animal models, tumor cells selected for the ability to attach via laminin produced 10-fold more metastases following intravenous injection (Terranova et al., 1982). Whole laminin bound to the tumor cell surface will stimulate hematogenous metastases; this effect requires the presence of the globular end regions of the molecule. Treating the cells with the receptor binding fragment of laminin markedly inhibits or abolishes metastases in a nontoxic fashion (Barsky et al., 1984b). Thus, we think that the laminin receptor can "stimulate" hematogenous metastases by at least two mechanisms: (1) The unoccupied receptor can be used by the cell to bind directly to host laminin; and (2) if the receptor is occupied, the cell can utilize cell surface laminin as an attachment bridge through the globular end regions. The fragment of laminin that binds to the receptor, but lacks the globular end regions, inhibits both of these mechanisms (Figure 6).

CONCLUSIONS

The laminin receptor is the first cell surface receptor specific for an extracellular matrix component to be isolated and partially characterized. The binding region on the laminin molecule has been determined. The laminin receptor has been shown experimentally to play a role in cell adhesion and cancer metastasis. However, there may be a multitude of other functions in which the laminin receptor is involved, including cell migration, cytoskeletal organization, and mitogenesis. A number of unanswered questions remain concerning internalization of the receptor, its transducer function,

LAMININ RECEPTOR MODEL

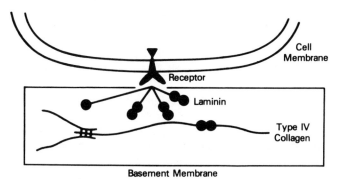

Figure 6. *Hypothetical schematic representation of the cell–basement membrane interface.* The cell surface laminin receptor binds to the rod-shaped position of the disulfide-bonded intersection of the three short arms of laminin. The globular end domains in turn bind to collagen type IV approximately 1/3 of the distance from the globular amino terminus (Rao et al., 1985). Thus the laminin forms a link between the cell and the collagen backbone of the basement membrane.

and the second messenger involved. The availability of monoclonal antibodies to the receptor should now make it possible to address some of these questions.

REFERENCES

Albrechtsen, R., M. Nielsen, U. Wewer, E. Engvall, and E. Ruoslahti (1981) Basement membrane changes in breast cancer detected by immunohistochemical staining for laminin. *Cancer Res.* **41**:5076–5081.

Anderson, R.G.W., M. S. Brown, and J. L. Goldstein (1977) Role of the coated endocytic vesicle in the uptake of receptor-bound low density lipoprotein in human fibroblasts. *Cell* **10**:351–364.

Barsky, S. H., G. P. Siegal, F. Jannotta, and L. A. Liotta (1983) Loss of basement membrane components by invasive tumors but not by their benign counterparts. *Lab. Invest.* **49**:140–147.

Barsky, S.H., C. N. Rao, D. Hyams, and L. A. Liotta (1984a) Characterization of a laminin receptor from human breast carcinoma tissue. *Breast Cancer Res. Treat.* **4**:181–188.

Barsky, S. H., C. N. Rao, J. E. Williams, and L.A. Liotta (1984b) Laminin molecular domains which alter metastasis in a murine model. *J. Clin. Invest.* **74**:843–848.

Bridges, K., J. Harford, G. Ashwell, and R. D. Klausner (1982) Fate of receptor and ligand during endocytosis of asialoglycoproteins by isolated hepatocytes. *Proc. Natl. Acad. Sci. USA* **79**:350–354.

Brown, S. S., H. L. Malinoff, and M. S. Wicha (1983) Connectin: Cell surface protein that binds both laminin and actin. *Proc. Natl. Acad. Sci. USA* **80**:5927–5930.

Carpenter, G., and S. Cohen (1976) ^{125}I-labeled human epidermal growth factor. Binding, internalization, and degradation in human fibroblasts. *J. Cell. Biol.* **71**:159–171.

Chung, A. E., R. Jaffe, I. L. Freeman, J. P. Vergnes, J. E. Braginski, and B. Carlin (1979) Properties of a basement membrane-related glycoprotein synthesized in culture by a mouse embryonal carcinoma-derived cell line. *Cell* **16**:277–287.

Couchman, J. R., M. Höök, D.A. Rees, and R. Timpl (1983) Adhesion, growth, and matrix production by fibroblasts on laminin substrates. *J. Cell. Biol.* **96**:177–183.

Das, M., and C. F. Fox (1978) Molecular mechanism of mitogen action: Processing of receptor induced by epidermal growth factor. *Proc. Natl. Acad. Sci. USA* **75**:2644–2648.

Dautry-Varsat A., A. Ciechanover, and H. F. Lodish (1983) pH and the recycling of transferrin during receptor-mediated endocytosis. *Proc. Natl. Acad. Sci. USA* **80**:2258–2262.

Engel, J., E. Odermatt, A. Engel, J. A. Madri, H. Furthmayr, H. Rohde, and R. Timpl (1981) Shapes, domain organizations and flexibility of laminin and fibronectin, two multifunctional proteins of the extracellular matrix. *J. Mol. Biol.* **150**:97–120.

Engvall, E., T. Krusius, U. Wewer, and E. Ruoslahti (1983) Laminin from rat yolk sac tumor: Isolation, partial characterization, and comparison with mouse laminin. *Arch. Biochem. Biophys.* **222**:649–656.

Farquhar, M. G., P. J. Courtoy, M. C. Lemkin, and Y. S. Kanwar (1982) Current knowledge of the functional architecture of the glomerular basement membrane. In *New Trends in Basement Membrane Research*, K. Kuehn, H. Schoene, and R. Timpl, eds., pp. 9–29, Raven, New York.

Goldstein, J. L., R. G. W. Anderson, and M. S. Brown (1979) Coated pits, coated vesicles, and receptor mediated endocytosis. *Nature* **279**:679–685.

Haglund, C., S. Nordling, P. I. Roberts, and P. Ekblom (1984) Expression of laminin in pancreatic neoplasms and in chronic pancreatitis. *Am. J. Surg. Pathol.* **8**:669–676.

Hand, P., A. Thor, C. N. Rao, J. Sclom, and L. A. Liotta (1985) Monoclonal antibodies to laminin receptor: Immunohistological staining of human breast carcinoma. *Cancer Res.* (in press).

Hogan, B. L. M. (1980) High molecular weight extracellular proteins synthesized by endoderm cells derived from mouse teratocarcinoma cells and normal extraembryonic membranes. *Dev. Biol.* **76**:275–285.

Hogan, B. L. M. (1981) Laminin and epithelial cell attachment. *Nature* **290**:737–738.

Hollenberg, M. D., and E. Nexø (1981) Receptor binding assays. In *Receptors and Recognition: Membrane Receptors, Methods for Purification and Characterization,* Vol. 2., S. Jacobs and P. Cuatrecasas, eds., pp. 3–31, Chapman and Hall, London.

Hopkins, C. R. (1983) The importance of the endosome in intracellular traffic. *Nature* **305**:684–685.

Kefalides, N. A., R. Alper, and C. C. Clark (1979) Biochemistry and metabolism of basement membranes. *Int. Rev. Cytol.* **61**:167–228.

Klausner, R. D., G. Ashwell, J. V. Renswoude, J. B. Harford, and K. R. Bridges (1983) Binding of apotransferrin to K562 cells: Explanation of the transferrin cycle. *Proc. Natl. Acad. Sci. USA* **80**:2263–2266.

Lesot, H., U. Köhl, and K. von der Mark (1983) Isolation of a laminin binding protein from muscle cell membranes. *EMBO J.* **2**:861–865.

Liotta, L. A., C. N. Rao, and S. H. Barsky (1983) Tumor invasion and the extracellular matrix. *Lab. Invest.* **49**:636–649.

Liotta, L. A., P. H. Hand, C. N. Rao, G. Bryant, S. H. Barsky, and T. Schlom (1985) Monoclonal antibodies to the human laminin receptor recognize structurally distinct sites. *Exp. Cell. Res.* **156**:117–126.

Malinoff, H. L., and M. S. Wicha (1983) Isolation of a cell surface receptor protein for laminin from murine fibrosarcoma cells. *J. Cell. Biol.* **96**:1475–1479.

Martinez-Hernandez, A., and P. S. Amenta (1983) The basement membrane in pathology. *Lab. Invest.* **48**:656–677.

Martinez-Hernandez, A., E. J. Miller, I. Damjanov, and S. Gay (1982) Laminin-secreting yolk sac carcinoma of the rat. Biochemical and electron immunohistochemical studies. *Lab. Invest.* **47**:247–257.

Ott, U., E. Odermatt, J. Engel, H. Furthmayr, and R. Timpl (1982) Protease resistance and conformation of laminin. *Eur. J. Biochem.* **123**:63–72.

Pastan, I. H., and M. C. Willingham (1981) Journey to the center of the cell: Role of the receptosome. *Science* **214**:504–509.

Pearse, B. M. F. (1976) Clathrin: A unique protein associated with intracellular transfer of membrane by coated vesicles. *Proc. Natl. Acad. Sci. USA* **73**:1255–1259.

Rao, C. N., I. M. K. Margulies, R. H. Goldfarb, I. A. Madri, and D. T. Woodley (1982a) Differential proteolytic susceptibility of laminin α and β subunits. *Arch. Biochem. Biophys.* **219**:65–70.

Rao, C. N., I. M. K. Margulies, T. S. Tralka, V. P. Terranova, J. A. Madri, and L. A. Liotta (1982b) Isolation of a subunit of laminin and its role in molecular structure and tumor cell attachment. *J. Biol. Chem.* **257**:9740–9744.

Rao, N. C., S. H. Barsky, V. P. Terranova, and L. A. Liotta (1983) Isolation of a tumor cell laminin receptor. *Biochem. Biophys. Res. Commun.* **111**:804–808.

Rao, N. C., I. M. K. Margulies, and L. A. Liotta (1985) Binding domain for laminin on type IV collagen. *Biochem. Biophys. Res. Commun.* **128**:45–52.

Risteli, L., and R. Timpl (1981) Isolation and characterization of pepsin fragments of laminin from human placental and renal basement membranes. *Biochem. J.* **193**:749–755.

Schlessinger, J., Y. Shechter, M. C. Willingham, and I. Pastan (1978) Direct visualization of binding, aggregation, and internalization of insulin and epidermal growth factor on living fibroblastic cells. *Proc. Natl. Acad. Sci. USA* **75**:2659–2663.

Schlessinger, J., A. B. Schreiber, T. A. Libermann, I. Lax, A. Avivi, and Y. Yarden (1983) Polypeptide-hormone-induced receptor clustering and internalization. In *Cell Membranes: Methods and Reviews*, Vol. 1, E. Elson, W. Frazier, and L. Glaser, eds., pp. 117–149, Plenum, New York.

Sugrue, S. P., and E. D. Hay (1981) Response of basal epithelial cell surface and cytoskeleton to solubilized extracellular matrix molecules. *J. Cell. Biol.* **91**:45–54.

Terranova, V. P., D. H. Rohrbach, and G. R. Martin (1980) Role of laminin in the attachment of PAM 212 (epithelial) cells to basement membrane collagen. *Cell* **22**:719–726.

Terranova, V. P., L. A. Liotta, R. G. Russo, and G. R. Martin (1982) Role of laminin in the attachment and metastasis of murine tumor cells. *Cancer Res.* **42**:2265–2269.

Terranova, V. P., C. N. Rao, T. Kalebic, I. M. K. Margulies, and L. A. Liotta (1983) Laminin receptor on human breast carcinoma cells. *Proc. Natl. Acad. Sci. USA* **80**:444–448.

Timpl, R., H. Rohde, P. Gehron Robey, S. I. Rennard, J. M. Foidart, and G. R. Martin (1979) Laminin—A glycoprotein from basement membranes. *J. Biol. Chem.* **254**:9933–9937.

Timpl, R., J. Engel, and G. R. Martin (1983a) Laminin—a multifunctional protein of basement membranes. *Trends Biochem. Sci.* **8**:207–209.

Timpl, R., S. Johansson, V. Van Delden, I. Oberbaümer, and M. Höök (1983b) Characterization of protease-resistant fragments of laminin mediating attachment and spreading of rat hepatocytes. *J. Biol. Chem.* **258**:8922–8927.

Varani, I., E. J. Lovett, O. P. McCoy, S. Shibata, D. E. Maddox, I. J. Goldstein, and M. Wicha (1983) Differential expression of a lamininlike substance of high- and low-metastatic tumor cells. *Am. J. Pathol.* **111**:27–34.

Vracko, R. (1982) The role of basal lamina in maintenance of orderly tissue structure. In *New Trends in Basement Membrane Research*, K. Kuehn, H. Schoene, and R. Timpl, eds., pp. 1–7, Raven, New York.

Wewer, U., R. Albrechtsen, M. Manthorpe, S. Varon, E. Engvall, and E. Ruoslahti (1983) Human laminin isolated in a nearly intact, biologically active form from placenta by limited proteolysis. *J. Biol. Chem.* **258**:12654–12660.

Willingham, M.C., and I. H. Pastan (1983) Receptor-mediated endocytosis: General considerations and morphological approaches. In *Receptors and Recognition: Receptor-Mediated Endocytosis*, P. Cuatrecasas and T. Roth, eds., pp. 1–17, Chapman and Hall, London.

Chapter 16

Integral Membrane Protein Complexes in Cell–Matrix Adhesion

CLAYTON A. BUCK
KAREN A. KNUDSEN
CAROLINE H. DAMSKY
CINDI L. DECKER
RHONDA R. GREGGS
KIM E. DUGGAN
DONNA BOZYCZKO
ALAN F. HORWITZ

ABSTRACT

Cell–matrix adhesion plays a central role in morphogenesis. It is required for the migration of cells, the movement of sheets of cells, and the extension of cell processes. It is at points of cell–matrix interaction that cytoskeletal elements are organized in characteristic structures in response to the molecules in the extracellular matrix. The mediators of this response are constituents of the surface membrane. Using both polyclonal and monoclonal antibodies, we have been able to isolate the membrane molecules involved in this adhesive process. Our data show that these are glycoproteins organized as heterodimers or heterotrimers and suggest that the organization of adhesion sites is a dynamic process involving a family of molecules that can be organized in various combinations in response to the composition of the extracellular matrix.

Morphogenesis is primarily an epigenetic phenomenon directed largely by the interaction of the cell with its environment. Migration, adhesion, and morphology are considered the principal driving forces of this process. Our interest is in cell adhesion as a directing agent for cell migration and as a determinant of cell morphology and polarization. Studies of the role of differential cell adhesion in morphogenesis are regarded as a cornerstone of developmental biology, beginning with the reaggregation experiments of Townes and Holtfreter (1955) using amphibian embryos.

For convenience, it is useful to subclassify adhesion as cell–cell and cell–matrix. Cell–cell adhesion includes direct contact of a cell with its neighbor.

Gap junctions, tight junctions, and desmosomes are common examples of specialized cell–cell adhesive interactions. The fasciculation of individual nerve processes into bundles is another, perhaps less obvious, example. Cell–matrix adhesion is that of a cell with the extracellular material that resides between cells and serves to separate them from tissues and from the molecular scaffolds for migration.

The role of these two adhesive classes as determinants of morphology is clearly seen in the cells of epithelia. These cells are highly polarized. They have basal surfaces that contact extracellular constituents of the basal lamina and free dorsal surfaces. The cells of the epithelium are directly joined to those of their neighbors by characteristic junctional complexes. The dependence of epithelial structure on both cell–cell and cell–matrix adhesion is illustrated in Figure 1. To totally disrupt this simple but organized collection of cells, one must merely deny it of its cell–substratum adhesion apparatus. Here this is done by exposing the cells to an antibody that disrupts cell–substratum adhesion. The end result is that the cells are unable to maintain their morphology because they lose the ability to adhere to a substratum (Figure 1, middle panel). If, on the other hand, this same monolayer is treated with antibodies that disrupt cell–cell interactions, they assume a fibroblastlike morphology as shown in Figure 1 (right panel), and can no longer form a sheet of cells that allows them to communicate and to directionally transport material from one surface to another.

Another example of the morphogenetic importance of cell–cell and cell–matrix adhesion is evident in the nerve. Less adhesive matrices promote fasciculation, the bundling of the neuronal process (Nakai, 1960), whereas more adhesive matrices promote the extension of individual neurites (Letourneau, 1975; Collins, 1978). These phenomena can be revealed through the use of an antibody that inhibits cell–matrix adhesion, thus promoting fasciculation (Figure 2).

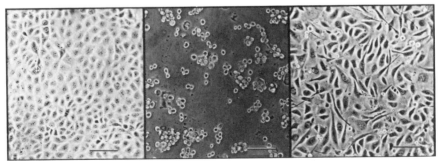

Figure 1. *Effect of antibodies on a sheet of epithelial cells.* Control cells exposed to preimmune serum (*left*). Cells exposed to anti-SFM I, which disrupts cell–matrix adhesion (*middle*). Cells exposed to anti-SFM II, which disrupts cell–cell adhesion (*right*). Calibration bars = 100 μm. Antisera described in Damsky et al., 1982.

Control

CSAT

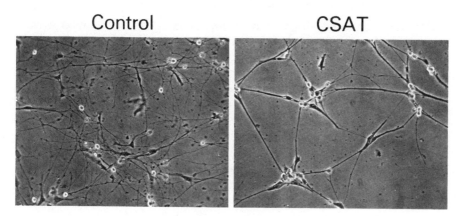

Ciliary ganglia

Figure 2. *Fasciculation promoted by antibody interference with cell–substratum adhesion.* Cultures of cells dissociated from ciliary ganglia were grown on polylysine-coated plates. The monoclonal antibody CSAT was added after 24 hours in culture and photographs were taken 24 hours later.

The focus of this chapter is the interaction of cells with the extracellular matrix, that is, cell–matrix adhesion. It is clear that the extracellular matrix plays a special role in morphogenesis. One of the best-studied examples of this involves the transitory neural crest cells, which include future peripheral nerves, Schwann cells, and melanocytes taking defined pathways to their target loci. These cells migrate through undifferentiated mesenchyme, following pathways devoid of cells and rich in fibronectin (Le Douarin, 1982; Sanes, 1983). *In vitro* studies have shown a clear preference of these cells for fibronectin during their migratory process. The precision of cell–matrix interactions is further demonstrated in reinnervation of degenerated adult frog muscle (Sanes et al., 1978). In this case, the muscle degenerates until only the basal lamina remains. With satellite cell division and differentiation blocked, axons extend along the basal lamina to reinnervate the original synaptic sites. The information directing this process rests in the basal lamina, with which the growth cones of the regenerating neurons interact. At the cellular level, it is also clear that the interaction of an individual cell with its extracellular matrix plays a major role in morphology, that is, in determining the shape of that cell.

The contact sites between cells and their matrices have been studied in detail at the structural level (Abercrombie et al., 1971; Izzard and Lochner, 1976; Heath and Dunn, 1978; Chen and Singer, 1982). They are regions characterized by a coordinated interaction of the cytoskeletal elements of the cell and the extracellular matrix, which is thought to be mediated by constituents of the cell surface. They are rich in such cytoskeletal elements as α-actinin, vinculin, and actin on the cytoplasmic side of the membrane, and such matrix molecules as fibronectin and collagen on the outer surface of the membrane. By providing sites of anchorage and organization of cellular filaments, these contact sites play a dominant role in determining cell morphology and cytoskeletal organization; when they are perturbed, as pointed out earlier, characteristic cell morphologies are destroyed. The integral membrane proteins thought to link extracellular materials with cytoskeletal elements deserve special attention since they may serve as cell surface receptors initiating these organizational events and helping to anchor cells to the matrix. They may also be important in the migratory behavior of cells, as this involves a dynamic forming and releasing of cell–matrix contacts, again requiring a coordinated interaction of matrix and cytoskeleton across the membrane.

USE OF BROAD-SPECTRUM ANTIBODIES TO IDENTIFY CELL–MATRIX ADHESION MOLECULES

Despite progress in identifying extracellular materials that serve as adhesive ligands, and in identifying cytosolic molecules that appear to link cytoskeletal constituents with the membrane, candidates for the integral membrane proteins that may serve as transmembrane links have been identified only recently. One approach is based on isolating molecules from the membrane through binding to extracellular ligands. A receptor for laminin has been purified in this fashion (Malinoff and Wicha, 1983). Our approach, on the other hand, has been patterned after that taken by Gerisch (1977) in studying the aggregation of cellular slime molds. Initially, polyclonal antisera were prepared against crude membranes or material shed into serum-free medium by mammalian cells. Typical effects of these antisera on hamster fibroblasts can be seen in Figure 3: A culture of BHK-21/C13 fibroblasts treated with preimmune serum is shown in the left panel; the right panel shows the same cells exposed to a complement-free antiserum that has caused them to become round and detach from the substratum. Because of the possibility that the antibodies might have caused changes in the cells' adhesive behavior by inducing nonspecific capping, by rearranging cell surface molecules, or perhaps by being toxic, the following control experiments were performed. First, to show that the altered morphology was not simply due to a general perturbation of the cell surface resulting from the cross-linking of membrane molecules with bivalent antibodies, cells were exposed to Fab fragments of the appropriate immunoglobulin. Precisely the same results were seen as shown in Figure 3. Second, quantitative radioimmune assays and immunochemical localization were carried out to demonstrate that mere ensheathment of cells with antisera could not, by itself, result in altered cell–substratum interactions (Damsky et al., 1979). Third, it was demonstrated that the effect of the antiserum on cell morphology was totally reversible (Figure 3, right panel). After removal

Effect of Anti-Membrane IgG on Hamster Fibroblasts

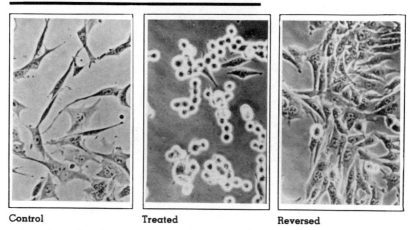

Control   Treated   Reversed

Figure 3. *Effect of broad-spectrum antibodies on cell–substratum adheson of BHK21/C13 hamster fibroblasts.* Control cells (*left*); cells exposed to anti-M, an antiserum that disrupts cell–substratum adhesion in mammalian cells (*middle*); cells treated as in the *middle* panel, 24 hours after removal of anti-M (*right*). Antiserum described in Wylie et al., 1979.

of the antibodies, the cells would again adhere to and spread on the substratum, demonstrating that the effects of the antisera were not due to cytotoxicity. Further evidence comes from the fact that certain affected cell lines could replicate in the presence of the "rounding" antisera (Wylie et al., 1979). Even more striking was the demonstration that myoblast fusion, cardiac and skeletal myogenesis, and chondrogenesis could occur in the presence of adhesion-perturbing antibodies (Decker et al., 1984). Thus, these antisera appeared to be altering cellular morphology by interacting with those elements of the surface membrane required for adhesion and not via some indirect or toxic effect on the cell.

Polyclonal antisera that rounded and detached cells in this manner were prepared against mouse, hamster, human, and chicken cells. These antisera were not species-specific; however, they were more effective in inducing rounding and detachment of their homologous cell type. In each case, the antigens could be extracted from cultured cells and detected by their ability to block the effect of the appropriate antibody on cell morphology. The antigens could only be extracted from the cell in an active form with nonionic detergents and hence are probably integral membrane components. The antigens could be fractionated by a series of procedures involving differential precipitation, ion exchange chromatography, lectin affinity chromatography, and antibody affinity chromatography (Knudsen et al., 1981). The antigens thus purified were found to be glycoproteins that appeared poorly resolved in the region of 120–160 kD on reduced SDS-polyacrylamide gels. That this material represented cell surface glycoproteins involved in cell substratum adhesion was shown by the following exper-

Control   +CSAT

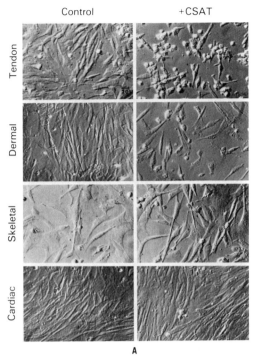

Tendon

Dermal

Skeletal

Cardiac

A

Figure 4. *Effect of monoclonal antibody CSAT on the cell–matrix adhesion of several types of chick embryo cells. A*: Fibroblast cultures derived from chick tendon, dermis, skeletal muscle, and cardiac muscle. *B(opposite page)*: Cardiac and skeletal myoblast cultures. In all cases, cells were cultured 24–48 hours prior to the addition of CSAT monoclonal antibody. The cells were observed under modulation contrast optics 4–24 hours after the addition of antibody. See text for details. (After Decker et al., 1984.)

iments. In the most highly purified material capable of blocking the morphological effects of the antisera, no other molecules could be detected by iodination, dansylation, metabolic labeling, or gel staining. Antibodies prepared against this material were even more effective in altering cell morphology than were the original antisera produced against crude antigen. Cellular variants that had lost their ability to adhere and spread on a substratum contained 10- to 100-fold less antigen than the parental adhesive cell type (Knudsen et al., 1982). Similarly, lymphoid cells seemed devoid of this antigen even though it could be detected in most other tissue examined (Damsky et al., 1982). The effects of the antibodies were not blocked by collagen, glycosaminoglycans, fibronectin, or laminin, further supporting the argument that the antigen is membrane-bound. Finally, fluorescence microscopy using antisera prepared against the purified antigen showed it to be localized in concentrated areas at the cell–substratum interface and at the edges of ruffled membranes. The antigen isolated from mouse, hamster, and human cells appeared to be structurally and serologically similar. Thus, studies using broad-spectrum antibodies pointed

Control   + CSAT

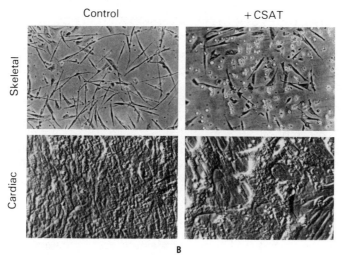

B

Figure 4. (*continued*)

to the existence of a class of glycoproteins (with molecular weights of 120–160 kD) involved in cell–substratum adhesion on the surface of most cell types.

MONOCLONAL ANTIBODY CSAT AS A PROBE FOR CELL–MATRIX ADHESION MOLECULES

Because of potential uncertainties concerning the specificity of polyclonal antisera, we turned to a second system, developed independently, for studying muscle morphogenes (Horwitz et al., 1982). This system used the monoclonal antibody CSAT, which was produced by injecting mice with plasma membrane vesicles shed from formaldehyde-treated chick myoblasts. This antibody will not cross-react with mammalian cells. The effects of CSAT on several types of primary chick cells are very different and range from complete rounding and detachment of myoblasts and chick tendon fibroblasts to the apparent refractility seen with cardiac fibroblasts (Figure 4). Chick skeletal fibroblasts and cardiac muscle cells exhibit an intermediate form of behavior when exposed to CSAT. Although they are not detached from the substratum, they are clearly less well spread, giving them a slightly plumped up appearance. The cardiac muscle cells stand out from the underlying fibroblastic layer as a result of their being less well spread in the presence of CSAT (Figure 4B). These islands of cells continue to contract synchronously as a single unit, showing further that CSAT has no effect on other cell surface functions. Possible reasons for these differences are discussed later. Both radioimmune assays and fluorescent localization studies show that all of these cells are expressing CSAT antigen.

CELL SURFACE GLYCOPROTEINS RECOGNIZED BY CSAT

Relying on the specificity of monoclonal antibodies, the antigen was purified from chick cells by antibody affinity chromatography using the monoclonal antibody CSAT. As in previous work with mammalian cells, the antigen could only be extracted from the cells by using nonionic detergents. Hypotonic shock, EDTA, high salt, and urea all failed to extract the antigen. Figure 5 is a composite showing the SDS-polyacrylamide gel electrophoresis profile of antigen extracted from chick cells and eluted from a monoclonal antibody affinity column. Lane A is an autoradiogram of the NP40 extract of chick cells before passage over the column. Lane B is a control lane showing a complete absence of the 120–160-kD range of material eluted from an antibody affinity column prepared with monoclonal antibody against a protein from hamster cell membranes. Lane C shows a typical profile seen on reduced SDS polyacrylamide gels of material eluted from

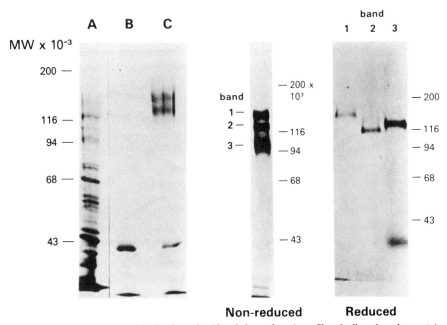

Figure 5. *Autoradiograms of SDS-polyacrylamide gel electrophoresis profiles of cell surface glycoproteins isolated by monoclonal antibody affinity chromatography.* Chick fibroblasts were grown in the presence of [^{35}S]methionine, extracted with the nonionic detergent NP40, and fractionated by monoclonal antibody affinity chromatography as described by Neff et al., 1982. *Lane A*: Unfractionated NP40 extract. *Lane B*: Eluate from control column. *Lane C*: Eluate from CSAT monoclonal antibody affinity column. These lanes were electrophoresed under reduced conditions. *Nonreduced lane:* Eluate as shown in C, electrophoresed in the absence of reducing agents. *Reduced lane:* Bands 1, 2, and 3 were cut from nonreduced gel and reelectrophoresed under reducing conditions.

the CSAT column. The material migrates in a poorly resolved fashion in the 120–160-kD region of the gel in much the same manner as that seen with antigen isolated from mammalian cells. Unlike N-CAM (Hoffman et al., 1982), this resolution could not be improved by treatment with neuraminidase or a mixture of glycosidases (Knudsen et al., 1985). Instead, when the sample was electrophoresed under nonreducing conditions, the antigen was resolved into three distinct bands with apparent molecular weights of approximately 160 kD (band 1), 140 kD (band 2), and 120 kD (band 3) (Figure 5, nonreduced). If the individual bands were cut from nonreduced gels and reelectrophoresed under reducing conditions, a marked change in their behavior was noted (Figure 5, bands 1–3). Bands 1 and 2 migrated more rapidly, displaying apparent molecular weights of 140 kD and 120 kD, respectively. Band 3, on the other hand, migrated more slowly, displaying a higher apparent molecular weight of 130 kD. These results explain why it has not been possible to resolve this class of cell–substratum adhesion molecules by reduced SDS-polyacrylamide gel electrophoresis. They also indicate that disulfide bonding is important to the configuration of these molecules. The apparent unfolding of band 3 indicates that internal disulfide bonds contribute characteristically to the structure of this molecule. The increased mobility of bands 1 and 2 on gels following reduction might be explained by the release of small disulfide-linked fragments.

Further evidence for the individuality of the three glycoproteins bound by CSAT comes from peptide mapping and studies with polyclonal antibodies made against individual bands. Cleveland peptide maps of the individual bands are different and tend to exclude the possibility that bands 2 and 3 are cleavage fragments of the higher molecular weight glycoprotein in band 1 (Knudsen et al., 1985). Polyclonal antisera prepared against band 1 do not cross-react with bands 2 or 3 on immunoblots. Conversely, a mouse antiserum that reacts on immunoblots with band 3 does not cross-react with either bands 1 or 2. Should any of these bands contain regions of extensive homology, it would be expected that these antisera would cross-react with glycoproteins in the other bands as all are equally denatured. From these data, we conclude that the three glycoproteins are composed of different polypeptides. Thus the antigen in NP40 extracts of cells that is recognized by the monoclonal antibody CSAT seems to consist of a molecular complex of three glycoproteins differing in their molecular size, conformation, polypeptide structure, and serological properties.

CSAT ANTIGEN BEHAVES AS AN OLIGOMER

Bands 1, 2, and 3 copurify on CSAT monoclonal antibody affinity columns whether the material passed over the column is a total, unfractionated NP40 extract of cells or whether it is material first fractionated as outlined

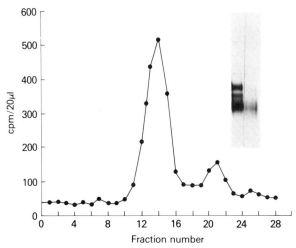

Figure 6. *Sucrose gradient centrifugation profile of glycoproteins eluted from CSAT monoclonal antibody affinity column.* Chick tendon fibroblasts were grown in [^{35}S]methionine and extracted with NP40; the extract was chromatographed on a monoclonal antibody affinity column (Neff et al., 1982). The eluate was concentrated and centrifuged through a gradient of 5–20% sucrose. 200-μl samples were collected and counted in a scintillation counter. Fractions 13–15 and 19–27 were pooled and subjected to SDS-polyacrylamide gel electrophoresis under nonreducing conditions. *Insert*: Autoradiogram of SDS-polyacrylamide gel electrophoresis profile of pooled fractions 13–15 (*left lane*) and pooled fractions 19–27 (*right lane*).

earlier. Thus we appear to be dealing with a complex of two or more molecules. This possibility has been further examined by centrifugation in 5–20% linear sucrose gradients, which fractionates macromolecules by differences in molecular weight, partial molecular volume, and effective radius. In the sucrose gradients, the main peak migrates at an $S_{20,w}$ value of 8.6 and contains all three peptides (Figure 6). However, a satellite peak at an $S_{20,w}$ value of 4.0 is also seen. This peak is enriched on SDS-polyacrylamide gels (band 3). Although the origin and significance of this satellite peak is unclear, its presence demonstrates that the gradient can resolve and separate monomers from oligomers.

Combining these gradient and sieving data, it was possible to estimate the molecular weight of the complex by using the methods described by Clarke (1975). The molecular weight for the antigen complex in detergent is approximately 235 kD. When this is corrected for bound detergent, the molecular weight is approximately 215 kD. The antigen is not a monomer or a higher order oligomer, but rather a dimer or a trimer. Our inability to clarify this arises from not knowing how overestimated the glycoprotein molecular weights indicated by SDS-polyacrylamide gel electrophoresis are.

LOCALIZATION STUDIES OF CSAT ANTIGENS

If bands 1, 2, and 3 are involved in adhesion, they should be concentrated at the cell–substratum interface at points thought to be characteristic of cell adhesive sites. Therefore, their distribution relative to actin-containing fibers, vinculin, and fibronectin was examined by fluorescence microscopy. Figure 7 is a composite showing chick tendon fibroblasts stained with CSAT antibody, detected by rhodamine-conjugated anti-mouse immuno-

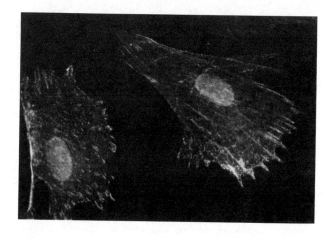

Figure 7. *Localization of CSAT antigen (top) and actin-containing fibers (bottom) on cultured chick tendon fibroblasts.* Cells were grown on cover slips, fixed, and permeabilized before staining with CSAT monoclonal antibody *(top)* or fluorescein-conjugated phalloidin *(bottom)*. CSAT antibody was detected by using rhodamine-conjugated anti-mouse serum.

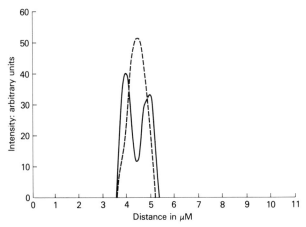

Figure 8. *Comparison of the spatial distribution of CSAT and vinculin staining in the region of putative focal contacts.* Negatives from micrographs of CSAT- and vinculin-stained cells were projected onto a screen and digitized using a video imagery system. A linear section of the image passing through putative adhesion sites on both micrographs was then generated. The intensity values are relative and are not scaled similarly for the two micrographs. (After Damsky et al., 1985.)

globulin, and stress fibers stained with fluorescein-conjugated phalloidin. CSAT antigen is enriched at sites of cell–substratum interaction in a very interesting and characteristic manner. Most striking is the "needle's eye" distribution seen in places where the cell is flattest, near the extended surface of the ruffled membrane. Here the antigen is concentrated in very elongated ovals with unstained centers (Figure 7, top). Comparison of their distribution with that of the actin-containing fibers (Figure 7, bottom) shows that these fibers superimpose directly on the unstained eye of these needles. If cells are stained with CSAT and anti-vinculin, the vinculin is seen concentrated in focal adhesion plaques and is considered to be a marker for these sites. In this case, CSAT is enriched at the periphery of the eye of the needle, which is now filled with antibody-stained vinculin. In order to compare precisely the distribution of the CSAT-antigen and vinculin, images from photographs of cells double-stained by anti-CSAT and vinculin have been digitized, superimposed, and the intensity of the staining of each antigen graphed as shown in Figure 8. In every case where a needle's eye is found, the CSAT staining is concentrated at the periphery, resulting in a biphasic distribution of fluorescence intensity. Vinculin and actin are found concentrated in one area (the needle's eye) in the same field, resulting in a curve with a single peak of fluorescence intensity. This peak is always in the center of the eye. Further controls, including the use of a monoclonal antibody known to stain focal contacts, demonstrate that these sites on the membrane are accessible to antibodies and are not devoid of CSAT staining because of poor penetration. A comparison of CSAT and fibronectin staining (Figure 9) shows that both molecules are localized along linear stress fibers and are enriched at the periphery of each needle's eye. In addition, fibronectin staining extends outside the margins of the cell where CSAT is absent, whereas CSAT staining is present in ruffled membranes and other regions of the ventral cell surface that lack fibronectin.

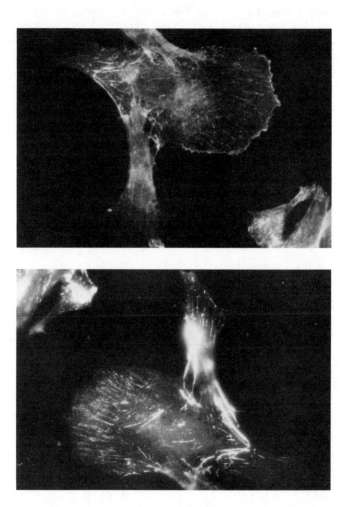

Figure 9. *Distribution of CSAT antigen (top) and fibronectin (bottom) on chick tendon fibroblasts.* Cells were grown on cover slips, fixed, and permeabilized before staining with CSAT monoclonal antibody (*top*) or anti-fibronectin antibody (*bottom*). Anti-fibronectin was detected by using fluorescein-conjugated anti-goat serum.

EFFECT OF CSAT ON DIFFERENT FIBROBLASTS

Knowing a bit more about the antigen we are dealing with, we have returned to the problem of why CSAT antibody affects cells differently. The most obvious possibility is that there is no CSAT antigen on the cells that appear refractile (Figure 4). This has been eliminated by showing that CSAT binds to all the cells discussed here. In fact, the amount of CSAT expressed on myoblasts and skeletal fibroblasts is about the same within a factor of two (Neff et al., 1982). The antigen isolated by antibody affinity chromatography from myoblasts, chick tendon fibroblasts, and cardiac fibroblasts is nearly identical and resembles that shown in Figure 5. All three SDS-gel bands are present in approximately the same ratio, and they appear to be of the same molecular weight. Thus, there is no obvious difference in the quality of the CSAT antigen expressed on these cells. This suggests that the more refractile cells may be adhering to the extracellular

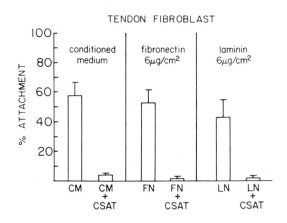

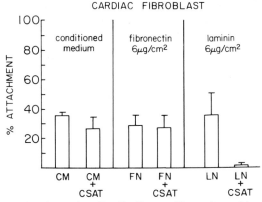

Figure 10. *Adhesion of tendon and cardiac fibroblasts to fibronectin- and laminin-coated substrata.* Adhesion assay is described in the text.

matrix through non-CSAT-mediated mechanisms. To test this possibility, the adhesion of two different types of fibroblasts to more defined substrata were compared. The two types of fibroblasts used were chick tendon fibroblasts, whose adhesive properties were markedly affected by CSAT, and chick cardiac fibroblasts, which appeared almost refractile. For this comparison, an assay based on that described by McClay et al. (1980) was developed in which the resistance of cells to a constant centrifugal force was used as a measure of their attachment to the substratum (Decker et al., 1984). Thus, a given number of cells was centrifuged onto surfaces coated with either fibronectin or laminin. The cells, now in contact with the substratum, were incubated for 30–45 minutes and the plates inverted and recentrifuged. The number of cells remaining on the substratum could then be quantitated. Tendon and cardiac fibroblasts adhered to both laminin and fibronectin (Figure 10). When CSAT was included in the assay, the attachment of tendon fibroblasts to both fibronectin and laminin was reduced by 80–90%. On the other hand, the binding of cardiac fibroblasts to fibronectin was refractile to CSAT, while binding to laminin was inhibited by 90%. These results were consistent even when the experiments were performed in the presence of cycloheximide to minimize the deposition of extracellular matrices by the cells during the 30–45-minute incubation period.

    These experiments suggest that the difference in effect of CSAT on various cell types is a reflection of different adhesive mechanisms. In this instance, tendon fibroblasts adhere to both fibronectin and laminin by a CSAT-sensitive mechanism, suggesting that some combination of bands 1, 2, and 3 may act as receptors for both of these molecules in tendon fibroblasts. Cardiac fibroblasts, on the other hand, use a similar CSAT-sensitive mechanism to adhere to laminin but use additional CSAT-insensitive mechanisms to adhere to fibronectin.

DISCUSSION

We believe that we are looking at molecules directly involved in cell–substratum adhesion for the following reasons: (1) The antigen was identified by using antibodies selected by their effect on cell–matrix adhesion; (2) the antigen localizes on the ventral surface of the cell in regions where cell–matrix interactions have been characterized through microscopy; (3) in assays of the initial events of cell adhesion, the antibody can inhibit virtually all adhesions to a defined component; (4) nonadherent cells contain considerably reduced quantities of these antigens; and (5) specific antisera raised against purified antigen disrupt cell–substratum adhesion.

    The glycoproteins described here, particularly those reacting with CSAT, appear different from certain molecules implicated in cell–substratum adhesion by other groups. Carter and Hakomori (1981) have described a

140-kD glycoprotein that remains associated with the extracellular matrix following detergent extraction of the cells. This molecule is similar to one described by Lehto (1983) as being associated with vinculin; it contains hydroxylysine and hydroxyproline (Carter, 1982), both of which are absent from the CSAT antigen complex reported here. Their molecular size and location distinguish CSAT glycoproteins from the 60-kD molecule observed in focal contacts by Oesch and Birchmeier (1982). Similarly, size would seem to distinguish the CSAT antigen from the 47-kD surface molecule that Aplin et al. (1981) were able to cross-link chemically to fibronectin, as well as from a laminin receptor with a molecular weight of 72 kD. Other molecules spanning ranges of molecular weight similar to CSAT-glycoproteins have been described, but there is insufficient detail reported to evaluate the extent of these similarities (reviewed in Damsky et al., 1984).

The CSAT antigen (bands 1, 2, and 3) is identical to the antigen that reacts with the monoclonal antibody JG22 isolated by Greve and Gottlieb (1982) and referred to in localization studies by Chen and Singer (1982). Both JG22 and CSAT were prepared by immunization with material from chick myoblasts. Both antibodies isolated identical components from cells when used in antibody affinity chromatography (Greve and Gottlieb, 1982; Neff et al., 1982), and they compete for the same binding sites in radioimmune assays (K. A. Knudsen and A. F. Horwitz, unpublished observations).

The CSAT antigen complex described here bears some resemblance to the 140-kD glycoprotein fibronectin receptor reported by Pytela et al. (1985) in that they are both glycoprotein complexes which are poorly resolved on reduced SDS-polyacrylamide gels, but are resolved into three glycoproteins when subjected to SDS-polyacrylamide gel electrophoresis under nonreduced conditions. Whether these similarities are merely fortuitous or can be taken to suggest that the CSAT antigen complex is indeed a fibronectin receptor will require further study. However, as shown here by immunofluorescence double-labeling, the CSAT antigen complex and fibronectin are found codistributed at the periphery of focal adhesion plaques, and chick tendon fibroblasts certainly adhere to fibronectin-coated surfaces by a mechanism sensitive to the CSAT monoclonal antibody. Further, very preliminary experiments show that the CSAT antigen complex can bind directly to fibronectin (A. F. Horwitz and C. A. Buck, unpublished observations). It is clear that in order to understand the role of the CSAT antigen in adhesion, we will need to determine the various elements of the extracellular matrix with which this glycoprotein complex can bind.

Our observations point to a multiplicity of adhesion mechanisms involved in the interactions of cells with extracellular matrices. The differential effects of the CSAT monoclonal antibody on the adhesion of different cell types are most readily reconciled, in the absence of an antigen gradient, by the presence of additional combinations of adhesion molecules not seen by our monoclonal antibody. This notion of multiple adhesion mechanisms is supported by the work of others. The genetic studies of Harper and

Juliano (1980), as well as interference reflection and immunoelectron microscopy, suggest multiple types of adhesion sites in fibroblasts. Of the adhesion contacts discussed by Chen and Singer (1982), the CSAT glycoprotein complex is most likely found in matrix and close contacts. It is not in focal contacts. This conclusion is based on the homology between the monoclonal antibody JG22 discussed by these workers and our monoclonal antibody CSAT. At close contact sites, the ventral surface of the cell comes to within 30–50 nm of the substratum. These contact sites tend to surround focal contacts when such are present and are regions where microfilament bundles interact laterally with the cell surface. The adhesion of highly motile cells is characterized by a predominance of close contacts. They are also the first adhesive structures to form in cell spreading. If CSAT antigen plays a major role in initial adhesive events, including cell migration and spreading, these are the structures where it should be found.

The discovery that cell–substratum adhesion, as described here, involves a complex of glycoproteins is both surprising and intriguing, and bears on our notion about cell–matrix adhesion. The arguments for the existence of the complex, as stated before, are that the glycoproteins copurify through several biochemical procedures: They bind and elute together from monoclonal antibody affinity columns; they behave as a complex during density gradient centrifugation under conditions that can resolve a complex of two or more of these glycoproteins from single glycoproteins of similar molecular weight. The oligomeric nature of the antigen is particularly interesting. The data suggest heterodimers or trimers as the most likely quaternary structure. We do not yet know the function of each glycoprotein or whether they function only as a complex through their quaternary structure. The simplest functional arrangement in the membrane is for all three glycoproteins to be associated with one another in such a way that the quaternary structure on either side of the complex forms the binding site required for the appropriate cytoskeletal or extracellular matrix interaction. An alternative possibility is based on the fact that the molecular weight of the complex, as determined by centrifugation studies, suggests the existence of heterodimers rather than heterotrimers. In this case, one glycoprotein could interact with the cytoskeleton, a second could serve to bind to the extracellular matrix, and the third could act as an "organizer" to coordinate the action of the other two. In this way, the "organizer" might interact with cell–substratum adhesion molecules in much the same manner as β_2-microglobulin does with members of the histocompatibility complex.

Our overall view of cell–matrix adhesion is that it involves a limited family of molecules which can be organized at any point on the cell surface in several configurations to accommodate either the requirements of various extracellular matrices or the organizational demands of the cytoskeleton. A combination of CSAT bands 1, 2, and 3 might be optimal for cell–fibronectin interactions at a close contact site; bands 2, 3, and "X" might be more effective at an extracellular matrix contact site. Adhesion sites organized so as to use a combination of membrane molecules would give the cell a dynamic and flexible means of adjusting its adhesive response to any matrix environment.

ACKNOWLEDGMENTS

This work was supported by National Institutes of Health Grants CA-19144-10, CA-27909-05, CA-10815, CA-32311, GM-23244, and HD/CA-15663 and by the H. N. Watts, Jr. Neuromuscular Disease Research Center. We wish to thank Mrs. Helen Schorr, J. M., for organizing, typing, and assembling the manuscript.

REFERENCES

Abercrombie, M. J., J. E. M. Heaysman, and S. M. Pegrum (1971) Locomotion of fibroblasts in culture. IV. Electron microscopy of leading lamella. *Exp. Cell Res.* **67**:359–367.

Aplin, J. D., R. C. Hughes, C. L. Jaffe, and N. Sharon (1981) Reversible cross-linking of cellular components of adherent fibroblasts to fibronectin and lectin-coated substrata. *Exp. Cell Res.* **134**:488–494.

Carter, W. G. (1982) The cooperative role of the transformation-sensitive glycoproteins gp140 and fibronectin in cell attachment and spreading. *J. Biol. Chem.* **257**:3249–3257.

Carter, W. G., and S.-I. Hakomori (1981) A new cell surface detergent-insoluble glycoprotein matrix of human and hamster fibroblasts. *J. Biol. Chem.* **256**:6953–6960.

Chen, W. T., and J. Singer (1982) Immunoelectron microscopic studies of the sites of cell–substratum and cell–cell contact in cultured fibroblasts. *J. Cell Biol.* **95**:205–226.

Clarke, S. (1975) Size and detergent binding of membrane proteins. *J. Biol. Chem.* **250**:5459–5469.

Collins, F. (1978) Induction of neurite outgrowth by a conditioned-medium factor bound to the culture substratum. *Proc. Natl. Acad. Sci. USA* **75**:5210–5213.

Damsky, C. H., D. E. Wylie, and C. A. Buck (1979) Studies on the function of cell surface glycoproteins. II. Possible role of surface glycoproteins in the control of cytoskeletal organization and surface morphology. *J. Cell Biol.* **80**:403–415.

Damsky, C. H., K. A. Knudsen, and C. A. Buck (1982) Integral membrane glycoproteins related to cell–substratum adhesion in mammalian cells. *J. Cell Biochem.* **18**:1–13.

Damsky, C. H., K. A. Knudsen, and C. A. Buck (1984) Integral membrane glycoproteins in cell–cell and cell–substratum adhesion. In *The Biology of Glycoproteins*, R. Ivatt, ed., pp. 1–64, Plenum, New York.

Damsky, C. H., K. A. Knudsen, D. Bradley, C. A. Buck, and A. F. Horwitz (1985) Distribution of the cell–substratum attachment (CSAT) antigen on myogenic and fibroblastic cells in culture. *J. Cell Biol.* **100**:1528–1539.

Decker, C. L., R. Greggs, K. E. Duggan, J. Stubbs, and A. F. Horwitz (1984) Adhesive multiplicity in the interaction of embryonic fibroblasts, myoblasts and neurons with extracellular matrices. *J. Cell Biol.* **99**:1398–1404.

Gerisch, G. (1977) Univalent antibody fragments as tools for analysis of cell interactions in Dictyostelium. *Curr. Top. Dev. Biol.* **14**:243–270.

Greve, J. M., and D. I. Gottlieb (1982) Monoclonal antibodies which alter the morphology of cultured chick myogenic cells. *J. Cell. Biochem.* **18**:221–230.

Harper, P. A., and R. L. Juliano (1980) Isolation and characterization of Chinese hamster ovary cell variants defective in adhesion to fibronectin coated collagen. *J. Cell Biol.* **87**:755–763.

Heath, J. P., and G. A. Dunn (1978) Cell to substratum contacts of chick fibroblasts and their relation to the microfilament system: A correlated interference-reflection and high voltage electron microscope study. *J. Cell Sci.* **29**:197–212.

Hoffman, S., B. C. Sorkin, P. C. White, R. Brackenbury, R. Mailhammer, U. Rutishauser, B. A. Cunningham, and G. M. Edelman (1982) Chemical characterization of a neural cell adhesion molecule purified from embryonic brain membranes. *J. Biol. Chem.* **257**:7720–7729.

Horwitz, A. F., N. Neff, A. Sessions, and C. Decker (1982) Cellular interactions in myogenesis. In *Muscle Development: Molecular and Cellular Control*, M. L. Pearson and H. F. Epstein, eds., pp. 291–299, Cold Spring Harbor Press, Cold Spring Harbor, New York.

Izzard, C., and L. R. Lochner (1976) Cell to substrate contacts in lung fibroblasts: An interference reflection study with an evaluation of technique. *J. Cell Sci.* **21**:129–159.

Knudsen, K. A., P. Rao, C. H. Damsky, and C. A. Buck (1981) Membrane glycoproteins involved in adherent vs. non-adherent melanoma cells. *J. Cell Biochem.* **18**:157–167.

Knudsen, K. A., A. F. Horwitz, and C. A. Buck (1985) A monoclonal antibody identifies a glycoprotein complex involved in cell–substratum adhesion. *Exp. Cell Res.* **157**:218–226.

Le Douarin, N. M. (1982) *The Neural Crest*, Cambridge Univ. Press, Cambridge, England.

Lehto, V. P. (1983) 140,000 dalton surface glycoproteins: A plasma membrane component of the detergent-resistant cytoskeletal preparations of cultured human fibroblasts. *Exp. Cell Res.* **143**:272–286.

Letourneau, P. C. (1975) Possible roles for cell-to-substratum adhesion in neuronal morphogenesis. *Dev. Biol.* **44**:77–91.

Malinoff, H. L., and M. S. Wicha (1983) Isolation of a cell surface receptor for laminin from murine fibrosarcoma cells. *J. Cell Biol.* **96**:1475–1479.

McClay, D. R., G. M. Wessel, and R. B. Marchase (1980) Intercellular recognition: Quantitation of initial binding events. *Proc. Natl. Acad. Sci. USA* **78**:4975–4979.

Nakai, J. (1960) Studies on the mechanism determining the course of nerve fibers in tissue culture. II. The mechanism of fasciculation. *Z. Zellforsch. Anat.* **52**:427–449.

Neff, N. T., C. Lowrey, C. L. Decker, A. Tovar, C. H. Damsky, C. A. Buck, and A. F. Horwitz (1982) A monoclonal antibody detaches embryonic skeletal muscle from extracellular matrices. *J. Cell Biol.* **95**:654–666.

Oesch, B., and W. Birchmeier (1982) A new surface component of fibroblasts' focal contacts identified by a monoclonal antibody. *Cell* **31**:671–679.

Sanes, J. R. (1983) Roles of extracellular matrix in neural development. *Annu. Rev. Physiol.* **45**:581–600.

Sanes, J. R., L. M. Marshall, and U. J. McMahan (1978) Reinnervation of muscle fiber basal lamina after removal of myofibers. Differentiation of regenerating axons at original synaptic sites. *J. Cell Biol.* **78**:176–198.

Townes, T. L., and J. Holtfreter (1955) Directed movements and selective adhesion of embryonic amphibian cells. *J. Exp. Zool.* **128**:53–120.

Wylie, D. E., C. H. Damsky, and C. A. Buck (1979) Studies on the function of cell surface glycoproteins. I. Use of antisera to surface membrane in the identification of proteins relevant to cell–substrate adhesion. *J. Cell Biol.* **80**:385–402.

Chapter 17

The Extracellular Matrix in Tissue Morphogenesis and Angiogenesis

PETER EKBLOM
IRMA THESLEFF
HANNU SARIOLA

ABSTRACT

The formation of tissues with several specialized cell types involves multiple cell–cell interactions. We have analyzed these events in the developing kidney: A key event is an inductive cell–cell interaction, which triggers the conversion of a mesenchyme to an epithelium. We have shown that this interaction has an immediate effect on the composition of the extracellular matrix. In addition, the cells become more adhesive, start to proliferate, and become responsive to the serum mitogen transferrin. These studies raise the possibility that inductive interactions affect a multitude of metabolic events. During this initial activation, there are no signs of overt cell polarity; the acquisition of polarity seems to be the next crucial event. The basement membrane matrix apparently influences polarization. Cells attach to the basement membrane, proliferate, gradually become terminally differentiated, and express segment-specific markers. The proliferation requires transferrin, which acts by transporting iron into the cells by receptor-mediated endocytosis. We do not know how proliferation and differentiation are coupled, but adhesive determinants of the basement membrane and the cell surface apparently restrict proliferation to the proper locations. The morphogenesis of the epithelium is linked to angiogenesis, although the two are separate events.

The developmental mechanisms that operate during epithelial morphogenesis do not apply to all stages of angiogenesis. Endothelial cells do not form in situ as a result of induction; instead, they migrate into the kidney. Both cell types nevertheless produce a similar matrix. Basement membrane matrices are therefore crucial in many morphogenetic events.

It has been known for years that cells *in vitro* require a proper substratum for growth and for maintenance of the differentiated state. It has also been convincingly demonstrated that the final density of the cells is determined by the serum concentration of the medium and by cell–cell interactions (Dulbecco and Stoker, 1970; Holley, 1975; Ross and Vogel, 1978). It is widely believed that these phenomena have their *in vivo* correlates. The substratum *in vivo* is the extracellular matrix, and the mitogens used in *in vitro* cultures are derived from serum or other biological fluids (Barnes and Sato, 1980), suggesting that they are involved in various biological events such as wound healing and other repair processes, and in embryogenesis.

In the embryo, tremendous growth occurs, but it is always tightly controlled and coupled to differentiation. It is therefore natural to assume that the extracellular matrix and various serum mitogens may represent crucial exogenous regulators of embryogenesis. During the past 10 years, our knowledge of the molecular composition of the extracellular matrix has increased dramatically. It has become apparent that cells deposit highly complex matrices, which often show a remarkable tissue specificity (Kleinman et al., 1981). Although both matrix molecules and soluble growth factors interact directly with the cell surface, there are reasons to believe that their effects on differentiation are mediated by completely different mechanisms. Mitogenic hormones and growth factors all seem to have certain common features not shared by the matrix molecules. Growth factors interact with their surface receptors and then are rapidly internalized and delivered to the intracellular compartment (for a review, see Brown et al., 1983). The matrix components are more immobile, do not become internalized, and instead provide suitable attachment sites for cells. This cell–matrix interaction nevertheless apparently affects cell behavior profoundly.

These processes have been studied in detail in monolayer cultures in a number of laboratories, and it is now becoming possible to analyze these exogenous regulators of cell function in more complex histogenetic processes as well.

Extracellular Matrix Components

In multicellular organisms, most cells produce an extracellular matrix. This is most readily apparent in mesenchymal cells such as those found in bone, cartilage, and fibroblasts. For many years, connective tissue research was therefore largely confined to analyses of these structures. Mesenchymal cells are surrounded on all sides by abundant matrices. It is now clear that they have many molecularly defined, tissue-specific features. These differences underlie the well-known physical differences between bone, cartilage, and loose mesenchyme. Fibroblasts are found not only in forming scar tissue and other "primitive" mesenchymes, but also in the stroma of parenchymal organs, where they are situated between the parenchymal epithelial cells and produce the easily recognizable interstitial matrix. What is not as easily noted is the basement membrane, which is a thinner, specialized matrix immediately surrounding the parenchymal cells. This extracellular matrix is not produced by the fibroblasts but by the parenchymal cells themselves (Pierce, 1966; Hay, 1982). In contrast to the interstitial type of matrices, the basement membranes of epithelial cells are not found on all sides of the cells, but are situated in a polar manner at the basal surface. Given the different anatomical (Figure 1) and physical characteristics of the various matrices, it is not surprising that their biochemical compositions differ greatly. All matrices contain collagens, glycoproteins, and proteoglycans, but they are often specific to each matrix. This principle is best illustrated by the collagens. Chondrocytes secrete collagen type II and other cartilage-type collagens, whereas fibroblasts produce collagen type I and type III. The basement membranes contain collagen type IV and perhaps some other collagenous peptides as well. More than 10 different collagen isotypes have been identified (Mayne, 1984). The distribution of the various isotypes is strictly controlled and the matrix composition is often tissue-specific.

The proteoglycans and the matrix glycoproteins are also to a large extent specific to each cell type. Chondrocytes deposit a large cartilage-specific proteoglycan, whereas basement membranes contain another type of proteoglycan with a distinct glycosaminoglycan composition. It has often been suggested that the proteoglycans found in epithelial basement membranes are of importance in morphogenesis (Bernfield et al., 1973). No proteoglycans specific for loose mesenchyme have been detected, but their matrices contain proteoglycans, and are rich in different glycosaminoglycans. There is some evidence that hyaluronate and its metabolism may influence the hydration level of the tissues, and thus the compactness of loose mesenchyme might be regulated by the amount of this glycosaminoglycan. Unlike the other glycosaminoglycans, hyaluronate is not covalently bound to proteins (Toole, 1973).

Recent studies have clearly demonstrated the presence of a number of matrix glycoproteins. Fibronectin, a glycoprotein found around normal fibroblasts, is typical for undifferentiated mesenchymal cells. However, it cannot be classified as a fibroblast-specific glycoprotein, since it is also

Cartilage   Mesenchyme   Epithelium

Figure 1. *Anatomical appearance of certain tissues and their extracellular matrices.* In cartilage, round cells are surrounded by a dense matrix. The cartilage matrix is shown as *small dots.* In embryonic mesenchyme, fibroblastlike cells are surrounded by a loose matrix. The mesenchymal matrix is shown as *short rods.* In epithelia, the polarized cells are surrounded by a matrix only on the basal side. This epithelial matrix, the basement membrane, is represented by the *black circle.*

present in blood and in close association with basement membranes (Vaheri and Mosher, 1978; Ruoslahti et al., 1981; Hynes and Yamada, 1982). It is not found in mature cartilage, which instead contains chondronectin, another matrix glycoprotein (Kleinman et al., 1981). The glycoprotein composition of basement membranes has been studied for many years (see Spiro, 1978; Kefalides et al., 1979) and several of their major molecular components have been detected and rapidly characterized. Laminin is a large basement membrane glycoprotein, initially found in a mouse tumor (Timpl et al., 1979). Its molecular weight is about 850 kD; it is apparently formed by several chains that interact to form a large, cross-shaped structure. The multidomain structure of laminin, first suggested by electron microscopy after rotary shadowing, is supported by proteolytic cleavage studies (Timpl and Martin, 1982). Laminin is found in all basement membranes, where it may be the major noncollagenous glycoprotein.

Several other, smaller basement membrane glycoproteins have now been detected. A sulfated glycoprotein, entactin, has a molecular weight of about 160 kD (Bender et al., 1981), and a similar glycoprotein, nidogen, consists of 80-kD chains that can aggregate to form large nests (Timpl et al., 1983). Many other minor components are known, and some are specific for certain parts of the body. It is very likely that other basement membrane components will be found. It is apparent from this brief introduction that the deposition of a basement membrane matrix is the result of a complex intermolecular interaction that may be instrumental in the formation of insoluble matrices (Kleinman et al., 1981). At present, it is somewhat unclear as to how basement membranes are layered. It is difficult to understand how such extraordinarily large components (100–400 nm) can fit into the thin membrane. The collagen type IV may represent a major structural component, as its chains are linked to each other through the 7S collagen region to form a chicken-wire network (Timpl and Martin, 1982). Laminin may have another function as a link between cells and the basement membrane (Terranova et al., 1980), whereas the self-aggregating nidogen could serve as a stabilizer of the membrane complex (Timpl et al., 1983).

Cells continuously interact with the matrices they secrete. This is strongly suggested by studies on the biological effects of various matrices (Adamson, 1982; Hay, 1982), and more recently it has been demonstrated that cells have defined surface components with affinities for different matrix components. Again, there seems to be some cell type specificity. The chondrocyte receptor for collagen is a small 32-kD glycoprotein distinct from the 68-kD receptor for laminin. These receptors, called anchorins, are to a certain extent located on different cells (Liotta et al., 1983; von der Mark, 1984). It is noteworthy that similar surface receptors for fibronectin have been difficult to find (Yamada, 1983). Some matrix components may thus react only weakly and transiently with many different surface components (Johansson and Höök, 1984). The receptors for laminin and other anchorin proteins may be responsible for the attachment of cells to the matrix. They may also mediate transmembrane signals between the matrix and the interior of the cells, but this has not yet been directly demonstrated.

We can conclude that adult cells elaborate matrices that reflect the differentiated state of the cells. The question is then: How do these profound differences arise during development? Certain mesenchymal cells become cartilage cells, others develop into fibroblasts, and some are geared toward epithelial differentiation. The continuous interaction between cells and the matrix is apparently important for the maintenance of the adult phenotype, but it does not explain how the cells become differentiated in the first place. There must be other forces at work during the segregation process of embryogenesis. A number of observations have convincingly demonstrated that the initial triggers for developmental transitions are due to interactions between nearby cell populations. The basic rules of these tissue interactions have been frequently reviewed (Rutter et al., 1964; Saxén et al., 1976; Wessells, 1977).

Embryological studies have established that inductive interactions regulate early differentiative events (Hogan and Tilly, 1981), the development of the central nervous system (Toivonen, 1979), and later development of epithelial organs such as the pancreas, the salivary and thyroid glands, the lungs, and the kidneys. The molecular mechanisms that underlie these events are largely unknown. There is now evidence from many systems that the interactions influence the composition of the matrix of the interacting cells (Ekblom, 1984). In some inductive systems, the interactions also alter responsiveness to the hormones and growth factors present in serum (Kratochwil and Schwartz, 1976; Cunha et al., 1983).

Because interactions between different cell lineages are so crucial, it would actually be desirable to use experimental differentiation models that allow an analysis of the cell interactions. Monolayer cultures are not well suited for this purpose because three-dimensional structures do not develop under such conditions. We have found that organ cultures of embryonic tissue rudiments are ideal systems in many respects. Here we review our studies on kidney development and focus on the role of extracellular matrix components and serum factors. Many other aspects already have been thoroughly reviewed (Kazimierzak, 1971; Lehtonen 1976).

Cell Lineages and Their Interactions

Kidney morphogenesis is driven by interactions between three cell lineages: the epithelial ureter bud, the metanephrogenic mesenchyme, and the endothelium (Figure 2). The trigger is a contact between the ureter bud and the mesenchyme (Grobstein 1956; Wartiovaara et al., 1974; Lehtonen, 1976). This inductive interaction is a vital reciprocal event that leads to a branching of the ureter bud and to a conversion of the mesenchyme into epithelial kidney tubules. Recombination studies have revealed that the

PROPOSAL FOR CELL LINEAGES

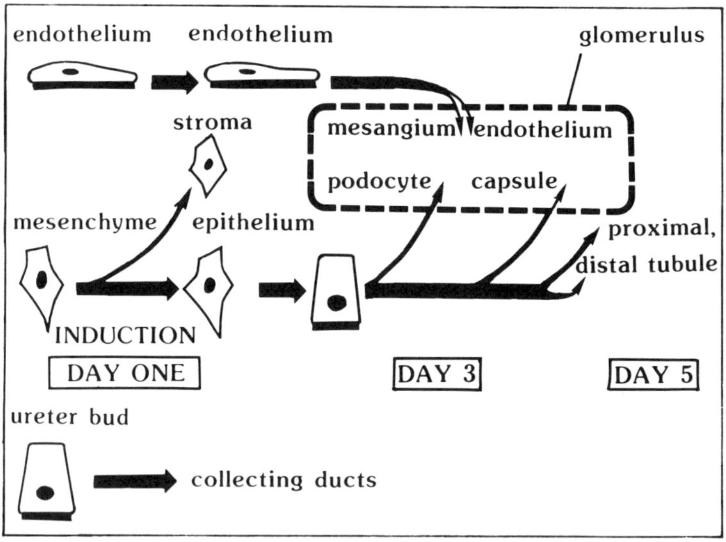

Figure 2. *Cell lineages during* in vivo *development*. Three cell lineages, the endothelium, the nephrogenic mesenchyme, and the ureter bud, interact to form the kidney. Induction by the ureter bud leads to the conversion of the mesenchyme to epithelium, and this in turn stimulates the migration of the endothelium. (After Ekblom et al., 1982b.)

mesenchyme can be induced by a variety of embryonic tissues, but of all the tissues tested, only the metanephric mesenchyme can stimulate branching and growth of the ureter. The metanephric mesenchyme is also the only mesenchyme that responds by forming tubules when exposed to the various inductor tissues (Grobstein, 1955, 1967).

Until recently, the third cell lineage, the endothelium, had not been much studied experimentally. Immunohistological studies on basement membrane formation clearly demonstrate its presence very early in *in vivo* development (Ekblom, 1981a,b). Its presence is not vital for the early differentiation of either the mesenchyme or the epithelial ureter (Berstein et al., 1981; Ekblom, 1981a). Onset of angiogenesis, on the other hand, is stimulated by mesenchymal–epithelial interaction. Once the mesenchyme has been induced, it starts to attract the endothelium, and the cell lineages eventually meet in the glomerulus, where they organize histiotypically (Figure 2).

These multiple cell–cell interactions between cell lineages would be difficult to analyze without using models of the individual steps. We have used two experimental models that allow us to study many events separately. The development of mesenchymal cells into epithelial kidney tubules has been analyzed in an organ culture system described by Grobstein (1956). In this system, the metanephrogenic mesenchyme is surgically separated

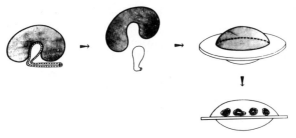

Figure 3. *Scheme of the* in vitro *model of Grobstein (1956).* The ureter bud is microsurgically removed, the mesenchyme is placed on top of the filter, and an inductor tissue attached under the filter. During culture, kidney tubules form in the mesenchyme.

from the ureter bud and then cultured on top of a filter. An inductor tissue is attached on the other side (Figure 3). It is now known that the endothelial cells are left out by the microsurgical procedure (Ekblom et al., 1982a), and therefore one monitors only the differentiation of the mesenchyme into epithelium (Figure 4). The filter acts as a convenient physical barrier that prevents intermixing of the inductor tissue with the mesenchyme, although proper filters allow the passage of crucial inducer signals (Lehtonen, 1976). In the *in vitro* system, precise timings are possible, and matrix deposition closely resembles the *in vivo* situation; studies on the hormonal requirements can be performed, as chemically defined media are available (Ekblom et al., 1981c).

Another culture technique has been used to follow the development of the endothelial cell lineage. Developing kidneys are separated prior to

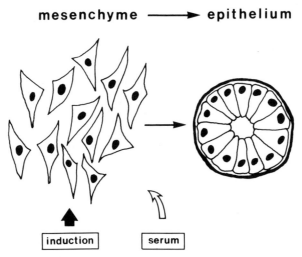

Figure 4. *Morphological changes during kidney differentiation* in vitro. In response to induction and serum, the mesenchyme converts into an epithelium.

vascularization and then placed on richly vascularized areas of chorioallantoic membranes of either chick or quail. In this system, many embryonic organs develop very well. Experimental manipulations are difficult, but it is still one of the best approaches for vascularization studies (Folkman, 1982; Schor and Schor, 1983); the differentiated tissues are composed of cells from two different species as the vessels are host-derived, whereas all other cells are of graft origin (Ekblom et al., 1982a).

EPITHELIAL MORPHOGENESIS

The morphological steps during differentiation of the mesenchyme *in vitro* are well known. A one-day exposure to the inductor tissue is sufficient to direct the mesenchyme cells irreversibly toward epithelial development. Morphologically, the induced cells are at that time indistinguishable from uninduced cells, yet the induced ones can now differentiate. On day two, aggregates become detectable, and on day three the first epithelial structures with a central lumen become evident (Wartiovaara, 1966). Later, a definite segregation of the epithelium takes place (Ekblom et al., 1981a,b,c). All morphological events occur after induction, and certain developmentally important metabolic changes preceding morphogenesis are therefore triggered by the inductive event. This has been known for a number of years (Grobstein, 1955, 1967), but apart from the finding that thymidine incorporation into cells increases (Vainio et al., 1965; for a review, see Saxén et al., 1968) there have been no reports on more specific changes during induction. However, in recent years it has been noted that induction leads to a whole set of molecularly defined changes. These changes are not directly linked to each other in any apparent way, but many are likely to be crucial for the subsequent steps of overt morphogenesis.

The Extracellular Matrix

The cells of the nephrogenic mesenchyme are morphologically similar to fibroblasts, and their epithelial bias is not in any way apparent. Immunohistology reveals a matrix phenotype typical for a loose mesenchyme. Collagens type I and III and fibronectin are found around the cells, and no basement membrane components can be detected. When the ureter bud begins to grow into the mesenchyme *in vivo*, a local dissolution of the original matrix occurs rapidly (Figure 5). This is seen only around the tips of the ureteric tree, where subsequent morphogenesis will take place. The same loss is seen *in vitro* within 24 hours. At this time, basement membrane components become detectable, but only in the induced areas. Initially, they are found in a punctate pattern, both intra- and extracellularly (Figure 6A–C). During the following stages, the deposition of the basement membrane components gradually becomes more polar, and finally they can be

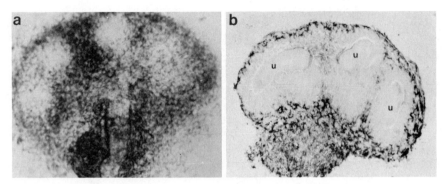

Figure 5. *Loss of interstitial collagens.* When the ureteric tree branches into the mesenchyme, cells of the mesenchyme become condensed around the tips of the ureteric tree. *a*: The tips and the surrounding mesenchyme are seen as a lighter area. *b*: Immunoperoxidase labeling reveals that collagen type III is lost in the mesenchyme around the tips of the ureteric (u) branches. (From Ekblom et al., 1981b.)

detected only at the basal surface of the cells (Figure 6D–E). For each segment of the nephron, this sequence is completed within approximately three days *in vitro* (Figure 7). This sequence was first reported for laminin, but it was soon found that collagen type IV and basement membrane proteoglycans are distributed in a similar fashion (Ekblom, 1981 a,b). The evidence for the matrix shift has been obtained by using specific antibodies prepared against the interstitial collagens (Timpl, 1982), fibronectin (Linder et al., 1975), and various basement membrane constituents (Timpl and Martin, 1982).

Control recombination experiments with other tissues have shown that this switch occurs only in an inductive situation. Thus, the inductors of the kidney mesenchyme cannot stimulate a loss of interstitial collagens in mesenchymes that do not have the capacity to become kidney tubules. Furthermore, tissues that do not induce the formation of kidney tubules do not switch the matrix phenotype of the nephrogenic mesenchyme (Ekblom et al., 1981a). Our results tell us that short-range interactions between dissimilar cell populations are the forces that reverse the matrix composition of differentiating cells. Without this influence, the cells would continue to deposit an interstitial matrix and would remain mesenchymal. It is noteworthy that only the initial switch is induction-dependent. The next stages, characterized by overt morphogenesis and a gradual formation of cell polarity, occur without the inductor tissue *in vitro*.

Based on the described sequence of extracellular matrix changes, a reasonably simple sequence for the molecular events that occur during the conversion of the mesenchyme to epithelium can be presented. The loss of the interstitial collagens and fibronectin may represent the first event. It has been suggested that this loss is due to an activation of tissue collagenase (Ekblom et al., 1981a; Ekblom, 1984), but no evidence for the

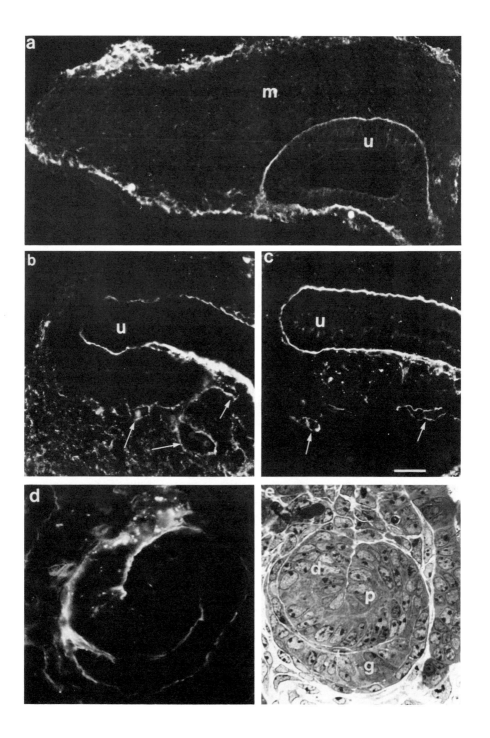

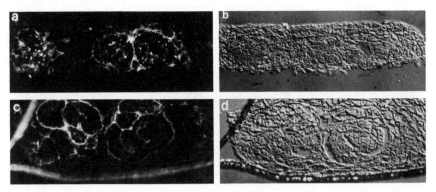

Figure 7.   *Basement membrane formation* in vitro. *a* and *b*: On day two, laminin is still found in a punctate pattern. *c* and *d*: On day three, laminin is largely confined to the basally located basement membrane. *a* and *c*, immunofluorescence labeling; *b* and *d*, corresponding Nomarski optics. (From Ekblom et al., 1980a.)

presence of collagenases has been reported so far. Collagenase activity is regulated by cell–cell interactions in other systems, suggesting a role for this enzyme in morphogenesis (Gross, 1974; Wirl, 1977; Johnson-Muller and Gross, 1978; Woolley et al., 1978).

The disappearance of the original mesenchymal matrix found between all cells may be directly related to the condensation of induced cells. The formation of condensed areas would thus, in part, be a rather passive event for the cells. It is likely, however, that the cells actively adhere to one another, and this requires the presence of adhesive proteins. Cell adhesion molecules (Edelman, 1983) and basement membrane proteins (Ekblom, 1981a) may serve this function. During kidney morphogenesis, both these components are present in the condensed areas, but their anatomical distribution is slightly different. We think that this difference is developmentally important. Whereas cell adhesion molecules are located on the cell surface in a nonpolar orientation (Thiery et al., 1982), laminin and other basement membrane components are found in a punctate pattern, and very soon become polar. We therefore postulate a two-step process for adhesion. The initial event after the loss of the interstitial matrix could be mediated by cell adhesion molecules. This cell–cell interaction does not, however, lead

Figure 6.   *Basement membrane formation during the conversion of the mesenchyme to epithelium.* Immunofluorescence labeling. *a*: Before branching of the ureter bud (u), basement membrane collagen is not found in the mesenchyme (m), and staining is detectable only in the capsule and in the ureteric basement membrane. *b* and *c*: When the ureteric tree branches and induces the mesenchyme, fibronectin is lost from the induced areas (*b*), while laminin appears in a punctate pattern (*c*). *Arrows* denote vessels. *d* and *e*: Basement membrane antigens first become polar in the proximal part of the nephron (*d*) when the mesenchyme-derived epithelium elongates to form an S-shaped structure (*e*). (From Ekblom, 1981a, b.)

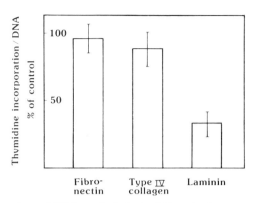

Figure 8. *Matrix proteins and DNA synthesis during kidney development* in vitro. On day two of *in vitro* development, the induced mesenchyme responds to transferrin by proliferation. Soluble fibronectin (50 μg/ml) or collagen type IV (50 μg/ml) do not alter the transferrin-dependent proliferation, but laminin (50 μg/ml) decreases proliferation by approximately 50% (I. Thesleff and P. Ekblom, unpublished observations.)

to formation of cell polarity. The second step, initiated very early, is the adhesion and attachment of cells to laminin and other basement membrane proteins. Since the deposition of the basement membrane complex is polar, it may influence the formation of morphological cell polarity.

Thus, we can see that the conversion of the mesenchyme to the epithelium involves a gradual specialization of the cell surface into basal, lateral, and luminal domains. The primary trigger for polarization is not known, but there are reasons to believe that the appearance of the basement membrane components is crucial. The anchorage of cells to a basement membrane could provide a sufficient signal. Accordingly, one would expect an inhibition of cell polarity when basement membrane proteins are present in a nonpolar fashion. Such inhibition can be accomplished experimentally by adding soluble exogenous laminin. It has now been shown, both in certain cell culture situations (Grover et al., 1983) and during kidney tubule development, that soluble laminin can disturb the formation of cell polarity. In our system, soluble laminin has an effect on proliferation (Figure 8). These studies support the model of Adamson and her associates (Grover et al., 1983) on the role of laminin in the polarization of epithelia. The model suggests a major role for an endogenous polar deposition of laminin in the acquisition of cell polarity (Figure 9).

Our scheme for the conversion of mesenchyme to epithelium ties together observations on matrix proteins (substrate adhesion molecules) and the findings of Thiery et al. (1982) on cell–cell adhesion molecules. In spite of this, the scheme is still incomplete in many respects. Direct interruption experiments, such as those performed with laminin, should be done for all matrix components. Data on the appearance of the laterally located proteins such as the desmoplakins (Franke et al., 1982), associated intracellular filaments, and adult-type cell adhesion molecules are almost completely lacking. We can now state, however, that this simple conversion involves a multitude of changes in many different protein classes, and each change may be important because it may be rate limiting.

Serum Growth Factors

Although adhesion between cells and attachment to the matrix are pre-requisites for the growth of monolayers and solid embryonic organ rudi-ments, they alone are not sufficient for growth and tissue histogenesis. Serum mitogens provide another kind of stimulus. Many types of mitogens have been found in serum, and unique fetal growth factors may be present in embryos. The differentiation of the kidney epithelium seems to be dependent on only one mitogen, transferrin. Transferrin is an abundant 80-kD serum protein that transports iron. Other hormones, such as epidermal and fibroblast growth factor or insulin, do not stimulate differentiating epithelium when tested in chemically defined media, whereas transferrin, regardless of the basal medium, supports morphogenesis (Ekblom et al., 1981c). It is likely that other growth factors cannot replace the effect of transferrin, because differentiation fails in a transferrin-depleted serum (Figure 10). As in other systems, transferrin acts by stimulating proliferation (Thesleff and Ekblom, 1984). Initially, the cells are completely unresponsive to transferrin. The uninduced mesenchyme cannot be stimulated by trans-ferrin, nor does transferrin affect proliferation on day one during induction; the effect becomes apparent on subsequent days (Figure 11). Inducer tissue is not responsive to transferrin in transfilter cultures (Figure 12). These results suggest that induction not only switches the matrix composition, but also alters a cell's capacity to respond to the serum mitogen (Ekblom et al., 1983; Thesleff et al., 1983b).

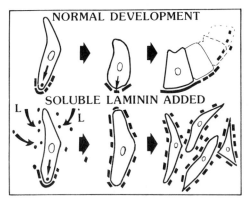

Figure 9. *Role of laminin in cell polarization.* Soluble laminin disturbs the formation of a polarized epithelium (Grover et al., 1983).

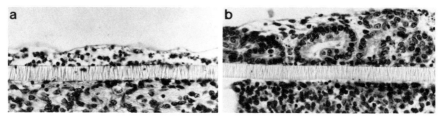

Figure 10.   *Role of serum transferrin in kidney morphogenesis.* In a transferrin-depleted serum, tubule formation fails (*a*), but when transferrin is added to the same serum, tubule formation is evident (*b*). (After Ekblom et al., 1983.)

The inductive event itself also has a direct effect on proliferation (Vainio et al., 1965; Saxén et al., 1983), but as mentioned above, the initial proliferation burst is not in any way stimulated by transferrin. There are therefore two proliferative events: an initial transferrin-independent one and a subsequent transferrin-dependent one. We do not know how the first stimulus is brought about, but it would seem logical to postulate that the onset of transferrin responsiveness is due to the appearance of receptors for transferrin. The transferrin receptor is a 180-kD integral membrane glycoprotein that has been detected in large quantities in rapidly growing cells, including malignant cells (Sutherland et al., 1981; Trowbridge and Omary, 1981).

Our scheme for the transferrin-mediated proliferation is as follows: In the embryo, transferrin is produced by the liver and the yolk sac. Of the many growth promoters secreted by the liver, only transferrin stimulates proliferation of the induced mesenchyme (Ekblom and Thesleff, 1985). Transferrin is not produced by inducer tissues such as the ureter bud and the spinal cord, which explains why it must be added to the medium in our experiments. In the embryo, transferrin is delivered to the differentiating

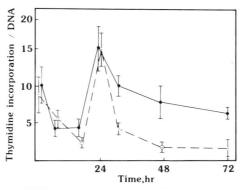

Figure 11.   *Time course of DNA synthesis during* in vitro *development.* Without transferrin (*broken line*), proliferation decreases on day two, but when transferrin is present (*solid line*) proliferation stays at a high level. (After Ekblom et al., 1983.)

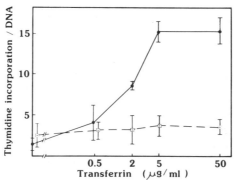

Figure 12. *Target tissue for transferrin.* A dose-dependent stimulation of proliferation by transferrin is seen on day two in the induced mesenchyme (*solid line*), but not in the inductor tissue (*broken line*). (After Thesleff and Ekblom, 1984.)

kidney through the blood vessels. When the cells become transferrin-dependent as a result of induction, transferrin reacts with the transferrin receptors present on the cell surface. The hormone–receptor complex is then rapidly internalized and delivered into intracellular acidic compartments. The acidity as such is sufficient to dissociate the iron from transferrin, and iron-free transferrin then rapidly returns to the extracellular space. Both morphological and biochemical evidence for this pathway has been reviewed elsewhere (Aisen and Listowsky, 1980; Brown et al., 1983).

For many growth factors, the initial interaction with the receptor is a sufficient signal for proliferation. This has been clearly demonstrated for the epidermal growth factor (EGF) (Schlessinger et al., 1983). The same rule does not apply for transferrin. Whereas monoclonal antibodies against the EGF receptor stimulate growth, antibodies against the transferrin receptor decrease growth (Trowbridge and Lopez, 1982). The antibodies against the mouse transferrin receptor also decrease the growth of embryonic organs (Thesleff et al., 1985). Hence, it does not seem likely that transferrin–receptor interactions would induce a signal for proliferation. Rather, the delivery of iron is the crucial event. In our system, free iron not bound to transferrin cannot enter the cells, because the only pathway is the receptor-mediated route. The existence of such pathways provides very delicate control mechanisms for the proliferation of cells. Accordingly, it is not surprising that free iron salts do not stimulate the proliferation of differentiating kidney epithelium (Landschulz et al., 1984).

It can be shown experimentally that iron is important for transferrin-mediated proliferation. Iron can be coupled to a variety of metal chelators. Most of these cannot enter cells, but when the chelator is lipophilic, it traverses the lipid membrane directly and delivers iron into the cell, thus bypassing the receptor-mediated pathway (Ponka et al., 1982; Landschulz et al., 1984). When iron is coupled to such a lipophilic chelator, proliferation

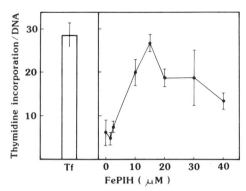

Figure 13. *Role of iron in transferrin-mediated proliferation.* Iron was coupled to a lipophilic chelator, pyridoxal isonicotoyl hydrazone (PIH). The bar to the *left* shows the level of proliferation when transferrin (50 µg/ml) is present. The dose–response curve to the *right* shows that 15 µM FePIH can stimulate proliferation as efficiently as can transferrin (Tf). (From Landschulz et al., 1984.)

of the cells is stimulated (Figure 13). In our model, the lipophilic iron chelator stimulates both differentiation and proliferation to the same degree as iron-containing transferrin. These experiments tell us that the sole role of transferrin in proliferation is the delivery of iron into the intracellular space. Many other growth factors are internalized in the same fashion as is transferrin, but the mechanisms that stimulate proliferation are quite different. We do not yet know how iron stimulates proliferation, but it has been suggested that many enzymes for DNA synthesis require it as a cofactor (Octave et al., 1983).

ANGIOGENESIS

Our studies on morphogenesis *in vitro* revealed that differentiating kidney epithelium segregates into defined nephron segments during later culture stages. The nephrogenic mesenchyme was thus shown to have several developmental options, all of which are triggered by the combined effects of the inducer tissue and the serum mitogen transferrin. However, in spite of prolonged cultures, we and others have failed to induce the development of endothelial cells from mesenchyme (Bernstein et al., 1981; Ekblom, 1981a, b, 1984). The glomeruli that form *in vitro* do not contain any endothelial cells, and the glomerular tufts are clusters of podocytes. It is noteworthy that the general histoarchitecture of the glomerulus could easily be seen in spite of the absence of the endothelium and the mesangium. The information that directs the morphogenesis of the glomerulus thus resides largely in epithelial cells.

The Origin of the Endothelium

Until recently, no simple theory for the origin of the endothelium had emerged (for a review, see Kazimierzak, 1971), and several authors thought that there was some conversion from mesenchyme to endothelium within the glomerulus at those stages when abundant vasculature and erythrocytes become detectable in the glomeruli (Reeves et al., 1980). This view was in accordance with the traditional theories of the vasculogenesis of other organs (Wagner, 1980). Based on our *in vivo* and *in vitro* results, we reasoned that endothelial cells do not derive from kidney mesenchyme, but arise instead from outside vessels that grow into the glomeruli (Ekblom, 1981a,b). This proposal was directly tested in an experimental model. The migration potential of outside vessels was studied by grafting avascular kidney rudiments onto chorioallantoic membranes of either quail or chick. This is a classic culture method still frequently used for studies on angiogenesis. We used the quail egg in these experiments because the quail nucleolar marker gave us an opportunity to study the origin of endothelial cells using ordinary histological preparations (Le Douarin, 1973). In the quail–mouse chorioallantoic cultures, the blood vessels were all from the quail, and they grew into the developing mouse glomeruli. In spite of the species differences, the cells organized histotypically, interacted properly, and formed hybrid glomeruli (Figure 14A,B). These studies provided clear evidence that neovascularization does not involve recruitment of mesenchymal cells to endothelium but instead involves solely a directed invasion of cells derived from another cell lineage.

Although the molecular mechanisms that initiate angiogenesis are unknown, some basic rules of the biology of this event have emerged for our studies. *In vivo*, vessels show a definite tendency to grow toward condensed areas, suggesting that condensation of the induced mesenchyme provides one trigger for angiogenesis. Experiments on the chorioallantoic membrane also suggest that the induction of differentiation activates the secretion of angiogenesis factors. We noted that induced kidney mesenchymes elicited an angiogenic response, but uninduced ones did not (Figure 15). The development of the epithelial part of the nephron thus influences and guides the angiogenetic process, and there is a constant epithelial–endothelial interaction throughout development (Sariola et al., 1983). A similar differentiation-dependent stimulation of vascularization has been noted by others (Castellot et al., 1982).

The Extracellular Matrix

The pathways taken by developing vessels follow a strictly controlled pattern, and each of the numerous glomeruli become properly vascularized. At present, both the nature of the initial stimuli and the factors that locally

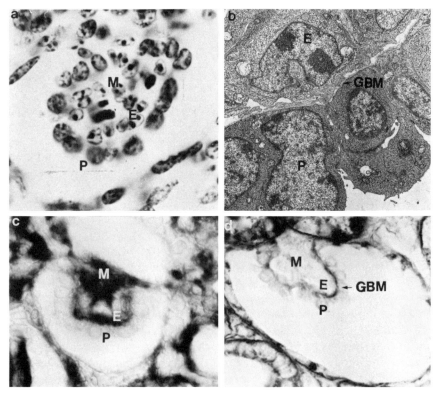

Figure 14. *Characterization of interspecies hybrid glomeruli. a*: In mouse kidneys grown on quail chorioallantoic membrane, quail endothelium has grown into the mouse glomerulus. The mesangium (M) and the endothelium (E) have a quail nucleolar marker, whereas the podocytes (P) are of mouse origin. *b*: Electron microscopy of the hybrid glomeruli shows a glomerular basement membrane (GBM) between the cells. *c*: In the hybrid glomeruli, the matrix of the mesangium and endothelium cells reacts positively with antibodies against avian collagen type IV. *d*: In the hybrid glomeruli, antibodies against mouse collagen type IV do not react with the mesangium or endothelium, but the glomerular basement membrane is positive (From Sariola et al., 1983, 1984.)

influence the direction of moving endothelium are largely unknown. In many other migration events, moving cells utilize preformed matrices as pathways, and these matrices may give the cells important developmental cues (Ebendal, 1977; Rovasio et al., 1983). Because rapid changes occur in the extracellular matrix composition of the kidney mesenchyme, it is tempting to postulate that these alterations play a role in the initial stimulus for angiogenesis. It could therefore be thought that the matrix of the target organ is instrumental for the attachment of moving endothelial cells. Endothelial cells attach well to the matrix, but the nature of the matrix is not important (Macarak and Howard, 1983). It is therefore crucial to analyze the *in vivo* situation. Our studies on matrix deposition during angiogenesis raise the possibility that the matrix produced by the endothelium itself may be important in the attachment of endothelial cells.

Matrix deposition has been studied reliably through the use of species-specific antibodies that react exclusively with either mouse- or avian-type basement membrane collagen. Such antibodies were obtained by screening several rabbit antisera by means of enzyme immunoassays. A few anti-collagen type IV antisera showed an exclusive reaction for either mouse or chick collagen type IV (Figure 16). These antibodies and monoclonal antibodies against chick collagen type IV (Mayne et al., 1983) were then used to study the histogenesis of basement membranes in the hybrid kidneys (Figures 17, 14C–D).

Several important conclusions have emerged from these immunolocalization studies. We found that the ingrowing vessels at all stages of development produced an endogenous matrix that was of the basement membrane type. No large discontinuities were detected at any stage of the migration process. The leading edge of the new vessels can be seen in the lower crevice of the S-shaped body, which is an early stage of nephron development. An endothelial-derived basement membrane was also detected. This suggests that basement membrane production is a constant and very early event in areas of neovascularization, and that discontinuities in the basement membranes at these sites (Folkman, 1982) may be smaller or more transient than previously thought. This proposal does not exclude a rapid remodeling of the basement membrane at the leading edge by enzymes that cleave collagen type IV (Folkman, 1982; Kalebic et al., 1983).

The availability of the species-specific antibodies also gave us an opportunity to study the origin of basement membranes in the glomeruli, the final destination of the endothelium. Within the glomerular tuft, vessel-

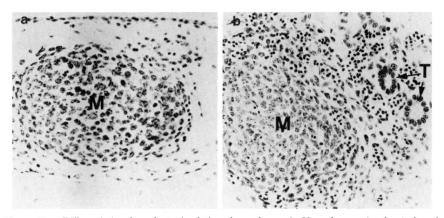

Figure 15. *Differentiation-dependent stimulation of vasculogenesis.* Vasculogenesis of uninduced (a) and partially induced (b) metanephric mesenchymes cultured on chorioallantoic membranes. The uninduced areas (M) remain avascular, whereas the differentiating areas (T) are vascularized by chorioallantoic vessels. (From Sariola et al., 1983.)

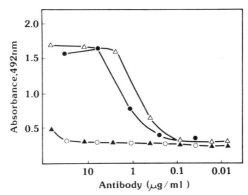

Figure 16. *Generation of species-specific antibodies against collagen type IV.* Reaction of rabbit antibodies against mouse (*open symbols*) and chick (*filled symbols*) collagen type IV in an enzyme immunoassay. Microtiter wells were coated with mouse (*triangles*) or chick (*circles*) collagen type IV. (From Sariola et al., 1984.)

derived cells (mesangium, endothelium) and mesenchyme-derived epithelial cells (podocytes) meet, organize histiotypically, and form a glomerular basement membrane, which is situated between the podocytes and the endothelium. This membrane serves as a major filter apparatus (Farquhar, 1982). Our studies demonstrate that all three cell types produce a basement membrane matrix; thus the glomerular basement membrane has a dual origin, as both the endothelium and the epithelium deposit basement membrane components (Figure 14C,D).

It is not known how the two basement membranes fuse. Ultrastructural evidence of development *in vivo* suggests that they are initially separate, and that during the next stage a fusion occurs (Reeves et al., 1980). This may involve noncovalent interactions between independent membranes

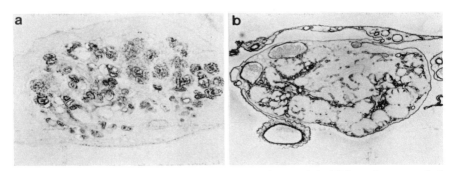

Figure 17. *Epithelial and vascular basement membrane collagen in hybrid kidneys.* In mouse/chick hybrids, the anti-mouse collagen type IV antibody reacts only with developing epithelial structures (*a*), whereas the anti-chick collagen type IV antibody reacts only with the developing vessels (*b*). The anti-chick collagen type IV antibody can be used to detect the extent of vascularization. (From Sariola et al., 1984.)

or covalent cross-linking between two or more parallel networks of type IV collagen molecules (Sariola et al., 1984). In the hybrid glomeruli, the fusion occurs but it is incomplete. The subtle species differences in the components are apparently just sufficient to disturb the normal fusion process (Sariola, 1984).

Possible Triggers for Angiogenesis

It is apparent that we cannot yet present a very detailed scheme of the molecular events of kidney angiogenesis. The initial event is somehow related to the onset of epithelial differentiation. The endogenous production of the basement membranes by the endothelium seems to be another crucial event. Many recent papers on angiogenesis in other systems have raised the possibility that angiogenesis is controlled by glycosaminoglycan metabolism. Hyaluronate turnover (Feinberg and Beebe, 1983) and heparin levels (Taylor and Folkman, 1982) strongly influence angiogenesis. It therefore seems that angiogenesis is stimulated by alterations of both collagenous components (Kalebic et al., 1983) and other matrix components. Data on these aspects are very limited. It has been shown that hyaluronidase activity changes during kidney morphogenesis, but it is not known whether this is related to angiogenesis or to epithelial differentiation (Belsky and Toole, 1983).

The migration of the endothelium must be accompanied by a replication of cells. It is uncertain whether the directional growth of the vessels represents a true movement of individual cells or a directional growth based on the replication of daughter cells derived from the original endothelium. The rapid appearance of a basement membrane probably immobilizes cells, and neovascularization thus requires a constant supply of new cells. The nature of the mitogens for angiogenesis are not well known, but there are probably both local and systemic mitogens. Perhaps some of the locally produced angiogenesis factors act directly as mitogens. Our knowledge of the mitogens for angiogenesis is in all systems rather limited. Various serum factors have been reported to stimulate angiogenesis. Foidart and Reddi found that hypophysectomy decreased angiogenesis (1980), and other factors from the central nervous system have also been shown to be stimulatory for angiogenesis (for a review, see Schor and Schor, 1983). We and others have found that EGF can influence angiogenesis of embryonic vessels in organ culture (Gospodarowicz and Tauber, 1980; Thesleff et al., 1983a). Our studies have raised the possibility that an EGF-like mitogen, sarcoma growth factor (SGF), may be an even more potent mitogen for endothelial cells. This polypeptide was initially found in tumor cells (Todaro and DeLarco, 1978). More recently, it has also been found in embryonic tissues; we have postulated that it may represent an embryonic form of EGF, which is also synthesized in large quantities in certain malignancies (Thesleff et al., 1983a).

Although we know the target cells for transferrin during kidney morphogenesis rather well, it is not known in detail how EGF and SGF stimulate angiogenesis. They act either directly on the endothelium or more indirectly by enhancing the proliferation of other cells in the organ cultures. Our preliminary studies on angiogenesis of the kidney suggest that both EGF and SGF have effects on angiogenesis in this tissue as well (P. Ekblom and I. Thesleff, unpublished observations).

No detailed schemes for the molecular steps constituting angiogenesis can be given at present. We still do not know enough of the guiding principles and have only a rough view of matrix composition during neovascularization. The role of serum mitogens also remains poorly understood.

CONCLUSIONS

The molecular nature of morphogenetic cell interactions remains a largely unsolved problem although the significance of such interactions for embryogenesis is obvious. Our studies do not clarify the nature of the inductive signals, but they give us a much clearer picture of the molecular changes that occur in response to induction. The presently known changes can be broadly classified into two main categories. On one hand, the cells in our system respond by altering the extracellular matrix composition and, on the other hand, cell proliferation is affected. Proliferation is directly stimulated, but the inductive event also leads to an acquisition of responsiveness to a serum mitogen. These induction-dependent alterations are developmentally meaningful, as the switch in matrix composition provides the proper attachment, and the onset of transferrin responsiveness is important for organ growth. Thus, the morphogenetic steps can be explained to a certain extent in terms of these developmentally regulated events.

Our results have led us to postulate that the proliferation that follows induction is largely stimulated by a very simple compound, iron. Its delivery, however, involves a multitude of complex events, and the regulation of iron intake is related to the appearance of transferrin responsiveness, which in turn is regulated by inductive interaction. The final shape of the tissue, again, is controlled by adhesive determinants that restrict proliferation to proper locations. The new generation of cells immediately becomes attached to the forming, endogenously synthesized basement membrane, and the kidney tubule elongates and branches appropriately. During this growth, there is also a gradual appearance of kidney-specific markers, which are typical for each segment of the nephron (Ekblom et al., 1980b, 1981b). The first terminally differentiated cells that we detect are the glomerular cells. Then proximal tubules start to express their luminal markers, and finally the distal tubules can be distinguished by using cell type-

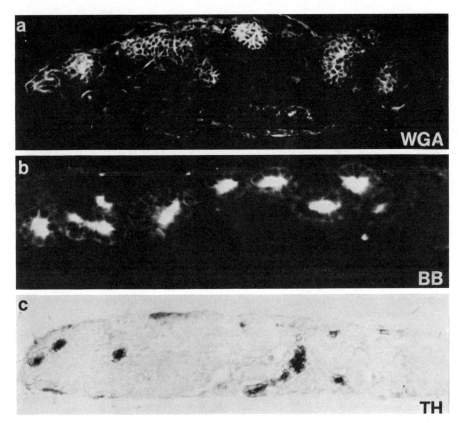

Figure 18.   In vitro *segregation of the nephron*. This was shown by using specific markers for
three different nephron segments. Glomerular podocytes react with the lectin wheat
germ agglutinin (WGA) (*a*); proximal tubules are revealed by antibodies against brush
border antigens (BB) (*b*); distal tubules are revealed by antibodies against the Tamm-
Horsfall (TH) glycoprotein (*c*). (From Ekblom et al., 1980b, 1981b).

specific markers (Figure 18). We do not know how this terminal differentiation is coupled to the proliferation of the cells. Iron cannot provide any
signals that would direct the segregation process. Several theories on the
relationship between differentiation and proliferation have been advanced
(Holzer, 1968), but no clear hypotheses can at present be given on this
aspect of the differentiation of the kidney tubules.

The scheme presented for kidney epithelium does not apply to the blood
vessels of the kidney. They use a totally different strategy for development
as they do not form *in situ*, but migrate into the kidney. It is noteworthy
that both the migrating endothelium and epithelium use a similar type of
matrix for attachment. The basement membrane matrix is thus of importance
in many different types of morphogenetic events.

REFERENCES

Adamson, E. D. (1982) The effect of collagen on cell division, cellular differentiation and embryonic development. In *Collagen in Health and Disease*, J. B. Weiss and M. I. V. Jayson, eds., pp. 218–243, Churchill Livingstone, London.

Aisen, P., and I. Listowsky (1980) Iron transport and storage proteins. *Annu. Rev. Biochem.* **49**:713–721.

Barnes, D., and G. Sato (1980) Serum-free culture: A unifying approach. *Cell* **22**:649–655.

Belsky, E., and B. P. Toole (1983) Hyaluronate and hyaluronidase in the developing chick embryo kidney. *Cell. Differ.* **12**:61–66.

Bender, B. L., R. Jaffe, B. Carlin, and A. Chung (1981) Immunolocalization of entactin, a sulfated basement membrane component in rodent tissues, and comparison with GP-2 (laminin). *Am. J. Pathol.* **103**:419–426.

Bernfield, M. R., R. H. Cohn, and S. D. Banerjee (1973) Glycosaminoglycan and epithelial organ formation. *Am. Zool.* **13**:1067–1083.

Bernstein, J., F. Cheng, and J. Roszka (1981) Glomerular differentiation in metanephric culture. *Lab. Invest.* **45**:183–190.

Brown, M. S., R. G. W. Anderson, and J. L. Goldstein (1983) Recycling receptors: The round trip itinerary of migrant membrane proteins. *Cell* **32**:663–667.

Castellot, J. J., Jr., M. J. Karnovsky, and B. M. Spiegelman (1982) Differentiation-dependent stimulation of neovascularization and endothelial cell chemotaxis by 3T3 adipocytes. *Proc. Natl. Acad. Sci. USA* **79**:5597–5601.

Cunha, G. R., L. W. K. Chung, J. M. Shannon, O. Taguchi, and H. Fujii (1983) Hormone-induced morphogenesis and growth: Role of mesenchymal–epithelial interactions. *Recent Prog. Horm. Res.* **30**:559–598.

Dulbecco, R., and M. G. P. Stoker (1970) Conditions determining initiation of DNA synthesis in 3T3 cells. *Proc. Natl. Acad. Sci. USA* **66**:204–208.

Ebendal, T. (1977) Extracellular matrix fibrils and cell contacts in the chick embryo. Possible roles in orientation of cell migration and axon extension. *Cell Tissue Res.* **175**:439–458.

Edelman, G. M. (1983) Cell adhesion molecules. *Science* **219**:450–457.

Ekblom, P. (1981a) Formation of basement membranes in the embryonic kidney: An immunohistological study. *J. Cell. Biol.* **91**:1–10.

Ekblom, P. (1981b) Determination and differentiation of the nephron. *Med. Biol.* **50**:139–160.

Ekblom, P. (1984) Basement membrane proteins and growth factors in kidney development. In *Role of the Extracellular Matrix in Differentiation*, R. L. Trelstad, ed., pp. 173–206, Alan R. Liss, New York.

Ekblom, P., and I. Thesleff (1985) Control of kidney differentiation by soluble factors secreted by the embryonic liver and the yolk sac. *Dev. Biol.* **110**:29–38.

Ekblom, P., K. Alitalo, A. Vaheri, R. Timpl, and L. Saxén (1980a) Induction of a basement membrane glycoprotein: Possible role of laminin in morphogenesis. *Proc. Natl. Acad. Sci. USA* **77**:485–489.

Ekblom, P., A. Miettinen, and L. Saxén (1980b) Induction of brush border antigens of proximal tubules in developing kidney. *Dev. Biol.* **74**:263–274.

Ekblom, P., E. Lehtonen, L. Saxén, and R. Timpl (1981a) Shift in collagen type as an early response to induction of the metanephric mesenchyme. *J. Cell. Biol.* **89**:276–283.

Ekblom, P., A. Miettinen, I. Virtanen, T. Wahlström, A. Dawnay, and L. Saxén (1981b) *In vitro* segregation of the metanephric nephron. *Dev. Biol.* **84**:88–95.

Ekblom, P., I. Thesleff, A. Miettinen, and L. Saxén (1981c) Organogenesis in a defined medium supplemented with transferrin. *Cell Differ.* **10**:281–288.

Ekblom, P., H. Sariola, M. Karkinen-Jääskeläinen, and L. Saxén (1982a) The origin of the glomerular endothelium. *Cell Differ.* **11**:35–39.

Ekblom, P., L. Saxén, and R. Timpl (1982b) The extracellular matrix in kidney differentiation. In *Membranes in Growth and Development*, G. Giebisch, J. Hoffman, and L. Bolis, eds., pp. 429–442, Alan R. Liss, New York.

Ekblom, P., I. Thesleff, L. Saxén, A. Miettinen, and R. Timpl (1983) Transferrin as a fetal growth factor: Acquisition of responsiveness related to embryonic induction. *Proc. Natl. Acad. Sci. USA* **80**:2651–2655.

Farquhar, M. G. (1982) The glomerular basement membrane. A selective macromolecular filter. In *Cell Biology of Extracellular Matrix*, E. D. Hay, ed., pp. 335–378, Plenum, New York.

Feinberg, R. N., and D. C. Beebe (1983) Hyaluronate in vasculogenesis. *Science* **220**:1177–1179.

Foidart, J. M., and A. H. Reddi (1980) Immunofluorescent localization of type IV collagen and laminin during endochondral bone differentiation and regulation by pituitary growth hormone. *Dev. Biol.* **75**:130–136.

Folkman, J. (1982) Angiogenesis: Initiation and control. *Ann. N.Y. Acad. Sci.* **401**:212–227.

Franke, W. W., R. Moll, D. L. Schiller, E. Schmid, J. Kartenbeck, and H. Müller (1982) Desmoplakins of epithelial and myocardial desmosomes are immunologically and biochemically related. *Differentiation* **23**:115–127.

Gospodarowicz, D., and J. P. Tauber (1980) Growth factors and the extracellular matrix. *Endocrin. Rev.* **1**:201–227.

Grobstein, C. (1955) Inductive interaction in the development of the mouse metanephros. *J. Exp. Zool.* **130**:319–340.

Grobstein, C. (1956) Trans-filter induction of tubules in mouse metanephrogenic mesenchyme. *Exp. Cell. Res.* **10**:424–440.

Grobstein, C. (1967) Mechanisms of organogenetic tissue interaction. *Natl. Cancer Inst. Monogr.* **26**:279–299.

Gross, J. (1974) Collagen biology: Structure, degradation and disease. *Harvey Lect. Ser.* **68**:351–432.

Grover, A., G. Andrews, and E. D. Adamson (1983) Role of laminin in epithelium formation by F9 aggregates. *J. Cell Biol.* **97**:137–144.

Hay, E. D. (1982) Interaction of embryonic cell surface and cytoskeleton with extracellular matrix. *Am. J. Anat.* **165**:1–12.

Hogan, B. L. M., and R. Tilly (1981) Cell interactions and endoderm differentiation in cultured mouse embryos. *J. Embryol. Exp. Morphol.* **62**:379–394.

Holley, R. (1975) Control of growth of mammalian cells in culture. *Nature* **253**:487–489.

Holzer, H. (1968) Induction of chondrogenesis: A concept in quest of mechanism. In *Epithelial–Mesenchymal Interactions*, R. Fleischmajer and R. E. Billingham, eds., pp. 152–164, Williams and Wilkins, Baltimore.

Hynes, R. O., and K. Yamada (1982) Fibronectins: Multifunctional modular glycoproteins. *J. Cell Biol.* **95**:369–377.

Johansson, S., and M. Höök (1984) Substrate adhesion of rat hepatocytes: On the mechanism of attachment of fibronectin. *J. Cell Biol.* **98**:810–817.

Johnson-Muller, B., and J. Gross (1978) Regulation of corneal collagenase production: Epithelial–stromal cell interactions. *Proc. Natl. Acad. Sci. USA* **75**:4417–4421.

Kalebic, T., S. Garbisa, B. Glaser, and L. A. Liotta (1983) Basement membrane collagen: Degradation by migrating endothelial cells. *Science* **221**:281–283.

Kazimierzak, J. (1971) Development of the renal corpuscle and the juxtaglomerular apparatus. A light and electron microscopic study. *Acta Pathol. Microbiol. Scand. (Suppl.)* **218**:1–61.

Kefalides, N. A., R. Alper, and C. C. Clark (1979) Biochemistry and metabolism of basement membranes. *Int. Rev. Cytol.* **61**:167–180.

Kleinman, H. K., R. J. Klebe, and G. R. Martin (1981) Role of collagenous matrices in the adhesion and growth of cells. *J. Cell Biol.* **88**:473–485.

Kratochwil, K., and P. Schwartz (1976) Tissue interaction in androgen response of embryonic mammary rudiment of mouse: Identification of target tissue for testosterone. *Proc. Natl. Acad. Sci. USA* **73**:4041–4045.

Landschulz, W., I. Thesleff, and P. Ekblom (1984) A lipophilic iron chelator can replace transferrin as a stimulator of cell proliferation and differentiation. *J. Cell Biol.* **98**:596–601.

Le Douarin, N. M. (1973) A biological cell labeling technique and its use in experimental embryology. *Dev. Biol.* **30**:217–222.

Lehtonen, E. (1976) Transmission and spread of embryonic induction. *Med. Biol.* **54**:108–128.

Linder, E., A. Vaheri, E. Ruoslahti, and J. Wartiovaara (1975) Distribution of fibroblast surface antigen in the developing chick embryo. *J. Exp. Med.* **142**:41–49.

Liotta, L. A., C. N. Rao, and S. H. Barsky (1983) Tumor invasion and the extracellular matrix. *Lab. Invest.* **49**:636–649.

Macarak, E. J., and P. S. Howard (1983) Adhesion of endothelial cells to extracellular matrix proteins. *J. Cell. Physiol.* **115**:76–86.

Mayne, R. (1984) Collagen isotypes. In *Role of Extracellular Matrix in Differentiation*, R.L. Trelstad, ed., pp. 33–42, Alan R. Liss, New York.

Mayne, R., R. D. Sanderson, H. Wiedemann, J. M. Fitch, K. von der Mark, and T. F. Linsenmayer (1983) The use of monoclonal antibodies to fragments of chicken type IV collagen in structural and localization studies. *J. Biol. Chem.* **258**:5794–5797.

Octave, J.-N., Y.-V. Schneider, A. Trouvet, and R. R. Crichton (1983) Iron uptake and utilization by mammalian cells. *Trends Biochem. Sci.* **8**:217–219.

Pierce, B. M. (1966) The development of basement membranes of the mouse embryo. *Dev. Biol.* **13**:231–249.

Ponka, P., H. M. Schulman, and A. Wilzynska (1982) Ferric pyridoxal isonicotinoyl hydrazone can provide iron for heme synthesis in reticulocytes. *Biochim. Biophys. Acta* **718**:151–156.

Reeves, W. H., Y. P. Kanwar, and M. G. Farquhar (1980) Assembly of the glomerular filtration surface. Differentiation of anionic sites in glomerular capillaries of newborn rat kidney. *J. Cell Biol.* **85**:735–753.

Ross, R., and A. Vogel (1978) The platelet derived growth factor. *Cell* **14**:203–210.

Rovasio, R. A., A. DeLouvée, K. M. Yamada, R. Timpl, and J.-P. Thiery (1983) Neural crest migration: Requirements for exogenous fibronectin and high cell density. *J. Cell Biol.* **96**:462–473.

Ruoslahti, E., E. Engvall, and E. Hayman (1981) Fibronectin: Current concepts of its structure and functions. *Collagen Relat. Res.* **1**:85–128.

Rutter, W. J., N. K. Wessells, and C. Grobstein (1964) Control of specific synthesis in the developing pancreas. *Natl. Cancer Inst. Monogr.* **13**:51–63.

Sariola, H. (1984) Incomplete fusion of endothelial and epithelial basement membranes in interspecies hybrid glomeruli. *Cell Differ.* **14**:189–195.

Sariola, H., P. Ekblom, E. Lehtonen, and L. Saxén (1983) Differentiation and vascularization of the metanephric kidney grafted on the chorioallantoic membrane. *Dev. Biol.* **96**:427–435.

Sariola, H., R. Timpl, K. von der Mark, R. Mayne, J. M. Fitch, T. F. Linsenmayer, and P. Ekblom (1984) Dual origin of glomerular basement membrane. *Dev. Biol.* **101**:86–96.

Saxén, L., O. Koskimies, A. Lahti, H. Miettinen, J. Rapola, and J. Wartiovaara (1968) Differentiation of kidney mesenchyme in an experimental model system. *Adv. Morphog.* **7**:251–293.

Saxén, L., M. Karkinen-Jääskeläinen, E. Lehtonen, S. Nordling, and J. Wartiovaara (1976) Inductive tissue interactions. In *The Cell Surface in Animal Embryogenesis and Development*, G. Poste and G. L. Nicolson, eds., pp. 331–407, North-Holland, Amsterdam.

Saxén, L., J. Salonen, P. Ekblom, and S. Nordling (1983) DNA synthesis and cell generation cycle during determination and differentiation of the metanephric mesenchyme. *Dev. Biol.* **98**:130–138.

Schlessinger, J., A. B. Schreiber, A. Levi, I. Lax, I. Libermann, and Y. Yarden (1983) Regulation of cell proliferation by epidermal growth factor. *CRC Crit. Rev. Biochem.* **14**:93–112.

Schor, A. M., and S. L. Schor (1983) Tumor angiogenesis. *J. Pathol.* **141**:385–414.

Spiro, R. G. (1978) Nature of the glycoprotein components of basement membranes. *Ann. N.Y. Acad. Sci.* **321**:106–120.

Sutherland, R., D. Delia, C. Schneider, R. Newman, J. Kemshead, and M. Greaves (1981) Ubiquitous cell-surface glycoprotein on tumor cells is proliferation-associated receptor for transferrin. *Proc. Natl. Acad. Sci. USA* **78**:4515–4519.

Taylor, S., and J. Folkman (1982) Protamine is an inhibitor of angiogenesis. *Nature* **297**:307–312.

Terranova, V. P., D. H. Rohrbach, and G. R. Martin (1980) Role of laminin in the attachment of PAM 212 (epithelial) cells to basement membrane collagen. *Cell* **22**:719–726.

Thesleff, I., and P. Ekblom (1984) Role of transferrin in branching morphogenesis, growth and differentiation of the embryonic kidney. *J. Embryol. Exp. Morphol.* **82**:147–161.

Thesleff, I., P. Ekblom, and J. Keski-Oja (1983a) Inhibition of morphogenesis and stimulation of vascularization by a sarcoma growth factor preparation. *Cancer Res.* **43**:5902–5907.

Thesleff, I., P. Ekblom, E. Lehtonen, P. Kuusela, and E. Ruoslahti (1983b) Exogenous fibronectin not required for organogenesis *in vitro*. *In Vitro* **19**:903–910.

Thesleff, I., A. M. Partanen, W. Landschulz, I. S. Trowbridge, and P. Ekblom (1985) Role of transferrin receptors and iron delivery in embryonic morphogenesis. *Differentiation* (in press).

Thiery, J.-P., J. L. Duband, U. Rutishauser, and G. M. Edelman (1982) Cell adhesion molecules in early chicken embryogenesis. *Proc. Natl. Acad. Sci. USA* **79**:6737–6741.

Timpl, R. (1982) Antibodies to collagens and procollagens. *Adv. Enzymol. Rel. Areas Mol. Biol.* **82**:472–498.

Timpl, R., and G. R. Martin (1982) Components of basement membranes. In *Immunochemistry of the Extracellular Matrix*, Vol. II, H. Furthmayr, ed., pp. 119–150, CRC Press, Boca Raton.

Timpl, R., H. Rohde, P. G. Robey, S. I. Rennard, J. M. Foidart, and G. R. Martin (1979) Laminin—a glycoprotein from basement membranes. *J. Biol. Chem.* **254**:9933–9937.

Timpl, R., M. Dziadek, S. Fujiwara, H. Nowak, and G. Wick (1983) Nidogen: A new self-aggregating basement membrane protein. *Eur. J. Biochem.* **137**:455–465.

Todaro, G., and J. E. DeLarco (1978) Growth factors produced by sarcoma virus-transformed cells. *Cancer Res.* **38**:4147–4151.

Toivonen, S. (1979) Transmission problem in primary induction. *Differentiation* **15**:177–181.

Toole, B. P. (1973) Hyaluronate and hyaluronidase in morphogenesis and differentiation. *Am. Zool.* **13**:1061–1065.

Trowbridge, I. S., and M. S. Omary (1981) Human cell surface glycoprotein related to cell proliferation is the receptor for transferrin. *Proc. Natl. Acad. Sci. USA* **78**:3039–3043.

Trowbridge, I. S., and F. Lopez (1982) Monoclonal antibody to transferrin receptor blocks transferrin binding and inhibits tumor cell growth *in vitro*. *Proc. Natl. Acad. Sci. USA* **79**:1175–1179.

Vaheri, A., and D. Mosher (1978) High molecular weight, cell surface glycoprotein (fibronectin) lost in malignant transformation. *Biochim. Biophys. Acta* **516**:1–25.

Vainio, T., J. Jainchill, K. Clement, and L. Saxén (1965) Studies on kidney tubulogenesis. VI. Survival and nucleic acid metabolism of differentiating mouse metanephrogenic mesenchyme *in vitro*. *J. Cell. Comp. Physiol.* **66**:311–317.

von der Mark, K., J. Mollenhauer, U. Kühl, J. Bee, and H. Lesot (1984) Anchorins: A new class of membrane proteins involved in cell–matrix interactions. In *Role of the Extracellular Matrix in Differentiation*, R. L. Trelstad, ed., pp. 67–88, Alan R. Liss, New York.

Wagner, R. C. (1980) Endothelial cell embryology and growth. *Adv. Microcirc.* **9**:45–75.

Wartiovaara, J. (1966) Cell contacts in relation to cytodifferentiation in metanephrogenic mesenchyme *in vitro*. *Ann. Med. Exp. Fenn.* **44**:469–503.

Wartiovaara, J., S. Nordling, E. Lehtonen, and L. Saxén (1974) Transfilter induction of kidney tubules: Correlation with cytoplasmic penetration into nucleopore filters. *J. Embryol. Exp. Morphol.* **31**:667–682.

Wessells, N. K. (1977) *Tissue Interactions and Development*, W. A. Benjamin, Menlo Park, Calif.

Wirl, G. (1977) Extractable collagenase and carcinogenesis of the mouse skin. *Connect. Tissue Res.* **7**:103–110.

Woolley, D. E., E. D. J. Harris, C. L. Mainardi, and C. E. Brinckerhoff (1978) Purification, characterization and inhibition of human skin collagenase. *Biochem. J.* **169**:265–276.

Yamada, K. M. (1983) Cell surface interactions with extracellular materials. *Annu. Rev. Biochem.* **52**:761–799.

Section 5

Specialized Junctions

Chapter 18

Gap Junctional Contact Between Cells

NORTON B. GILULA

ABSTRACT

The transmission of molecular information between cells (cell–cell communication) via gap junctional contacts is now documented extensively in most animal tissues. This chapter contains a synthesis of current information available on the structure, permeability, formation, and biochemical properties of gap junctions. In addition, the expression of gap junctional communication during development is discussed, along with some recent observations on the relationship of junctional communication to regulation of development in an amphibian embryo.

Communication between cells via specialized low-resistance pathways has been widely documented in most animal tissues during the last 10–15 years. This form of cell–cell communication is fundamentally responsible for the transmission of information, in the form of small organic molecules or inorganic ions, that can regulate and/or synchronize the activities of cells in multicellular systems (for reviews, see Hertzberg et al., 1981; Loewenstein, 1981; Spray et al., 1982).

In this chapter, an attempt is made to provide some background information that is important for understanding and appreciating the value of the recent progress on studies of communication in an embryonic system.

GAP JUNCTIONAL COMMUNICATION

This fundamental cellular property was initially described in the invertebrate nervous system as an electrical or electrotonic synapse (Furshpan and Potter, 1959; Bennett, 1977). Similar synapses have subsequently been identified in most metazoan tissues, most prominently in the myocardium (Weidmann, 1952; Dreifuss et al., 1966) where such communication provides the basis for synchronizing activity of cells by permitting the rapid propagation of action potentials between cells.

Independently, communication (transfer) of low molecular weight metabolites was detected between cultured cells in contact. This phenomenon was described as "metabolic cooperation between cells" (Subak-Sharpe et al., 1969).

In the early seventies, a study was published demonstrating that a specific specialization of the cell surface membrane, a gap junction, was present between cells that communicate by the transfer of small metabolites and by the transmission of electrical signals (current pulses) (Gilula et al., 1972). From this study, using both communication-competent and communication-defective cells, it was possible to conclude that the gap junction provides a structural pathway for both metabolic and ionic (electrical) communication. However, it was not possible to directly demonstrate that molecules move through the junction, nor that similar or different channels are involved in the transfer of different molecules (inorganic ions vs. metabolites).

In the last 10 years, a sizeable volume of literature documenting gap junctions, electrical coupling, and metabolic coupling between cells *in vivo* and in culture has developed (for reviews, see Hertzberg et al., 1981; Loewenstein, 1981; Spray et al., 1982; Meda et al., 1984). These observations have contributed to a widely accepted belief that junctional communication must have some functional relevance in nonexcitable and embryonic tissues since it is so widespread.

GAP JUNCTION STRUCTURE

Gap junctions have been studied extensively through the use of a number of structural techniques (for a review, see Peracchia, 1980). Because of the junctions' small size, most of the structural analyses have been carried out with the electron microscope. In thin sections, the junction is characterized by a close apposition of the cell surface membranes of adjacent cells, where the membranes are separated by a small space or "gap" of 2–4 nm (Figure 1; Revel and Karnovsky, 1967). The junctions exist as plaquelike elements that vary in size and distribution throughout tissues. The internal structure of the junction membrane contains a polygonal lattice of 8.5-nm intramembrane particles that can be visualized by freeze-fracture analysis and negative staining (of isolated junctions) (Figures 2,3). Subcellular fractions of isolated mammalian liver gap junctions have been structurally characterized by low-dose electron diffraction and X-ray diffraction (Makowski et al., 1977; Unwin and Zampighi, 1980; Baker et al., 1983). These analyses have indicated that the junction membrane consists of a bilayer lipid structure that is interrupted by a polygonal arrangement of protein. The protein is organized in oligomers to form the particles, termed connexons, that contain the

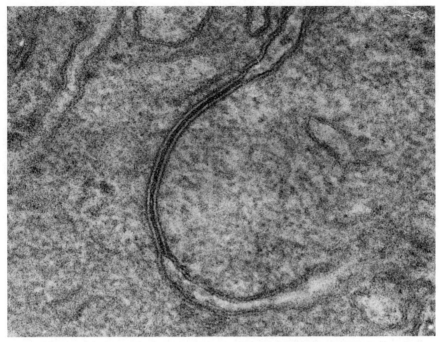

Figure 1. *Appearance of a gap junction between adjacent rat liver hepatocytes in a thin-section electron micrograph.* At the site of junctional contact, the surface membranes are closely apposed with a small intervening space or "gap." × 174,000.

polar channel for cell–cell transmission of molecules. Evidence for this channel was initially obtained from negative-stain penetration of the central region (about 1.2–1.5 nm) of the 8.5-nm connexon (Figure 3). Each cell contributes one-half to this bipartite structure (Figure 4). In a recent analysis, structural evidence has been obtained for a calcium-dependent regulation of the channel structure (Unwin and Ennis, 1983, 1984).

JUNCTIONAL PERMEABILITY

In general, junctional channels are permeable to most small molecules in the cytoplasm of cells. The channels appear to discriminate molecules based on size (molecular weight) rather than any other distinguishing properties. Consequently, the channels function much like a molecular sieve: Molecules less than about 1000 Daltons can pass, whereas molecules larger than 1000 Daltons cannot pass (Simpson et al., 1977). In biological terms, the size restriction on junctional permeability permits cells to retain their individuality of differentiation or expression. For example, cells can retain their individuality by retaining the large molecules, such as enzymes

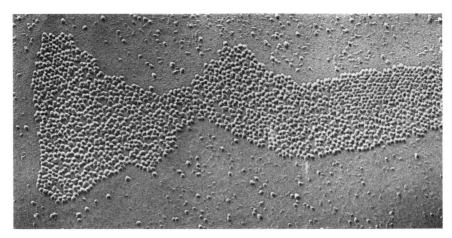

Figure 2. *Internal membrane structure at the site of gap junctional contact between adjacent rat ovarian granulosa cells.* The gap junctional membrane contains a polygonal arrangement of intramembrane particles of similar size that is distinct from the adjacent region of nonjunctional plasma membrane. × 102,500.

and nucleic acids, that are responsible for differentiation, whereas, all cells in a multicellular system can be "regulated" and/or synchronized by the cell–cell transmission of small molecules, that is, cyclic nucleotides, Ca^{2+}, and Na^+.

The regulation of junctional permeability has been analyzed in a number of systems, both embryonic and adult. Thus far, previous studies have focused on the following mechanisms for regulation: (1) intracellular pH (Turin and Warner, 1977, 1980; Spray et al., 1981); (2) intracellular Ca^{2+} concentration (Rose and Loewenstein, 1975; Rose and Rick, 1978); and (3) cell metabolism. In general, most cellular activities that influence cellular homeostasis have effects on junctional permeability that are nonspecific and most likely indirect. The three mechanisms listed above can certainly be included in such a category. It is of interest to note that lowering intracellular pH has a highly reversible effect on junctional permeability (Turin and Warner, 1980); other treatments frequently disrupt junctional permeability irreversibly. Finally, in certain vertebrate embryonic systems, the junctional channels have a detectable voltage dependence (Spray et al., 1979). This voltage dependence has not been observed in most other cellular systems that have been examined (in cell cultures or in intact tissues).

FORMATION OF JUNCTIONS

Gap junctional channels can form quickly between cells *in vivo* and in culture (for a review, see Sheridan, 1978). Formation occurs so quickly (within minutes, and perhaps within seconds) that in cell culture systems, channels appear to form without any significant dependence on protein synthesis (Epstein et al., 1977). Consequently, protein for junctional channels must be available in some form that can be rapidly reorganized, modified, or recruited for channel formation. The half-life of the major junction protein in mouse liver has been reported to be extremely short, about five hours (Fallon and Goodenough, 1981). Since the gap junction is a bipartite structure, formation should result from each cell contributing material to form half of the complete channel. This possibility has been strengthened by experimental observations demonstrating that heterologous cell types can readily form junctional channels (Michalke and Loewenstein, 1971; Epstein and Gilula, 1977), presumably because they are capable of producing homologous junction channel-forming molecules.

Figure 3. *Negative-stain image of a gap junction isolated from rat liver.* The 8.5-nm particles (connexons) contain a central electron-dense region that is the proposed location of the cell–cell junctional channel. × 326,700.

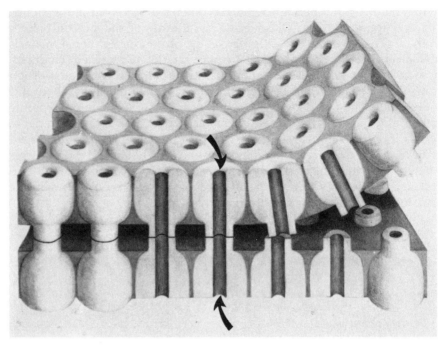

Figure 4. *Three-dimensional representation of the gap junction as visualized by negative staining.*
The junction is a bipartite structure that can be separated (by treatment with hypertonic
sucrose) as illustrated in the *right-hand* portion of the drawing. Each cell membrane
contributes one-half of the complete channel-bearing unit, both halves forming a channel
(*arrows*) that provides cytoplasmic continuity between two cells.

The structural events associated with junction formation have been
analyzed in several systems by using freeze-fracture analysis. A common
sequence of structural events can be detected in many tissues (Johnson et
al., 1974). These include: (1) the appearance of a flattened formation plaque
containing large 10-nm intramembrane "precursor" particles; (2) the subsequent appearance of small aggregates of 8.5-nm "mature" junctional
particles; and (3) finally, an enlargement of the "mature" particle aggregates
with a disappearance of the "precursor" particles. Thus far, no chemical
information has been obtained that can be applied constructively to understanding this set of structural changes associated with junction formation.

The most extensive molecular characterization of junction protein biosynthesis has been carried out on a unique protein present in the vertebrate
eye lens. Studies on the cell-free translation of this protein (Ramaekers et
al., 1980; Paul and Goodenough, 1983) and its subsequent processing
(Anderson et al., 1983; Paul and Goodenough, 1983) indicate that no
detectable modification of the protein is required for its integration into a
membrane system. In essence, the primary translation product is identical
in size and antigenicity as the protein that exists in the membrane *in vivo*.

ISOLATION AND CHARACTERIZATION OF JUNCTION PROTEINS

Three types of junctions have been isolated via subcellular fractionation procedures in relatively enriched fractions. Gap junction fractions have been obtained from mammalian liver (Henderson et al., 1979; Hertzberg and Gilula, 1979), mammalian heart (Kensler and Goodenough, 1980; Manjunath et al., 1982; Gros et al., 1983), lens fiber junctions from bovine eye lenses (Benedetti et al., 1976; Goodenough, 1979; Hertzberg et al., 1982), and gap junctions from crustacean (arthropod) tissues (Berdan and Gilula, 1982; Finbow et al., 1984).

In general, gap junctions isolated from vertebrate tissue contain homologous protein. This fact is based, in part, on chemical, immunological, and biological data. The major protein identified in vertebrate liver junction fractions has an apparent molecular weight of 27 kD. Recent studies using antibodies specific for this protein have reported homologous antigens in a variety of tissues (Dermietzel et al., 1984; Hertzberg and Skibbens, 1984). Some exceptions have been reported (Dermietzel et al., 1984); however, these negative findings may result from technical or reagent difficulties. Finally, cells from different vertebrate tissues and organisms can readily form junction channels with each other (Epstein and Gilula, 1977), strongly indicating a molecular homology. The existing observations provide a reasonable basis for concluding that the gap junction protein in most vertebrate tissues is quite similar.

At present, the best characterized junction protein exists exclusively in the vertebrate eye lens (Bloemendal, 1977). This protein is the major protein in the cell surface membrane of lens fibers, and it is located in both junctional and nonjunctional regions of the plasma membrane. Comparative chemical and immunological analyses have provided strong evidence that the lens junction protein is distinctly different from the gap junction protein present in other tissues (Hertzberg et al., 1982; Nicholson et al., 1983). The tissue-specific uniqueness of the lens junction protein has been further strengthened by recent sequence information obtained from lens junction cDNA (Gorin et al., 1984).

Arthropod gap junctions have been regarded as unique since they have different ultrastructural and biological properties (Epstein and Gilula, 1977; Peracchia, 1980). The recent isolation of enriched arthropod gap junction fractions has provided stronger evidence for this uniqueness. Thus far, the major protein in these junctions is smaller in size (about 16–17 kD) (Berdan and Gilula, 1982; Finbow et al., 1984), and there is no detectable antigenic homology between the arthropod and the vertebrate gap junction proteins (R. Berdan and N. B. Gilula, unpublished observations). Clearly, the arthropod gap junctions must represent a unique evolutionary divergence in gene structure, since serological evidence for homologous gap junction protein has been recently obtained for organisms in a number of invertebrate (Mollusca, Annelida, Nematoda, Coelenterata) and vertebrate phyla (C. Green and N. B. Gilula, unpublished observations).

GAP JUNCTIONAL COMMUNICATION IN DEVELOPMENT

Gap junctional communication has been well documented in a number of developing vertebrate organisms from oogenesis through organogenesis (for reviews, see Gilula, 1980; Spray et al., 1982; Warner, 1983; Schultz, 1985). In most cases, junctional communication has been detected as heterologous cell interactions by either electrical coupling, dye coupling, or the presence of gap junction structures.

In addition to gap junctional communication, several other types of cell–cell interactions frequently coexist during development. Two of the most common cell contacts are the tight junction (*zonula occludens*) and the desmosome (*macula adherens*) (for a review, see Gilula, 1978). The tight junction normally joins epithelial cells to provide a transepithelial permeability barrier for regulating the flow of substances (via the extracellular space) across the epithelium. This structure is characterized by a true fusion of the plasma membrane from adjacent cells. In freeze-fracture replicas, the sites of fusion contain an internal membrane specialization, a strand or ridge, that joins other strands to form a "barrier" or network. The desmosome is a broad category of special cell contacts that contain modified extracellular matrix and cytoplasmic surfaces. These structures are involved in stabilizing cytoskeletal elements (by serving as insertion sites) and cell–cell interactions via adhesion. In the last few years, substantial progress has been made in characterizing some of the molecular components associated with the desmosome (Franke et al., 1981). In general, all of these cell contacts (gap junctions, tight junctions, and desmosomes) contribute to the adhesive stabilization that must exist between cells, in addition to their other functional activities.

During vertebrate oogenesis, the oocyte interacts with a heterologous population of follicular cells, and this interaction has a defined regulatory influence on the maturational events of the oocyte that ultimately prepare the follicle for ovulation and the oocyte for fertilization (for a review, see Schultz, 1985). In the rat ovarian preovulatory follicle, a specialized population of follicular cells, the cumulus granulosa cells, communicate via gap junctions with the oocyte (Gilula et al., 1978). This communication occurs, in spite of the protective zona pellucida, through granulosa cell processes that penetrate the zona to form boutonlike endings that establish gap junctional contact with the oocyte plasma membrane; it persists until the final stage of oocyte maturation. At that stage, communication is disrupted and ovulation occurs (Gilula et al., 1978). The disruption of communication appears to be under hormonal control, and the signal for triggering this event may, in turn, be transmitted via junctional communication between the granulosa cells and the oocyte. Thus, gap junctional communication participates in the maturational stages of oogenesis, and is modulated (disrupted) at the end of this differentiation event to permit the oocyte to interact with a new cell, a sperm, thereby initiating the subsequent process of embryogenesis. Similar gap junction events have been described for oogenesis in frogs, chickens, sheep, and mice (for a review, see Schultz, 1985).

Following fertilization, the initial two cleavage stages usually occur in vertebrates without detectable gap junction expression. However, during the third cleavage, the eight-cell stage, gap junctional communication can be detected as an exclusive cell–cell pathway (Lo and Gilula, 1979a; Goodall and Johnson, 1984). Analysis of communication in early embryos is frequently complicated because of the persistence of cytoplasmic bridges resulting from incomplete cytokinesis. In the mouse embryo, both electrical coupling and dye transfer via junctions occurs from the late eight-cell stage until the postimplantation blastocyst stage. At that time, electrical coupling exists between all cells, but dye coupling is restricted to "compartments" (Lo and Gilula, 1979b). For example, dye will spread between cells of the inner cell mass region, but it will not spread from inner cell mass cells to trophectoderm cells. This qualitative change in the dye-coupling property of gap junctions has been observed in several other developing systems (Warner and Lawrence, 1982; Weir and Lo, 1982; Blennerhassett and Caveney, 1984). This property may have a profound effect on establishing gradients of information that might contribute to patterning in development. Thus far, there is no specific information available to explain the molecular or regulatory basis for "switching" the junction channels from a dye-permeable to a dye-impermeable state at a specific compartmental boundary.

Throughout embryogenesis, gap junctions occur with variable frequencies, depending on several factors. Junctions are usually small and difficult to detect between migrating embryonic cells. Also, as cells become committed to differentiate, the junctional population closely mimics the character that exists in the differentiated tissue. The initial widespread distribution of gap junctions in the early embryo is increasingly restricted with time, to reflect the regional specialization of cells participating in organogenesis. Physical boundaries, that is, connective tissue matrices, and spatial factors provide the most obvious restrictions for junctional communication in the developing embryo. The expression of gap junctional communication is clearly developmentally regulated, and evidence for this has come from studies in several systems. For example, during vertebrate myogenesis, mononucleated muscle precursor cells communicate via gap junctions prior to the formation of differentiated, multinucleated muscle fiber (Kalderon et al., 1977). However, muscle fibers (myotubes) do not form gap junctions with other cells, except perhaps transiently with neurons during the early stages of neuromuscular junction formation (Fischbach, 1972). The differentiation of the embryonic avian otocyst is also accompanied by a dramatic modulation of gap junctional communication (Ginzberg and Gilula, 1979, 1980). The undifferentiated epithelium in this organ is triggered to differentiate into a sensory epithelium containing hair cells and support cells by interactions with a population of invading neurites. Initially, all cells in the epithelium are joined by extensive gap junctions, and presumptive sensory cells are soon uncoupled from the nonsensory cells by an internalization of the junctions that exist between the two cell types. The internalization of the junctions effectively creates a population of sensory epithelial cells that are electrically insulated from the population of electrically coupled support cells. This "determination" of the epithelium is accompanied by the appearance of small transient gap junctions between the neurites and the basal region of some of the epithelial cells.

Presumably, developmentally regulated transient gap junctional events, such as the ones identified in the embryonic otocyst described above, will be the most important form of junctional interactions that occurs during the "inductive" stages of development. Such transient gap junctional interactions can be responsible for transmitting important molecular information that, together with inputs from other components and genetic factors, ultimately regulate the course of embryonic development.

GAP JUNCTIONAL COMMUNICATION IN THE AMPHIBIAN EMBRYO

In a recent study with Sarah Guthrie and Anne Warner of University College London, an effort was made to utilize gap junction-specific antibodies to more precisely determine the relationship of junctional communication to early development (Warner et al., 1984).

The experimental rationale for this study was based on the fact that affinity-purified rabbit antibodies prepared against the 27-kD gap junction protein from rat liver interact with homologous antigens in homogenates of *Xenopus* oocytes and early cleavage-stage embryos. Thus, rat liver gap junction antibodies could be used to examine the possible influence of anti-junction antibodies on junctional communication in the amphibian embryo. The experimental strategy for this determination resulted directly from the initial observations made by Sarah Guthrie on patterns of junctional communication (dye spread) in the early *Xenopus* embryo (Guthrie, 1984). From her analysis, it was possible to select a specific cell in the eight-cell stage embryo, located in the grey crescent region, that has defined dye transfer capabilities two cleavages later (32-cell stage). Thus, reagents for testing could be microinjected into the selected cell at the eight-cell stage, and the progeny of this cell (four cells) could be examined for junctional communication (dye transfer and electrical coupling) at the 32-cell stage.

The following observations were made using the strategy outlined above. Microinjection of immunoglobulins from preimmune serum, several buffers, and an anti-β-fibrinogen (fibrinogen is present in *Xenopus* at this early stage) into the selected cell in the eight-cell stage had no significant effect on the junctional communication properties of the progeny cells at the 32-cell stage. Conversely, microinjection of two different specific antibodies for the 27-kD junction protein had significant effects on junctional communication in the 32-cell stage progeny. The incidence of both dye transfer and electrical coupling were significantly reduced by treatment with the immune reagents. In the treated organisms, there was no appreciable alteration of cleavage and growth in the subsequent embryonic stages. Consequently, it was possible to make the following conclusions: (1) Antibodies specific for the 27-kD junction protein can selectively disrupt junctional communication upon microinjection by interacting with the antigen *in vivo*. The antigenic determinant(s) is localized on the cytoplasmic surface of the junction (data from immunolocalization studies), and it is likely to be closely associated with the junctional channel since monovalent fragments of the immunoglobulin will also block junctional channel permeability. (2) The 27-kD protein that is isolated from rat liver gap junctions is directly involved in the channel-conducting property that has been ascribed to these junctions. Therefore, the 27-kD protein now has *both* a structural and functional relationship to the gap junction.

Selective disruption of gap junctional communication in early *Xenopus* embryos has profound consequences on subsequent development. Initially, it was determined (through lineage analysis) that the specific cell utilized in the eight-cell embryo contributes to ectodermal and mesodermal derivatives on the right-hand side of the embryo. Analysis of stage-36 tadpoles treated with nonimmune reagents revealed normal asymmetries of structures on both sides of the embryo. However, in stage-36 tadpoles that were treated with junction antibodies, striking asymmetries were present. In some of these organisms, the eye on the right-hand side failed to form, and this condition was usually accompanied by a distinct underdevelopment of the brain on the right-hand side. Although it is not possible at the present time to provide a precise mechanistic understanding for the observed developmental defects, the initial observations do provide a strong indication that junctional communication participates in regulating embryonic development. Clearly, in the absence of junctional communication, cells within the embryo can survive, but they will probably fail to receive or transmit the appropriate information that is essential for permitting the "triggering" or induction of proper structures. Current efforts are focused on determining if the developmental consequences result from a long-term (hours to days) interference of the injected antibody on the synthesis and assembly of new junctional protein.

Studies in the next few years utilizing specific probes to selectively disrupt gap junctional communication should prove valuable in clarifying the precise biological role of this cell–cell information transfer mechanism in embryonic development. Inductive events in early embryogenesis will undoubtedly be a fertile area to investigate.

ACKNOWLEDGMENTS

The studies described from the author's laboratory have been supported by the Welch Foundation (Q-930) and by National Institutes of Health

Grants HL-28446 and GM-32230. The author is grateful for the expert assistance of Ms. Suzanne Saltalamacchia in preparing this manuscript.

REFERENCES

Anderson, D. J., K. E. Mostov, and G. Blobel (1983) Mechanisms of integration of de novo-synthesized polypeptides into membranes: Signal recognition particle is required for integration into microsomal membranes of calcium ATPase and of lens MP26 but not of cytochrome b5. Proc. Natl. Acad. Sci. USA 80:7249–7253.

Baker, T. S., D. L. D. Caspar, C. J. Hollingshead, and D. A. Goodenough (1983) Gap junction structures. IV. Asymmetric features revealed by low-irradiation microscopy. J. Cell Biol. 96:204–216.

Benedetti, E. L., I. Dunia, C. J. Bentzel, A. J. M. Vermorken, M. Kibbelaar, and H. Bloemendal (1976) A portrait of plasma membrane specializations in the eye lens epithelium and fibers. Biochim. Biophys. Acta 457:353–384.

Bennett, M. V. L. (1977) Electrical transmission: A functional analysis and comparison to chemical transmission. In Handbook of Physiology, Vol. I, The Nervous System, Sect. I, Cellular Biology of Neurons, E. R. Kandel, ed., pp. 357–416, American Physiological Society, Bethesda, Maryland.

Berdan, R. C., and N. B. Gilula (1982) Isolation and preliminary biochemical analysis of an invertebrate gap junction. J. Cell Biol. 95:94a.

Blennerhassett, M. G., and S. Caveney (1984) Separation of developmental compartments by a cell type with reduced junctional permeability. Nature 309:361–364.

Bloemendal, H. (1977) The vertebrate eye lens. Science 197:127–138.

Dermietzel, R., A. Liebstein, U. Frixen, U. Janssen-Timmen, O. Traub, and K. Willecke (1984) Gap junctions in several tissues share antigenic determinants with liver gap junctions. EMBO J. 3:2261–2270.

Dreifuss, J. J., L. Girardier, and W. G. Forssman (1966) Étude de la propagation de l'excitation dans le ventricule de rat du moyen des solutions hypertoniques. Pfluegers Arch. 292:13–33.

Epstein, M. L., and N. B. Gilula (1977) A study of communication specificity between cells in culture. J. Cell Biol. 75:769–787.

Epstein, M. L., J. D. Sheridan, and R. G. Johnson (1977) Formation of low resistance junctions in vitro in the absence of protein synthesis and ATP production. Exp. Cell Res. 104:25–30.

Fallon, R. F., and D. A. Goodenough (1981) Five-hour half-life of mouse liver gap junction protein. J. Cell Biol. 90:521–526.

Finbow, M. E., T. E. J. Buultjens, N. J. Lane, J. Shuttleworth, and J. D. Pitts (1984) Isolation and characterization of arthropod gap junctions. EMBO J. 3:2271–2278.

Fischbach, G. D. (1972) Synapse formation between dissociated nerve and muscle cells in low density cell cultures. Dev. Biol. 28:407–429.

Franke, W. W., E. Schmid, C. Grund, H. Mueller, I. Engelbrecht, R. Moll, J. Stadler, and E. D. Jarasch (1981) Antibodies to high molecular weight polypeptides of desmosomes: Specific localization of a class of junctional proteins in cells and tissues. Differentiation 20:217–241.

Furshpan, E. J., and D. D. Potter (1959) Transmission at giant motor synapses of the crayfish. *J. Physiol. (Lond.)* **143**:289–325.

Gilula, N. B. (1978) Structure of intercellular junctions. In *Intercellular Junctions and Synapses: Receptors and Recognition*, Vol. 2, J. Feldman, N. B. Gilula, and J. D. Pitts, eds., pp. 1–19, Chapman and Hall, London.

Gilula, N. B. (1980) Cell-to-cell communication and development. In *The Cell Surface: Mediator of Developmental Processes*, S. Subtelny and N. K. Wessels, eds., pp. 23–41, Academic, New York.

Gilula, N. B., O. R. Reeves, and A. Steinbach (1972) Metabolic coupling, ionic coupling, and cell contacts. *Nature* **235**:262–265.

Gilula, N. B., M. L. Epstein, and W. H. Beers (1978) Cell-to-cell communication and ovulation. A study of the cumulus–oocyte complex. *J. Cell Biol.* **78**:58–75.

Ginzberg, R. D., and N. B. Gilula (1979) Modulation of cell junctions during differentiation of the chicken otocyst sensory epithelium. *Dev. Biol.* **68**:110–129.

Ginzberg, R. D., and N. B. Gilula (1980) Synaptogenesis in the vestibular sensory epithelium of the chick embryo. *J. Neurocytol.* **9**:405–424.

Goodall, H., and M. H. Johnson (1984) The nature of intercellular coupling within the preimplantation mouse embryo. *J. Embryol. Exp. Morphol.* **79**:53–76.

Goodenough, D. A. (1979) Lens gap junctions: A structural hypothesis for nonregulated low-resistance intercellular pathways. *Invest. Ophthalmol. Visual Sci.* **18**:1104–1122.

Gorin, M. B., S. B. Yancey, J. Cline, J.-P. Revel, and J. Horwitz (1984) The major intrinsic protein (MIP) of the bovine lens fiber membrane: Characterization and structure based on cDNA cloning. *Cell* **39**:49–59.

Gros, D. B., B. J. Nicholson, and J.-P. Revel (1983) Comparative analysis of the gap junction protein from rat heart and liver: Is there a tissue specificity of gap junctions? *Cell* **35**:539–549.

Guthrie, S. C. (1984) Patterns of junctional communication in the early amphibian embryo. *Nature* **311**:149–151.

Henderson, D., H. Eibl, and K. Weber (1979) Structure and biochemistry of mouse hepatic gap junctions. *J. Mol. Biol.* **132**:193–218.

Hertzberg, E. L., and N. B. Gilula (1979) Isolation and characterization of gap junctions from rat liver. *J. Biol. Chem.* **254**:2138–2147.

Hertzberg, E. L., and R. V. Skibbens (1984) A protein homologous to the 27,000 dalton liver gap junction protein is present in a wide variety of species and tissues. *Cell* **39**:61–69.

Hertzberg, E. L., T.-S. Lawrence, and N. B. Gilula (1981) Gap junctional communication. *Annu. Rev. Physiol.* **43**:479–491.

Hertzberg, E. L., D. J. Anderson, M. Friedlander, and N. B. Gilula (1982) Comparative analysis of major polypeptides from liver gap junctions and lens fiber junctions. *J. Cell Biol.* **92**:53–59.

Johnson, R. G., M. Hammer, J. Sheridan, and J.-P. Revel (1974) Gap junction formation between reaggregated Novikoff hepatoma cells. *Proc. Natl. Acad. Sci. USA* **71**:4536–4540.

Kalderon, N., M. L. Epstein, and N. B. Gilula (1977) Cell-to-cell communication and myogenesis. *J. Cell Biol.* **75**:788–806.

Kensler, R. W., and D. A. Goodenough (1980) Isolation of mouse myocardial gap junctions. *J. Cell Biol.* **86**:755–764.

Lo, C. W., and N. B. Gilula (1979a) Gap junctional communication in the preimplantation embryo. *Cell* **18**:399–409.

Lo, C. W., and N. B. Gilula (1979b) Gap junctional communication in the postimplantation mouse embryo. *Cell* **18**:411–422.

Loewenstein, W. R. (1981) Junctional intercellular communication: The cell-to-cell membrane channel. *Physiol. Rev.* **61**:829–913.

Makowski, L., D. L. D. Caspar, W. C. Phillips, and D. A. Goodenough (1977) Gap junction structures. II. Analysis of the X-ray diffraction data. *J. Cell Biol.* **74**:629–645.

Manjunath, C., G. E. Goings, and E. Page (1982) Isolation and protein composition of gap junctions from rabbit hearts. *Biochem. J.* **205**:189–194.

Meda, P., A. Perrelet, and L. Orci (1984) Gap junctions and cell-to-cell coupling in endocrine glands. *Mod. Cell Biol.* **3**:131–196.

Michalke, W., and W. R. Loewenstein (1971) Communication between cells of different types. *Nature* **232**:121–122.

Nicholson, B. J., L. J. Takemoto, M. W. Hunkapiller, L. E. Hood, and J.-P. Revel (1983) Differences between liver gap junction protein and lens MIP 26 from rat: Implications for tissue specificity of gap junctions. *Cell* **32**:967–978.

Paul, D. L., and D. A. Goodenough (1983) *In vitro* synthesis and membrane insertion of bovine MP 26, an integral protein from lens fiber plasma membrane. *J. Cell Biol.* **96**:636–638.

Peracchia, C. (1980) Structural correlates of gap junction permeation. *Int. Rev. Cytol.* **66**: 81–146.

Ramaekers, F. C. S., A. M. E. Selten-Versteegen, E. L. Benedetti, I. Dunia, and H. Bloemendal (1980) *In vitro* synthesis of the major lens membrane protein. *Proc. Natl. Acad. Sci. USA* **77**:725–729.

Revel, J.-P., and M. J. Karnovsky (1967) Hexagonal array of subunits in intercellular junctions of the mouse heart and liver. *J. Cell Biol.* **33**:C7–C12.

Rose, B., and W. R. Loewenstein (1975) Permeability of cell junction depends on local cytoplasmic calcium activity. *Nature* **254**:250–252.

Rose, B., and R. Rick (1978) Intracellular pH, intercellular free Ca and junctional cell–cell coupling. *J. Membr. Biol.* **44**:377–415.

Schultz, R. M. (1985) Roles of cell-to-cell communication in development. *Biol. Reprod.* **32**:27–42.

Sheridan, J. D. (1978) Junctional formation and experimental modification. In *Intercellular Junctions and Synapses: Receptors and Recognition*, Vol. 2, J. Feldman, N. B. Gilula, and J. D. Pitts, eds., pp. 37–60, Chapman and Hall, London.

Simpson, I., B. Rose, and W. R. Loewenstein (1977) Size limit of molecules permeating the junctional membrane channels. *Science* **195**:294–296.

Spray, D. C., A. L. Harris, and M. V. L. Bennett (1979) Voltage dependence of junctional conductance in early amphibian embryos. *Science* **204**:432–434.

Spray, D. C., A. L. Harris, and M. V. L. Bennett (1981) Gap junctional conductance is a simple and sensitive function of intracellular pH. *Science* **211**:712–715.

Spray, D. C., A. L. Harris, and M. V. L. Bennett (1982) Control of intercellular communication via gap junctions. In *Cellular Communication During Ocular Development*, J. B. Sheffield and S. R. Hilfer, eds., pp. 57–84, Springer-Verlag, New York.

Subak-Sharpe, J. H., R. R. Bürle, and J. D. Pitts (1969) Metabolic cooperation between biochemically marked mammalian cells in tissue culture. *J. Cell Sci.* **4**:353–367.

Turin, L., and A. E. Warner (1977) Carbon dioxide reversibly abolishes ionic communication between cells of early amphibian embryo. *Nature* **210**:56–57.

Turin, L., and A. E. Warner (1980) Intracellular pH in early *Xenopus* embryos: Its effect on current flow between blastomeres. *J. Physiol. (Lond.)* **300**:489–504.

Unwin, P. N. T., and P. D. Ennis (1983) Calcium-mediated changes in gap junction structure: Evidence from the low angle X-ray pattern. *J. Cell Biol.* **97**:1459–1466.

Unwin, P. N. T., and P. D. Ennis (1984) Two configurations of a channel-forming membrane protein. *Nature* **307**:609–613.

Unwin, P. N. T., and G. Zampighi (1980) Structure of the junction between communicating cells. *Nature* **283**:545–549.

Warner, A. E. (1983) Physiological approaches to development. *Recent Adv. Physiol.* **10**:87–123.

Warner, A. E., and P. A. Lawrence (1982) Permeability of gap junctions at the segmental border in insect epidermis. *Cell* **28**:243–252.

Warner, A. E., S. C. Guthrie, and N. B. Gilula (1984) Antibodies to gap junctional protein selectively disrupt junctional communication in the early amphibian embryo. *Nature* **311**:127–131.

Weidmann, S. (1952) The electrical constants of Purkinje fibers. *J. Physiol. (Lond.)* **118**:348–360.

Weir, M. P., and C. W. Lo (1982) Gap junctional communication compartments in the *Drosophila* wing disk. *Proc. Natl. Acad. Sci. USA* **79**:3232–3235.

Chapter 19

Biology of Gap Junction Molecules and Development

JEAN-PAUL REVEL
S. BARBARA YANCEY
BRUCE J. NICHOLSON

ABSTRACT

We describe the morphology and chemistry of gap junctions with special reference to developing systems. Recent results on the sequencing of junction protein are presented, including a discussion of a partial sequence of heart junction protein and the complete sequence of lens MIP, as deduced from DNA. The possible roles that junctions might play in the embryo are reviewed.

One of the earliest papers on the topic of cell–cell communication described patterns of coupling and dye transfer in a developing system, the squid embryo (Potter, et al., 1966). Data accumulated since then suggest that gap junctions provide one of the many ways in which cells can interact with each other during embryogenesis, allowing for diffusion of informational or permissive molecules between cytoplasmic compartments of adjacent cells. The argument for the role gap junctions play in development is based on the following observations: (1) There are changes in gap junction complement that coincide temporally with important stages of development (see, for example, Fujisawa et al., 1976; Revel and Brown, 1976; Ginzberg and Gilula, 1979; Lo and Gilula, 1979 a,b); (2) molecules of a size which might diffuse through gap junctions (such as cAMP and other as yet undetermined small molecules) are believed to play roles in controlling differentiation (Crick, 1970; Schaller, 1973; Lawrence et al., 1978; Meinhardt, 1983); and (3) direct evidence is emerging that interfering with junction permeability causes abnormalities in embryos (Warner et al., 1984).

Gap junctions are nearly ubiquitous in multicellular organisms. They are found in animals from coelenterates (Hand and Gobel, 1972; Wood

411

and Kuda, 1980) and fish (Robertson, 1963) to mammals (Revel and Karnovsky, 1967) and in some of the earliest stages of embryogenesis (Ducibella et al., 1975; Lo and Gilula, 1979 a,b) through adult life (Kelley and Perdue, 1980). One important step in assessing the role of gap junctions in adults and embryos is to gain a clearer understanding of the chemical nature of their constituents and of their molecular organization at various times. At present, much of this knowledge is restricted to junctions in adult organisms. Hopefully, future work will show that the ideas and concepts derived from analysis of these junctions will also be relevant to junctions between embryonic and differentiating cells.

STRUCTURE OF GAP JUNCTIONS

Gap junctions, also referred to as nexus (Dewey and Barr, 1962) or maculae communicantes (Simionescu et al., 1975), are clusters of matching channels that penetrate the membranes of two adjacent cells. Each channel ["connexon" (Goodenough, 1975)] in one cell membrane docks in the intercellular space with an identical channel in the membrane of the apposing cell. The channels are often arranged in closely packed arrays. In some instances (Peracchia, 1980), they are quasi-crystalline. They can be detected by freeze cleaving (Kreutziger, 1968; Goodenough and Revel, 1970) and other morphological approaches (Revel and Karnovsky, 1967).

Morphology of Gap Junctions in Embryos

The appearance of gap junctions in embryos is usually similar to that in adult organisms (see, e.g., Revel et al., 1973). There are, however, a few exceptions to this rule, instances of tissues in which there are differences (Bellairs et al., 1975) that can sometimes be quite dramatic and may be accompanied by changes in permeability.

In the developing heart (Gros et al., 1978, 1979), and in heart cell cultures (Griepp et al., 1978; Williams and DeHaan 1981), gap junctions form linear strands 2–5 particles wide. These strands meander and form bizarre shapes, including loops. In the adult, they are replaced by typical junctional plaques (see Page, 1978; Shibata and Yamamoto, 1979) composed of closely packed particles. It is of interest to note that the "embryonic" appearance of the junction is maintained in the adult amphibian heart (Kensler et al., 1977).

Another example of dramatic morphological changes occurs in the developing lens. Here there is a developmental gradient (Schuetze and Goodenough, 1982) from anterior lens epithelium to lens fibers. This gradient duplicates the developmental temporal sequence. Gap junctions of ordinary appearance and permeability properties are replaced by structures in which the packing of particles is much looser. Concomitantly, the coupling between lens cells (now lens fibers) seems to become less sensitive to changes in pCO_2. The new structures appearing on the lens fiber membranes consist of a different junction protein (Vermorken et al., 1977; Hertzberg et al., 1982; Nicholson et al., 1983) and have different properties (Goodenough, 1979). It is actually not clear that the new structures are really gap junctional in nature (Zampighi et al., 1982). Paul and Goodenough (1983), using immunocytochemical approaches, failed to detect the main intrinsic protein (MIP) of the lens fiber (believed by many to be the junction protein of mature lens fibers) in gap junction-like structures. Other workers, using different antibodies, or other criteria, find support for the idea that MIP is part of lens fiber junctions (Bok et al., 1982; Kuszak et al., 1982; Keeling et al., 1983). There is recent evidence, based on sequencing of the lens protein (Gorin et al., 1984) and on introduction of the lens junction membrane into liposomes (Girsch and Peracchia, 1983), which does suggest that there is indeed a junctional structure between lens fibers, albeit apparently different from that found between anterior epithelial cells.

Molecular Anatomy of Gap Junctions

Models of the gap junction derived by X-ray analysis of isolated junctional membranes (Caspar et al., 1977; Makowski et al., 1977) and by low-dose electron microscopy (Unwin and Zampighi, 1980: Baker et al., 1983) indicate that each connexon in liver consists of six protein rodlets which delimit a central pore. Matching of connexons across the intercellular space allows the formation of a continuous aqueous channel through which ions and small molecules can diffuse between neighboring cells. Evidence for the size and nature of molecules that can be exchanged through the junctions suggests a general lack of selectivity except for one based on molecular size (Simpson et al., 1977; Flagg-Newton et al., 1979; Flagg-Newton, 1980). In some instances, there might be selectivity based on charge, as shown in the case of molecules close to the exclusion limit for penetration through the channel (Brink and Dewey, 1980). Junction morphology (Bernardini and Peracchia, 1981) and permeability can be altered by changes in Ca^{2+} concentration (Rose and Loewenstein, 1975; Rose and Loewenstein, 1976; Rose and Rick, 1978), pCO_2 (Turin and Warner, 1977), and pH (Turin and Warner, 1980; Spray et al., 1981), or by changes in membrane potential (Spray et al., 1979; Harris et al., 1983). A publication by Unwin and Ennis (1984) suggests a mechanism for closing the pore: In the presence of Ca^{2+}, the subunit constituents of the connexons tilt and this configurational change obstructs the channel. A somewhat different model had been proposed previously by Makowski et al. (1977). Changes in particle packing (Bernardini and Peracchia, 1981) do not always accompany uncoupling (Hanna et al., 1984).

THE CHANNEL PROTEIN

Gap junctions can be recognized morphologically and physiologically in a wide range of animals and developmental stages. However, evidence gathered to date suggests that the protein component may vary widely. Gap junction-associated proteins in different tissues seem to be less homologous than gap junction proteins derived from the same tissues in different animal species (Gros et al., 1983; Nicholson et al., 1983). The differences observed could reflect the germ layer of origin and/or different proteins could be expressed under different physiological conditions or at different developmental stages. (Vermorken et al., 1977; Schuetze and Goodenough, 1982). Even though they differ, junction protein molecules must often be functionally equivalent. They have been shown to participate *in vitro* in the formation of physiologically functional, heterologous junctions (Michalke and Lowenstein, 1971; Gaunt and Subak-Sharpe, 1979; Flagg-Newton and Loewenstein, 1980; but see also Pitts and Burk, 1976; Epstein and Gilula, 1977). Junctions between different cell types also occur *in vivo*. One is thus justified in expecting to find some similarity in configuration in spite of the differences in primary sequence (after all, they must behave like gap junctions in some respects, in order to be recognized!).

Size of the Junction Protein

A possible expression of this similarity in configuration is to be found in the rather uniform molecular weight of the known vertebrate junctional proteins. Even though a variety of molecular weights have been proposed on the basis of early work, there now seems to be a consensus for a molecular weight of about 25–30 kD, although a major dissenting view has recently been put forward by Finbow and collaborators (1983), and there are some data suggesting the possibility of a higher molecular weight (Henderson and Weber, 1982). There are several reports giving a molecular weight of 28 kD for liver protein (Henderson et al., 1979; Hertzberg and Gilula, 1979; Finbow et al., 1980; Nicholson et al., 1981). In the heart, multiple molecular weights had also been assigned in the pioneering work of Kensler and Goodenough (1980) and Manjunath et al. (1982); there now seems to be some evidence for a major entity of 28 kD or perhaps 30 kD (Gros et al., 1983). The case of the eye lens is particularly interesting: The apparent molecular weight of the lens protein MIP26 on SDS-polyacrylamide gels (Benedetti, 1974; Bloemendal et al., 1977; Takemoto and Hansen, 1981; Hertzberg et al., 1982; Nicholson et al., 1983) is about 26 kD, dropping to 21 kD in older portions of the lens. A molecular weight of 28.2 kD has now been calculated on the basis of the amino acid sequence obtained by analysis of the cDNA derived from lens MIP26 mRNA (Gorin et al., 1984).

Arrangement of Junction Protein in the Membrane

A number of investigators have used proteolytic enzymes to study the arrangement of the junctional polypeptide in isolated plaques, where presumably it is still in its "native" configuration. Even though a number of enzymes with different specificities (trypsin, chymotrypsin, V8 protease) were used, in each case a rather large portion of the junctional molecule was protected. However, treatment of SDS-extracted junctional proteins with the same enzyme produces much smaller peptides.

In the case of the lens protein MIP26, a polypeptide of 21–22 kD is obtained in all instances examined after trypsin treatment (Keeling et al., 1983; Nicholson et al., 1983). Sequencing by Edman degradation (Nicholson et al., 1983) suggests that five amino acids (with a molecular weight of 750 Daltons) are removed from the amino terminus of the molecule and the rest (approximately 4 kD) from the carboxyl terminus of the molecule (Gorin et al., 1984). If MIP26 is indeed part of a junction, where the intercellular space is too narrow (2 nm) to allow for penetration by trypsin (5 nm), one can argue that both ends of the molecule appear to be exposed at the cytoplasmic surface.

In the liver, fragments obtained by proteolysis of isolated junctions have a size of approximately 10 kD. In fact, the evidence indicates that there are two such peptides, so that fragments totaling 20 kD are spared proteolysis. Edman degradation shows no loss from the amino terminus, and one assumes that all of the peptide(s) removed are farther toward the carboxyl end of the molecule. The situation has not been investigated as thoroughly in the case of the heart, but here again, two molecules of approximately 10 kD can be isolated under similar conditions, suggesting that liver and heart may have a very similar organization, and except for the availability of appropriate sites to give peptides of 10 kD instead of one of 21 kD, could be organized in the same way as lens junction protein.

Assuming that the extent of proteolysis reflects not only the number of cleavable sites but also their disposition in the membrane allows rough guesses about the arrangement of the junction protein. It would appear that in all three cases studied the carboxyl end of the molecule is exposed to easy attack, perhaps because it is at the cytoplasmic side of the junctions. The data available in the case of the lens suggest that the amino terminal is on the same side as the carboxyl end. There are few other proteolytic sites exposed, except most likely cytoplasmic loops which, when split, leave intact peptides of about 10 kD in the case of liver and heart.

DIVERSITY OF JUNCTION PROTEINS

Although, as we have seen, there are similarities between them, the diversity of junctional protein becomes obvious when one examines two-

dimensional peptide maps of proteolytic fragments or compares the partial
primary sequence data that have been obtained for several of the proteins.

Peptide Maps

Purified gap junction polypeptides obtained by eluting the protein from
bands after separation by SDS-polyacrylamide gel electrophoresis are treated
with proteolytic enzymes, the resulting peptides are run in a first dimension
by electrophoresis and in a second by chromatography in organic solvents
that sort the peptides according to their hydrophobicity (Elder et al., 1977).
By this approach, very little if any homology is seen between gap junctions
of different organs. This is true whether the peptides are labeled at tyrosine
groups (chloramine T) or at amino groups (Bolton-Hunter reagent). Comparison of the maps obtained for heart, liver, and lens do not show more
peptides in common than either has with bacteriorhodopsin, another
transmembrane protein, which a priori is expected to be unrelated to the
gap junction protein. However, the peptide maps obtained cannot be used
to deny categorically the existence of common sequences in the different
membrane proteins. In contrast to intertissue differences, comparison of
peptide maps for the same organ in different species shows a great deal
of homology, suggesting that the differences may well have functional
significance. There is as yet no data on comparison of the protein extracted
from the same organ at different stages of development.

Amino Acid Sequencing

The availability of microsequencing techniques (Hunkapiller and Hood,
1980) and improvements in gap junction isolation techniques to yield
sufficient amounts of protein have allowed us to begin amino acid sequencing
of junction proteins from different tissues. So far it has been possible to
analyze the amino terminal portions of lens, heart, and liver proteins, as
well as a cyanogen bromide peptide from the lens. Comparison of the
sequence of mouse to that of rat liver shows that of the 20 or so residues
which form the overlapping sequence only three are changed. An even
closer homology is found between rat and bovine lens proteins. Quite to
the contrary, comparison of sequences obtained for rat liver and rat lens
show only two amino acids in common. Recently, it has been possible to
obtain sequence data for the amino terminal portion of rat heart junctions.
The results of this preliminary analysis show homology with rat liver
protein in about 40% of the residues, more if one includes functionally
conservative residues. Residues that are identical in both sequences are
distributed evenly along each sequence, leading one to believe that junction
proteins may not be made of large domains which are identical and others
which are completely different, but rather that divergences and similarities
may occur throughout the molecule. Of course, it is only by complete

sequencing that it will become possible to determine the correctness of this view, presently based on a very small sample.

In contradistinction to the "diversity" we report, the first observations made using antibodies indicate a great deal of similarity between junctional proteins. There is cross-reactivity of the liver antibody with proteins of different tissues over a wide range of different species (Hertzberg, 1980; Ziegler and Horwitz, 1981; Traub et al., 1982). There is even a report of a low but positive cross-reaction between liver and lens proteins (Traub et al., 1982). The comparison of the amino acid sequence of portions of the liver and of heart suggests sufficient homology to allow for the same immunological determinants to be present on different protein molecules, which could, at the same time, display enough diversity to account for the differences seen by other approaches.

GAP JUNCTIONS AND DEVELOPMENT

Gap junctions, and electrical as well as dye coupling, have been reported in even the earliest embryonic stages. They have been seen in squid embryos (Potter et al., 1966), in early stages of chick development (Trelstad et al., 1967; Sheridan, 1968; Revel et al., 1973), even before the eggs are laid (Bellairs et al., 1975), in early Xenopus (Slack and Palmer, 1969), in early amphibian embryos (Warner, 1973; Hanna et al., 1980), in imaginal disks, (Ryerse, 1982), between mammalian blastomeres (Goodall and Johnson, 1982), and at the compaction stage in mammalian embryos (Ducibella et al., 1975; Lo and Gilula, 1979a). The role of junctions during development (Bennett, 1973) may also be changing, as judged from changes in physiological properties reported by Lo and Gilula (1979b), Schuetze and Goodenough (1982) and others. Caution in interpreting such results is needed, however (Bennett et al., 1978). We have already seen that gap junctions are not unique structures found only in the embryo; one must also assume that the general role that they play in the embryo resembles those known from the study of adult organisms.

Precisely what those roles are is difficult to define at the present time. They have been implicated in the exchange of second messenger molecules (Lawrence et al., 1978) and may well exert controlling effects in this fashion. In turn, gap junctions may also be under cyclic AMP control (Azarnia et al., 1981). It has now been shown that L cells, long believed to be devoid of gap junctions and to be incapable of intercellular coupling, acquire both under conditions of up regulation of adenylate cyclase. Hormonal treatments also seem to affect the level of junction observed morphologically (Merk et al., 1972; Decker, 1976; Dahl and Berger, 1978; Garfield, 1980; Decker, 1981). Other agents and conditions also affect the junction complement. Witness the great increases in gap junctions that can be found after treatment with vitamin A and derivatives (Elias and Friend, 1976; Elias et al., 1981). The changes observed in gap junction protein and in the number of gap junctions, as well as in electrical coupling, after partial hepatectomy (Yee and Revel, 1978; Yancey et al., 1979; Meyer et al., 1981) could also be the results of variations in hormonal or messenger levels in the cells.

The idea that gap junctions may play a role in development has been very appealing. Especially interesting are reports which suggest that in fact there may well be changes in gap junction permeability at the boundary between developmental compartments (Warner and Lawrence, 1982; Weir and Lo, 1982). It has been proposed that gradients of morphogens (Crick, 1970) in cell populations linked by gap junctions such as those seen in *Hydra* (Fraser and Bode, 1981) or in chick limb bud could provide positional (Summerbell et al., 1973; Tickle et al., 1975) and other kinds of information (see for example Meinhardt, 1983). At the present time, critical tests of various hypotheses are difficult to make. Perhaps the availability of antibodies against junction proteins (Hertzberg, 1980; Traub et al., 1982; Warner et al., 1984) will provide for more direct experimental intervention. It is unlikely, however, that the answer is going to be very simple. The recent results of Harris et al. (1983) on the bistable modulation of electrical conductivity in gap junctions by changes in membrane potential are an indication of the possible complexity. Experiments on rectifying synapses, structures which look like gap junctions, also suggest complexity. Here electrical coupling can be eliminated while dye coupling is preserved. These results are reminiscent of the experiments of Flagg-Newton and Loewenstein (1980), who found that the extent of coupling in heterologous pairs depended on which member of the pair was impaled.

It is difficult at present to extract a coherent pattern that would explain differentiation, development, morphogenesis, etc. Understanding the control of synthesis or expression of gap junction proteins (Fallon and Goodenough, 1981; Yancey et al., 1981) and the modulation of the gap junction permeability could well allow experiments that would explain presently obscure aspects of development.

COMPLETE SEQUENCE OF THE LENS PROTEIN

Knowing the structure and arrangement of junction proteins in the membrane would be a big step toward understanding how they work. This goal is now within reach with the sequencing of the lens MIP26 (Gorin et al., 1984). Unfortunately, the junctional status of MIP26 is disputed (Zampighi et al., 1982). On the one hand, a number of investigators (Kuszak et al., 1982; Keeling et al., 1983; Fitzgerald et al., 1983) have pointed to the similarity between gap junctions and lens junctions and suggested that they are the same structures. Others have focused on different aspects and have concluded that MIP26 is not part of gap junctions (Zampighi et al., 1982; Paul and Goodenough, 1983). As indicated earlier in this chapter, we have been able to sequence cDNA corresponding to the gene for the bovine lens junction. Our analysis has not proceeded far enough to discuss the details of the amino acid sequence derived from the DNA, but a general picture of the possible secondary structure can already be guessed at from the primary sequence obtained. The most interesting result at present is that the polypeptide has a sufficient number of hydrophobic sequences to cross the bilayer six times. The model derived predicts trypsin-sensitive sites where they are expected to be on the basis of the molecular weights obtained by experimentation. Assuming that the putative transmembrane segments are in α-helical configuration, at least one of these has an amphiphilic character, with one hydrophilic and one hydrophobic face. This is precisely what is to be expected should MIP26 be a junction protein, since similar amphiphilic helices have already been detected in acetylcholine receptors (Finer-Moore and Stroud, 1984) and other pore-forming polypeptides. While such a finding does not prove that the lens protein is indeed a gap junction protein, it is one of the necessary steps. With the sequence of a potential junction protein at hand, it will become possible to carry out direct tests of the role of junctions in development and throughout the life of the organism.

CONCLUSIONS

The picture that emerges is one of embryos endowed with gap junctions morphologically and physiologically similar to those of the adult. Although the junctions seem homologous by these criteria we might expect them to differ in primary amino acid sequence since the other known junctional proteins are so diverse. These differences may, if in fact they do exist, support subtle but developmentally significant alterations in junction behavior. Other types of modulation of junction function may also play roles in determining the future of various cells in the embryo (Harris et al., 1983). The evidence available at present, suggesting the presence of passively permeable pores, may be giving us much too simple an image and is likely an "artifact" of our methods of testing.

ACKNOWLEDGMENTS

The authors thank Susan Mangrum for her expert and cheerful typing. The original work described was supported in part by the Institute of General Medical Sciences of the U.S. Public Health Service.

REFERENCES

Azarnia, R., G. Dahl, and W. R. Loewenstein (1981) Cell junctions and cyclic AMP. III. Promotion of junctional membrane permeability and junctional membrane particles in a junction deficient cell type. *J. Membr. Biol.* **63**:133–146.

Baker, T. S., D. L. D. Caspar, C. J. Hollingshead, and D. A. Goodenough (1983) Gap junction structures. IV. Asymmetric features revealed by low-irradiation microscopy. *J. Cell Biol.* **96**:204–216.

Bellairs, R., A. S. Breathnach, and M. Gross (1975) Freeze fracture replication of junctional complexes in unincubated and incubated chick embryos. *Cell Tiss. Res.* **162**:235–252.

Benedetti, E. L., I. Dunia, and H. Bloemendal (1974) Development of junctions during differentiation of lens fibers. *Proc. Natl. Acad. Sci. USA* **71**:5073–5077.

Bennett, M. V. L. (1973) Function of electrotonic junctions in embryonic and adult tissues. *Fed. Proc.* **32**:65–75.

Bennett, M. V. L., M. E. Spira, and D. C. Spray (1978) Permeability of gap junctions between embryonic cells of *Fundulus*: A reevaluation. *Dev. Biol.* **65**:114–125.

Bernardini, G., and C. Peracchia (1981) Gap junction crystallization in lens fibers after an increase in cell calcium. *Invest. Opthalmol. Vis. Sci.* **21**:291–299.

Bloemendal, H. (1977) The vertebrate eye lens. A useful system for the study of fundamental biological processes on a molecular level. *Science* **197**:127–138.

Bok, D., J. Dockstader, and J. Horwitz (1982) Immunocytochemical localization of the lens main intrinsic polypeptide (MIP26) in communicating junctions. *J. Cell Biol.* **92**:213–220.

Brink, P. R., and M. M. Dewey (1980) Evidence for fixed charge in the nexus. *Nature* **285**:101–102.

Caspar, D. L. D., D. A. Goodenough, L. Makowski, and W. C. Phillips (1977) Gap junction structures. I. Correlated electron microscopy and X-ray diffraction. *J. Cell Biol.* **74**:605–628.

Crick, F. (1970) Diffusion in embryogenesis. *Nature* **225**:420–422.

Dahl, G., and W. Berger (1978) Nexus formation in the myometrium during parturition and induced by estrogen. *Cell Biol. Int. Rep.* **2**:381–387.

Decker, R. S. (1976) Hormonal regulation of gap junction differentiation. *J. Cell Biol.* **69**:669–685.

Decker, R. S. (1981) Gap junctions and steroidogenesis in the fetal mammalian adrenal cortex. *Dev. Biol.* **82**:20–31.

Dewey, M. M., and L. Barr (1962) Intercellular connection between smooth muscle cells: The nexus. *Science* **137**:670–672.

Ducibella, J., D. Albertini, E. Anderson, and J. D. Biggers (1975) The preimplantation mammalian embryo: Characterization of intercellular junctions and their appearance during development. *Dev. Biol.* **47**:231–250.

Elder, J. H., R. A. Pickett, J. Hampton, and R. A. Lerner (1977) Radioiodination of proteins in single polyacrylamide gel slices. *J. Biol. Chem.* **252**:6510–6515.

Elias, P. M., and D. S. Friend (1976) Vitamin A-induced mucus metaplasia: An *in vitro* system for modulating tight and gap junction differentiation. *J. Cell Biol.* **68**:173–188.

Elias, P. M., S. Grayson, T. M. Caldwell, and N. S. McNutt (1980) Gap junction proliferation in retinoic acid-treated human basal cell carcinoma. *Lab. Invest.* **42**:469–474.

Elias, P. M., S. Grayson, E. G. Gross, G. L. Peck, and N. S. McNutt (1981) Influence of topical and systemic retinoids on basal cell carcinoma cell membranes. *Cancer* **48**:932–938.

Epstein, M. L., and N. B. Gilula (1977) A study of communication specificity between cells in culture. *J. Cell Biol.* **75**:769–787.

Fallon, R. F., and D. A. Goodenough (1981) Five hour half-life of mouse liver gap junction protein. *J. Cell Biol.* **90**:521–526.

Finbow, M. E., S. B. Yancey, R. Johnson, and J.-P. Revel (1980) Independent lines of evidence suggesting a major gap junctional protein with a molecular weight of 26,000. *Proc. Natl. Acad. Sci. USA* **77**:970–974.

Finbow, M. E., J. Shuttleworth, A. E. Hamilton, and J. D. Pitts (1983) Analysis of vertebrate gap junction protein. *EMBO J.* **2**:1479–1486.

Finer-Moore, J., and R. M. Stroud (1984) Amphipathic analysis and possible formation of the ion channel in an acetylcholine receptor. *Proc. Natl. Acad. Sci. USA* **81**:155–159.

Fitzgerald, P. G., D. Bok, and J. Horwitz (1983) Immunocytochemical localization of the main intrinsic polypeptide (MIP) in ultrathin frozen sections of rat lens. *J. Cell Biol.* **97**:1491–1499.

Flagg-Newton, J. L. (1980) The permeability of the cell-to-cell membrane channel and its regulation in mammalian cell junctions. *In Vitro* **16**:1043–1048.

Flagg-Newton, J. L., and W. R. Loewenstein (1980) Asymmetrically permeable membrane channels in cell junction. *Science* **207**:771–773.

Flagg-Newton, J. L., I. Simpson, and W. R. Loewenstein (1979) Permeability of the cell-cell membrane channels in mammalian cell junction. *Science* **205**:404–409.

Fraser, S. E., and H. R. Bode (1981) Epithelial cells of *Hydra* are dye coupled. *Nature* **294**:356–358.

Fujisawa, H., H. Morioka, K. Watanabe, and M. Nakamura (1976) A decay of gap junctions in association with cell differentiation of neural retina in chick embryo development. *J. Cell Sci.* **22**:585–596.

Garfield, R. E., D. Merrett, and A. K. Grover (1980) Gap junction formation and regulation in myometrium. *Am. J. Physiol.* **239**:C217–C228.

Gaunt, S. J., and J. H. Subak-Sharpe (1979) Selectivity in metabolic cooperation between cultured mammalian cells. *Exp. Cell Res.* **120**:307–320.

Ginzberg, R. D., and N. B. Gilula (1979) Modulation of cell junctions during differentiation of the chicken otocyst sensory epithelium. *Dev. Biol.* **68**:110–129.

Girsch, S. J., and C. Peracchia (1983) Lens junction protein (MIP26) self-assembles in liposomes forming large channels regulated by calmodulin. *J. Cell Biol.* **97**:83a.

Goodall, H., and M. H. Johnson (1982) Use of carboxyfluorescein diacetate to study the formation of permeable channels between mouse blastomeres. *Nature* **295**:524–526.

Goodenough, D. A. (1975) Methods for the isolation and structural characterization of hepatocyte gap junctions. In *Methods in Membrane Biology*, Vol. 3, *Plasma Membranes*, E. D. Korn, ed., pp. 51–80, Plenum, New York.

Goodenough, D. A. (1979) Lens gap junctions: A structural hypothesis for nonregulated low-resistance intercellular pathways. *Invest. Ophthalmol.* **18**:1104–1122.

Goodenough, D. A., and J.-P. Revel (1979) A fine structural analysis of intercellular junctions in the mouse liver. *J. Cell Biol.* **45**:272–290.

Gorin, M., S. B. Yancey, J. Cline, J.-P. Revel, and J. Horwitz (1984) The major intrinsic protein (MIP) of the bovine lens fiber membrane: Characterization and structure based upon cDNA cloning. *Cell* **38**:49–59.

Griepp, E. B., J. Peacock, M. Bernfield, and J.-P. Revel (1978) Morphological and functional correlates of synchronous beating between embryonic heart cell aggregates and layers. *Exp. Cell Res.* **113**:273–282.

Gros, D., J. P. Mocquard, C. E. Challice, and J. Schrevel (1978) Formation and growth of gap junctions in mouse myocardium during ontogenesis: A freeze-cleave study. *J. Cell Sci.* **30**:45–61.

Gros, D., J. P. Mocquard, C. E. Challice, and J. Schrevel (1979) Formation and growth of gap junctions in mouse myocardium during ontogenesis: Quantitative data and their implications on the development of intercellular communication. *J. Mol. Cell. Cardiol.* **11**: 543–554.

Gros, D. B., B. J. Nicholson, and J.-P. Revel (1983) Comparative analysis of the gap junction protein from rat heart and liver: Is there a tissue specificity of gap junctions? *Cell* **35**: 539–549.

Hand, A. R., and S. Gobel (1972) The structural organization of the septate and gap junctions of the *Hydra*. *J. Cell Biol.* **52**:397–408.

Hanna, R. B., P. G. Model, D. C. Spray, M. V. L. Bennett, and A. L. Harris (1980) Gap junctions in early amphibian embryos. *Am. J. Anat.* **158**:111–114.

Hanna, R. B., G. D. Pappas, and M. V. L. Bennett (1984) The fine structure of identified electrotonic synapses following increased coupling resistance. *Cell Tiss. Res.* **235**:243–249.

Harris, A. L., D. C. Spray, and M. V. L. Bennett (1983) Control of intercellular communication by voltage-dependent gap junction conductance. *J. Neurosci.* **3**:79–100.

Henderson, D., and K. Weber (1982) Immunological analysis of gap junction proteins from liver, lens and heart muscle. *Biol. Cell* **45**:229a.

Henderson, D., H. Eibl, and K. Weber (1979) Structure and biochemistry of mouse hepatic gap junctions. *J. Mol. Biol.* **132**:193–218.

Hertzberg, E. L. (1980) Biochemical and immunological approaches to the study of gap junctional communication. *In Vitro* **16**:1057–1067.

Hertzberg, E. L., and N. B. Gilula (1979) Isolation and characterization of gap junctions from rat liver. *J. Biol. Chem.* **254**:2138–2147.

Hertzberg, E. L., D. J. Anderson, M. Friedlander, and N. B. Gilula (1982) Comparative analysis of the major polypeptides from liver gap junctions and lens fiber junctions. *J.Cell Biol.* **92**:53–59.

Hunkapiller, M. W., and L. E. Hood (1980) New protein sequenator with increased sensitivity. *Science* **207**:523–525.

Keeling, P., K. Johnson, D. Sas, K. Klukas, P. Donahue, and R. Johnson (1983) Arrangement of MIP26 in lens junctional membranes: Analysis with proteases and antibodies. *J. Membr. Biol.* **74**:217–228.

Kelley, R. D., and B. D. Perdue (1980) Development of the aging cell surface: Structural patterns of gap junction assembly between metabolic mutants and progressively sub-cultivated human diploid fibroblasts (IMR-90). *Exp. Gerontol.* **15**:407–421.

Kensler, R. W., and D. A. Goodenough (1980) Isolation of mouse myocardial gap junctions. *J. Cell Biol.* **86**:755–764.

Kensler, R. W., P. Brink, and M. M. Dewey (1977) Nexus of frog ventricle. *J. Cell Biol.* **73**: 768–781.

Kreutziger, G. O. (1968) Freeze etching of intercellular junctions of mouse liver. *Proc. EMSA* **26**:138.

Kuszak, J. R., J. L. Rae, B. U. Pauli, and R. S. Weinstein (1982) Rotary replication of lens gap junctions. *J. Ultrastruct. Res.* **81**:249–256.

Lawrence, T. S., W. H. Beers, and N. B. Gilula (1978) Transmission of hormonal stimulation by cell-to-cell communication. *Nature* **272**:501–506.

Lo, C. W., and N. B. Gilula (1979a) Gap junctional communication in the preimplantation mouse embryo. *Cell.* **18**:399–409.

Lo, C. W., and N. B. Gilula (1979b) Gap junctional communication in the post-implantation mouse embryo. *Cell* **18**:411–422.

Makowski, L., D. L. D. Caspar, W. C. Phillips, and D. A. Goodenough (1977) Gap junction structures. II. Analysis of the X-ray diffraction data. *J. Cell Biol.* **74**:629–645.

Manjunath, C. K., G. E. Goings, and E. Page (1982) Isolation and protein composition of gap junctions from rabbit hearts. *Biochem. J.* **205**:189–194.

Meinhardt, H. (1983) Cell determination boundaries as organizing regions for secondary embryonic fields. *Dev. Biol.* **96**:375–385.

Merk, F. B., C. R. Botticelli, and J. T. Albright (1972) An intercellular response to estrogen by granulosa cells in the rat ovary: An electron microscope study. *Endocrinology* **90**: 992–1007.

Meyer, D. J., S. B. Yancey, and J.-P. Revel (1981) Intercellular communication in normal and regenerating rat liver: A quantitative analysis. *J. Cell Biol.* **91**:505–523.

Michalke, W., and W. R. Loewenstein (1971) Communication between cells of different types. *Nature* **232**:121–122.

Nicholson, B. J., M. W. Hunkapiller, L. B. Grim, L. E. Hood, and J.-P. Revel (1981) The rat liver gap junction protein: Properties and partial sequence. *Proc. Natl. Acad. Sci. USA* **78**:7594–7598.

Nicholson, B. J., L. J. Takemoto, M. W. Hunkapiller, L. E. Hood, and J.-P. Revel (1983) Differences between the liver gap junction protein and lens MIP26 from rat: Implications for tissue specificity of gap junctions. *Cell* **32**:967–978.

Page, E. (1978) Quantitative ultrastructural analysis in cardiac membrane physiology. *Am. J. Physiol.* **235**:C147–C158.

Paul, D. L., and D. A. Goodenough (1983) Preparation, characterization and localization of antisera against bovine MIP26, an integral membrane protein of the lens fiber. *J. Cell Biol.* **96**:625–632.

Peracchia, C. (1980) Structural correlates of gap junction permeation. *Int. Rev. Cytol.* **66**: 81–146.

Pitts, J. D., and R. R. Burk (1976) Specificity of junctional communication between animal cells. *Nature* **264**:762–764.

Potter, D. D., E. J. Furshpan, and E. S. Lennox (1966) Connections between cells of the developing squid as revealed by electrophysiological methods. *Proc. Natl. Acad. Sci. USA* **55**:328–335.

Revel, J.-P., and S. S. Brown (1976) Cell junctions in development with particular reference to the neural tube. *Cold Spring Harbor Symp. Quant. Biol.* **40**:443–455.

Revel, J.-P., and M. J. Karnovsky (1967) Hexagonal array of subunits in intercellular junctions of the mouse heart and liver. *J. Cell Biol.* **33**:C7–C12.

Revel, J.-P., P. Yip, and L. L. Chang (1973) Cell junctions in the early chick embryo—a freeze-etch study. *Dev. Biol.* **35**:302–317.

Robertson, J. D. (1963) The occurrence of a subunit pattern in the unit membranes of club endings in Mauthner cell synapse in goldfish brains. *J. Cell Biol.* **19**:201–221.

Rose, B., and W. R. Loewenstein (1975) Permeability of cell junction depends on local cytoplasmic calcium activity. *Nature* **254**:250–252.

Rose, B., and W. R. Loewenstein (1976) Permeability of a cell junction and the local cytoplasmic free ionized calcium concentration: A study with aequorin. *J. Membr. Biol.* **28**:87–119.

Rose, B., and R. Rick (1978) Intracellular pH, intracellular free calcium, and junctional cell–cell coupling. *J. Membr. Biol.* **44**:377–415.

Ryerse, J. S. (1982) Gap junctions are non-randomly distributed in Drosophila wing discs. *Roux's Arch. Dev. Biol.* **191**:335–339.

Schaller, C. H. (1973) Isolation and characterization of a low molecular weight substance activating head and bud formation in *Hydra*. *J. Embryol. Exp. Morphol.* **29**:27–38.

Schuetze, S. M., and D. A. Goodenough (1982) Dye transfer between cells of embryonic chick lens becomes less sensitive to CO_2 treatment with development. *J. Cell Biol.* **92**:694–705.

Sheridan, J. D. (1968) Electrophysiological evidence for low-resistance intercellular junctions in the early chick embryo. *J. Cell Biol.* **37**:650–659.

Shibata, Y., and T. Yamamoto (1979) Freeze-fracture studies of gap junctions in vertebrate cardiac muscle cells. *J. Ultrastruct. Res.* **67**:79–88.

Simionescu, M., N. Simionescu, and G. Palade (1975) Segmental variations of cell junctions in the vascular endothelium: The microvasculature. *J. Cell Biol.* **67**:863–885.

Simpson, I., B. Rose, and W. R. Loewenstein (1977) Size limit of molecules permeating the junctional membrane channels. *Science* **195**:294–296.

Slack, C., and J. P. Palmer (1969) The permeability of intercellular junctions in the early embryo of *Xenopus laevis* studied with a fluorescent tracer. *Exp. Cell Res.* **55**:416–431.

Spray, D. C., A. L. Harris, and M. V. L. Bennett (1979) Voltage dependence of junctional conductance in early amphibian embryos. *Science* **204**:432–434.

Spray, D. C., A. L. Harris, and M. V. L. Bennett (1981) Gap junctional conductance is a simple and sensitive function of intracellular pH. *Science* **211**:712–715.

Summerbell, D., J. H. Lewis, and L. Wolpert (1973) Positional information in chick limb morphogenesis. *Nature* **244**:492–496.

Takemoto, L. J., and J. S. Hansen (1981) Gap junctions from the lens: Purification and characterization by chemical cross-linking reagent. *Biochem. Biophys. Res. Commun.* **99**:324–331.

Tickle, C., D. Summerbell, and L. Wolpert (1975) Positional signaling and specification of digits in chick limb morphogenesis. *Nature* **254**:199–202.

Traub, O., U. Janssen-Timmen, P. M. Druge, R. Dermietzel, and K. Willecke (1982) Immunological properties of gap junction protein from mouse liver. *J. Cell. Biochem.* **19**:27–44.

Trelstad, R. L., J.-P. Revel, and E. D. Hay (1967) Cell contact during early morphogenesis in the chick embryo. *Dev. Biol.* **16**:78–106.

Turin, L., and A. E. Warner (1977) Carbon dioxide reversibly abolishes ionic communication between cells of early amphibian embryo. *Nature* **270**:56–57.

Turin, L., and A. E. Warner (1980) Intracellular pH in early *Xenopus* embryos: Its effect on current flow between blastomeres. *J. Physiol. (Lond.)* **300**:489–504.

Unwin, P. N. T., and P. D. Ennis (1984) Two configurations of a channel-forming membrane protein. *Nature* **307**:609–613.

Unwin, P. N. T., and G. Zampighi (1980) Structure of the junctions between communicating cells. *Nature* **283**:545–549.

Vermorken, A. J. M., J. M. H. Hilderink, I. Dunia, E. L. Benedetti, and H. Bloemendal (1977) Changes in membrane protein pattern in relation to lens cells differentiation. *FEBS Lett.* **83**:301–306.

Warner, A. E. (1973) The electrical properties of the ectoderm in the amphibian embryo during induction and early development of the nervous system. *J. Physiol. (Lond.)* **235**:267–286.

Warner, A. E., and P. A. Lawrence (1982) Permeability of gap junctions at the segmental border of the insect epidermis. *Cell* **28**:243–252.

Warner, A. E., S. Guthrie, and N. B. Gilula (1984) Antibodies to gap junctional protein selectively disrupt junctional communication in the early amphibian embryo. *Nature* **311**:127–131.

Weir, M. P., and C. W. Lo (1982) Gap junction communication compartments in the *Drosophila* wing disc. *Proc. Natl. Acad. Sci. USA* **79**:3232–3235.

Williams, E. H., and R. L. DeHaan (1981) Electrical coupling among heart cells in the absence of ultrastructurally defined gap junctions. *J. Membr. Biol.* **60**:237–248.

Wood, R. L., and A. M. Kuda (1980) Formation of junctions in regenerating *Hydra*: Gap junctions. *J. Ulstrastruct. Res.* **73**:350–360.

Yancey, S. B., D. Easter, and J.-P. Revel (1979) Cytological changes in gap junctions during liver regeneration. *J. Ultrastruct. Res.* **67**:229–242.

Yancey, S. B., B. N. Nicholson, and J.-P. Revel (1981) The dynamic state of liver gap junctions. *J. Supramol. Struct. Cell. Biochem.* **16**:221–232.

Yee, A. G., and J.-P. Revel (1978) Loss and reappearance of gap junctions in regenerating liver. *J. Cell Biol.* **78**:544–564.

Zampighi, G., S. Simon, J. Robertson, T. McIntosh, and M. Costello (1982) On the structural organization of isolated bovine lens fiber junctions. *J. Cell Biol.* **93**:175–189.

Ziegler, J. S., and J. Horwitz (1981) Immunochemical studies on the major intrinsic polypeptides from human lens membrane. *Invest. Ophthalmol. Vis. Sci.* **21**:46–51.

Chapter 20

The Desmosome–Intermediate Filament Complex

PAMELA COWIN
WERNER W. FRANKE
CHRISTINE GRUND
HANS-PETER KAPPRELL
JUERGEN KARTENBECK

ABSTRACT

The desmosome is a prominent plasma membrane domain, specific to certain kinds of cells (epithelia, myocardium, arachnoidea), which is involved in both intercellular adhesion and intracellular organization of cytoskeletal components, most notably the intermediate-sized filaments. This type of junction can be isolated as a stable symmetrical complex of two membrane domains connected by extraneous structures and retaining, on their cytoplasmic sides, dense plaques, which in turn are attached to bundles of intermediate-sized filaments (tonofilaments). Progress in elucidating the molecular organization of the desmosome has been made, and several major proteins and glycoproteins have been identified and localized by the use of specific antibodies. These permit a more sensitive and accurate means of classifying desmosomes and related structures (hemidesmosomes, intermediate-sized filament-associated intracytoplasmic vesicles), and of distinguishing them from other junctional structures, including those of similar morphology. In addition, they provide a means of following the processes of junction formation and disintegration. The possible biological functions of the desmosome–tonofilament complex and the role played by the individual protein constituents of these structures in establishing and maintaining cell and tissue architecture are discussed.

OCCURRENCE

True desmosomes (maculae adherentes), in contrast to other intercellular junctions, are characteristic of vertebrate epithelial cells (Farquhar and Palade, 1963; for reviews, see Campbell and Campbell, 1971; Overton, 1974; Staehelin, 1974; Hull and Staehelin, 1979). In addition, junctions with the typical structure of desmosomes have been reported to occur within the intercalated disks of vertebrate myocardiac tissue, in Purkinje fiber cells, and in cultured myocardiac cells (Fawcett and McNutt, 1969; Goshima, 1970; DeHaan and Sachs, 1972; Gross and Mueller, 1977; Eriksson and Thornell, 1979; Moses and Claycomb, 1982). They have also been described at the boundaries between arachnoidal cells of the brain and in meningioma tumors derived therefrom (e.g., Cervós-Navarro and Vazques, 1969; Tani et al., 1974; Copeland et al., 1978; Goldman et al., 1980; Kepes, 1982; Kartenbeck et al., 1984). Through the use of both ultrastructural and biochemical critieria, desmosomes have been identified in all vertebrate species examined. Structurally similar junctions have occasionally been described in certain invertebrates (see Lane, 1978; Blanquet and Riordan, 1981; see also Schnepf and Maiwald, 1970; Friedman, 1971; Lawrence and Green, 1975), but whether these structures are indeed homologous in organization and composition to true desmosomes of vertebrates remains to be seen. (For reports of an absence of intermediate-sized filament proteins homologous to cytokeratins and desmin in invertebrates, see Fuchs and Marchuk, 1983; Quax et al., 1984.)

FUNCTION

The desmosomal domains of the plasma membrane exhibit a specialized morphology characterized by two forms of interactions with other cellular structures. First, they form membrane anchorage sites for intermediate-sized filaments (IF) of at least three different types (cytokeratins, vimentin, and desmin). This feature clearly distinguishes them from the adherens junctions (zonulae and fasciae adherentes, puncta adherentia) that form the membrane anchorage sites for actin microfilaments. Second, desmosomes are regions in which a specific membrane domain interacts with a corresponding domain of the plasma membrane of an adjacent cell, apparently mediating intercellular adhesion in a stable way. This is evident in the ability of these domains to remain adherent during isolation procedures that dissolve and disrupt cell membranes (e.g., Skerrow and Matoltsy, 1974a,b; Drochmans et al., 1978). Through the architectural co-incidence of these two interactions, the desmosome–IF complex seems ideally suited to impart a tensile strength and resiliency to the tissue as a whole.

STRUCTURE

Desmosomes from different epithelia vary greatly in their sizes and cell type-specific patterns of distribution and abundance. Nevertheless, desmosomes from all sources share a basic common morphology (Farquhar and Palade, 1963; Kelly, 1966; Campbell and Campbell, 1971; Staehelin, 1974; Matoltsy, 1975).

En face, they occupy oval or circular areas of the cell membrane, varying in diameter from approximately 0.1 to 1.5 μm, depending on the tissue and the specific form of the desmosome (e.g., from maculae adherentes diminutae to giant, probably fused, desmosomes). In most cases, mature desmosomes measure 0.3–0.5 μm in diameter. In cross sections (Figure 1), they appear as a pair of straight and parallel plasma membranes, separated from each other by an intercellular space of 20–30 nm. This relatively wide intercellular space is bisected by a 5–9-nm thick electron-dense stratum ("central disk;" Farquhar and Palade, 1963), or midline, which in transverse sections often has a beaded appearance and has been described from freeze-fracture specimens as consisting of globular particles with a diameter of 5–10 nm (Kelly and Shienvold, 1976; Leloup et al., 1979). In many instances, threadlike cross-bridge structures have been resolved between this central disk and the plasma membranes (Figure 1), that is, within the intermembranous material of the "desmoglea" (Gorbsky and Steinberg, 1981). Distinct, regular 8–15-nm projections from the membrane surface, which seem to be synonymous with these cross-bridges, have been resolved after contrasting the aqueous space with externally added lanthanum salts (Rayns et al., 1969; Franke et al., 1983b), and after splitting the desmosome by treatment with denaturing agents (Franke et al., 1983b; see Figure 4E).

The plasma membrane of the desmosomal domain (5–7-nm thick) usually reveals a pronounced trilaminar appearance and is indistinguishable from interdesmosomal regions of the plasma membrane. Freeze-fracture images of the desmosomal membrane reveal clusters of particles ranging from 8 to 20 nm in diameter, which locate, depending on the type of fixation employed, either to the P or E face of fractured membrane leaflets (e.g., Orwin et al., 1973; Staehelin, 1974; Elias and Friend, 1975, 1976; Kelly and Shienvold, 1976; Leloup et al., 1979). Characteristically, a pair of electron-dense, rigid plaques of 14–20-nm thickness are found subjacent and closely applied to the desmosomal membranes. Bundles of IF (tonofilaments) converge toward and laterally abut the plaques (Kelly, 1966). It is not yet clear, however, whether the filaments enter the plaque structure or approach it superficially and then loop back into the cytoplasm. Sometimes one gains the impression of a second, plaquelike, fuzzy densification at some distance (10–20 nm) from the primary plaque, and in such situations the IF seem to abut upon the "secondary plaque" (Figure 1B). Some authors further suggest the presence of a second set of thinner (4–5 nm) filaments that may connect the tonofilaments to the plaques and/or the plaques to the plasma membrane (Kelly, 1966; Kelly and Shienvold, 1976; Hull and Staehelin, 1979; Kelly and Kuda, 1981). It is possible that such thin filaments are protofilamentous subcomponents fraying out from the IF (cf. Leloup

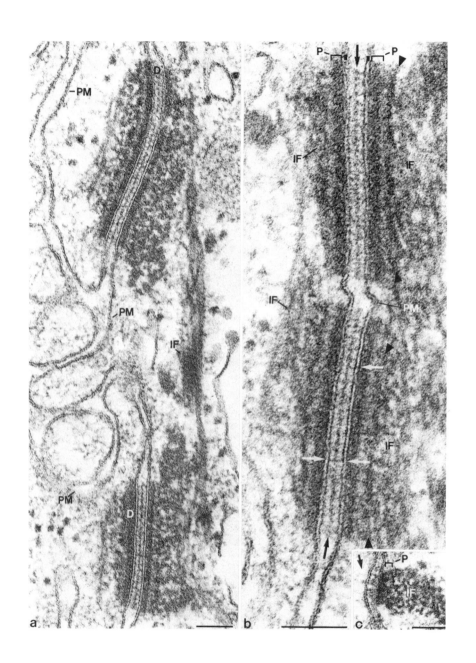

et al., 1979; for protofilament structure observed during IF reconstitution *in vitro*, see Renner et al., 1981; Franke et al., 1982c; Steven et al., 1982; Aebi et al, 1983; Quinlan et al., 1984).

In several stratified epithelia, notably cornifying ones, the structure of the desmosome changes as the cells differentiate and move into the strata. These changes include a thickening of the desmoglea (e.g., Orwin et al., 1973; Allen and Potten, 1975; Elias and Friend, 1975; White and Gohari, 1984). In cornified envelopes and squames of fully cornified epithelia, it is difficult to identify distinct desmosomal substructures.

Junctions with an ultrastructural organization and protein components similar to those of epithelial desmosomes have so far been recognized only in myocardiac cells and Purkinje fibers of the heart (Franke et al., 1981a, 1982b; Cowin and Garrod, 1983; Thornell et al., 1985), in arachnoidal cells of the meninges, and in meningiomas (see Kartenbeck et al., 1984; Schwechheimer et al., 1984). However, these cells are devoid of cytokeratins and instead express IF of the desmin (in myocardium and Purkinje fibers) or vimentin (in arachnoidea) type. In addition, attachment of bundles of vimentin IF to true desmosomes has been found in certain clones of cultured cells selected for the presence of little, if any, cytokeratin IF and an abundance of vimentin IF. For example, certain clones selected from originally cytokeratin-positive cells of the human lung carcinoma-derived cell line A-427 (American Type Culture Collection: Human tumor cell line bank No. HTB53) are abundant in vimentin but devoid of cytokeratin (J. Kartenbeck and W. W. Franke, unpublished observations). Desmosomes of these cells are usually rather small ($\leq$ 0.2 μm) and the IF have the loosely bundled appearance characteristic of vimentin and desmin filaments, in contrast to the familiar densely bundled swathes of cytokeratin filaments. Again in this type of interaction, vimentin filaments appear to approach

Figure 1. *Electron micrographs of cross sections through desmosomes from fetal (20 weeks) human foot sole epidermis showing typical examples of desmosome–intermediate filament complexes. a:* Two desmosomes (D) are separated in this region by a relatively long and convoluted intercept of intradesmosomal plasma membrane (PM). Intermediate filament (IF) bundles of the cytokeratin type can be seen both in close association with the cytoplasmic sides of the desmosomes and linking the desmosomes together. *b:* Two desmosomes separated from each other by a very short interdesmosomal plasma membrane (PM) are shown at higher magnification. They display classic desmosomal features: a pair of straight and parallel membranes (denoted by *black bars* at *top*) in which the inner and outer leaflets can be clearly seen. Two nearly vertical *black arrows* denote the central, electron-dense, midline structure that appears to be connected to the plasma membrane by small cross-bridge structures (*white arrows*). Electron-dense rigid plaques (P) lie subjacent to the plasma membrane and appear to be associated with intermediate filaments (IF), either directly or through a secondary plate (denoted by *black arrowheads* in the cell on the *right*); the plate may include cross-sectional views of IF coursing parallel to the plaques. *c:* Asymmetrical desmosome showing a plaque (P) and an associated IF bundle present on only one side of the desmosome (the absence of these structures in the other cell is seen in the region denoted by the *arrow*). Calibration bars = 0.1 μm.

the desmosomal plaques laterally, as shown in Figure 2A and C. Correspondingly, cells of the myocardium (Franke et al., 1982b; Kartenbeck et al., 1983; Tokuyasu et al., 1983), Purkinje fibers (Franke et al., 1982b; Thornell et al., 1985), and cultured cardiac myocytes (Kartenbeck et al., 1983) show desmin filaments attached to desmosomes lying amid numerous fasciae adherentes (Figure 2D, E); the desmin IF approach the desmosomal

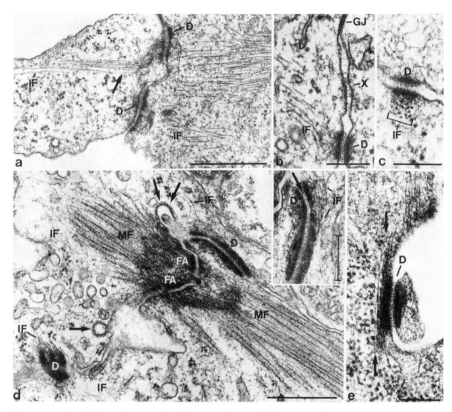

Figure 2. *Association of noncytokeratin types of IF with desmosomal plaques. a*: A cytokeratin-depleted clone of cultured human A-427 cells derived from lung in which only vimentin IF is detectable (see text). Two desmosomes (D) with attached IF bundles that show the typical loosely fasciated (*arrow*) appearance of vimentin filaments. *b*: Same cell clone as in *a* showing a very small desmosome (D), with loosely associated IF, clustered together with a gap junction (GJ) and an unclassified junction structure (X). *c*: Same cell clone as in *a* showing a cross section of a vimentin–IF bundle laterally associated with a desmosome (D). *d*: Cardiac myocytes growing in primary culture (cf. Kartenbeck et al., 1983) showing desmosomes (D) in cross section (*upper right; inset* shows higher magnification; *arrow* in *inset* denotes midline structure) and in grazing section (D, *lower left*), interspersed with fasciae adherentes (FA) and coated pits (*arrows*). Note that desmin IF approach desmosomes, whereas myofibrils (MF) terminate at fasciae adherentes. *e*: Same cell culture as in *d* showing the lateral association of the desmin IF (*arrows*) with a desmosome (D). Calibration bars = 0.5 μm (*a,d*); 0.2 μm (*b,c,e*).

plaque laterally. This association of three different classes of IF proteins with desmosomes points to their homology, in terms of interaction with this membrane domain, and suggests the existence of a common sequence or conformational feature (for a conspicuously homologous sequence at the carboxy-terminal end of the α-helical rod portion in these proteins, see Weber and Geisler, 1984).

Both epithelial and myocardiac desmosomes do not occlude the extra-cellular space. For example, lanthanum ions can perfuse first the central disk region and later also penetrate most of the intermembranous desmoglea, as shown by various authors (e.g., Rayns et al., 1969; Elias and Friend, 1975; Franke et al., 1983b). Therefore, it is reasonable to assume that the desmoglea-associated water and ions are in equilibrium with the fluid surrounding the cell.

HEMIDESMOSOMES, ABNORMAL DESMOSOMES, AND INTRACYTOPLASMIC VESICLES ASSOCIATED WITH DESMOSOMAL COMPONENTS

Hemidesmosomes are domains characteristic of regions where the cell surface membrane borders on extracellular material (see Krawczyk and Wilgram, 1973; Hay, 1977; Gipson et al., 1983). In most cases, these structures are very small (0.05–0.5 μm) regions of surface membrane characterized by the association of a dense cytoplasmic plaque and a punctate extracellular plate reminiscent of the desmosomal central disk. The hemidesmosomal plaques are frequently associated with tufts or bundles of IF. As the name implies, the organization of hemidesmosomes suggests that they are equivalent to one half of a desmosome. This view appears to be supported by demonstrations of desmoplakins in hemidesmosomal regions of certain epithelia (Müller and Franke, 1983), but it is complicated by negative findings in the hemidesmosomes of Purkinje fibers (Thornell et al., 1985) and the absence of immunocytochemical reaction for some other desmosomal components in both epithelial and myocardiac tissue (Cowin et al., 1984a). In addition, Kelly and colleagues have reported that hemidesmosomes are similar to desmosomes in freeze-fracture analyses, but that their intra-membranous particles are somewhat larger (20–30 nm diameter; Kelly and Shienvold, 1976; Shienvold and Kelly, 1976; Kelly and Kuda, 1981).

The typical appearance and frequency of hemidesmosomes at the basal surface of stratified epithelia and cultured epithelial cells is shown in Figure 3A. In some tissues, one also gains the impression, depending on the type of fixation used, that the basal lamina shows localized thickenings or densifications at places corresponding to hemidesmosomes. These den-sifications are often termini of extracellular-anchoring fibrils on the one side and finely filamentous threads between basal lamina and hemides-mosomal membrane on the other. Both structures are assumed to be

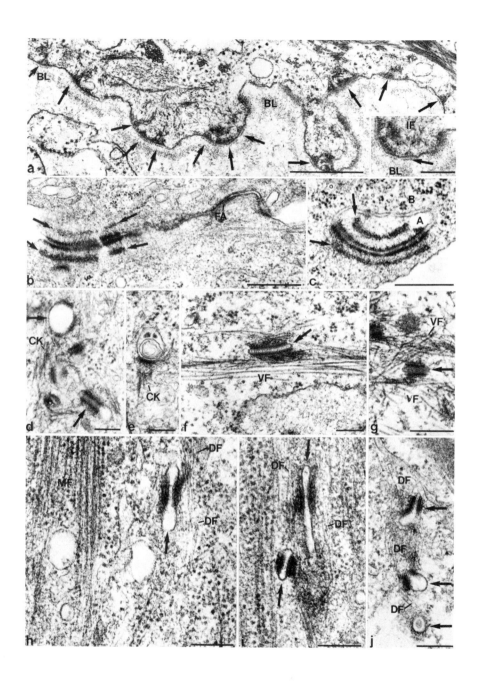

involved in cell anchorage to the basal lamina, and possibly also in the nucleation of hemidesmosome formation (Fawcett, 1981; Gipson et al., 1983). Such hemidesmosomes have also been shown to be present in cultured cells of myocardiac and meningiomal origin. In culture, they are usually found on cell surface areas facing the substratum, although they occasionally occur at other regions of the cell surface (e.g., Schmid et al., 1983a; for increased frequencies of similar asymmetrical structures under the influence of certain inhibitors, see Overton and DeSalle, 1980).

The positioning of hemidesmosomes is suggestive of an adhesive role; however, experimental evidence for this is lacking. The formation of hemidesmosomes, per se, demonstrates that the assembly of the typical desmosomelike domain is not exclusive to the interaction between two cells but can be induced by extracellular components, probably the basal lamina.

A special morphological variant characterized by the appearance of an additional dense plaque structure at some distance (0.1–0.15 μm) from the primary desmosomal plaque structure has been described in certain epithelial cell cultures ("accessory plaques;" cf. Schmid et al., 1983a), as well as in the myocardiac cells of mammalian embryos (e.g., the "imagined desmosomes" of Forbes and Sperelakis, 1975), and those of lower vertebrates (Cowin et al., 1984a). In these cells, the IF usually terminate at the accessory plaque. Figure 3B and C present typical examples of this kind of accessory

---

Figure 3. *Electron micrographs of desmosome-related structures of various kinds.* a: Small hemidesmosomes at the basal surface (BL, basal lamina) of human fetal epidermis (as in Figure 1a) are frequent (some are denoted by *arrows*). At higher magnification (*inset* at lower right) they reveal a dense plaque, often associated with IF bundles, and a dense punctate line resembling the midline structure of the desmosome at some distance from the plasma membrane. b and c: Desmosomes of a cultured epithelial cell line from bovine mammary gland epithelium (BMGE + HM; cf. Schmid et al., 1983a) showing secondary (accessory) plaques (*arrows*) at some distance from the primary plaques. Like primary plaques, secondary plaques contain desmoplakins. Desmosomes shown are located (b) at a relatively straight cell boundary and (c) at a deep invagination of one cell (A) into another (B). d and e: Intracellular vesicles associated with short dense plaque structures and IF of the cytokeratin type (CK) in epithelial cells of the BMGE + HM line shown in b. These vesicles, which are probably derived from endocytotic events such as those that occur during trypsinization, appear in different shapes: for example, a tennis-racket shape (*arrow* at *lower right* in d), or spheroidal shapes, with (e) or without (*arrow* at *upper left* in d) membranous inclusions. Flattened vesicles with a symmetrical structure resembling that of true, that is, intercellular, desmosomes are also seen. f and g: Intracellular vesicles (*arrows*) related to desmosomal structures found in cultured human meningioma cells (cf. Kartenbeck et al., 1984), again with different sizes and shapes, are associated with IF of the vimentin type (VF). Note typical dense plaques and lateral association of IF and midlinelike structure within the flattened vesicles (f). h–j: Intracellular vesicles (*arrows*) of various sizes and shapes in cultured rat myocardiac cells (cf. Kartenbeck et al., 1983). Note dense plaques and laterally associated bundles of IF of the desmin type (DF), as well as indications of midline structure in flattened vesicles. The presence of myofibrils (MF) indicates the myogenic nature of these cells. Calibration bars = 0.5 μm (a–c); 0.2 μ (inset in a; g–j); 0.1 μm (f).

plaque structure, which has been immunologically demonstrated to contain desmoplakin proteins (Schmid et al., 1983a) and appears to be connected to the primary plaque by thin filaments 3–5 nm in diameter. Remarkably, hemidesmosomes in these cells are also accompanied by such additional plaques.

The appearance of desmosomelike structures is not restricted to the cell periphery. A number of reports have shown the occurrence of intracyto-plasmic vesicles in certain normal tissues and pathological conditions (see Caputo and Prandi, 1972; Komura and Watanabe, 1975; Petry, 1980; Schenk, 1980). They have also been observed frequently in cell cultures of epithelial and carcinoma origin (e.g., Franke et al., 1981a; Schmid et al., 1983a), and their formation can be induced or enhanced by treatment with trypsin or chelation of Ca^{2+} from the extracellular medium (Overton, 1968, 1974, 1975; Kartenbeck et al., 1982). By labeling the extracellular space with colloidal gold during desmosome splitting induced by Ca^{2+} depletion, Kartenbeck et al. (1982) have shown that at least part of the intracellular vesicles formed under these conditions are the products of endocytosis of the desmosomal halves. Such vesicles, which are usually characterized by a short dense plaque positive for desmoplakins and associated with IF, can appear in various sizes and forms, including spherical vesicles, tennis-racket shapes, and flattened sacs. Such vesicles are shown in Figure 3D–J and include examples of cytokeratin-associated (Figure 3D, E), vimentin-associated (Figure 3F, G), and desmin-associated (Figure 3H–J) types. While an endocytotic origin is easily imagined for the circular vesicles possessing a cap of desmosomal plaque material, the frequent observation of tennis-racket and sac-like vesicles with midlinelike structures implies a folding over and self-zipping of the half-domain. Such a process indicates that the adhesive molecules of one half of a desmosome can interact and adhere to themselves, in a fashion similar to that undertaken with the domain of another cell. It is not yet known whether these endocytosed vesicles can be reutilized in desmosome formation.

ISOLATION AND FRACTIONATION OF DESMOSOMES

As desmosomes are particularly abundant in stratified tissues, occupying up to 50% of the plasma membrane, these tissues are particularly suitable sources from which to isolate desmosomes in quantities suitable for bio-chemical analyses. Basically, three procedures have been used for the biochemical isolation and purification of desmosomal material.

1. Extraction of tissue homogenates with low and high salt buffers containing nondenaturing detergents, followed by sucrose gradient cen-trifugation, results in fractions of structurally well-preserved desmosomes with attached bundles of tonofilaments (see Figure 4A-D; cf. Drochmans

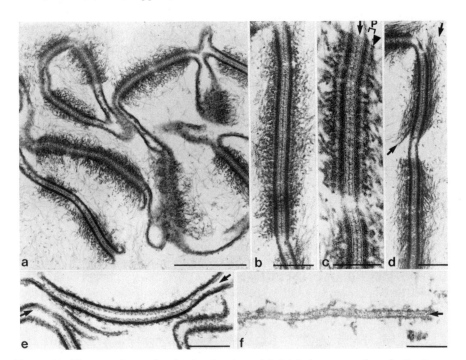

Figure 4. *Electron micrographs showing fractions of isolated desmosomes from bovine tongue (a,b,d) and snout epidermis (c,e,f) prepared by different procedures. a–d*: Isolation involving extraction at pH 9 and high salt buffer (cf. Müller and Franke, 1983; Kapprell, 1983) showing desmosome–tonofilament complexes. The survey micrograph (*a*) illustrates the purity of the fraction, and higher magnification pictures (*b* and *c*) show good preservation of central disk ("midline") and plaque structures. *Arrows* in *d* denote lateral approach of tonofilaments. *e*: Isolated desmosomes extracted in buffer containing 9 *M* urea (cf. Franke et al., 1983b). Note splitting along the "midline" plane (*arrows*) and the appearance of regular projections on outer membrane surface. *f*: Desmosomal appearance after isolation in citric acid (cf. Franke et al., 1981a; Müller and Franke, 1983). Note that most of the plaque material is removed and the central disk structure (*arrow*) is partly disorganized. Calibration bars = 0.5 (*a*); 0.2 μm (*b–f*).

et al., 1978). In biochemical studies, such fractions have the disadvantage of containing more IF protein material than desmosomal components.

2.   Fractions obtained as described in the preceding paragraph can be depleted of IF by extraction with such strong denaturing agents as high concentrations of urea or guanidinium hydrochloride (Franke et al., 1983b). Fractions thus obtained show little residual IF protein material but retain well-preserved plaques and membrane substructure. However, frequent splitting or separation of the desmosomal halves along the plane of the central disk (Figure 4E) with partial extraction of glyco-polypeptides 4a and 4b (Franke et al., 1983b) is found.

3.   The low pH citrate buffer extraction procedure developed by Skerrow and Matoltsy (1974a,b) and modified with additional metrizamide

gradient fractionation by Gorbsky and Steinberg (1981) results in very clean preparations greatly enriched in "desmosomal cores." The disadvantage of this procedure is that the desmosomal plaques tend to distintegrate and parts of the plaque material are removed (Figure 4F; Skerrow and Matoltsy, 1974a,b; Franke et al., 1981a; Gorbsky and Steinberg, 1981). In addition, there is some derangement of central disk structure.

Most of the fractions of desmosomes described in the literature are derived from bovine snout epidermis (Drochmans et al., 1978; Franke et al., 1981a; Gorbsky and Steinberg, 1981). However, good fractions have also been obtained from other stratified epithelia such as bovine tongue mucosa and human and rat epidermis (see Kapprell, 1983; Müller and Franke, 1983). Observations that the desmosomal structure is maintained in plasma membrane fractions from rat liver (e.g., Franke et al., 1979b), in cytoskeletal fractions from rat brush border (Franke et al., 1979a, 1981b), and in various cultured cells (Schmid et al., 1983a) should encourage more systematic fractionation work to isolate desmosomes from simple epithelia and cultured cells.

Similarly, it can be seen from published work on the isolation of gap junctions and fasciae adherentes from myocardiac tissue (Kensler and Goodenough, 1980; Colaco and Evans, 1981, 1983) that myocardiac desmosomes are also stable to rigorous extractive procedures and should be suitable for isolation.

BIOCHEMICAL COMPOSITION OF DESMOSOMES

Analysis of the lipid composition of isolated desmosomes from cow snout epidermis has not revealed any significant difference from the lipid pattern of total plasma membranes (Drochmans et al., 1978).

Preparations of bovine snout epidermal desmosomes can be resolved into eight major polypeptide bands, and at least seven individual proteins have been identified as true desmosomal constituents (Figure 5A–C). These

---

Figure 5. *Major polypeptides of isolated desmosomes from cow snout epidermis,* as revealed on one-dimensional SDS-polyacrylamide gels (*a*) and on two-dimensional gels that underwent either isoelectric focusing (IEF; *b*) or nonequilibrium pH gradient electrophoresis (NEPHGE; *c*). Two-dimensional electrophoresis in *b* and *c* was in the presence of SDS; higher pH values are to the *left* in *b* and to the *right* in *c*. Desmosomal polypeptides are indicated by arabic numerals (cf. Franke et al., 1981a; Müller and Franke, 1983) and major keratin polypeptides by roman numerals, as introduced by Franke et al. (1978). Reference proteins coelectrophoresed are: A, α-actin; BSA, bovine serum albumin; PGK, phosphoglycerokinase. Note that desmoplakins I and II (polypeptides of bands 1 and 2) are not at isoelectric equilibrium in *c*, and that little of the protein of component 3 has entered the focusing in *b*.

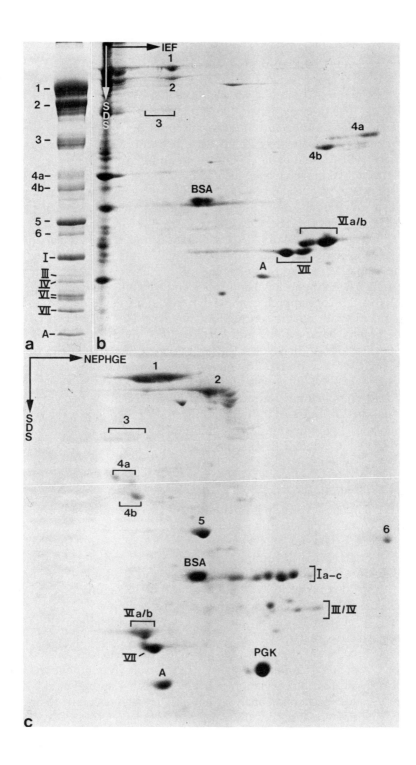

are the nonglycosylated proteins desmoplakin I (with a molecular weight of approximately 250 kD), desmoplakin II (approximately 215 kD), polypeptide band 5 (approximately 83 kD), and polypeptide band 6 (approximately 75 kD), as well as the glycosylated proteins of band 3 (approximately 150 kD), band 4a (approximately 130 kD), and band 4b (approximately 115 kD). In addition, minor and variable amounts of a smaller glycosylated component of approximately 22 kD have been noted (e.g., Gorbsky and Steinberg, 1981; Cohen et al., 1983; Cowin and Garrod, 1983).

The desmoplakins have been located by immunoelectron microscopy in the desmosomal plaques and by fluorescence microscopy in a large number of tissues, carcinomas, and cultured cells from a wide variety of species (Franke et al., 1981a, 1983c; Cowin and Garrod, 1983; Müller and Franke, 1983; Cowin et al., 1984a,b). Desmoplakins I and II are similar in their range of isoelectric pH values (6.8–7.2), amino acid compositions, and peptide maps (Müller and Franke, 1983; Kapprell et al., 1985), and show cross-reaction with monoclonal antibodies and conventional antisera (Franke et al., 1981a, 1982b; Kapprell et al., 1985). In 9.5 M urea, polypeptide band 5 is isoelectric with bovine serum albumin (pH 6.3–6.4) and band 6 polypeptide is very basic (Müller and Franke, 1983). Both have been located to the desmosome at the level of the light microscope (Cowin and Garrod, 1983; Cowin et al., 1984a). However, their precise location within the desmosome has yet to be established. These proteins are dissimilar in amino acid composition (Kapprell et al., 1985) and have been shown to be the distinct products of individual mRNAs (Franke et al., 1983d).

A minor polypeptide component (antigen "DP1") in a molecular weight region similar to that of the desmoplakins has also been located to the desmosomal plaque (Franke et al., 1981a).

The glycoproteins of band 3 and bands 4a and 4b are biochemically and immunologically distinct from each other (Cohen et al., 1983; Cowin and Garrod, 1983). Band 3 protein is often resolved into a number of polypeptides differing slightly in isoelectric pH and molecular weight values; however, a proteolytic origin for this heterogeneity can so far not be excluded (Kapprell et al., 1985). Its amino acid composition and sugar content are clearly distinct from those of glycoproteins 4a and 4b (Table 1; cf. Kapprell et al., 1985). In contrast, polypeptides 4a and 4b are similar to each other in isoelectric values (Müller and Franke, 1983) and amino acid compositons (Kapprell et al., 1985), and in all cases reported so far have shown immunological cross-reactivity (Cohen et al., 1983; Cowin and Garrod, 1983). However, analysis of their cleavage peptides products, amino acid compositions, and sugar contents suggests that these two polypeptides differ significantly in their glycosylation pattern (Table 1; Kapprell et al., 1985).

Most detailed biochemical analyses have so far been restricted to fractions from bovine epidermal tissues (Skerrow and Matoltsy, 1974a,b; Drochmans et al., 1978; Gorbsky and Steinberg, 1981). However, the presence of desmoplakins of similar molecular weight and isoelectric pH values has

Table 1. Carbohydrate Contents of Gel-Electrophoretically Purified Desmosomal Polypeptides from Bovine Muzzle Epidermis[a]

	Band 1 (Desmoplakin I)	Band 2 (Desmoplakin II)	Band 3	Band 4a	Band 4b	Band 5	Band 6
Fucose	—	—	—	—	—	—	—
Sialic acid	—	—	1.9 (2)	3.3 (3)	2.3 (2)	—	—
Mannose	—	—	1.35 (1)	4.5 (5)	7.2 (7)	—	—
Glucose	—	—	1.75 (2)	—	—	—	—
Galactose	—	—	5.2 (5)	4.5 (5)	3.5 (4)	0.17 (0)	—
Glucosamine	—	—	0.94 (1)	1.85 (2)	7.1 (7)	—	0.41 (0)
Galactosamine	—	—	0.8 (1)	1.1 (1)	2.2 (2)	—	0.2 (0)

Source: Kapprell et al., 1985.

[a] Values are presented as molar ratios of carbohydrate residues per polypeptide; numbers in parentheses give approximated whole number ratios.

been shown in electrophoretically purified polypeptides from epidermal desmosomes of other species and from bovine tongue (Müller and Franke, 1983). The presence of at least one desmoplakin component in human and bovine myocardial tissue has been demonstrated in immunoblot experiments (Franke et al., 1982b).

Desmosomal proteins 5 and 6 also seem to occur in tissues other than epidermis. Polypeptides with a similar molecular weight and isoelectric pH have been detected in cytoskeletal preparations from tongue mucosa and cultured bovine mammary gland epithelial cells (Kapprell, 1983). Evidence for the occurrence of the two groups of glycoproteins of polypeptide bands 3, 4a, and 4b is so far based exclusively on observations of immunological cross-reaction with desmosomal structures in some but not all tissues (Cowin and Garrod, 1983; Cowin et al., 1984a,b).

IMMUNOLOCALIZATION OF DESMOSOMAL COMPONENTS

Antibodies to purified desmosomal proteins can be used specifically to label or stain desmosomal structures and thereby distinguish them from other membrane-associated domains including dense plaque-containing junctions such as the zonulae adherentes and fasciae adherentes (Franke et al., 1981a, 1982b; Geiger et al., 1983). Typical results of immunofluorescence microscopy, showing the reaction of antibodies against constituent proteins of the desmosomal plaque from cow snout on the tissue of origin, are presented in Figure 6A–C. The reaction is restricted to epithelial cells and is preferential for, if not exclusive to, cell boundaries. At a higher resolution, it is resolved into distinct fluorescent dots that are especially evident in grazing sections (Figure 6B) and in the simple epithelia and ducts of the epidermal glands (Figure 6C), where the desmosomes are more sparsely distributed. Correspondingly, these antibodies have been immunolocalized, at the electron microscope level, to the desmosomal plaque (examples of immunoperoxidase and immunogold staining are shown in Figure 7A, B; Franke et al., 1981a). They have also been localized to the hemidesmosomal plaques of stratified epithelium (Franke et al., 1981a). Antibody reaction is also found in punctate patterns along the cell boundaries of other epithelial cell types, including transitional epithelia of bladder, polarized simple epithelia (Figure 8 presents examples of chicken and bovine intestinal cells; cf. Müller and Franke, 1983), hepatocytes (Franke et al., 1981b; Cowin and Garrod, 1983; Müller and Franke, 1983), and the epithelial reticular cells of the thymus (Franke et al., 1981a, 1982b). Desmosomal staining by such antibodies demonstrates the cell type-specific distribution of desmosomes *in situ* and in isolated cells (Figure 8C–E). It should also facilitate study of the altered desmosomal patterns found in some disease states, including certain tumors (Franke et al., 1982b, 1983c; for reports on altered desmosomal frequencies in certain carcinomas, see

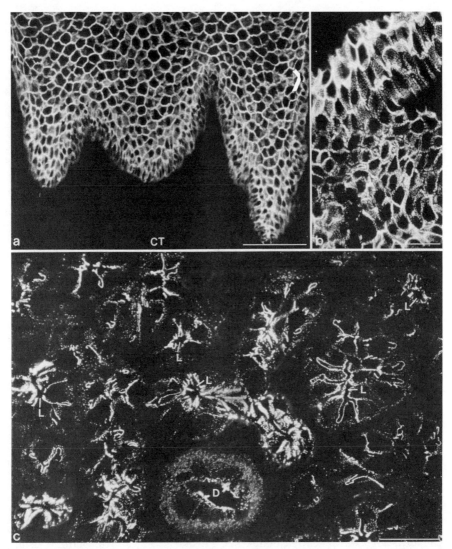

Figure 6. *Immunofluorescence microscopy of guinea pig antibodies to desmoplakins* (Müller and Franke, 1983) on frozen tissue sections through bovine snout epidermis showing specific reaction at the cell boundaries of stratified epidermal cells (*a* and *b*) and the simple epithelia of mucous glands (*c*; L, lumina) and ducts (D) (cf. Franke et al., 1980). At higher magnification (*b* and *c*), especially in grazing sections (*b*), the staining is resolved into small fluorescent dots representing individual desmosomes. CT, connective tissue. Calibration bars = 100 μm (*a*); 50 μm (*c*); 25 μm (*b*).

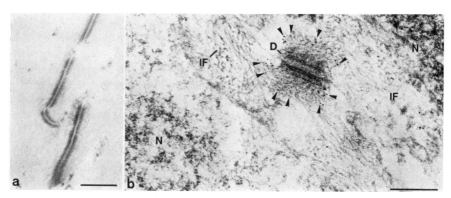

Figure 7. *Immunoelectron micrographs showing the specific reaction of desmoplakin antibodies* (as in Figure 6) with desmosomal plaques of bovine tongue tissue, as revealed with peroxidase (*a*) and colloidal gold (*b*; region decorated by 5-nm gold particles is denoted by *arrowheads*) techniques (for methods, see Franke et al., 1981a, 1982b, Müller and Franke, 1983). D, desmosome; IF, intermediate filament bundles; N, nuclei. Calibration bars = 0.5 μm.

also Pauli et al., 1978; Alroy et al., 1981). Characteristically, polarized cells show dispersed patterns of desmosomes on their lateral cell–cell boundaries, as well as a conspicuous subapical ring of fluorescent dots representing the individual maculae adherentes (Figure 8C–E).

On densely grown monolayer cell cultures, a punctate decoration of cell boundaries can be visualized by using antibodies to desmosomal proteins. (Figure 9A presents an example of an epithelial cell in which adjacent cells are in contact over most of their lateral surface.) Antibodies against desmoplakin introduced into cultured cells by microinjection also specifically bind to the desmosomal plaques, resulting in a similar "dotted line pattern" in the injected cells (Figure 9B), thus confirming the specific location of this protein. Certain epithelial cell cultures do not form extended contacts but are connected by bridges produced on the ends of cytoplasmic projections of neighboring cells. In these cells, staining is found exclusively in the bridges, each site representing one or several desmosomes. (Figure 9C presents an example; for details, see Franke et al., 1981a; Schmid et al., 1983b; for human cell line A-431, see Franke et al., 1983c.) In mitotic cells, a significant number of desmosomes continue to connect the dividing cell with its neighboring interphase cells (Figure 9D).

Punctate patterns corresponding to the distribution of desmosomes along cell boundaries have also been described, following the reaction of antibodies to bovine epidermal desmoplakins with cultures of certain non-cytokeratin-containing, nonepithelial cells, notably rat cardiac myocytes, meningioma cells, and the aforementioned A-427 cell clones (cf. Figure 10; Kartenbeck et al., 1983, 1984).

Using antibodies against desmosomal plaque proteins in combination with antibodies against IF proteins enables the anchorage of IF bundles at desmosomes and the integration of desmosomes into the IF cytoskeletal meshwork to be visualized by double-label immunofluorescence microscopy. Figure 10 presents examples of cytokeratin IF (Figure 10a, A') and vimentin IF (Figure 10B, B'). Similar double-labeling experiments in which antibodies against desmin IF were used have been described by Kartenbeck et al. (1983).

In certain cultured cells, particularly after trypsinization or treatment with Ca^{2+}-chelating agents, the desmosome-positive reaction is found not

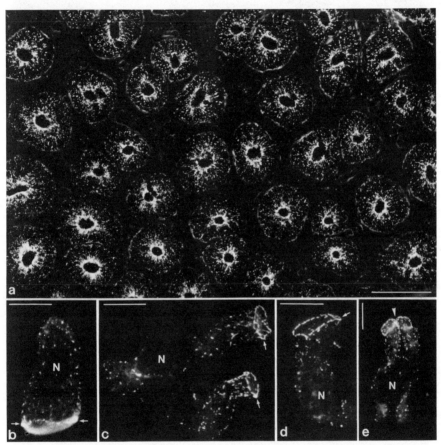

Figure 8. *Immunofluorescence microscopy showing the reaction of guinea pig antibodies to desmoplakin* isolated from bovine snout epidermis with mucosal cells of the small intestine of chicken (*a* and *b*) and cow (*c–e*) in frozen tissue section (*a*, survey picture of cross-sectioned crypts) and on detached cells (*b–e*; for preparations, see Franke et al., 1981a; Geiger et al., 1983; Müller and Franke, 1983). Desmosomes are revealed as individual fluorescent dots. Note the high density of desmosomes in the subapical ring of maculae adherentes (*short arrows* in *b–e*) as opposed to a more dispersed and less regular distribution of desmosomes on the more basal regions of lateral cell surfaces. N, nuclei. Calibration bars = 50 μm (*a*); 10 μm (*b–e*).

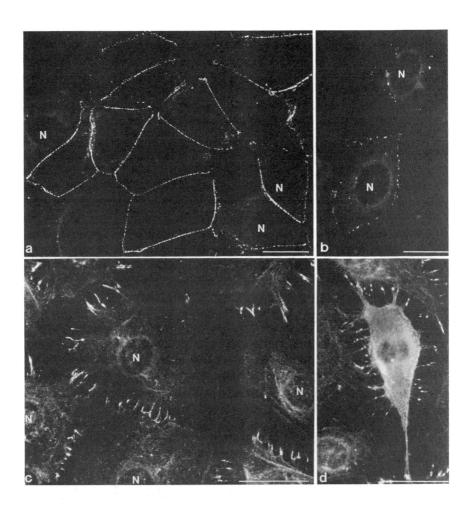

only at cell boundaries but also in places apparently deep in the cytoplasm, usually in smaller "dots" (e.g., see Figures 9A, C; 10A). At least some of these intracellular "desmoplakin-positive" dots correspond to the intracytoplasmic, desmosome-related vesicles described above and to hemidesmosomes. However, without electron microscopy it is difficult to decide whether a given dot represents one or the other structure.

The occurrence of desmosomes in diverse cell lines indicates that the expression of desmoplakins, and probably also of other desmosomal components, is maintained during the culturing of many cells, independent of IF protein expression. In other cell lines, however, desmosomes or particular desmosomal constituents are no longer detectable, for example in the cytokeratin-positive marsupial kidney epithelial cell line PtK$_2$ (Zerban and Franke, 1978; Franke et al., 1981a, 1982b; Cowin, 1984) and in certain cytokeratin-positive and -negative rat hepatoma-derived cell clones (Venetianer et al., 1983). Present evidence therefore indicates that the expression of desmosomal components is not necessarily linked to that of a specific category of IF proteins, although in living organisms such a correlation is common.

Several studies have suggested that desmosomal components are highly conserved throughout the vertebrates (Franke et al., 1982b; Cowin and Garrod, 1983; Cowin et al., 1984a), although certain species-restricted antigens such as desmosomal plaque component "DP1" have been described

---

Figure 9. *Immunofluorescence microscopy showing the reaction of guinea pig antibodies against desmoplakins* with desmosomal structures at cell boundaries in monolayer cultures of bovine cell lines derived from kidney epithelium (MDBK line; *a* and *b*) or mammary gland (BMGE + H line, described by Schmid et al., 1983b; *c* and *d*). *a*: MDBK cell cultures grown on cover slips were fixed and permeabilized by treatment with cold methanol and acetone and processed for indirect immunofluorescence microscopy (cf. Franke et al., 1981a, 1982b; Geiger et al., 1983; Müller and Franke, 1983). Note fluorescent dot staining in linear assays, most of which represent desmosomes at boundaries of cells; the desmosomes in this line are in close contact over most of the cell surface. Some "dots" that are smaller and less intense are also seen, in what appear to be intracellular regions. These sites might represent hemidesmosomes and/or intracellular desmosome-related vesicles (see text). *b*: A dense monolayer culture of MDBK cells, two of which had been microinjected with desmoplakin antibodies (purified IgG; 2 mg/ml). After incubation for four hours, cells were rinsed with phosphate-buffered saline (PBS), treated with methanol–acetone, rinsed again in PBS, treated with fluorescein-labeled secondary antibodies (goat anti-guinea pig Ig), washed repeatedly in PBS, and mounted in Mowiol (see the references mentioned above). Note that punctate staining at the cell periphery is restricted to the two injected cells (N, nuclei), whereas the surrounding cells are negative. *c*: BMGE + H cells stained as described for MDBK cells in *a*. These cells do not contact each other over the whole surface but form contacts that are situated at cell–cell bridges (Schmid et al., 1983b); desmosomes, or groups of desmosomes, are positioned in the central parts of these bridges (N, nuclei). *d*: Same culture as in *c*, showing a mitotic cell that is still connected to its neighbors by a number of desmoplakin-positive, that is, desmosome-containing, bridge connections. Calibration bars = 25 μm (*a,c,d*); 20 μm (*b*).

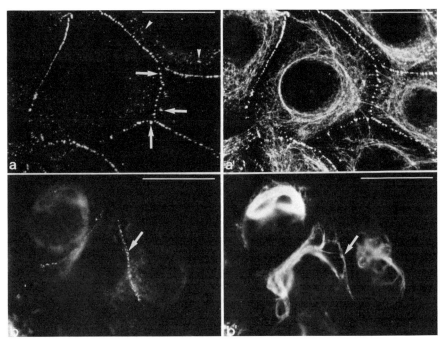

Figure 10. *Double-label immunofluoresence microscopy of cultured cells,* by using guinea pig
antibodies to desmosomal plaque antigen "DP1" (*a*) or to desmoplakin (*b*), in combination
with antibodies to cytokeratins (*a'*; for details of the specific type of antibody labeling
applied, see Geiger et al., 1983) and rabbit antibodies to vimentin (*b'*). *a* and *a'*: Bovine
eptihelial (MDBK) cells showing the display of desmosome–IF complexes at cell boundaries
(some are denoted by *arrows* in *a*), illustrating the anchorage of cytokeratin filament
bundles at desmosomes. The cell in the *upper right* contains a number of small desmoplakin-
positive dots; they seem to be located deeper within the cytoplasm and may represent
hemidesmosomes and/or intracytoplasmic desmosome-related vesicles (see text). Note
that associations with cytokeratin filaments are also seen at such internal dots (some are
denoted by *arrowheads*). *b* and *b'*: Human A427 cells of a clone selected for absence of
cytokeratin and abundance of vimentin IF (as in Figure 2a–c) after treatment with
colcemid (for conditions, see Franke et al., 1979c) in order to induce formation of
juxtanuclear aggregates of vimentin IF, followed by double-label immunofluorescence
microscopy (simultaneous addition of guinea pig antibodies to desmoplakins and rabbit
antibodies to vimentin; for general conditions, see Geiger et al., 1983; Franke et al.,
1984). Note that most of the vimentin IF is seen in the juxtanuclear aggregates ("whorls"),
but some vimentin IF bundles are connected to desmosomes (one is denoted by *arrow*),
indicating the stable anchorage of vimentin IF at desmosomes in this cell. Calibration
bars = 25 μm.

(Franke et al., 1981a). Immunological cross-reactivity ranging from fish to
humans has been observed by Cowin et al. (1984a). In studies using
polyclonal (Cowin and Garrod, 1983; Cowin et al., 1984a) and monoclonal
(P. Cowin and W. W. Franke, unpublished results) antibodies, a more
restricted tissue distribution has been noticed, notably for the glycoprotein
groups 3 and 4 ("desmogleins;" Shida et al., 1983). This is particularly

pronounced with both polyclonal and monoclonal antibodies against gly-copolypeptides 4a and 4b ("desmocollins" according to Cowin et al., 1984b). These show a greatly diminished staining intensity or, as in the case of several monoclonal antibodies produced so far, an absence of staining of a number of nonstratified tissues. The results suggest that in some cell types these glycopolypeptides are absent or represented by non-cross-reacting members of a polypeptide family.

Molecules detected by antibodies directed against the 4a and 4b com-ponents of bovine snout desmosomes have been located by fluorescence and immunoelectron microscopy to the cell surface of dispersed cultured bovine epithelial cells (MDBK line) prior to cell contact (Cowin et al., 1984b). Upon cell–cell contact, these molecules appear to distribute on the basolateral surface of the cell in regions known subsequently to form desmosomes. Furthermore, when MDBK cells are cultured in the presence of monovalent fragments (Fab') of these antibodies, desmosome formation, but not intercellular adhesion, is prevented (Cowin et al., 1984b). These observations, together with the reduction of band 4a and 4b glycoproteins from preparations of isolated desmosomes that are split by treatment with 9.5 M urea (Kapprell et al., 1985), strongly implicates them ("desmocollins") as true desmosomal adhesives.

DEFINITION OF JUNCTIONS BY BIOCHEMICAL AND IMMUNOLOGICAL CRITERIA

Originally, intercellular junctions were classified by morphological criteria, notably by their appearance in the electron microscope (Farquhar and Palade, 1963; Staehelin, 1974). While junctional structures displaying the classical ultrastructural features are easy to identify, it is often difficult to classify other more rudimentary forms that may be missing certain char-acteristic features. This is especially problematic with the two categories of junctions and "half-junctions" that are associated with dense cytoplasmic plaque structures and bundles of cytoskeletal filaments. These are, on the one hand, the desmosomes and the related structures described above, which contain a specific set of desmosomal proteins and display specific attachment to IF, and on the other hand, the junctions of the fascia adherens type. The latter also include the zonulae adherentes, the puncta adherentia, and the asymmetrical focal adhesions prominent at the bottom surface of cultured cells, which are associated with a dense cytoplasmic plaque and actin microfilaments. Considerable difficulty can be encountered in deciding whether a given junctional or "hemijunctional" structure falls into the desmosome or the fascia adherens category (also referred to as "intermediate junctions") or into other, as yet poorly characterized, plaque-associated junctions such as the pre- and postsynaptic densities of nervous tissues (Pfenninger, 1973; Peters et al., 1976). Hence, many structures have been described as "desmosomelike" in diverse cell types, including nonepithelial cells and a broad range of tumors (e.g., Altorfer et al., 1974; Russell, 1976a,b; Nagano and Suzuki, 1978; Romrell and Ross, 1979; Ghadially, 1980, 1982; Fawcett, 1981; Mirra and Miles, 1982). The availability of an increasing number of junction type-specific, biochemically defined markers, and of antibodies that allow their visualization, have led to a new definition and classification of junctions in molecular terms. In fact, structures of the desmosome group can be distinguished from those of the fascia adherens group by a number of criteria. Membrane domains related to the fascia adherens category are positively identified by combinations of antibodies to vinculin, α-actinin, a cell adhesion protein ("uvomorulin"), and a 135-kD amphiphilic protein. In addition, they are often the termini of bundles of actin microfilaments (Tokuyasu et al., 1981, 1983; Geiger et al., 1983; Volk and Geiger, 1984; Boller et al., 1985; and references quoted therein). They do not contain detectable amounts of the desmosomal components immunologically related to proteins of bands 1–6 of epidermal desmosomes (see above) and are not associated with IF. Therefore they can be distinguished from structures of the desmosome category at high resolution, that is, even when these two types of junctions are present within a junctional complex (Geiger et al., 1983). Such distinctions are also of importance for cell typing in the diagnostic classification of tumors (cf. Ghadially, 1980, 1982; Erlandson, 1981).

DESMOSOME FORMATION

Desmosome formation ("desmogenesis") has been studied in diverse systems, including embryos of chick (Overton, 1962; Hay, 1968) and fish (Lentz and Trinkaus, 1971), ovine, bovine, and human epidermis (Breathnach, 1971; Orwin et al., 1973; Leloup et al., 1979), epidermal wound healing in mice (Krawczyk and Wilgram, 1973), and various cell cultures (Patrizi, 1967; Dembitzer et al., 1980; Hennings and Holbrook, 1983).

Desmosomes are structures that appear early in embryonic differentiation and, at least in the mouse embryo, their advent correlates with the demonstrable appearance of cytokeratin filaments, that is, at the morula–blastocyst transition, when small ("nascent," "incipient") desmosomes are found in the trophectodermal cells, both with and without short cytokeratin IF tufts attached (Ducibella et al., 1975; Jackson et al., 1980; for bovine blastocysts, see Massip et al., 1981). Similarly, Lentz and Trinkaus (1971) have noted the first appearance of desmosome-related structures in blastulae of the fish, *Fundulus heteroclitus*. Subsequently, desmosomes with cytokeratin IF are found in all ecto- and endodermal cells (for the mouse see, e. g., Jackson et al., 1981), but they disappear in those ectoderm-derived cells that convert into primary mesenchymal cells (Franke et al., 1982a, 1983a), indicating that expression of desmosomal components is coupled to cell differentiation during embryo development.

A number of studies have specifically dealt with the sequence of assembly of desmosomal structural components during desmosome or hemidesmosome formation, both in tissues and in cell cultures. At present, it is still a matter of controversy whether clustering of the intrinsic desmosomal membrane components precedes the association of plaque structures, or whether the interaction of plaque components induces the local accumulation of the membrane components with their prominent desmogleal specializations. Several authors studying various tissue and cell culture systems have reported parallel alignment of plasma membrane regions, and/or the condensation of finely filamentous material on the extracellular surface of what appears to be a site of nucleation of desmosome formation, as the initial feature (Krawczyk and Wilgram, 1973; Orwin et al., 1973; Leloup et al., 1979). Correspondingly, the extracellular fibrils associated with the basal lamina, notably the "anchoring fibrils," have been said to nucleate hemidesmosome formation (Gipson et al., 1983; see also references discussed therein). Other reports have described a simultaneous occurrence of both extracellular and intracellular (plaque) densities or have emphasized that plaque formation defines the site of ensuing desmosome formation or growth (Lentz and Trinkaus, 1971; Dembitzer et al., 1980; Hennings and Holbrook, 1983; for reviews, see Overton, 1974, 1975; Garrod and Cowin, 1984). Obviously, morphological criteria alone are not sufficient to define the sequence of events in this complex process.

Desmosome and hemidesmosome stability and formation are clearly dependent, in a number of tissues and cell systems, on Ca^{2+}, as has been shown in experiments employing chelators or media with low Ca^{2+} content (Muir, 1967; Borysenko and Revel, 1972; Overton, 1974; Cereijido et al., 1978; Ducibella and Anderson, 1979; Hennings et al., 1980; Jones et al., 1982; Kartenbeck et al., 1982; Brysk et al., 1984; Trinkaus-Randall and Gipson, 1984). As isolated desmosomes are not split by Ca^{2+} depletion (e.g., in up to 20 mM EDTA; P. Cowin, W. W. Franke, C. Grund, H.-P. Kapprell, and J. Kartenbeck, unpublished observations), it is questionable whether the positive effect of Ca^{2+} is exerted on the desmosome structure as such; it may well be an indirect effect. Desmosome formation and stability may be controlled by other intracellular factors, as suggested by experiments utilizing cytochalasin B (Overton and Culver, 1973), but here again it is difficult to unravel these findings in molecular terms.

Glycoproteins (desmogleins, desmocollins) are widely thought to be involved in desmosome-mediated cell adhesion. However, when tunicamyin, a potent inhibitor of the addition of asparagine-linked oligosaccharide moieties, is added, the frequency of desmosomes formed in aggregating, cultured chick corneal cells is reduced by a factor of two or three (Overton, 1982). This finding of a moderate reduction in the number of desmosomes suggests that the formation of desmosomes in principle does not depend on the N-linkage type of glycosylation of desmosomal proteins. This, of course, does not exclude the possibility that the carbohydrate moieties may serve other functions.

Desmosome formation appears to be independent of cell type specificity. Several studies (Armstrong, 1970; Overton and Kapmarski, 1975) have shown that desmosomes can occur between cells of different species and between cells of different tissues. In this respect, the functional adhesive domains of desmosomes from different sources seem to be compatible with each other. Desmosomal adhesives therefore cannot be seen as contributing to cell–cell recognition through cell type-specific molecules. However, it is possible that they may contribute to cell–cell selectivity by conveying a quantitative differential of adhesion of the type described by Steinberg (1964).

Information on the lifespan of desmosomes and the turnover of desmosomal components is presently not available. It is clear, however, that the maintenance of desmosomes does not depend on its association with the IF bundle meshwork in a specific cell. In various epithelium-derived cells growing in culture, and also in certain tissues or tumors, cytokeratin IF are transiently detached from the desmosomes, unraveled and rearranged into large spheroidal aggregates (Franke et al., 1982d; Geiger et al., 1984; see there for further references). Similarly, cytokeratin IF material is disconnected from desmosomes during certain forms of toxic damage in hepatocytes, followed by rearrangement of the IF protein material into large aggregates ("Mallory bodies") deeper in the cytoplasm. Nevertheless, in these cells desmosomes are found in normal frequency, with typical morphology and in typical distribution, particularly along the bile canaliculi (Denk and Franke, 1981). This illustrates that both parts of the IF–tonofilament complex can dissociate from one another without considerable effects on desmosome stability and positioning.

ACKNOWLEDGMENTS

We thank Caecilia Kuhn, Helga Mueller, and Hans-Josef Opferkuch for valuable contributions, and Irmgard Purkert for typing. The work has been supported in part by the Deutsche Forschungsgemeinschaft.

REFERENCES

Aebi, U., W. E. Fowler, P. Rew, and T.-T. Sun (1983) The fibrillar substructure of keratin filaments unraveled. *J. Cell Biol.* **97**:1131–1143.

Allen, T. D., and C. S. Potten (1975) Desmosomal form, fate, and function in mammalian epidermis. *J. Ultrastruc. Res.* **51**:94–105.

Alroy, J., B. U. Pauli, and R. S. Weinstein (1981) Correlation between numbers of desmosomes and the aggressiveness of transitional cell carcinoma in human urinary bladder. *Cancer* **47**:104–112.

Altorfer, J., T. Fukuda, and C. Hedinger (1974) Desmosomes in human seminiferous epithelium. An electron microscopic study. *Virchows Arch. B. Cell Pathol.* **16**:181–194.

Armstrong, P. B. (1970) A fine structural study of adhesive cell junctions in heterotypic cell aggregates. *J. Cell Biol.* **47**:197–210.

Blanquet, R. S., and G. P. Riordan (1981). An ultrastructural study of the subumbrellar musculature and desmosomal complexes of *Cassiopea xamachana* (Cnidaria: Scyphozoa). *Trans. Am. Microsc. Soc.* **100**:109–119.

Boller, K., D. Vestweber, and R. Kemler (1985) Cell-adhesion molecule uvomorulin is localized in the intermediate junctions of adult intestinal epithelial cells. *J. Cell Biol.* **100**:327–332.

Borysenko, J. Z., and J.-P. Revel (1972) Experimental manipulation of desmosome structure. *Am. J. Anat.* **137**:403–422.

Breathnach, A. S. (1971) Embryology of human skin. A review of ultrastructural studies. *J. Invest. Dermatol.* **57**:133–143.

Brysk, M. M., J. Miller, and G. K. Walker (1984) Characteristics of a human epidermal squamous carcinoma cell line at different extracellular calcium concentrations. *Exp. Cell Res.* **150**:329–337.

Campbell, R. D., and J. H. Campbell (1971) Origin and continuity of desmosomes. In *Origin and Continuity of Cell Organelles*, J. Reinert and H. Ursprung, eds., pp. 261–298, Springer Verlag, Berlin.

Caputo, R., and G. Prandi (1972) Intracytoplasmic desmosomes. *J. Ultrastruct. Res.* **41**: 358–368.

Cereijido, M., E. S. Robbins, W. J. Dolan, C. A. Rotunno, and D. D. Sabatini (1978) Polarized monolayers formed by epithelial cells on a permeable and translucent support. *J. Cell Biol.* **77**:853–880.

Cervós-Navarro, J. and J. J. Vazquez (1969) An electron microscopic study of meningiomas. *Acta Neuropathol.* **13**:301–323.

Cohen, S. M., G. Gorbsky, and M. S. Steinberg (1983) Immunochemical characterization of related families of glycoproteins in desmosomes. *J. Biol. Chem.* **258**:2621–2627.

Colaco, C. A. L. S., and W. H. Evans (1981) A biochemical dissection of the cardiac intercalated disk: Isolation of subcellular fractions containing fascia adherentes and gap junctions. *J. Cell Sci.* **52**:313–325.

Colaco, C. A. L. S., and W. H. Evans (1983) Plasma membrane intercellular junctions. Morphology and protein composition. In *Electron Microscopy of Proteins*, Vol. 4, J. R. Harris, ed., pp. 331–363, Academic, New York.

Copeland, D. D., S. W. Bell, and J. D. Shelburne (1978) Hemidesmosomelike intercellular specializations in human meningiomas. *Cancer* **41**:2242–2249.

Cowin, P. (1984) Analysis of vertebrate desmosomes. Unpublished doctoral dissertation, University of Southampton, England.

Cowin, P., and D. R. Garrod (1983) Antibodies to epithelial desmosomes show wide tissue and species cross-reactivity. *Nature* **302**:148–150.

Cowin, P., D. Mattey, and D. R. Garrod (1984a) Distribution of desmosomal components in the tissues of vertebrates, studied by fluorescent antibody staining. *J. Cell Sci.* **66**: 119–132.

Cowin, P., D. Mattey, and D. R. Garrod (1984b) Identification of desmosomal surface components (desmocollins) and inhibition of desmosome formation by specific Fab'. *J. Cell Sci.* **70**:41–60.

DeHaan, R. L., and H. G. Sachs (1972) Cell coupling in developing systems: The heart-cell paradigm. *Curr. Top. Dev. Biol.* **7**:193–228.

Dembitzer, H. M., F. Herz, A. Schermer, R. C. Wolley, and L. G. Koss (1980) Desmosome development in an *in vitro* model. *J. Cell Biol.* **85**:695–702.

Denk, H., and W. W. Franke (1981) Rearrangement of the hepatocyte cytoskeleton after toxic damage: Involution, dispersal and peripheral accumulation of Mallory body material after drug withdrawal. *Eur. J. Cell Biol.* **23**:241–249.

Drochmans, P., C. Freudenstein, J.-C Wanson, L. Laurent, T. W. Keenan, J. Stadler, R. Leloup, and W. W. Franke (1978) Structure and biochemical composition of desmosomes and tonofilaments isolated from calf muzzle epidermis. *J. Cell Biol.* **79**:427–443.

Ducibella, T., and E. Anderson (1979) The effects of calcium deficiency on the formation of the zonula occludens and blastocoel in the mouse embryo. *Dev. Biol.* **73**:46–58.

Ducibella, T., D. F. Albertini, E. Anderson, and J. D. Biggers (1975) The preimplantation mouse embryo: Characterization of intercellular junctions and their appearance during development. *Dev. Biol.* **45**:231–250.

Elias, P. M., and D. S. Friend (1975) The permeability barrier in mammalian epidermis. *J. Cell Biol.* **65**:180–191.

Elias, P. M., and D. S. Friend (1976) Vitamin-A-induced mucous metaplasia. An *in vitro* system for modulating tight and gap junction differentiation. *J. Cell Biol.* **68**:173–188.

Eriksson, A., and L.-E. Thornell (1979) Intermediate (skeleton) filaments in heart Purkinje fibers. *J. Cell Biol.* **80**:231–247.

Erlandson, R. A. (1981) *The Interpretation of Submicroscopic Structures in Neoplastic Cells*, Masson Publishing, New York.

Farquhar, M. G., and G. E. Palade (1963) Junctional complexes in various epithelia. *J. Cell Biol.* **17**:375–412.

Fawcett, D. W. (1981) *The Cell*, W. B. Saunders, Philadelphia.

Fawcett, D. W., and N. S. McNutt (1969) The ultrastructure of the cat myocardium. I. Ventricular papillary muscle. *J. Cell Biol.* **42**:1–29.

Forbes, M. S., and N. Sperelakis (1975) The "imaged-desmosome": A component of intercalated discs in embryonic guinea pig myocardium. *Anat. Rec.* **183**:243–258.

Franke, W. W., K. Weber, M. Osborn, E. Schmid, and C. Freudenstein (1978) Antibody to prekeratin. Decoration of tonofilament-like arrays in various cells of epithelial character. *Exp. Cell Res.* **116**:429–445.

Franke, W. W., B. Appelhans, E. Schmid, C. Freudenstein, M. Osborn, and K. Weber (1979a) The organization of cytokeratin filaments in the intestinal epithelium. *Eur. J. Cell Biol.* **19**:255–268.

Franke, W. W., E. Schmid, J. Kartenbeck, D. Mayer, H.-J. Hacker, P. Bannasch, M. Osborn, K. Weber, H. Denk, J.-C. Wanson, and P. Drochmans (1979b) Characterization of the intermediate-sized filaments in liver cells by immunofluorescence and electron microscopy. *Biol. Cell* **34**:99–110.

Franke, W. W., E. Schmid, S. Winter, M. Osborn, and K. Weber (1979c) Widespread occurrence of intermediate-sized filaments of the vimentin-type in cultured cells from diverse vertebrates. *Exp. Cell Res.* **123**:25–46.

Franke, W. W., E. Schmid, C. Freudenstein, B. Appelhans, M. Osborn, K. Weber, and T. W. Keenan (1980) Intermediate-sized filaments of the prekeratin type in myoepithelial cells. *J. Cell Biol.* **84**:633–654.

Franke, W. W., E. Schmid, C. Grund, H. Müller, I. Engelbrecht, R. Moll, J. Stadler, and E.-D. Jarasch (1981a) Antibodies to high molecular weight polypeptides of desmosomes: Specific localization of a class of junctional proteins in cells and tissues. *Differentiation* **20**:217–241.

Franke, W. W., S. Winter, C. Grund, E. Schmid, D. L. Schiller, and E.-D. Jarasch (1981b) Isolation and characterization of desmosome-associated tonofilaments from rat intestinal brush border. *J. Cell Biol.* **90**:116–127.

Franke, W. W., C. Grund, C. Kuhn, B. W. Jackson, and K. Illmensee (1982a) Formation of cytoskeletal elements during mouse embryogenesis. III. Primary mesenchymal cells and the first appearance of vimentin filaments. *Differentiation* **23**:43–59.

Franke, W. W., R. Moll, D. L. Schiller, E. Schmid, J. Kartenbeck, and H. Müller (1982b) Desmoplakins of epithelial and myocardial desmosomes are immunologically and biochemically related. *Differentiation* **23**:226–237.

Franke, W. W., D. L. Schiller, and C. Grund (1982c) Protofilamentous and annular structures as intermediate during reconstitution of cytokeratin filaments *in vitro. Biol. Cell* **46**:257–268.

Franke, W. W., E. Schmid, C. Grund, and B. Geiger (1982d) Intermediate filament proteins in non-filamentous structures: Transient disintegration and inclusion of subunit proteins in granular aggregates. *Cell* **30**:103-113.

Franke, W. W., C. Grund, B. W. Jackson, and K. Illmensee (1983a) Formation of cytoskeletal elements during mouse embryogenesis. IV. Ultrastructure of primary mesenchymal cells and their cell–cell interactions. *Differentiation* **25**:121–141.

Franke, W. W., H.-P. Kapprell, and H. Müller (1983b) Isolation and symmetrical splitting of desmosomal structures in 9 *M* urea. *Eur. J. Cell Biol.* **32**:117–130.

Franke, W. W., R. Moll, H. Müller, E. Schmid, C. Kuhn, R. Krepler, U. Artlieb, and H. Denk (1983c) Immunocytochemical identification of epithelium-derived human tumors with antibodies to desmosomal plaque proteins. *Proc. Natl. Acad. Sci. USA* **80**:543–547.

Franke, W. W., H. Müller, S. Mittnacht, H.-P. Kapprell, and J. L. Jorcano (1983d) Significance of two desmosomal plaque-associated polypeptides of molecular weights 75,000 and 83,000. *EMBO J.* **2**:2211–2215.

Franke, W. W., E. Schmid, S. Mittnacht, C. Grund, and J. L. Jorcano (1984) Integration of different keratins into the same filament system after microinjection of mRNA for epidermal keratins into kidney epithelial cells. *Cell* **36**:813–825.

Friedman, M. H. (1971) Arm-bearing microtubules associated with an unusual desmosome-like junction. *J. Cell Biol.* **49**:916–920.

Fuchs, E., and D. Marchuk (1983) Type I and type II keratins have evolved from lower eukaryotes to form the epidermal intermediate filaments in mammalian skin. *Proc. Natl. Acad. Sci. USA* **80**:5857–5861.

Garrod, D. R., and P. Cowin (1985) Desmosome structure and function. In *Receptors in Tumor Immunology*, C.M. Chadwick, ed., Marcel Dekker, New York (in press).

Geiger, B., E. Schmid, and W. W. Franke (1983) Spatial distribution of proteins specific for desmosomes and adherens junctions in epithelial cells demonstrated by double immuno-fluorescence microscopy. *Differentiation* **23**:189–205.

Geiger, B., T. E. Kreis, O. Gigi, E. Schmid, S. Mittnacht, J. L. Jorcano, D. B. von Bassewitz, and W. W. Franke (1984) Dynamic rearrangements of cytokeratins in living cells. In *Cancer Cells*, Vol. 1, *The Transformed Phenotype*, A. Levine, G.F. Vande Wode, W.C. Topp, and J. D. Watson, eds., pp. 202–215, Cold Spring Harbor Laboratory, Cold Spring Harbor, New York.

Ghadially, F. N. (1980) *Diagnostic Electron Microscopy of Tumours*, Butterworth, London.

Ghadially, F. N. (1982) *Ultrastructural Pathology of the Cell and Matrix*, 2nd Ed., Butterworth, London.

Gipson, I. K., S. M. Grill, S. J. Spurr, and S. J. Brennan (1983) Hemidesmosome formation *in vitro*. *J. Cell Biol.* **97**:849–857.

Goldman, J. E., D. S. Horoupian, and A. B. Johnson (1980) Granulofilamentous inclusions in a meningioma. *Cancer* **46**:156–161.

Gorbsky, G., and M. S. Steinberg (1981) Isolation of the intercellular glycoproteins of desmosomes. *J. Cell Biol.* **90**:243–248.

Goshima, K. (1970) Formation of nexuses and electrotonic transmission between myocardial and fl cells in monolayer culture. *Exp. Cell Res.* **63**:124–130.

Gross, W. O., and C. Mueller (1977) A mechanical momentum in ultrastructural development of the heart. *Cell. Tissue Res.* **178**:483–494.

Hay, E. D. (1968) Organization and fine structure of epithelium and mesenchyme in the developing chick embryo. In *Epithelial–Mesenchymal Interactions*, R. Fleischmajer and R.E. Billingham, eds., pp. 31–55, Williams and Wilkins, Baltimore.

Hay, E. D. (1977) Epithelium. In *Histology*, L. Weiss and R. O. Greep, eds., pp. 113–144, McGraw-Hill, New York.

Hennings, H., and K. A. Holbrook (1983) Calcium regulation of cell–cell contact and differentiation of epidermal cells in culture. An ultrastructural study. *Exp. Cell Res.* **143**:127–142.

Hennings, H., D. Michael, C. Cheng, P. Steinert, K. A. Holbrook, and S. H. Yuspa (1980) Calcium regulation of growth and differentiation of mouse epidermal cells in culture. *Cell* **19**:245–254.

Hull, B. E., and L. A. Staehelin (1979) The terminal web. A reevaluation of its structure and function. *J. Cell Biol.* **81**:67–82.

Jackson, B. W., C. Grund, E. Schmid, K. Bürki, W. W. Franke, and K. Illmensee (1980) Formation of cytoskeletal elements during mouse embryogenesis. Intermediate filaments of the cytokeratin type and desmosomes in preimplantation embryos. *Differentiation* **17**:161–179.

Jackson, B. W., C. Grund, S. Winter, W. W. Franke, and K. Illmensee (1981) Formation of cytoskeletal elements during mouse embryogenesis. II. Epithelial differentiation and intermediate-sized filaments in early postimplantation embryos. *Differentiation* **20**:203–216.

Jones, J. C. R., A. E. Goldman, P. M. Steinert, S. Yuspa, and R. D. Goldman (1982) Dynamic aspects of the supramolecular organization of intermediate filament networks in cultured epidermal cells. *Cell Motil.* **2**:197–213.

Kapprell, H. P. (1983) Biochemische und immunologische Charakterisierung von Desmosomenproteinen. Unpublished doctoral dissertation, University of Heidelberg, Federal Republic of Germany.

Kapprell, H. P., P. Cowin, W. W. Franke, H. Ponstingl, and H. J. Opferkuch (1985) Biochemical characterization of desmosomal proteins isolated from bovine muzzle epidermis: Amino acid and carbohydrate composition. *Eur. J. Cell Biol.* **36**:217–229.

Kartenbeck, J., E. Schmid, W. W. Franke, and B. Geiger (1982) Different modes of internalization of proteins associated with adhaerens junctions and desmosomes: Experimental separation of lateral contacts induces endocytosis of desmosomal plaque material. *EMBO J.* **1**: 725–732.

Kartenbeck, J., W. W. Franke, J.G. Moser, and U. Stoffels (1983) Specific attachment of desmin filaments to desmosomal plaques in cardiac myocytes. *EMBO J.* **2**:735–742.

Kartenbeck, J., K. Schwechheimer, R. Moll, and W. W. Franke (1984) Attachment of vimentin filaments to desmosomal plaques in human meningiomal cells and arachnoidal tissue. *J. Cell Biol.* **98**:1072–1081.

Kelly, D. E. (1966) Fine structure of desmosomes, hemidesmosomes and an adepidermal globular layer in developing newt epidermis. *J. Cell Biol.* **28**:51–72.

Kelly, D. E., and A. M. Kuda (1981) Traversing filaments in desmosomal and hemidesmosomal attachments: Freeze-fracture approaches toward their characterization. *Anat. Rec.* **199**: 1–14.

Kelly, D. E., and F. Schienvold (1976) The desmosome: Fine structural studies with freeze-fracture replication and tannic acid staining of sectioned epidermis. *Cell Tissue Res.* **173**:309–323.

Kensler, R. W., and D. A. Goodenough (1980) Isolation of mouse myocardial gap junctions. *J. Cell Biol.* **86**:755–764.

Kepes, J. J. (1982) *Biology, Pathology, and Differential Diagnosis,* Masson Publishing, New York.

Komura, J., and S. Watanabe (1975) Desmosome-like structures in the cytoplasm of normal human keratinocyte. *Arch. Dermatol. Res.* **253**:145–149.

Krawczyk, W. S., and G. F. Wilgram (1973) Hemidesmosome and desmosome morphogenesis during epidermal wound healing. *J. Ultrastr. Res.* **45**:93–101.

Lane, N. J. (1978) Intercellular junctions and cell contacts in invertebrates. *Proc. Int. Congr. Electron Microsc.* **3**:673–688.

Lawrence, P. A., and S. M. Green (1975) The anatomy of a compartment border. The intersegmental boundary in *Oncopeltus. J. Cell Biol.* **65**:373–382.

Leloup, R., L. Laurent, M.-F. Ronveaux, P. Drochmans, and J.-C. Wanson (1979) Desmosomes and desmogenesis in the epidermis of calf muzzle. *Biol. Cell* **34**:137–152.

Lentz, T. L., and J. P. Trinkaus (1971) Differentiation of the junctional complex of surface cells in the developing fundulus blastoderm. *J. Cell Biol.* **48**:455–472.

Massip, A., J. Mulnard, R. Huygens, C. Hanzen, P. van der Zwalmen, and F. Ectors (1981) Ultrastructure of the cow blastocyst. *J. Submicrosc. Cytol.* **13**:31–40.

Matoltsy, A. G. (1975) Desmosomes, filaments, and keratohyaline granules: Their role in the stabilization and keratinization of the epidermis. *J. Invest. Dermatol.* **65**:127–142.

Mirra, S. S., and M. L. Miles (1982) Subplasmalemmal linear density: A mesodermal feature and a diagnostic aid. *Hum. Pathol.* **13**:365–380.

Moses, R. L., and W. C. Claycomb (1982) Ultrastructure of terminally differentiated adult rat cardiac muscle cells in culture. *Am. J. Anat.* **164**:113–131.

Müller, H., and W. W. Franke (1983) Biochemical and immunological characterization of desmoplakins I and II, the major polypeptides of the desmosomal plaque. *J. Mol. Biol.* **163**:647–671.

Muir, A. R. (1967) The effects of divalent cations on the ultrastructure of the perfused rat heart. *J. Anat.* **101**:239–261.

Nagano, T., and F. Suzuki (1978) Cell to cell relationships in the seminiferous epithelium in the mouse embryo. *Cell Tissue Res.* **189**:389–401.

Orwin, D. F. G., R. W. Thomson, and N. E. Flower (1973) Plasma membrane differentiation of keratinizing cells of the wool follicle. *J. Ultrastruct. Res.* **45**:15–29.

Overton, J. (1962) Desmosome development in normal and reassociating cells in the early chick blastoderm. *Dev. Biol.* **4**:532–548.

Overton, J. (1968) The fate of desmosomes in trypsinized tissue. *J. Exp. Zool.* **168**:203–214.

Overton, J. (1974) Cell junctions and their development. *Prog. Surf. Membr. Sci.* **8**:161–208.

Overton, J. (1975) Experiments with junctions of the adhaerens type. *Curr. Top. Dev. Biol.* **10**:1–34.

Overton, J. (1982) Inhibition of desmosome formation with tunicamycin and with lectin in corneal cell aggregates. *Dev. Biol.* **92**:66–72.

Overton, J., and N. Culver (1973) Desmosomes and their components after cell dissociation and reaggregation in the presence of cytochalasin B. *J. Exp. Zool.* **185**:341–356.

Overton, J., and R. DeSalle (1980) Control of desmosome formation in aggregating embryonic chick cells. *Dev. Biol.* **75**:168–176.

Overton, J., and R. Kapmarski (1975) Hybrid desmosomes in aggregated chick and mouse cells. *J. Exp. Zool.* **192**:33–42.

Patrizi, G. (1967) Desmosomes in tissue cultures: Their reconstruction after trypsinization. *J. Cell Biol.* **35**:182A.

Pauli, B. U., S. M. Cohen, J. Alroy, and R. S. Weinstein (1978) Desmosome ultrastructure and the biological behavior of chemical carcinogen-induced urinary bladder carcinomas. *Cancer Res.* **38**:3276–3285.

Peters, A., S. L. Palay, and H. deE. Webster (1976) *The Fine Structure of the Nervous System: The Neurons and Supporting Cells*, W. B. Saunders, Philadelphia.

Petry, G. (1980) "Autodesmosomen," desmosomale Kontakte von Teilen derselben Zelle im menschlichen Chorion laeve und Amnion. *Eur. J. Cell Biol.* **23**:129–136.

Pfenninger, K. H. (1973) *Synaptic Morphology and Cytochemistry*, Gustav Fischer Verlag, Stuttgart.

Quax, W., R. van den Heuvel, W. Vree Egberts, Y. Quax-Jeuken, and H. Bloemendal (1984) Intermediate filament cDNAs from BHK-21 cells: Demonstration of distinct genes for desmin and vimentin in all vertebrate classes. *Proc. Natl. Acad. Sci. USA* **81**:5970–5974.

Quinlan, R. A., J. A. Cohlberg, D. L. Schiller, M. Hatzfeld, and W. W. Franke (1984) Heterotypic tetramer (A_2D_2) complexes of non-epidermal keratins isolated from cytoskeletons of rat hepatocytes and hepatoma cells. *J. Mol. Biol.* **178**:365–388.

Rayns, D. G., F. O. Simpson, and J. M. Ledingham (1969) Ultrastructure of desmosomes in mammalian intercalated disc; appearances after lanthanum treatment. *J. Cell Biol.* **42**:322–326.

Renner, W., W. W. Franke, E. Schmid, N. Geisler, K. Weber, and E. Mandelkow (1981) Reconstitution of intermediate-sized filaments from denatured monomeric vimentin. *J. Mol. Biol.* **149**:285–306.

Romrell, L. J., and M. H. Ross (1979) Characterization of sertoli cell–germ cell junctional specializations in dissociated testicular cells. *Anat. Rec.* **193**:23–42.

Russell, L. (1976a) Desmosome-like junctions between sertoli and germ cells in the rat testis. *Am. J. Anat.* **148**:301–312.

Russell, L. (1976b) Movement of spermatocytes from the basal to the adluminal compartment of the rat testis. *Am. J. Anat.* **148**:313–328.

Schenk, P. (1980) Intracytoplasmic desmosomes in malignant keratinocytes of laryngeal carcinoma. *Arch. Oto-rhino-laryngol.* **226**:219–223.

Schmid, E., W. W. Franke, C. Grund, D. L. Schiller, H. Kolb, and N. Paweletz (1983a) An epithelial cell line with fusiform myoid morphology derived from bovine mammary gland: Expression of cytokeratins and desmoplakins in abnormal arrays. *Exp. Cell Res.* **146**:309–328.

Schmid, E., D. L. Schiller, C. Grund, J. Stadler, and W. W. Franke (1983b) Tissue type-specific expression of intermediate filament proteins in a cultured eptihelial cell line from bovine mammary gland. *J. Cell Biol.* **96**:37–50.

Schnepf, E., and M. Maiwald (1970) Halbdesmosomen bei Phytoflagellaten. *Experientia (Basel)* **26**:1343.

Schwechheimer, K., J. Kartenbeck, R. Moll, and W. W. Franke (1984) Vimentin filament–desmosome cytoskeleton of diverse types of human meningiomas. A distinctive diagnostic feature. *Lab. Invest.* **51**:584–591.

Shida, M., S. M. Cohen, G. J. Giudice, and M. S. Steinberg (1983) Quantitative electron microscopic immunocytochemistry of desmosomal antigens. *J. Cell Biol.* **97**:325.

Shienvold, F. L., and D. E. Kelly (1976) The hemidesmosome: New fine structural features revealed by freeze-fracture techniques. *Cell Tissue Res.* **172**:289–307.

Skerrow, C. J., and A. G. Matoltsy (1974a) Isolation of epidermal desmosomes. *J. Cell Biol.* **63**:515–523.

Skerrow, C. J., and A. G. Matoltsy (1974b) Chemical characterization of isolated epidermal desmosomes. *J. Cell Biol.* **63**:524–531.

Staehelin, L. A. (1974) Structure and function of intercellular junctions. *Int. Rev. Cytol.* **39**: 191–283.

Steinberg, M. S. (1964) The problem of adhesive selectivity in cellular interactions. In *Cellular Membranes in Development*, M. Locke, ed., pp. 321–366, Academic, New York.

Steven, A. C., J. Wall, J. Hainfeld, and P. M. Steinert (1982) Structure of fibroblastic intermediate filaments: Analysis by scanning transmission electron microscopy. *Proc. Natl. Acad. Sci. USA* **79**:3101–3105.

Tani, E., K. Ikeda, S. Yamagata, M. Nishiura, and N. Higashi (1974) Specialized junctional complexes in human meningioma. *Acta Neuropathol.* **28**:305–315.

Thornell, L.-E., A. Eriksson, B. Johansson, and U. Kjörell (1985) Intermediate filament and associated proteins in heart purkinje fibers. A membrane–myofibril-anchored cytoskeletal system. *Ann. N.Y. Acad. Sci.* (in press).

Tokuyasu, K. T., A. H. Dutton, B. Geiger, and S. J. Singer (1981) Ultrastructure of chicken cardiac muscle as studied by double immunolabeling in electron microscopy. *Proc. Natl. Acad. Sci. USA* **78**:7619–7623.

Tokuyasu, K. T., A. H. Dutton, and S. J. Singer (1983) Immunoelectron microscopic studies of desmin (skeletin) localization and intermediate filament organization in chicken cardiac muscle. *J. Cell Biol.* **96**:1736–1742.

Trinkaus-Randall, V., and I. K. Gipson (1984) Role of calcium and calmodulin in hemidesmosome formation *in vitro*. *J. Cell Biol.* **98**:1565–1571.

Venetianer, A., D. L. Schiller, T. Magin, and W. W. Franke (1983) Cessation of cytokeratin expression in a rat hepatoma cell line lacking differentiated functions. *Nature* **305**: 730–733.

Volk, T., and B. Geiger (1984) A 135-kd membrane protein of intercellular adherens junctions. *EMBO J.* **3**:2249–2260.

Weber, K., and N. Geisler (1984) Intermediate filaments from wool α-keratins to neurofilaments: A structural overview. In *Cancer Cell*, Vol. 1, *The Transformed Phenotype*, A. Levine, G. F. Vande Wode, W. C. Topp, and J. D. Watson, eds., pp.153–159, Cold Spring Harbor Laboratory, Cold Spring Harbor, New York.

White, F. H., and K. Gohari (1984) Some aspects of desmosomal morphology during differentiation of hamster cheek pouch epithelium. *J. Submicrosc. Cytol.* **16**:407–422.

Zerban, H., and W. W. Franke (1978) Modified desmosomes in cultured epithelial cells. *Cytobiologie* **18**:360–373.

Chapter 21

Molecular Domains of Adherens Junctions

BENJAMIN GEIGER
ZAFRIRA AVNUR
TOVA VOLBERG
TALILA VOLK

ABSTRACT

Adherens junctions are cell contacts which are associated at their endofacial surfaces with actin filament bundles. In this chapter we describe the domain substructure of these cellular contacts, their molecular composition, and the dynamic rearrangements of their constituents during junction formation or modulation. Four major domains are discussed: an extracellular domain consisting of the surfaces to which the cell binds; an integral membrane domain with "contact receptors" and components involved in transmembrane linkage to the cytoplasmic elements; a plaque domain, which is associated with the membrane; and microfilaments, which are attached to the plaque domain.

We describe a series of attempts to dissect the junction between neighboring domains and show that the plaque and the membrane contact are largely independent of actin. We also show that plaque-bound actin and α-actinin are relatively stable during in vivo *treatment with cytochalasin B or azide. Detachment of cell contacts by EGTA, on the other hand, leads to a rapid detachment from the membrane of plaque components, along with the attached filaments. A junctional protein with an apparent molecular weight of 135 kD that is specifically localized in intercellular adherens junctions is described. At the extracellular surface, we show that adhesion-promoting matrix proteins such as fibronectin may initially be associated with vinculin-rich contacts but are subsequently removed from these sites.*

Analysis of the dynamic properties of various cellular junctional components indicates that each component is associated with two pools: a diffusible extrajunctional pool and a junction-associated, immobile pool. We discuss the possibility that controlled exchange of components between these two pools is involved in the biogenesis and modulation of adherens junctions.

Adherens junctions are a specialized class of cell contacts formed with extracellular substrates or with the membranes of neighboring cells that are associated with actin filaments at their endofacial surfaces. Intercellular

461

adherens junctions of polarized epithelia were defined according to their position and fine ultrastructure, as revealed by transmission electron microscopy more than two decades ago (Farquhar and Palade, 1963; Staehelin, 1974). The intercellular adherens junction in the polar epithelium of the intestine, kidney, pancreas, and so on, is termed the zonula adherens and forms a continuous subapical belt with a characteristic intercellular gap of about 150–200 Å. Near the cytoplasmic faces of the junction, numerous bundled microfilaments were identified running along the junctional membrane. In other types of cells, such as cardiac myocytes and cultured fibroblasts, intercellular adherens junctions are abundant, though their topology is quite different from that of the epithelial junctions. Fascia adherens is one of the junctional elements within the intercalated disks of cardiac muscle and is often flanked by desmosomes (maculae adherentes), gap junctions, and possibly tight junction-like contacts. Transmission electron microscopy indicates that thin filaments of the terminal sarcomeres terminate in the membrane-bound, electron-dense plaques of these adherens junctions. In cultured fibroblastic or epithelial cells, numerous contact areas were detected, which according to our definition are adherens junctions. They are often smaller than the junctions mentioned above, but are laminated symmetrical structures and are clearly associated, at their cytoplasmic faces, with cortical bundles of microfilaments (Heaysman and Pegrum, 1973). Examples of some of the adherens junctions, as seen by transmission electron microscopy, are shown in Figure 1. In addition to classic adherens junctions, we have also included junctions in lens tissues (Figure 1E) that do not have the typical appearance of adherens junctions, but were identified as such by immunohistochemical labeling for vinculin, as is shown below.

Adherens junctions may also be formed with extracellular matrices and substrates. In intact tissues, "hemi-adherens junctions" with intercellular connective tissue or with basement membrane are common and widespread structures. In smooth muscle, there are numerous membrane-bound dense plaques (Figure 1A), the majority of which are not associated with cell–cell contacts but rather with areas enriched in intercellular connnective

---

Figure 1. *Transmission electron microscope morphology of intercellular adherens junctions in intact tissues and cultured cells. a*: The apical region of mouse intestinal epithelium showing microvilli (MV), tight junction (TJ), zonula adherens (AJ), and desmosome (D). *b*: A tangential section through the junctional complex of centroacinar epithelial cells of chick pancreas. *c*: The subapical adherens junction (AJ) of epithelial cells in the proximal tubules of chicken kidney. *d*: Fascia adherens (FA) and desmosomes (D) along the intercalated disk of chicken heart. *e*: Intercellular junctions between chicken lens fibers. These cells contain numerous adherens junctions, as defined by immunolabeling for actin and vinculin (see text). The junctions are probably located between gap junctions (GJ) and are often characterized by electron-dense (submembrane) material (*arrow*). *f*: Intercellular adherens junctions (AJ) formed between chick lens cells in culture. Notice the filament bundles attached to the membrane in the junction area. Calibration bars = 0.1 μm.

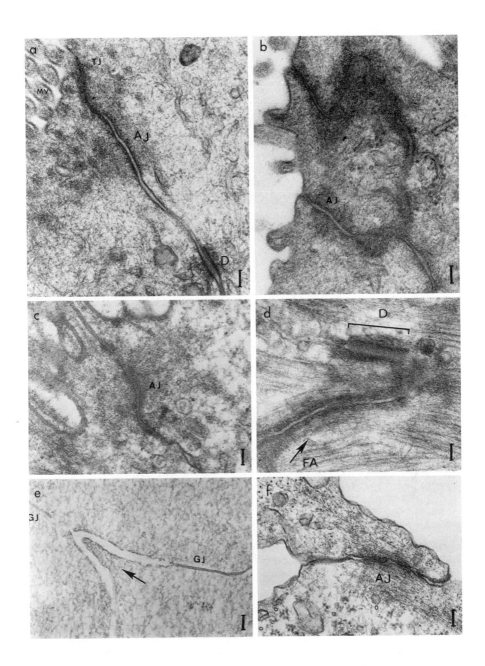

463

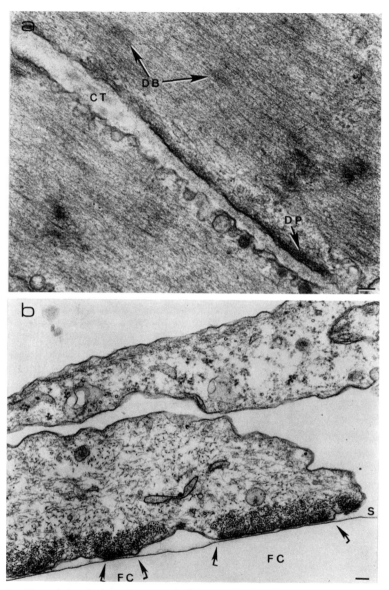

Figure 2. *Transmission electron microscopy of adherens junctions formed with intercellular connective tissue or tissue culture substrate. a:* Longitudinal section through mouse intestine smooth muscle, showing an abundance of intercellular connective tissue (CT), membrane-bound dense plaque (DP), and cytoplasmic dense bodies (DB). *b:* Cell–substrate focal contacts (FC, marked by the *twin arrows*) formed by cultured chick lens cells. Notice the abundance of microfilaments near the endofacial surfaces of the contact. S, the plane of the substrate. Calibration bars = 0.1 μm.

tissue. Most adherent cultured cells form tight adhesions to the substrate through small areas along their ventral surfaces, areas known as focal contacts or focal adhesion plaques. These areas, identified by transmission electron microscopy (Figure 2B) or by interference reflection microscopy, were shown to be associated with the termini of microfilament bundles, or stress fibers (Abercrombie et al., 1971; Heath and Dunn, 1978). Visualization of focal contacts in living motile fibroblasts indicated that these substrate contacts are relatively static structures and that, during cell locomotion, focal contacts are continuously formed under the leading lamellipodium, while the posterior ones in the trailing edge are disrupted (Abercrombie, 1980; Chen, 1981).

In view of the structural variability among different adherens junctions and the absence of distinctive intramembrane organization, it was important to identify and characterize the molecular constituents of the various adherens junctions and determine their fine topology. The identification of specific junctional components may be a key step toward understanding junction biogenesis and characterizing mechanisms involved in the transmission of transmembrane signals at sites of cell contact.

Experiments performed several years ago suggested that a reliable molecular marker for essentially all types of adherens junctions might be the cytoskeletal protein vinculin. This protein was initially shown to be associated with focal contacts, and later it was detected at the cytoplasmic surfaces of all adherens junctions tested (Geiger, 1979, 1982).

DOMAIN SUBSTRUCTURES OF ADHERENS JUNCTIONS

Adherens junctions are complex laminated structures. Electron microscope studies and various other approaches suggest that there are at least four distinct molecular domains in these junctions, positioned next to one another (Figure 3). The outermost domain contains the extracellular surfaces to which the membrane is anchored. These may consist of extracellular matrix elements, glass or plastic surfaces, or membrane components of neighboring cells. It is conceivable that this domain is different in different cells and tissues, displaying a wide molecular variability.

The second domain consists of integral membrane elements. The integral protein or proteins should be involved in at least two molecular interactions: at the cell exterior, with components of the extracellular domain mentioned above; and at the cell interior, with proteins of the cytoskeleton. In principle, the two interactions in a given junction may be mediated by one transmembrane protein or by two or more proteins which are laterally interacting.

The third domain, which may be defined as the junctional plaque, consists of peripheral membrane proteins which presumably are engaged in two types of interactions: toward the cell exterior, they interact with elements of the integral domain; toward the cell interior, they interact with

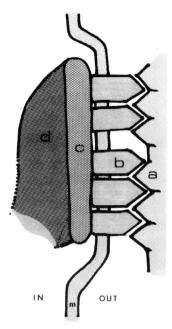

Figure 3. *A scheme showing the proposed domain substructure of adherens junctions (both intercellular and cell–substrate contacts). a*: The extracellular domain consisting of the exogenous surfaces to which the cell binds (in the case of intercellular junction this domain may consist of membrane components of the neighboring cell). *b*: The integral membrane domain containing "contact receptor(s)" and mediating the transmembrane linkage to the plaque. *c*: The membrane-associated plaque domain to which the microfilaments attach. *d*: The cytoskeletal domain attached to the plaque, consisting of actin and actin-associated proteins. The proportional dimensions of the different domains in the scheme do not represent actual proportions in adherens junctions.

actin-containing microfilament bundles. The typical plaque protein is vinculin and is discussed later.

The last and innermost domain consists of cytoplasmic microfilaments associated with the membrane-bound plaque that extend into the cortical cytoplasm. At considerable distances from the membrane these junctional filaments become intermeshed and associated with the rest of the cortical cytomatrix.

The delineation of the domain substructures of adherens junctions described here was very helpful in developing experimental approaches to elucidate the molecular properties of this class of junctions and their biogenesis. It has been possible to study the composition and dynamics of the various domains, their structural interdependence, and their interactions during the formation, maintenance, and reorganization of the junctions.

DISSECTION OF ADHERENS JUNCTIONS
INTO THEIR MOLECULAR DOMAINS

To determine the specific composition of each of the junctional domains and their interdependence, we explored the possibility of dissecting the junctions into their domains by selective perturbation of domain organization (Figure 4). A first step in this direction was made several years ago with the development of the $ZnCl_2$ method for isolating ventral membranes that retain both contact with the substrate (focal and close contacts) and association with vinculin and actin (Avnur and Geiger, 1981a). In other words, this system provided us with apparently intact adherens junctions, allowed free access to their endofacial surfaces, and avoided the use of detergents or other permeabilizing agents. We prepared such ventral membranes and tried to remove actin and vinculin selectively from their cytoplasmic faces. For the removal of actin, we used a Ca^{2+}-dependent actin-severing protein named fragmin. As reported earlier (Avnur et al., 1983), we found that short treatment (0.5–5 minutes) with fragmin in concentrations of 10–40 µg/ml caused an essentially complete removal of actin from the membranes or from Triton-permeabilized cells without affecting the cell–substrate contacts. Labeling of the same specimens for a variety of actin-associated proteins enabled us to distinguish between actin-dependent and actin-independent cytoskeletal proteins. The actin-dependent proteins, including myosin, tropomyosin, filamin, and α-actinin, were removed by fragmin along with actin. The actin-independent proteins included vinculin and talin and were not significantly affected by the fragmin treatment (Figure 4G, H). We concluded that vinculin and talin are components of the junctional plaque domain, whereas the other proteins are associated primarily with the bundle of actin microfilaments (see Figure 5). It should be mentioned that the close association of vinculin with the membrane was noticed several years ago when its organization was investigated by immunoelectron microscopy (Geiger et al., 1981; Tokuyasu et al., 1981). The single and double immunolabeling of vinculin and other actin-associated proteins on ultrathin frozen sections indicated that vinculin was considerably closer to the cytoplasmic faces of the membrane in the junction areas than any other protein tested.

Additional attempts were made to disconnect the cytoskeletal domain from the plaque domain *in vivo* by treating cells with cytochalasin B (Figure 4E, E′) or with azide (Figure 4F). The former treatment brought about considerable deformation of the microfilament system and the loss of most stress fibers, but initially had relatively limited effect on focal contacts. More specifically, double immunolabeling for actin and vinculin showed that in focal contacts actin was relatively stable and remained associated with the membrane long after the stress fibers were deformed or destroyed. Prolonged azide treatment had detrimental effects on stress fibers (repre-

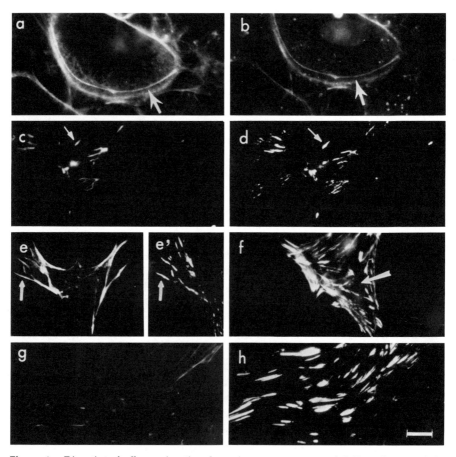

Figure 4. *Dissection of adherens junctions by various treatments. a* and *b*: Detachment of the vinculin-rich plaque (*b*) and associated actin-containing belt (*a*), labeled with fluorescent phalloidin, from the junctional membrane of MDBK cells after 30 minutes treatment with 2 mM EGTA. The *matched arrows* point to the detached belt. *c* and *d*: Extraction of actin (*c*) and vinculin (*d*) from ventral membrane of chicken fibroblasts isolated by 10 minutes treatment with 1 M KI. Notice that most of the residual actin left after extraction was associated with vinculin-containing sites. In many instances actin was removed and vinculin remained associated with the membrane. *e* and *e'*: Actin and vinculin, respectively, in cells incubated before fixation for 20 minutes with 5 μg/ml cytochalasin B. Most of the stress fibers disappear and only their terminal portions attached to the vinculin plaques are retained. *f*: Immunofluorescent labeling of chick fibroblast for α-actinin after 30 minutes incubation with 20 mM NaN₃. Notice that stress fiber-associated labeling disappears while the focal contact-associated fluorescence is apparently unaffected. *g* and *h*: Labeling for actin and vinculin of ventral membranes pretreated with fragmin. Notice the extensive removal of actin from the membrane and the marked stability of vinculin toward this treatment. Calibration bar = 10 μm.

sented by α-actinin in Figure 4F), but again did not seem to dissociate actin and α-actinin from the vinculin-rich plaques. High salt extraction, which effectively removed actin from ventral membranes (Figure 4C, D), indicated that focal contact-bound actin was relatively resistant to such chemical extractions. These treatments resulted in the selective removal of actin and vinculin from focal contacts and revealed the substructure of the extraction-resistant residues of the two proteins in the contact areas. The high salt-resistant proteins often displayed a dotted distribution within focal contacts. These subfoci exhibited linear alignment along the axis of the plaque. The significance of these resistant clusters is yet to be elucidated. It should be mentioned that the association of actin with the plaque did not seem to depend on the presence of intracellular divalent ions, since treatment of ventral membranes with EGTA or EDTA had limited effect on focal contact-bound actin.

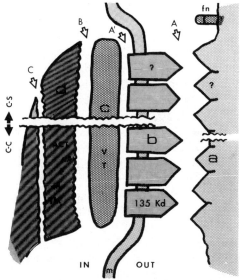

Figure 5. *A scheme summarizing the dissection of adherens junctions into their domains (a–d as in Figure 3).* The *labeled arrows* at the top point to the specific perturbation of the interactions between domains. A and A' represent the dissociation of the membrane receptors from the extracellular domain, and the subsequent detachment of the plaque from the membrane, respectively. B represents the removal of actin by fragmin. C represents the effect of microfilament-disrupting treatments such as cytochalasin B, azide, and various extractions. In this scheme we distinguish between cell–cell (C–C) and cell–substrate (C–S) contacts, which may differ in their molecular constituents. We indicate the locations of some of the relevant molecules in various domains. Fibronectin (fn) is shown here outside the contact area, as is discussed later in the text. A membrane component of intercellular adherens junctions (only) is the 135-kD protein. The plaque contains vinculin (V); the cytoskeletal domain contains actin, α-actinin, filamin, myosin, tropomyosin, and probably additional actin-associated proteins.

Additional experiments were carried out in an attempt to dissect the chain of domains further away between the plaque and the integral membrane constituents (domains c and b in Figure 3, respectively). Exposure of isolated ventral membranes to a large variety of buffer conditions did not result in a specific removal of vinculin. As pointed out above, the treatments that removed vinculin also removed actin, and usually had a marked effect on the substrate contacts themselves. Apparently, a dissociation of the membrane from the vinculin–actin complexes in intercellular adherens junctions can be induced in living cultured cells by the removal of extracellular Ca^{2+}. Madin-Darby bovine kidney (MDBK) cells form sheets of polarized epithelium in culture. As has been shown (Kartenbeck et al., 1982), the addition of EGTA to such cultures results in a rapid deterioration of the subapical junctional complex formed between the cells. The transmembrane conductance rapidly increases, suggesting the breakdown of tight junctions. In addition, marked changes occur in the structure of both desmosomes and adherens junctions. The former exhibit a marked sensitivity to Ca^{2+} withdrawal and split into two "hemidesmosomes" that are rapidly endocytosed by the cells. A similar phenomenon was noticed when cell monolayers were dispersed by trypsin treatment (Overton, 1974). The hemidesmosome-bound vesicles could be detected within the cytoplasm for long periods after the endocytosis, and there still is no direct indication as to whether they may be reutilized, that is, recirculate and reappear on the cell surface.

The zonula adherens in the same cells undergoes a very different mode of reorganization. Within a few minutes after removal of extracellular Ca^{2+}, the entire junctional belt dissociates from the endofacial surfaces of the membrane and contracts toward the cell center. (This is represented by the arrows A and A' in the scheme in Figure 5). Immunofluorescent labeling of EGTA-treated cells indicates that actin and vinculin remain associated and detach together from the membrane, suggesting that exogenous Ca^{2+} and the maintenance of the intercellular contacts are essential for continuous association between the plaque and the cell membrane. Further studies on the fate of the detached cytoskeletal belt, which could be detected in the medullary cytoplasm by transmission electron microscopy, indicate that the process is largely irreversible—the vinculin- and actin-containing bundles can not reassociate with the membrane, even if the cells are supplemented with Ca^{2+}-containing medium shortly after the EGTA treatment. This suggests that the reassembly of the junctional domains is polar and depends primarily on the reestablishment of new intercellular contacts. Further details and a discussion of these experiments will be provided elsewhere (T. Volberg, W. W. Franke, J. Kartenbeck, and B. Geiger, in preparation).

The dissection of the various types of adherens junctions into their four major domains is summarized schematically in Figure 5. Arrows A, A', B, and C represent different treatments that cleave neighboring domains (see legend). In the subsequent sections we briefly discuss the molecular properties and interactions of the four domains.

THE CYTOSKELETAL DOMAIN

Several microfilament-associated proteins have been detected at some distance from the junctional plasma membrane. Among these are actin, myosin, tropomyosin, filamin, and α-actinin. Localization experiments by immunoelectron and immunofluorescence microscopy indicate that each of these proteins exhibits specific and distinctive organization along the membrane-bound bundles. In most of the adherens junctions examined (including focal contacts, dense plaque of smooth muscle, and fascia adherens), actin apparently terminated end-on in the membrane-bound plaque. Less clear is the mode of association of actin with the junctional plaque in the zonulae adherentes of polarized epithelia. Ultrastructural observations indicate that the junctional belt runs largely parallel to the membrane, although the possibility can not be excluded that individual actin filaments or groups of them diverge from the bundle and anchor end-on in the plaque.

Decoration of actin filaments that terminate in the junctional plaque with such myosin fragments as heavy meromyosin or S_1 indicate that while actin filaments along the bundle show mixed polarity, the filaments attached to the plaque are uniformly oriented, with their barbed ends facing the membrane (Begg et al., 1978; Goldman et al., 1979).

The terminal area of the microfilament bundle contains α-actinin and often filamin. Both proteins have been detected in the neighborhood of several types of adherens junctions by use of immunofluorescence and immunoelectron microscopy. The latter studies have shown that α-actinin was particularly enriched immediately next to the junctional plaque (Geiger et al., 1981). It should be pointed out that recent studies have shown that α-actinin can bind to diacylglycerol and palmitic acid, raising the possibility that the protein may be directly linked to the membrane bilayer (Meyer et al., 1982; P. Burn, A. Rotman, R. K. Meyer, and M. M. Burger, *Nature*, 1985, in press). Immunoelectron microscope studies of adherens junctions in cells and tissues did not favor such association, but they may exist at other cellular locations such as leading lamella and membrane ruffles. Other actin-associated proteins, including myosin and tropomyosin, were associated with filaments at a considerable distance from the membrane, suggesting that their involvement in the anchorage of the filaments is limited and probably indirect.

A related topic is the dynamic rearrangement of the various cytoskeletal elements in living cells. This aspect was discussed in detail elsewhere (Geiger et al., 1983) and is therefore only briefly considered here. As evident from numerous immunocytochemical and biochemical studies, actin and its associated proteins display different forms of organization and levels of assembly in the various spatial domains within cells. In addition to the tightly packed F-actin bundles, actin may form meshworks or webs of individual filaments in the cortical cytoplasm and in lamellipodia and ruffles. Microfilaments may also form polygonal baskets in the perinuclear regions and may exist in a soluble, diffusible state in the form of either monomeric G-actin or relatively short oligomers. It is conceivable that the dynamic molecular transformations of actin from one form of assembly to the other, the involvement of actin-associated proteins, the effect of specific environmental stimuli, and the roles of intrinsic constraints are complex processes that play major roles in cellular dynamics.

In our studies we have tried to characterize the dynamic state of cytoskeletal elements in or near cell–substrate adherens junctions. As described previously (Kreis et al., 1982; Geiger et al., 1983; Kreis et al., 1985), we have microinjected fluorescently labeled actin, α-actinin, and vinculin into living fibroblasts, and by following their incorporation into stress fibers and/or focal contacts we have measured their gross translational movements by fluorescence photobleaching recovery (FPR) techniques (see Schlessinger and Elson, 1982). The results indicate that both α-actinin and actin are organized in two distinct forms: a diffusible cytoplasmic pool, and a relatively immobile pool associated with the stress fibers along their length, including their plaque-bound termini. The diffusion coefficients of the "soluble" α-actinin and actin is about 3×10^{-9} cm^2/sec. These values are comparable to those obtained with the same experimental system for such irrelevant proteins as bovine serum albumin or goat gammaglobulin (diffusion coefficients of both are about 6×10^{-9} cm^2/sec). It was thus concluded that the diffusible actin pool contains either G-actin monomers or short oligomers. Both actin and α-actinin appeared essentially immobile in their cytoskeleton-bound forms. In stress fibers, both exhibited a slow, though highly significant, recovery of fluorescence after photobleaching with a half-time in the order of minutes. The rate of recovery was largely unrelated to the diameter of the bleaching beam (providing the latter was small relative to the surface area of the cell). This indicates that the "slow" recovery is not controlled by diffusion (for further discussion, see Geiger et al., 1983).

A similar indication for the binding or exchange of cytoskeletal components between soluble and junctional pools was obtained by a different experimental approach. Fluorescently labeled α-actinin, when incubated with unfixed ventral membranes (purified by the ZnCl$_2$ method), was avidly bound to the α-actinin-rich striations and plaques along stress fibers and at their termini. This binding could be inhibited by an excess of unlabeled α-actinin or by heavy meromyosin, suggesting that there are spatially restricted sites along the actin bundle which are free and capable of binding exogenously added soluble α-actinin. Similar experiments with fluorescently labeled, polymerization-competent actin yielded inconclusive results, since the added G-actin tended to polymerize from many sites, including the cell-free surfaces of the coverslips. It is nevertheless conceivable that careful selection of polymerization conditions, as well as modification in the mode of membrane isolation, may shed light on the possible role of adherens junctions and their membrane-bound plaques in the nucleation of actin polymerization.

The various experiments described indicate that the cytoskeletal domain of adherens junction is a molecularly defined structure and that its major known components, actin and α-actinin, maintain a dynamic exchange with a soluble cytoplasmic pool of the respective components. That such an equilibrium exists suggests the possibility that changes in the size or form of organization of microfilament bundles may result from a perturbation to the dynamic equilibrium that exists between the mobile and immobile pools of the related cytoskeletal components.

THE MEMBRANE-BOUND PLAQUE DOMAIN

The existence of a plaque as an independent junctional domain has been demonstrated by several experimental approaches. Electron microscope examinations of a variety of adherens-type junctions indicated that an electron-dense layer was present immediately subjacent to the junctional membrane (Figures 1, 2; Staehelin, 1974). The density of this plaque and its dimensions varied from one junction to another, and the extent of molecular homology between them could not be directly appreciated from structural studies only.

Immunocytochemical labeling at the electron microscope level provided the first indications of the spatial segregation of the cytoskeletal domain and the plaque domain. Vinculin was associated with the plaque, whereas α-actinin, tropomyosin, and myosin were clearly excluded from the immediate vicinity of the membrane and were present further away from it. A rough quantitative analysis of the distribution of immunoferritin molecules as a function of distance from the endofacial surfaces of the junction have shown that most (approximately 60%) of the label for vinculin was restricted to a relatively narrow zone of about 500 Å next to the plasma membrane. More than 80% of the labeling for α-actinin in the same sections of chicken intestine was detected further away from the membrane, concentrated predominantly in the 500–1500 Å zone. Unfortunately, the level of resolution of indirect immunofluorescence used in those studies was 200–300 Å (the maximal distance from the antigen-binding site on the primary antibody to the core of ferritin bound to the secondary antibody). This resolution seems to be too close to the thickness of the plaque in many systems and thus does not allow for an accurate localization of specific proteins in defined subregions across the plaque.

So far, two proteins have been localized in the membrane-bound plaques of adherens junctions. Vinculin was initially isolated from chicken smooth muscle and subsequently localized in all types of adherens junctions, both cell–cell and cell–substrate. In fact, vinculin still seems to be the best-characterized marker for adherens junctions. The use of vinculin as a junctional marker was helpful in identifying "new" adherens junctions with atypical morphology. One such example is the junction formed between lens fibers or between adjacent lens-forming cells. These cells contain, beside many gap junctions, extended regions with irregular submembrane densities that could not be conclusively defined as adherens junctions on the basis of electron microscope morphology. Immunolabeling for vinculin and other junctional proteins indicates that these regions correspond to vinculin-positive intercellular junctions, namely, adherens junctions. Another plaque component present in focal contacts is a recently described 215-kD protein named talin (Burridge and Connell, 1983). Careful double immunofluorescence experiments show that both plaque proteins have an essentially identical distribution along focal contact areas. In other sites in the same cells some differences have been noted, including the presence of talin in vinculin-free areas along the leading lamella. In addition, it is not clear whether talin is present only in cell–substrate or extracellular matrix adherens junctions or is also found in intercellular junctions. Undoubtedly the availability of antibodies to both components will make it possible to clarify the gross spatial relationships of the two proteins in many cell types and in various types of adherens junctions (see note added in proof at the end of this chapter).

The dynamic properties of vinculin were determined by using the combination of microinjection and FPR measurements described above for actin and α-actinin. Although there were some differences in the diffusion coefficients and rate of slow recovery between vinculin and α-actinin or actin, the general behavior of all three proteins was similar. We have concluded that vinculin can be either soluble in the cytoplasm or anchored in the plaque domain and that these two pools maintain a continuous exchange (for further details, see Kreis et al., 1985).

To further analyze the binding of soluble vinculin to the plaque, we have used rhodamine- or fluorescein-coupled vinculin for the "decoration" of isolated ventral membranes. The labeled protein shows highly specific binding to focal contacts and has no apparent affinity to nonjunctional membranes. To our surprise, we noticed that, unlike the binding of α-actinin, the binding of vinculin is not readily saturable, suggesting that it might undergo local aggregation within the plaque domain.

On the basis of biochemical analyses (Jockusch and Isenberg, 1981; Burridge and Feramisco, 1982; Wilkins and Lin, 1982), as well as the *in situ* binding experiments and the dissection of the various junctional domains, one may assume that proteins of the plaque domain may be involved in several molecular interactions: (1) Vinculin may bind to F-actin and probably induces the assembly of filaments into bundles; (2) vinculin can bind to an additional protein with an apparent molecular weight of about 200 kD, probably talin (Burridge and Connell, 1983; Otto, 1983); (3) the binding of vinculin to the plaque is largely independent of actin and is not affected significantly by the removal of actin with fragmin, suggesting that vinculin interacts avidly with nonactin components in the plaque; (4) the association of soluble vinculin with the plaque, as determined by direct binding assay to isolated focal contacts, is Ca^{2+}-independent and occurs over a wide range of pH (5.5–8.5) and ionic strength (KCl concentrations 5–300 mM); (5) the plaque probably represents a multilayered structure of vinculin, as suggested by immunoelectron microscopy and by the nonsaturating kinetics obtained by the *in situ* binding analysis mentioned above; (6) the association of the entire plaque with the membrane depends on the maintenance of cell contact. Upon the disruption of the cell contact by EGTA or trypsin, the plaque may dissociate from the junctional membrane.

THE MEMBRANE DOMAIN

Integral membrane proteins are expected to play key roles in adherens junction formation, since they presumably mediate the transmembrane linkage between the external and internal domains. Moreover, it is conceivable that the specificity of cellular recognition leading to the formation of adherens junctions depends on the availability of specific "contact receptors" on the cell surface. It should be pointed out that until recently, essentially no information was available on the identity of junctional receptors specific for adherens junctions. The characterization of desmosomal and gap junctional components (see chapters by Cowin et al. and Gilula, this volume) was made possible by the development of reliable and relatively simple procedures for the purification and isolation of these two junctions. Unfortunately, there still is no comparable procedure for the isolation of adherens junctions. Moreover, functional assays, such as inhibition of contact formation, were found useful for the identification and isolation of contact-related molecules, as discussed in detail by several authors in this volume. Most of these molecules, however, were found to be involved in cell attachment and aggregation, but they were not the specific residents of a particular class of junctions.

Two aspects of junctional membranes have emerged from our studies: (1) translational movements of membrane lipids and proteins in cell–substrate adhesions; and (2) identification of a new membrane protein apparently specific for intercellular adherens junctions. The first topic was approached several years ago by using the $ZnCl_2$ method to isolate ventral membranes (Avnur and Geiger, 1981a) combined with FPR techniques. Two types of fluorescence markers were introduced into the plasma membranes of cultured fibroblasts—a lipid probe (WW591) and a nonpenetrating fluorophore, lissamine–rhodamine sulfonylchloride, which was bound covalently to membrane proteins (for details, see Geiger et al., 1982, 1983). FPR analysis has shown that the lipid probe is free to move through focal contacts, though at a reduced rate, compared to neighboring unattached membranes. Different information was obtained from FPR experiments with membranes whose proteins were fluorescently labeled *in vivo*. These experiments indicated that there are two comparable populations of membrane proteins with markedly different dynamic properties; about 50% of the fluorescence recovers rapidly after photobleaching, with an apparent diffusion coefficient of 0.8×10^{-9} cm^2/sec. This value is lower by about 50% than the diffusion coefficients measured in nonattached areas along the ventral membrane. The other half of the fluorescence is largely immobile and does not recover within the time scale of diffusion ($D < 10^{-12}$ cm^2/sec). While this approach could not provide direct information on the molecular nature of components (lipids or proteins) that are specifically associated with these adherens junctions, it provides an insight into the general biophysical state of this membrane microdomain. It indicates that a significant population of membrane proteins within the contact is not free to diffuse laterally, possibly due to "vertical" interactions either with the substrate, with the immobile plaque domain, or with both. The existence of a mobile pool of proteins and lipids within the focal contacts suggests that these regions do not contain major diffusional barriers. Thus, proteins which are not anchored can diffuse into, away from, or through the junction. The reduced rate of movement observed for lipids and proteins in the contact may reflect merely a reduction in the space free for diffusion. From the functional point of view, this suggests that membrane receptors within the junction which become temporarily free may diffuse away from the junction, leading to weakening of the contact, while extrajunctional contact receptors may diffuse randomly into the junctions and may be trapped and immobilized in them, leading to strengthening of the cell contact.

As for the molecular nature of the contact receptors, only limited information is available. Most attempts to identify contact receptors by functional assays have led to the identification of membrane molecules that are involved in recognition and contact formation but are not structural elements of the adherens junctions themselves. We have identified a membrane protein specifically associated with the intercellular subclass of adherens junctions. Our experiments included the following steps: (1) purification of intercalated disks from chicken cardiac muscle; (2) isolation of membrane proteins from this fraction by extraction with Triton X-114 and detergent partitioning; (3) preparation of monoclonal antibodies to the extracted membrane proteins; (4) selection of clones producing antibodies that react specifically with adherens junctions. Details of the cloning were published in Volk and Geiger (1984). Here we discuss the major findings and observations obtained with one monoclonal antibody called ID 7.2.3. In a radioimmunoassay, this antibody reacted with intercalated disk-rich membrane fractions of cardiac muscle and only poorly with nonjunctional membranes. Immunofluorescence labeling of frozen thin sections of chicken heart showed that the antigen was particularly enriched in intercalated disks and in sparsely spaced dots along the lateral cell borders. Among the various additional tissues tested, lens fibers were most extensively labeled along the cell membranes. In cultured cells, the spatial distribution of the new protein could be demonstrated at a higher level of resolution, even at the level of immunofluorescence. In permeabilized and fixed cells, extensive labeling was found along areas of intercellular contact. In heart myocyte cultures, large areas of positively labeled contacts were observed; fibroblastic cells in the same cultures or in chick gizzard cell cultures formed contacts through many processes that apparently contained the antigen recognized by ID 7.2.3 antibodies. Extensive labeling was observed along the intercellular junctions formed by cultured eye lens epithelial cells (Figure 6). Electron microscope examination of these cells indicated that they contained a most prominent subapical bundle of microfilaments that invariably terminated in a junctional plaque, as well as very prominent stress fibers bound to focal contacts at their ventral surfaces (Figure 6A). When double immunofluorescence labeling of the same cells with ID 7.2.3 and fluorescent anti-vinculin or phalloidin was used, the labeling was exclusively associated with the junctional membrane in close association with vinculin-rich plaques (Figure 6B, C) and termini of actin bundles (Figure 6D, E). An important difference was noted, however, between the staining pattern with ID 7.2.3 and anti-vinculin: the latter stained extensively both intercellular and cell–substrate contacts; the former reacted with intercellular adherens junctions exclusively.

As mentioned above, the positive immunostaining of cultured cells required permeabilization prior to labeling. It was essential to determine whether the antigen was actually intracellular or was facing the junctional gap and thus not accessible to the immune reagents. A direct indication that the antigen was indeed exposed on the exofacial surfaces of the membrane was obtained by immunolabeling of fixed or viable cells after short dissociation of the junctions by the Ca^{2+} chelator EGTA. We have found that after relatively short incubation with EGTA (0.5–10 minutes), the intercellular adherens junctions were split open, fully exposing the antigenic determinants of their outer surfaces. Further support was obtained from immunoelectron microscope labeling of EGTA-treated cells by monoclonal antibodies ID 7.2.3 and gold-conjugated anti-mouse IgG. Since the ultrastructure of the junction was not retained after this treatment and the antigen apparently diffused away from its original site, the labeling was broadly distributed over the cell surface. Nevertheless, all of the label was tightly associated with the plasma membrane. It would obviously be desirable to obtain complementary immunoelectron microscope labeling for the new antigen on frozen sections of intact tissue. Such experiments are now being carried out, and preliminary results are consistent with the notions specified above.

Immunoblotting analysis of extracts prepared from a variety of tissues indicate that the antigen is present in all tissues tested, including various muscles, liver, brain, and lens, though its relative concentrations in cardiac muscle and in lens are clearly elevated. The band labeled by ID 7.2.3

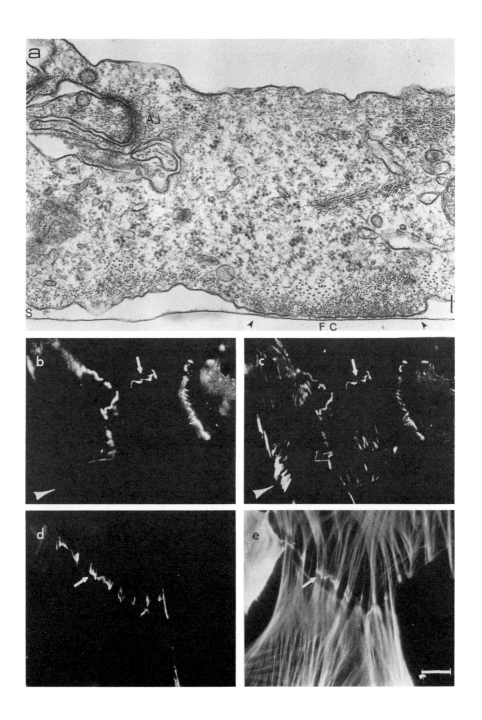

antibodies has an apparent molecular weight of 135 kD. The physiological roles of the 135-kD protein are still not clear. Preliminary experiments suggest that while ID 7.2.3 antibodies do not have dramatic effects on cultured chick lens cells, they interfere with the reformation of intercellular adherens junctions during the recovery of intercellular contacts following EGTA treatment. The limited available information clearly indicates that the 135-kD protein is a specific junctional protein which appears to be closely related to intercellular adherens junctions.

An important aspect that should be reemphasized is the absence of the 135-kD protein from focal contacts or cell–matrix contacts such as dense plaques of smooth muscle. It appears that the molecular homology between cell–cell and cell–matrix junctions is partial and that the two may differ in their specific membrane constituents. It is conceivable that interactions of cell membranes with different surfaces, extracellular matrix components, or neighboring membranes may be mediated by different receptors, though the aggregation or immobilization of each of these receptors in a newly formed junction may subsequently lead to a similar assembly of the subjacent plaque and cytoskeletal domains.

How many different membrane proteins are involved in the construction of a single adherens junction? How large is the variability between the membrane receptors of different adherens junctions? The limited amount of data available is clearly insufficient to permit answers to these questions. In fact, the notion that intercellular and cell–substrate junctions involve different membrane molecules is based on our preliminary observations, which in turn are based on our experience with only one protein. Further progress in this line of studies will require the identification of a complete battery of contact receptors specific for different adherens junctions. In view of the difficulties of purifying these receptors and the possibility that the junctional membranes contain nonjunctional proteins in addition to contact receptors, it seems that a monoclonal antibody approach, similar to the one used here, might be suitable for the identification and characterization of additional receptors. Indeed, by using a similar approach, Oesch and Birchmeier (1982) have identified a protein that apparently is related to focal contacts and is reported to be a surface component, because antibodies reacting with it inhibit cell attachment. The detailed tissue distribution of this antigen has not yet been reported.

Figure 6. *Localization of the 135-kD protein in cultured chick lens cells. a*: Transmission electron micrograph showing both intercellular adherens junctions (AJ) and focal contacts (FC) in the same transverse section. *b* and *c*: Double immunofluorescent labeling with ID 7.2.3 for the 135-kD protein and with anti-vinculin. Notice the colocalization of the two in intercellular contacts (*arrows*) and the absence of the former from cell–substrate focal contacts (*arrowheads*). *d* and *e*: Double immunofluorescence localization of the 135-kD protein and actin in the same cultured cells. The filament bundles apparently terminate in the 135-kD protein-containing junction. Calibration bars = 0.1 μm (*a*); 10 μm (*b–e*).

THE EXTRACELLULAR DOMAIN

The outermost compartment of the junction is the extracellular domain, which consists of those structures or molecules to which the contact receptors bind. In view of the diverse modes of attachment in different adherens junctions (different artificial or natural substrates, neighboring cells, basement membranes), one may consider two alternative possibilities for the specificity of recognition in these junctions: (1) The junctional membrane may contain many different receptors with broad substrate specificities, each of which interacts directly with specific extracellular substrates; (2) the contact in the different adherens junctions may be mediated through one or only a few receptors reacting with specific "adhesive molecule(s)" which are adsorbed on the different substrates and may even be deposited there by the cells.

Unfortunately, in contrast to the wealth of information available on adhesion-promoting molecules, very little is known about the exogenous molecules involved in the formation of any junction, including adherens junctions. In this section, we shall not discuss the possible direct inducers of focal contact formation, but rather describe studies carried out in our laboratory that characterize the relationships between known adhesive macromolecules and focal contacts.

A few years ago, we tried to correlate the distribution of specific extracellular matrix components to focal contacts by using double immunofluorescence labeling for fibronectin or collagen and vinculin. To our surprise, we found that focal contacts were usually devoid of labeling for fibronectin and collagen. This apparent exclusion was most prominent when the cells were fixed prior to permeabilization. We were therefore concerned with the possibility that this pattern resulted (at least in part) from lack of accessibility of the immunological reagents, rather than from true absence of the matrix proteins (Avnur and Geiger, 1981b). To clarify this point, we took two alternative experimental approaches: (1) extensive permeabilization prior to immunolabeling; (2) plating of cells on rhodamine-conjugated fibronectin.

Permeabilization revealed two distinctly different relationships between fibronectin and vinculin. Large and apparently "mature" focal contacts were usually devoid of fibronectin, whereas significant correlation could be detected between threadlike, vinculin-containing contacts and small fibronectin cables (both patterns can be seen in Figure 7A, B). This observation was subsequently corroborated by immunoelectron microscope labeling of ultrathin frozen sections of cultured fibroblasts for vinculin and fibronectin (Chen and Singer, 1982). It was shown that although fibronectin could be detected in close substrate contacts in an apparent association with vinculin, the "true" focal contacts (as defined by electron microscope criteria) were usually devoid of this matrix protein. The coexistence of fibronectin-containing and fibronectin-free contacts, and the possible in-

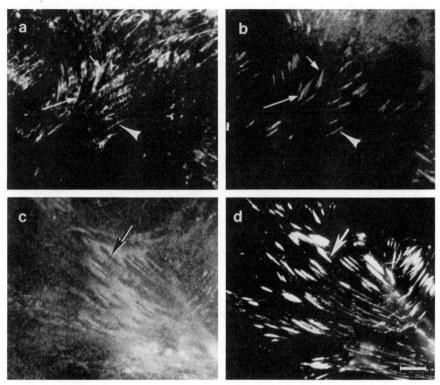

Figure 7.   *Colocalization of fibronectin (a) or chondroitin sulfate proteoglycans (c) and vinculin (b and d) in cultured chick fibroblasts.* Notice the exclusion of labeling for fibronectin from large focal contacts (*arrows* in *a* and *b*) and the presence of the two in "fibrillary" contacts (*arrowheads*). The cells in *c* and *d* were extensively permeabilized before fixation. Nevertheless, the exclusion of proteoglycan labeling from many focal contacts (*arrows*) is clearly apparent. Calibration bar = 10 μm.

terconversion from one to the other, may be the reason for some discrepancy among different reports on the interrelationships between vinculin and fibronectin (Birchmeier et al., 1980; Avnur and Geiger, 1981b; Singer and Paradiso, 1981; Hynes et al., 1982). As an extension to these studies, we examined the distributions of other extracellular matrix components and their relationships to focal contacts. Collagen, as pointed out, shows similar distributions to those of fibronectin and is largely absent from defined vinculin-rich focal contacts. Laminin is considerably more sparse than fibronectin, is found predominantly on the dorsal surfaces of the cells, and shows no spatial relationships to focal contacts. Recently, we prepared monoclonal antibodies reactive with chondroitin sulfate-containing proteoglycan. Staining of fibroblast cultures with these antibodies indicates that focal contacts, especially those formed by relatively stationary cells, are not labeled for the proteoglycan, whereas neighboring regions are highly enriched with it (see Figure 7C, D).

A general scheme for the involvement of fibronectin in contact formation may be as follows: Fibronectin, being an "adhesive" matrix protein, may promote cell attachment and spreading via interaction with specific surface receptors. Thus, cell contacts may primarily and preferentially be formed with fibronectin-rich areas on the substrates. These newly formed contacts may induce local organization of vinculin and may correspond to those threadlike adhesions described above (arrowheads in Figure 7A, B). However, at later stages when these contacts extend their areas and develop into large focal contacts, fibronectin may be excluded, either because of lack of accessibility for newly secreted fibronectin or because of active removal of preexisting fibronectin matrices from these areas.

A second set of experiments was directed toward the mechanism of exclusion of fibronectin from mature focal contacts. Cells were plated onto cover slips coated with rhodamine-labeled fibronectin, the cultures were fixed after different periods of time, and vinculin or actin was localized in them by immunofluorescence microscopy. The results (Figure 8; Avnur and Geiger, 1981b) indicated that shortly after cells adhered to the fibronectin "carpet," the underlying matrix protein was progressively removed from the nascent contacts and reorganized into extracellular matrix cables similar to those commonly found in fibroblast cultures. It is conceivable that the cables of fibronectin around cells are formed by this active reorganization. Interestingly, the displacement of fibronectin had a fixed directionality, and the matrix protein was displaced centripetally from the periphery of the cells toward their center. This is reminiscent of the centripetal movement of adsorbed particles on the surfaces of cells (Harris and Dunn, 1972), and suggests that a centripetal flow of membrane components occurs in focal contacts at the ventral cell surfaces.

Our final topic in relation to the extracellular matrix–focal contact relationships is the requirement for exogenous extracellular matrix components to induce focal contact formation. We found that chick fibroblasts plated on coverslips in serum-free medium fail to develop normal focal contacts (Figure 9B; cf. the control in A). We have tried to determine what serum component (s) are essential or sufficient for the induction of focal contacts. We found that the addition of fibronectin to serum-free medium or precoating of the coverslips with the matrix protein (Figure 9C) completely restored and often facilitated the formation of focal contacts (compared to cultures

---

Figure 8. *Removal of rhodamine-conjugated fibronectin from the substrate by cultured chick fibroblasts. a and b: A group of cells during early stages of spreading, fixed and photographed with fluorescent (a) or phase contrast (b) optics. Various degrees of removal of the substrate-attached fibronectin are seen; the matched arrows point to very early stages of fibronectin removal. In all cases the substrate-attached fibronectin is displaced unidirectionally toward the cell center. c: Pattern of distributions of rhodamine fibronectin after longer incubation time (1 hour) showing removal of the matrix protein from focal contact-like regions and the formation of extracellular fibronectin cables. Calibration bar = 10 μm.*

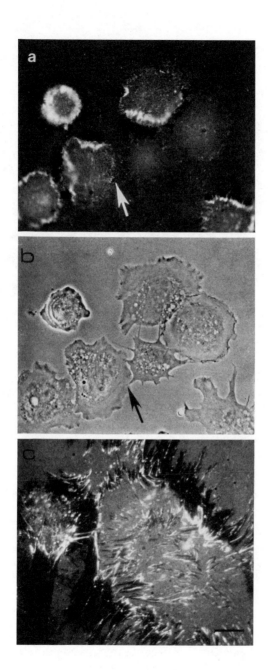

483

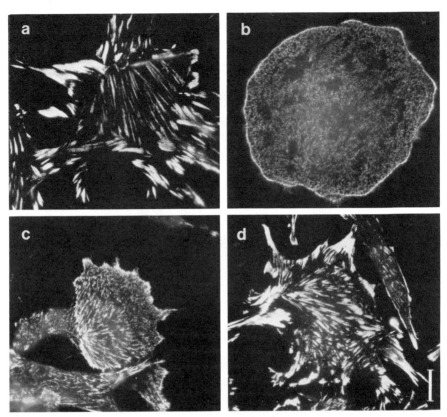

Figure 9. *Effect of fibronectin on focal contact formation and the assembly of vinculin in chick fibroblasts.* a: Cells maintained in regular 10% fetal calf serum-containing medium, fixed two hours after plating and stained for vinculin. b: A typical distribution of vinculin in cells fixed two hours after plating in serum-free Dulbecco's modified Eagle's medium containing 2 mg/ml insulin and 5 μg/ml transferrin. c: The pattern of distribution of vinculin in cells plated as above in medium containing 10% fibronectin-depleted fetal calf serum. Notice the absence of normal focal contacts and the abundance of small "fibrillary" contacts similar to those shown in b. d: The pattern of distribution of vinculin in cells plated in serum-free medium two hours earlier on fibronectin-coated coverslips. Notice the effect of the fibronectin-carpet on the formation of vinculin-rich focal contacts (cf. b and d). Calibration bar = 10 μm.

seeded in serum-containing medium). Removal of fibronectin (and possibly also other gelatin-binding proteins) from fetal calf serum by adsorption to Sepharose-bound gelatin column causes a marked inhibition of focal contact formation (Figure 9C). Focal contacts are formed only after several hours, probably as a result of the production and secretion of endogenous fibronectin by the cells.

Thus it appears that fibronectin plays conflicting roles in different phases of focal contact formation. The protein is required for the formation of initial contacts with the substrate, but as soon as these adhesions develop into typical focal contacts fibronectin is displaced to extrajunctional areas. Two additional observations that deserve reemphasis are the apparent absence of secretory activity in focal contact areas and the directional, centripetal displacement of the extracellular matrix carpet.

TRANSMEMBRANE COMMUNICATION IN ADHERENS JUNCTIONS: CONCLUDING COMMENTS

We have reported on the domain substructure of different adherens junctions and elucidated some of their molecular properties. In summary, we have shown that:

1. Adherens junctions are a family of cell contacts with similar modes of attachments to cytoskeletal microfilaments.
2. All adherens junctions are associated with actin filaments through a vinculin-containing plaque.
3. The adherens junctions are laminated structures consisting of four major domains, including anchoring substrates, membrane domains, plaque domains, and cytoskeletal filaments.
4. Each domain contains a unique set of proteins and displays a unidirectional interdependence: microfilament may be artificially removed without affecting the more distal plaque, but dissociation of the membrane contact leads to complete detachment of all other junctional domains.
5. All known cellular components of adherens junctions maintain an equilibrium between two pools: a junctional pool of relatively immobile components and a diffusible pool of extrajunctional proteins.
6. The formation of adherens junctions is a polar process triggered by contact and proceeds by a sequential assembly of the more central domains.
7. In contrast to the molecularly uniform plaque and cytoskeletal domains, there appears to be heterogeneity in the integral domain of the membrane and the extracellular adhesive surfaces.
8. A protein exclusively associated with intercellular adherens junctions was described. This protein, with an apparent molecular weight of 135 kD, appears to be a membrane protein.
9. Among the various adhesive proteins and proteoglycans tested, none was found to be specifically and reproducibly associated with the various adherens junctions. Moreover, some extracellular matrix proteins such as fibronectin were usually actively displaced from underneath focal contacts.

The biological role of adherens junctions and the nature of the transmembrane signals transmitted in these areas are extremely complex, in view of the wealth of junctional components, their interactivity, modifications, and rearrangements. Experiments based on the perturbation of specific junctional domains are very difficult, if not impossible, to interpret. Nevertheless, we outline here several cellular processes that seem to depend on the formation of adherens junctions. These may be classified in two major categories: (1) the formation of mechanical transmembrane or transcellular linkages; (2) the transmission of signals that affect and regulate cell behavior and growth.

The formation of linkages seems to be directly related to the assembly of the junction. In motile cells, focal contacts are formed under the anterior protrusions of leading lamella and serve as new organizing centers for microfilament bundles. These bundles elongate and intermesh with the rest of the microfilament-related contractile system, and usually apolar focal contacts lead to the massive assembly of actin. In other systems, such as the subapical junctions of epithelium, transmembrane interaction between the cytoskeletal junctional belts of neighboring cells provides a mechanically integrated contractile network through the entire sheet of epithelium. In cardiac muscle, the transmembrane linkage between the ends of myofibrils of neighboring cells, mediated through adherens junctions, contributes to the efficiency of contraction and prevents deformation of the cells.

The second role of adherens junctions relates to their possible involvement in the regulation of cell growth and motility. It is well established that cell contacts may have two, apparently conflicting, effects on cell growth and behavior. Most normal cells require an attachment to a solid substrate and will not grow unless they interact with a suitable adhesive surface. This phenomenon is often referred to as anchorage dependence. On the other hand, when cells reach high density they often become immobilized and stop growing, a phenomenon termed contact inhibition. It is conceivable that signals transmitted through specific cell–substrate or cell–cell contacts turn cell growth either on or off.

Two major indirect observations suggest that, among the different cellular contacts, adherens junctions might be the ones involved in these control mechanisms. The first observation concerns the wide occurrence of adherens junctions in cells of different origins. They are commonly detected in fibroblasts, epithelia, muscle, neurons, astrocytes, endothelium, macrophages, and probably many other cells that so far have not been studied. In this respect, they appear to exhibit the widest range of distribution, compared to all other junctional specializations.

The second observation is based on the deterioration of adherens junctions in transformed cells. It has been reported that such cells usually contain reduced substrate or intercellular contacts. It has been directly demonstrated in several systems that focal contacts in transformed cells and the associated cytoskeleton are highly deteriorated (for further discussion, see Geiger et al., 1983). The molecular nature of these apparent interrelationships between adherens junction formation and the normal contact-induced regulation of growth is not clear at present. Is there a primary defect or modification of a contact receptor or of one of the other junctional domains in transformed cells? Is this a block in the transmission of the signal from the exterior to the interior of the cell at sites of contact formed by these cells? Are the physiological manifestations of transformation directly related to the abnormal contact formation or are the two affected independently by yet another factor? These fundamental questions have no conclusive answers. It is anticipated that further research on the molecular constituents of the various domains and their interrelationships will shed light on the involvement of adherens junctions in the regulation of cell behavior.

ACKNOWLEDGMENTS

The studies discussed in this chapter were supported in part by a grant from the Muscular Dystrophy Association and a joint grant from the National Council for Research Development in Israel and the Deutches Krebsforschungszentrum in the Federal Republic of Germany. Some of the studies were performed in collaboration with Victor Small's group from the Austrian Academy of Science in Salzburg and with Werner Franke's group from the Deutches Krebsforschungszentrum in Heidelberg. It is our pleasant duty to acknowledge with gratitude the excellent technical assistance of H. Sabanay in the electron microscope studies.

Note added in proof: Since this chapter was written, we have shown that talin is present in the junctional plaques of cell–matrix contacts only and not in cell–cell adherens junctions.

REFERENCES

Abercrombie, M. J. (1980) The crawling movement of metazoan cells. *Proc. R. Soc. Lond. (Biol.)* **207**:129–147.

Abercrombie, M. J., J. E. M. Heaysman, and S. M. Pegrum (1971) The locomotion of fibroblasts in culture. IV. Electron microscopy of the leading lamella. *Exp. Cell Res.* **67**:359–367.

Avnur, Z., and B. Geiger (1981a) Substrate-attached membranes of cultured cells. Isolation and characterization of ventral cell membranes and the associated cytoskeleton. *J. Mol. Biol.* **153**:361–379.

Avnur, Z., and B. Geiger (1981b) The removal of extracellular fibronectin from areas of cell–substrate contact. *Cell* **25**:121–132.

Avnur, Z., J. V. Small, and B. Geiger (1983) Actin-independent association of vinculin with the cytoplasmic aspects of the plasma membrane in cell–substrate contacts. *J. Cell Biol.* **96**:1622–1630.

Begg, D. A., R. Rodewald, and L. I. Rebhun (1978) The visualization of actin filament polarity in thin sections. Evidence for the uniform polarity of membrane-associated filaments. *J. Cell Biol.* **79**:846–852.

Birchmeier, C., T. E. Kreis, H. M. Eppenberger, K. H. Winterhalter, and W. Birchmeier (1980) Corrugated attachment membrane in WI-38 fibroblasts: Alternating fibronectin fibers and actin-containing focal contacts. *Proc. Natl. Acad. Sci. USA* **77**:4108–4112.

Burridge, K., and L. Connell (1983) A new protein of adhesion plaques and ruffling membranes. *J. Cell Biol.* **97**:359–367.

Burridge, K., and J. R. Feramisco (1982) α-Actinin and vinculin from nonmuscle cells: Calcium-sensitive interactions with actin. *Cold Spring Harbor Symp. Quant. Biol.* **46**:587–597.

Chen, W.-T. (1981) Surface changes during retraction-induced spreading of fibroblasts. *J. Cell Sci.* **49**:1–13.

Chen, W.-T., and S. J. Singer (1982) Immunoelectron microscopic studies of the sites of cell–substratum and cell–cell contacts in cultured fibroblasts. *J. Cell Biol.* **95**:205–222.

Farquhar, M. G., and G. E. Palade (1963) Junctional complexes in various epithelia. *J. Cell Biol.* **17**:375–409.

Geiger, B. (1979) A 130-K protein from chicken gizzard: Its localization at the termini of microfilament bundles in cultured chicken cells. *Cell* **18**:193–205.

Geiger, B. (1982) Involvement of vinculin in contact-induced cytoskeletal interaction. *Cold Spring Harbor Symp. Quant. Biol.* **46**:671–682.

Geiger, B., A. H. Dutton, K. T. Tokuyasu, and S. J. Singer (1981) Immunoelectron microscope studies of membrane-microfilament interactions: Distributions of α-actinin, tropomyosin and vinculin in intestinal epithelial brush border and chicken gizzard smooth muscle cells. *J. Cell Biol.* **91**:614–628.

Geiger, B., Z. Avnur, and J. Schlessinger (1982) Restricted mobility of membrane constituents in cell–substrate focal contacts of chicken fibroblasts. *J. Cell Biol.* **93**:495–500.

Geiger, B., Z. Avnur, T. E. Kreis, and J. Schlessinger (1983) The dynamics of cytoskeletal organization in areas of cell contact. In *Cell and Muscle Motility*, Vol. 5, J. W. Shay, ed., pp. 195–234, Plenum, New York.

Goldman, R. D., J. Talian, A. Goldman, and B. Chojnacki (1979) Adhesion plaques of embryonic fibroblasts are analogous to Z lines. *J. Cell Biol.* **83**:324a.

Harris, A. K., and G. Dunn (1972) Centripetal transport of attached particles on both surfaces of moving fibroblasts. *Exp. Cell Res.* **73**:519.

Heath, J. P., and G. A. Dunn (1978) Cell-to-substratum contact of chick fibroblasts and their relation to the microfilament system: A correlated interference reflexion and high voltage electron microscopy study. *J. Cell Sci.* **29**:197–212.

Heaysman, J. E. M., and S. M. Pegrum (1973) Early contacts between fibroblasts. An ultrastructural study. *Exp. Cell Res.* **78**:71–78.

Hynes, R. O., A. T. Destree, and D. D. Wagner (1982) Relationships between microfilaments, cell–substratum adhesion and fibronectin. *Cold Spring Harbor Symp. Quant. Biol.* **46**:659–670.

Jockusch, B. M., and G. Isenberg (1981) Interaction of α-actinin and vinculin with actin: Opposite effect on filament network formation. *Proc. Natl. Acad. Sci. USA* **78**:3005–3009.

Kartenbeck, J., E. Schmid, W. W. Franke, and B. Geiger (1982) Different modes of internalization of proteins associated with adherens junctions and desmosomes: Experimental separation of lateral contacts induces endocytosis of desmosomal plaque material. *EMBO J.* **1**:725–732.

Kreis, T. E., B. Geiger, and J. Schlessinger (1982) The mobility of microinjected rhodamine–actin within living chicken gizzard cells determined by fluorescence photobleaching recovery. *Cell* **29**:835–845.

Kreis, T. E., Z. Avnur, J. Schlessinger, and B. Geiger (1985) Dynamic properties of cytoskeletal proteins in focal contacts. *Cold Spring Harbor Symp. Quant. Biol.* (in press).

Meyer, R. K., H. Schindler, and M. M. Burger (1982) α-actinin interacts specifically with model membranes containing glycerides and fatty acids. *Proc. Natl. Acad. Sci. USA* **79**:4280–4284.

Oesch, B., and W. Birchmeier (1982) New surface component of fibroblast's focal contacts identified by a monoclonal antibody. *Cell* **34**:671–679.

Otto, J. J. (1983) Detection of vinculin-binding proteins with a ^{125}I-vinculin gel overlay technique. *J. Cell Biol.* **97**:1283–1287.

Overton, J. (1974) Cell junctions and their development. *Prog. Surf. Membr. Sci.* **8**:161–208.

Schlessinger, J., and E. L. Elson (1982) Fluorescence methods for studying membrane dynamics. In *Methods of Experimental Physics*, Vol. 20, H. Ehrenstein and H. Lecar, eds., pp. 197–227, Academic, New York.

Singer, I. I., and P. R. Paradiso (1981) Transmembrane relationships between fibronectin and vinculin (130-kD protein): Serum modulation in normal and transformed hamster fibroblasts. *Cell* **24**:481–492.

Staehelin, L. A. (1974) Structure and function of intercellular junctions. *Int. Rev. Cytol.* **39**:191–283.

Tokuyasu, K. T., A. H. Dutton, B. Geiger, and S. J. Singer (1981) Ultrastructure of chicken cardiac muscle as studied by double immunolabeling in electron microscopy. *Proc. Natl. Acad. Sci. USA* **78**:7619–7623.

Volk, T., and B. Geiger (1984) A 135-kD membrane protein of intercellular adherens junctions. *EMBO J.* **3**:2249–2260.

Wilkins, J. A., and S. Lin (1982) High affinity interaction of vinculin with actin filaments *in vitro*. *Cell* **28**:83–90.

Contributors

W. Steven Adair
Departments of Biology & Biophysics
Washington University
St. Louis, Missouri 63130

Steven K. Akiyama
Laboratory of Molecular Biology
National Cancer Institute
Bethesda, Maryland 20205

Zafrira Avnur
Department of Chemical Immunology
The Weizmann Institute of Science
Rehovot 76100
Israel

Virginia Bayer
Departments of Physiology & Biophysics
University of California/Irvine
Irvine, California 92717

Donna Bozyczko
Department of Biochemistry
University of Pennsylvania School of
   Medicine
Philadelphia, Pennsylvania 19104

Gail Bryant
Laboratory of Pathology
National Cancer Institute
Bethesda, Maryland 20205

Clayton A. Buck
The Wistar Institute of Anatomy
   & Biology
Philadelphia, Pennsylvania 19104

Wen-Tien Chen
Howard University Cancer Center
Washington, D.C. 20060

Patricia Collin-Osdoby
Departments of Biology & Biophysics
Washington University
St. Louis, Missouri 63130

Pamela Cowin
Institute of Cell & Tumor Biology
German Cancer Research Center
D-6900 Heidelberg
Federal Republic of Germany

Bruce A. Cunningham
The Rockefeller University
1230 York Avenue
New York, New York 10021

Ivan Damjanov
Department of Pathology
Hahnemann Medical College
Philadelphia, Pennsylvania 19104

Caroline H. Damsky
Departments of Anatomy & Oral
   Biology
University of California/San Francisco
San Francisco, California 94143

Cindi L. Decker
Department of Biochemistry
University of Pennsylvania School of
   Medicine
Philadelphia, Pennsylvania 19104

Annie Delouvée
Institut d'Embryologie
Centre National de la Recherche
   Scientifique
94130 Nogent-sur-Marne
France

Jean-Loup Duband
Institut d'Embryologie
Centre National de la Recherche
  Scientifique
94130 Nogent-sur-Marne
France

Kim E. Duggan
Department of Biochemistry
University of Pennsylvania School of
  Medicine
Philadelphia, Pennsylvania 19104

Gerald M. Edelman
The Rockefeller University
1230 York Avenue
New York, New York 10021

Peter Ekblom
Friedrich Miescher Laboratorium
Max Planck Gesellschaft
D-7400 Tübingen
Federal Republic of Germany

Andreas Faissner
Department of Neurobiology
University of Heidelberg
D-6900 Heidelberg
Federal Republic of Germany

Günther Fischer
Department of Neurobiology
University of Heidelberg
D-6900 Heidelberg
Federal Republic of Germany

Werner W. Franke
Institute of Cell & Tumor Biology
German Cancer Research Center
D-6900 Heidelberg
Federal Republic of Germany

Scott E. Fraser
Departments of Physiology & Bio-
  physics
University of California/Irvine
Irvine, California 92717

Benjamin Geiger
Department of Chemical Immunology
The Weizmann Institute of Science
Rehovot 76100
Israel

Norton B. Gilula
Department of Cell Biology
Baylor College of Medicine
Houston, Texas 77030

Ursula W. Goodenough
Departments of Biology & Biophysics
Washington University
St. Louis, Missouri 63130

Rhonda R. Greggs
Department of Biochemistry
University of Pennsylvania School of
  Medicine
Philadelphia, Pennsylvania 19104

Christine Grund
Institute of Cell & Tumor Biology
German Cancer Research Center
D-6900 Heidelberg
Federal Republic of Germany

Etsuko Hasegawa
Laboratory of Molecular Biology
National Cancer Institute
Bethesda, Maryland 20205

Takayuki Hasegawa
Laboratory of Molecular Biology
National Cancer Institute
Bethesda, Maryland 20205

Kohei Hatta
Department of Biophysics
Faculty of Science
Kyoto University
Kyoto 606
Japan

John E. Heuser
Departments of Biology & Biophysics
Washington University
St. Louis, Missouri 63130

Alan F. Horwitz
Department of Biochemistry
University of Pennsylvania School of
  Medicine
Philadelphia, Pennsylvania 19104

Martin J. Humphries
Laboratory of Molecular Biology
National Cancer Institute
Bethesda, Maryland 20205

Antone G. Jacobson
Center for Developmental Biology
University of Texas
Austin, Texas 78712

Martin H. Johnson
Department of Anatomy
University of Camibridge
Cambridge CB2 3DY
England

Hans-Peter Kapprell
Institute of Cell & Tumor Biology
German Cancer Research Center
D-6900 Heidelberg
Federal Republic of Germany

Juergen Kartenbeck
Department of Cell Biology
Yale University School of Medicine
New Haven, Connecticut 06510

Gerhard Keilhauer
Department of Neurobiology
University of Heidelberg
D-6900 Heidelberg
Federal Republic of Germany

Karen A. Knudsen
The Wistar Institute of Anatomy
  & Biology
Philadelphia, Pennsylvania 19104

Jan Kruse
Department of Neurobiology
University of Heidelberg
D-6900 Heidelberg
Federal Republic of Germany

Volker Künemund
Department of Neurobiology
University of Heidelberg
D-6900 Heidelberg
Federal Republic of Germany

Jürgen Lindner
Department of Neurobiology
University of Heidelberg
D-6900 Heidelberg
Federal Republic of Germany

Lance A. Liotta
Laboratory of Pathology
National Cancer Institute
Bethesda, Maryland 20205

Bruce J. Nicholson
Division of Biology
California Institute of Technology
Pasadena, California 91125

Björn Öbrink
Department of Medical & Physiological
  Chemistry
University of Uppsala
S-751 23 Uppsala
Sweden

Carin Ocklind
Department of Medical & Physiological
  Chemistry
University of Uppsala
S-751 23 Uppsala
Sweden

Per Odin
Department of Medical & Physiological
  Chemistry
University of Uppsala
S-751 23 Uppsala
Sweden

Kenneth Olden
Laboratory of Molecular Biology
National Cancer Institute
Bethesda, Maryland 20205

George E. Palade
Department of Cell Biology
Yale University School of Medicine
New Haven, Connecticut 06510

Pasko Rakic
Section on Neuroanatomy
Yale University School of Medicine
New Haven, Connecticut 06510

C. Nageswara Rao
Laboratory of Pathology
National Cancer Institute
Bethesda, Maryland 20205

Jean-Paul Revel
Divison of Biology
California Institute of Technology
Pasadena, California 91125

Jean Richa
The Wistar Institute of Anatomy
  & Biology
Philadelphia, Pennsylvania 19104

Kristofer Rubin
Department of Medical & Physiological
  Chemistry
University of Uppsala
S-751 23 Uppsala
Sweden

Hannu Sariola
Friedrich Miescher Laboratorium
Max Planck Gesellschaft
D-7400 Tübingen
Federal Republic of Germany

Mellita Schachner
Department of Neurobiology
Univeristy of Heidelberg
D-6900 Heidelberg
Federal Republic of Germany

Yasuaki Shirayoshi
Department of Biophysics
Faculty of Science
Kyoto University
Kyoto 606
Japan

Masatoshi Takeichi
Department of Biophysics
Faculty of Science
Kyoto University
Kyoto 606
Japan

Irma Thesleff
Department of Pathology
University of Helsinki
SF-00290 Helsinki
Finland

Jean-Paul Thiery
Institut d'Embryologie
Centre National de la Recherche
  Scientifique
94130 Nogent-sur-Marne
France

Tova Volberg
Department of Chemical Immunology
The Weizmann Institute of Science
Rehovot 76100
Israel

Talila Volk
Department of Chemical Immunology
The Weizmann Institute of Science
Rehovot 76100
Israel

Heide Wernecke
Department of Neurobiology
University of Heidelberg
6900 Heidelberg
Federal Republic of Germany

Ulla M. Wewer
Laboratory of Pathology
National Cancer Institute
Bethesda, Maryland 20205

Margaret Wheelock
The Wistar Institute of Anatomy
  & Biology
Philadelphia, Pennsylvania 19104

Kenneth M. Yamada
Laboratory of Molecular Biology
National Cancer Institute
Bethesda, Maryland 20205

S. Barbara Yancey
Divison of Biology
California Institute of Technology
Pasadena, California 91125

Chikako Yoshida-Noro
Department of Biophysics
Faculty of Science
Kyoto University
Kyoto 606
Japan

Index